Vincent Jowett

THE AMERICAN FORESTRY SERIES
WALTER MULFORD, Consulting Editor

FOREST MENSURATION

THE AMERICAN FORESTRY SERIES
WALTER MULFORD, Consulting Editor

Allen—
  An Introduction to American Forestry
Baker—
  The Theory and Practice of Silviculture
Boyce—
  Forest Pathology
Brown, Panshin, and Forsaith—
  Textbook of Wood Technology, Volume I
Bruce and Schumacher—
  Forest Mensuration
Chapman and Meyer—
  Forest Mensuration
  Forest Valuation
Doane, Van Dyke, Chamberlin, and Burke—
  Forest Insects
Guise—
  The Management of Farm Woodlands
Harlow and Harrar—
  Textbook of Dendrology
Kittredge—
  Forest Influences
Marquis—
  Economics of Private Forestry
Matthews—
  Cost Control in the Logging Industry
  Management of American Forests
Preston—
  Farm Wood Crops
Stoddart and Smith—
  Range Management
Trippensee—
  Wildlife Management
Wackerman—
  Harvesting Timber Crops

FOREST MENSURATION

BY

Herman H. Chapman, M.F., Sc.D.
Professor Emeritus, Yale University

AND

Walter H. Meyer, M.F., Ph.D.
*Harriman Professor of Forest Management
Yale School of Forestry*

FIRST EDITION

NEW YORK   TORONTO   LONDON
McGRAW-HILL BOOK COMPANY, INC.
1949

FOREST MENSURATION

Copyright, 1949, by the McGraw-Hill Book Company, Inc. Printed in the United States of America. All rights reserved. This book, or parts thereof, may not be reproduced in any form without permission of the publishers.

THE MAPLE PRESS COMPANY, YORK, PA.

PREFACE

This text supersedes "Elements of Forest Mensuration" by Chapman and Demeritt, second edition, 1936. The original American text on the subject was "Forest Mensuration" by Henry S. Graves, John Wiley & Sons, Inc., 1906. This was followed by "Forest Mensuration," by H. H. Chapman, John Wiley & Sons, Inc., 1921. The text by Chapman and Demeritt first appeared in 1931 and was revised in 1936. Although certain fundamental techniques have been carried over from previous texts, the content has been completely rewritten, rearranged, and extended to include the more essential newer techniques developed during the last 15 years. The greatest amplification lies in the study of growth and increment for all classes of stands and forest.

The economic approach has been retained, *viz.*, this book deals first with uses and products, which gives the basis for all other steps in mensuration. Hence cubic, board, and piece measures for cords, logs, and dimensions, including scaling and log rules, are first developed (Chaps. 3 to 7). The construction and use of volume tables for the entire tree follow (Chaps. 8 to 12). The third stage is the measurement of standing trees and the determination of area and volume, commonly called "timber estimating," or "cruising" (Chaps. 13 to 20). Then comes the treatment of growth (Chaps. 22 to 31). In Chaps. 2 and 21 the elementary graphical, mathematical, and statistical techniques required in forest mensuration are discussed, but the more complicated and theoretical treatment of these subjects is avoided.

It is assumed throughout that the student wants to know why certain methods are required, not merely how to do the work. Hence much attention is paid to correlating the various mensuration techniques with the silvicultural characteristics of trees and stands on the one hand and with the demands of forest management on the other.

The text is designed to satisfy the need for an elementary course of instruction in forest mensuration, as well as that for an advanced course. To serve the first class of need, the instructor is advised to make a selection of chapters and sections that will fit his desire, because the text is probably too full for the usual first course in the subject.

For advanced students and research workers, the complete text will be required plus the list of selected references which accompanies each chapter. It is impossible in a text of this scope to include detailed discussion of all the advanced processes advocated by various authors. For example, the rapidly developing field of aerial photography in its application to mensuration has been covered by only a brief summary. This does not mean that the importance of aerial photographic techniques is minimized but rather that the limitations of space preclude full and adequate treatment of this important technique. Further instruction is left to specially prepared texts.

Special attention is called to Chaps. 22 to 31, inclusive, in which the subject of tree growth and stand increment has been brought up to date by the inclusion of recently accepted methods. Emphasis is placed on the importance and application of normal yield tables in predicting the growth of partially stocked even-aged stands. The general methods applicable to many-aged stands, based on stand tables and past periodic growth in diameter, are treated in detail. In Chap. 33, several case examples are given of the application of methods of timber estimating and growth determination to different types of problems encountered in the forest survey. Judgment in the choice of mensurational methods for the forest survey depends not only upon the physical condition of the forest, but even more on the utility and value of the products and the requirements as to time and cost. In solving this fundamental problem, the science of statistics, dealing as it does with sampling errors and their control, is of especial importance.

Grateful acknowledgment is made to Prof. Randolph M. Brown, School of Forestry, University of Minnesota, for his painstaking review and criticism of many of the chapters, especially those dealing with growth.

HERMAN H. CHAPMAN
WALTER H. MEYER

NEW HAVEN, CONN.
July, 1949

CONTENTS

PREFACE v

CHAPTER 1

ELEMENTS OF FOREST MENSURATION 1

    1. The Scope of Forest Mensuration—2. The Measurement of Products and of Logs—3. Measurement in the Tree—4. The Measurement of Stands of Timber—5. The Measurement of Growth

CHAPTER 2

ELEMENTARY COMPUTATIONS. 8

    6. Introductory—7. Accuracy of Measurement—8. Significant Figures—9. Rounding Off—10. Appearance of Data—11. Proficiency in Calculation—12. Population and Sample—13. Frequency Distributions—14. Graphs—15. Measurements of Central Tendency—16. Measures of Dispersion—17. Standard Error of the Mean

CHAPTER 3

CORD OR STACKED MEASURE. 34

    18. Utility of Cord Measure—19. Measurement by Weight—20. Units of Stacked or Cord Measure—21. Solid Cubic Contents of Stacked Wood—22. The Causes of Variation in Solid Contents of Cords—23. Deduction for Defect in Cordwood—24. The Volume of Bark—25. Board-foot Equivalents of Cordwood

CHAPTER 4

CUBIC VOLUME OF LOGS. 45

    26. Logs and Bolts—27. The Cylinder as a Basis for the Cubic Volume of Logs and Bolts—28. The Form and Cubic Volume of Logs—29. Scientific Standards of Accuracy in Log Measurements—30. Commercial Methods of Log Measurement—31. Bark Measurements and Scaling Diameter—32. Log Rules Based on Cubic Contents of Logs—33. Cubic Volume of Squared Timbers

CHAPTER 5

THE MEASUREMENT OF BOARD FEET IN THE LOG. LOG SCALING. 56

    34. The Necessity for Log Measurement—35. Log Rules as a Standard for Log Measure—36. Scaling Sound Straight Logs. Diameters—37. Scaling Sound Straight Logs. Lengths—38. Overrun—39. Deductions from Gross Scale. Sound Scale—40. Interior Defects. Deductions—41. Side or Exterior Defects. Deductions—42. Crook or Sweep. Deductions—43. Merchantable and Cull Logs—44. Sample Scaling—45. Scaling for Cubic Contents. Quebec—46. Scaling Records and Check Scaling

CONTENTS

CHAPTER 6
CONSTRUCTION OF LOG RULES 72

47. Factors Causing Variation in Sawed Output of Logs—48. Standardizing by Means of Diagrams. Diagram rules—49. Constructing Log Rules by Formula. The Board Foot–Cubic Foot Ratio—50. Waste from Saw Kerf and Slabs—51. Log Rules Based on Formulas. International Log Rule—52. Log Rules Based on Formulas. The Doyle Rule—53. A Universal Log Rule for Board Feet—54. Log Rules Based on Mill Tallies

CHAPTER 7
PIECE PRODUCTS. MEASUREMENT 84

55. Classes of Piece Products and Standards of Measurement—56. Round Products—57. Mine Timbers—58. Crossties. Specifications—59. Crossties. Tie "Log Rules"

CHAPTER 8
DETERMINATION OF VOLUME OF FELLED TREES FOR THE CONSTRUCTION OF VOLUME TABLES 89

60. Standard Tree Measurements—61. Tree Form as a Variable in Tree Volumes. Form Classes—62. Total versus Merchantable Height—63. Rules for Measuring Standard Merchantable Heights—64. Additional Measurements for Butt Swell—65. Diameter Outside and Inside Bark. Thickness of Bark—66. The Swedish Bark Gauge—67. Procedure for Measuring Felled Trees—68. The Cubic Content of Felled Trees—69. Graphic Determination of the Cubic Volume of Felled Trees—70. The Board-foot Volume of Felled Trees

CHAPTER 9
LOCAL TREE VOLUME TABLES. CONSTRUCTION. LIMITATIONS. 105

71. The Volume Table Problem—72. Construction of a Local Volume Table Based on D.B.H. Classes—73. The Graphic Plotting of Averages—74. Reading and Rechecking the Curve—75. Importance of the Sample, or Measured Values—76. The Board-foot Unit. Log Rules—77. Merchantable Diameter Limits—78. Log Lengths—79. Tree Heights and Standard Volume Tables—80. Total versus Merchantable Heights—81. Tree Form and Bark Thickness. Form Class

CHAPTER 10
THE CONSTRUCTION OF STANDARD VOLUME TABLES FOR CUBIC FEET 120

82. Standard Volume Tables—83. Standard Volume Tables for Cubic Feet. Basis—84. Volume Tables Constructed from Harmonized Curves—85. Checking the Accuracy of the Curves—86. Derivation of Local from Standard Volume Tables—87. Cordwood Volume Tables—88. Cubic Form Factors.–Basis—89. The Construction of Standard Volume Tables by the Method of Least Squares

CONTENTS

CHAPTER 11
THE CONSTRUCTION OF ALINEMENT CHARTS FOR STANDARD VOLUME TABLES FOR CUBIC FEET 139

90. Alinement Charts—91. Addition Charts—92. Multiplication Charts—93. Alinement Chart Volume Tables for Cubic Feet. Forest Service Method—94. The First Chart Correction—95. The Second Chart Correction—96. The Third Chart Correction—97. Completion of Chart and Test for Aggregate Difference—98. Average Percentage Deviation—99. Cubic Volume Tables for Partial Utilization of the Stem

CHAPTER 12
THE CONSTRUCTION OF STANDARD VOLUME TABLES FOR BOARD FEET 160

100. Board-foot Volume Tables. Basis—101. Standard and Local Volume Tables Based on Butt Log Form Quotients. Girard Method—102. Determination of Butt Log Form Class—103. Alinement Chart Volume Tables for Board Feet—104. Data Which Should Accompany Volume Tables—105. Adaptation and Conversion of Volume Tables—106. Checking the Accuracy of Standard Volume Tables for Board Feet—107. The Volume-diameter Ratio Method

CHAPTER 13
THE CONSTRUCTION AND APPLICATION OF TAPER TABLES 180

108. Application of Taper Tables for Different Form Classes to Trees of Different Species—109. Construction of Tables for Percentile Tapers

CHAPTER 14
THE MEASUREMENT OF DIAMETERS OF STANDING TREES 185

110. Tree Diameters. Tree Calipers—111. The Diameter Tape—112. The Biltmore Stick—113. Errors of Diameter Instrumentation

CHAPTER 15
THE MEASUREMENT OF HEIGHTS OF STANDING TREES 194

114. The Determination of Heights of Trees—115. Height Determination by Tangents of Angles. The Abney Hand Level. Forest Service and Faustmann Hypsometers—116. Height Determination by Similar Isosceles Triangles. Staff Hypsometer—117. Height Determination by the Ratio of Legs of Two Similar Triangles, Not Isosceles. Merritt Hypsometer. Christen Hypsometer—118. The Chapman Hypsometer—119. The Preparation of a Curve of Height on Diameter—120. The Principle of Sampling as Applied to Height Measurements—121. Errors of Height Instrumentation

CHAPTER 16
THE BOUNDARY SURVEY AND DETERMINATION OF TOTAL AREA 213

122. Retracement or Reestablishment of Surveys. Identification of Monuments—123. The Field Notes—124. The United States Public Land Survey. Essential Data—125. Meander Surveys and Township Plats—126. The Marking of United States Land Survey Corners and

x CONTENTS

Accessories—127. Reestablishment of Lost or Obliterated Corners—128. Retracement of Old Lines—129. Determination of Areas

CHAPTER 17
TIMBER ESTIMATING FOR CUBIC AND CORD MEASURE 227

130. Timber Estimating in General. Ocular Estimating—131. Measurement of Every Tree, or 100 Per Cent Estimate with Volume Table—132. Factors Determining the Cost of Timber Estimating—133. Estimating by Sample Areas. Partial Measurement—134. Units of Land Measurement. Linear Measurement. The Strip System of Timber Estimating—135. Area Measurement—136. Choice of Percentage of Area to Be Covered—137. Determination of Width of Strips—138. Errors of Measurement on Strips—139. Organization of a Strip Survey for Relatively Small Areas. Instruments Used—140. The Measuring of Distances by Pacing—141. Determination of Total Area and Estimate—142. Substitution of Plots for Strip Estimate—143. Plots Arbitrarily Chosen

CHAPTER 18
TIMBER ESTIMATING FOR BOARD FEET BY USE OF LOG RULES 247

144. The Log as a Unit in Timber Estimating—145. Rough Methods of Log Estimate—146. Tally of Log Diameter—147. Training in Ocular Estimates of Log Diameters—148. Deduction for Defect and Cull

CHAPTER 19
TIMBER FOR BOARD FEET ESTIMATING BY USE OF VOLUME TABLES. 251

149. Factors in the Choice of a Volume Table—150. The Tree Tally—151. Heights—152. Ocular Tally of Dimensions—153. Checking the Accuracy of Estimates—154. Deduction for Defect—155. Forest Types as Units for Timber Estimating—156. Site Qualities—157. The Stand or Forest Cover—158. Volume or Density Correction—159. Varying the Percentage Covered by Strips in Different Cover Types—160. Maps Showing Stocking or Stand per Acre, and Topographic Maps

CHAPTER 20
MEASUREMENT OF PIECE PRODUCTS IN THE TREE. 269

161. Estimation of Poles and Piling—162. Dendrometer for Top Diameters of Piling or Poles—163. Measurement of Sweep in Standing Trees—164. Tree Volume Tables for Ties, by Grades—165. Estimation of Mine Props and Mine Timbers—166. Estimating Post Timber

CHAPTER 21
FUNDAMENTAL STATISTICAL TECHNIQUES 275

167. Estimating by Sampling—168. The Normal Frequency Distribution—169. Fitting of the Normal Curve of Error to Observational Data—170. Probability of Occurrence—171. Standard Error of the Mean—172. Regression—173. Standard Error of Estimate from a Regression Equation—174. Standard Error of a Difference and of a Sum—175. Correla-

tion—176. Coefficient of Alienation—177. Multiple and Curvilinear Relationships

CHAPTER 22
INFLUENCES AFFECTING THE GROWTH OF TREES 298
178. Purpose of Studying the Growth of Trees—179. The Life Pattern of Tree Species—180. Normal Mortality. Its Control—181. Abnormal Mortality. Calamities—182. Characteristics of Height Growth—183. Characteristics of Diameter Growth—184. Slowing Down, or Deceleration, of Diameter Growth—185. Effect of Logging and of Mortality on Diameter Growth—186. Effect of Climatic Cycles on Periodic Growth

CHAPTER 23
THE AGE OF TREES AND STANDS 318
187. The Age of Trees—188. Seedling Age and Total Age of Trees—189. The Age of Even-aged Stands—190. The Increment Borer—191. Correction of Age for Suppressed Trees

CHAPTER 24
THE GROWTH OF TREES AND OF STANDS. 324
192. Nature of Forest Crops—193. Even-aged versus Many- or All-aged Stands—194. Current and Periodic Annual Increment for Even-aged Stands—195. Average (Mean) Annual Increment—196. Periodic and Current Increment of Many- or All-aged Stands—197. The Purpose of Growth Studies

CHAPTER 25
THE GROWTH OF TREES IN TOTAL HEIGHT BASED ON AGE. 335
198. Objectives of the Study of Height Growth—199. Height Growth of a Single Tree Based on Ages of Sections—200. Curves of Average Height on Age

CHAPTER 26
THE GROWTH OF TREES IN DIMENSION AND FORM 341
201. Growth of Trees in Diameter—202. Influences Affecting the Diameter Growth of Trees—203. Diameter Growth on the Stump—204. Compilation and Averaging of Diameter Growth on the Stump—205. Correlation of Stump Diameter Growth with D.B.H. outside Bark—206. Plotting the Curve of Growth in D.B.H.—207. The Derivation of Growth at B.H. from Current Growth in Diameter on the Stump—208. Purpose of Determining Growth in Upper Diameters or Form—209. Measurement of Diameter Growth at Upper Sections of the Bole—210. Averaging the Growth Data for Trees of a Given Group—211. The Graph of Tree Form Based on Age—212. Growth for Shorter Periods—213. Volume Growth of the Average Tree—214. Differentiation by Site Class

CHAPTER 27
NORMAL YIELD TABLES FOR EVEN-AGED STANDS AND THEIR CONSTRUCTION . 361
215. Yield Tables, Definition—216. The Normal Standard of Stocking for Yield Tables—217. The Effect of Site Quality on Yields—218. Classi-

fication of Forest Soils and Environment into Site Classes—219. The Site Index Graph—220. Alternate Methods of Using Plots in Constructing Yield Tables—221. Normal or Fully Stocked versus Under- and Overstocked Stands—222. Progress of Subnormal Stands toward Normality—223. The Need for Normal Yield Tables for Thinned Stands—224. Selection of Plots for a Yield Table—225. Measurements Taken on Plots—226. Office Work. Plot Computations—227. Construction of a Site Index Chart or Table—228. Determination of Site Index for Each Plot—229. Construction of a Table of Total Basal Area Based on Age—230. Yield Tables for Board Feet—231. Merchantable Yields for Partial Stands—232. Stand and Stock Tables for Fully Stocked Even-aged Stands

CHAPTER 28

PREDICTION OF THE PERIODIC ANNUAL GROWTH OF "AVERAGE" EVEN-AGED STANDS BY USE OF NORMAL YIELD TABLES 392

233. The Application of Normal Yield Tables to Average Stands—234. Empirical, or Subnormal, Yield Tables—235. Understocked Stands—236. Numerical Measures of Stocking-density Ratio—237. A Measure of Stocking Independent of Age and Site. Stand Density Index—238. Field Procedure for Predicting Growth of Even-aged Stands by Use of Normal Yield Tables—239. Prediction by Use of Present Stocking Percentage—240. Prediction Based on Normality Changes Measured from Permanent Sample Plots—241. Prediction by Use of a Density Percentage Based on Gerhardt's Empirical Formula—242. Prediction Based on Growth Variables. Gevorkiantz—243. Prediction Based on Comparative Cumulative Tallies of Stocking by Numbers of Trees per Acre

CHAPTER 29

PERIODIC AND CURRENT GROWTH IN DIAMETER BASED ON STAND TABLES . . 411

244. Substitution of Diameters for Age Classes in All-aged Stands—245. Objectives of Stand Table Method—246. The Stand Table as Related to Forest Types, Site Classes, and Age Classes—247. Mortality as Affecting Future Stand Tables—248. Requirements of Sampling for Diameter Growth—249. Modifying the Stand Table for Use in Growth Studies. Ingrowth—250. Application of Past Diameter Growth to Stand Tables—251. Application of Stand Tables and Growth Data to Total Area—252. Reproduction Surveys. Stocked Quadrats

CHAPTER 30

PREDICTING THE PERIODIC ANNUAL GROWTH OF STANDS BY THE STAND TABLE METHOD 434

253. Measurements of Diameter Growth at B.H.—254. Correction for Growth in Thickness of Bark—255. Periodic Growth in Height, Based on D.B.H. Classes—256. Prediction of Growth in Volume, Based on Growth in D.B.H., Height, and Use of Volume Tables. Influence of Changing Form Factors—257. Correction for Mortality—258. Method of Comparison of Present with Past Growth—259. Measuring Directly the Effect of Release Cutting

CONTENTS

CHAPTER 31

GROWTH PERCENTAGE.. 456

260. Definition and Character of Growth Percentage—261. Base of Growth Percentage—262. Methods of Determining Growth Percentage—263. Current versus Mean Annual Growth Percentage—264. Periodic Growth Percentages—265. Utility of Growth Percentages in Growth Predictions

CHAPTER 32

SAMPLE PLOTS.. 466

266. Growth Predictions Based on Historical Records. Recurring Inventories—267. Purpose of Sample Plots—268. Choice and Location—269. Size and Number—270. Establishment of Plots—271. Identification of Trees—272. Measurement of Trees—273. Listing Sheets versus Source Tables—274. The Charting of Sample Plots

CHAPTER 33

THE FOREST SURVEY OR INVENTORY...................................... 478

275. Objectives of the Forest Survey—276. General Outlines of Procedure to Be Followed in a Forest Survey—277. Aerial Photographs as an Aid in Forest Survey—278. Survey Procedure on a Large Area—279. Survey Procedure on a Small Area—280. Case Examples of Volume and Growth Surveys in Various Forest Regions

APPENDIX... 503

INDEX... 507

CHAPTER 1

ELEMENTS OF FOREST MENSURATION

1. The Scope of Forest Mensuration. Managers of any commercial enterprise whose objective is to supply the demand for goods must know approximately at any time the quantity of the various items of stock on hand. The periodic inventory, supplemented by records of purchases and sales, supplies the need. Of equal importance is a knowledge of the capacity of the plant and the rate at which these goods are being produced.

The forest may be compared with a factory, a farm, or a mine. As an enterprise in growing crops of trees, the volume produced and the time required for production are essential factors. As a wood mine, or reserve of standing timber, the estimated quantity of this timber constitutes the inventory of raw materials, corresponding to unmined ore.

The farmer may estimate his probable yield of grain before harvest, or he may sell his apple crop on the trees. More frequently he harvests these crops himself annually and prefers to measure them in the bin. The forest crop requires many years to mature, during which time it may be bought and sold while still in the tree. The farmer's crops are usually sold in the form in which they are harvested. Their value may depend upon their final use as flour, bread, or other manufactured foodstuffs, but they are sold as wheat, corn, or animals on the hoof. The forest crop may in turn be sold as logs, bolts, or cordwood, but the price paid will depend upon the value of the ultimate product such as boards, paper, or manufactured articles. In the forest, to a far greater extent than on the farm, the manufactured products for which the timber is to be used, especially if lumber, will influence the form in which the volume of the crop is measured. Cubic volume of merchantable wood offers a universal standard of measurement, but this is seldom preferred in dealing with forests or their products. Purchasers prefer instead other standards based upon the manufactured product.

Thus the factory or manufacturing phase of the forest industry exerts a strong influence on forest mensuration. As a result, the units of measure used to obtain the volume of forest crops are usually board feet, ties, poles, or cordwood rather than cubic volume.

In an industry engaged in the manufacture of forest products, the quantity of timber in the forest normally exceeds manyfold the value of products which have been cut and remain on hand. For this reason, the measurement of volume of standing timber rather than of logs or products is the usual form of forest inventory. In addition, estimates of the rate of growth or forest production are commonly expressed in the above units of measure.

2. Measurement of Products and of Logs. In the operations of securing timber products from the forest, such products may require measurement at three stages, *viz.*, in the finished or manufactured form, in the log or bolt after cutting and before manufacture, and finally in the standing tree or forest.

In each of these cases, whenever a sale is made, as of lumber, logs, or standing timber, the material purchased must be measured to furnish the basis for the transaction. A second reason for measurement is to permit of direct payment for piecework, since it is customary to pay contractors and even saw crews on the basis of the quantity of timber handled. When a forest owner cuts and manufactures his own timber, employing labor on a day basis, the only measurement he needs is the final tally of the sawed lumber. But when a sawmill man purchases logs cut and delivered, neither party yet knows what these logs will saw out. Settlement for the logs on the basis of the final lumber tally would involve delay. Further, if several purchases are made, segregation of each lot from the products of other logs passing through the mill and into the yard would be necessary. Only when dealing with a single owner who can wait for his settlement can the purchase of the logs or even of standing timber be made on the basis of lumber or product tally, which gives an accurate measurement of the quantity acquired.

In the majority of cases, it is evident that payment must be made promptly and the ownership of logs acquired before manufacture. It is thus necessary to measure in the log, before sawing, the *probable* quantities of the kinds of products into which these logs will be manufactured. The exact basis of the sale must be agreed upon in advance by both buyer and seller. Round peeled products (such as piling, poles, or posts) and hewn ties are completed and shaped for use before transportation. For such products, a final accurate tally is possible on the "landing." For logs and bolts, however, uncertainty exists as to the quantities of lumber, ties, shingles, or other products that these pieces may contain. This uncertainty is due to the varying methods, machinery, and skill used in manufacturing and to the variety of products which, at the option of the purchaser, may be cut from identical logs; for a given log may be cut into planks and boards of different thicknesses, ties, dimension stock for remanufacture, or billets for special purposes such as staves or tool handles.

Because of this condition, the basis of log measurement which will be accepted in lieu of actual manufactured output is arbitrarily standardized or fixed in advance. When lumber is the designated product of the logs, standard log rules are adopted, giving fixed contents in board measure for logs of each different standard length and diameter. The thickness of the boards supposed to be contained in the log is standardized in the rule adopted for log measure. Settlement is on the basis of the log rule values, and possible departures in manufacture from the standardized thickness, and assumed quantity of sawed output thus indicated, are disregarded as not concerning the seller. When the entire substance of the log is utilized, as in the case of paper pulp, wood distillation, or firewood, the cubic contents of these logs or bolts, either with or without bark, is the real basis of the final standard for purchase. Even here, the wood is usually measured in stacked form as cords, rather than for net cubic contents.

Log rules and other standards of measure assume that the pieces to be measured are straight and sound or come within certain specified requirements. In the application of these accepted standards or rules, the physical measurement requires only dimensions, *i.e.*, diameters in inches and lengths in feet. The contents corresponding to these sizes as given in the log rule are then recorded. For lumber, defects which render the wood mechanically unsound call for a deduction from the standard contents. Such deduction requires judgment and experience on the part of those engaged in log measurement. Piles, poles, crossties, and other products are bought on specifications as to their dimensions, quality, and defects. The disqualifying defects in such products depend upon shapes, sizes, species, soundness, and whether cut living or dead. This results in the rejection of many pieces which may have the required standard lengths and diameters but are otherwise disqualified.

In this manner both standardized rules and uniform practices for measuring forest products apply to the log, bolt, or cord, and permit of prompt settlement. Such rules do not pretend to foretell accurately the quantity that will be finally secured in manufacture, since this is variable. They do undertake to give the volume of products that would be obtained from sound straight logs if manufactured according to the exact specifications for which the rule was made. These conditions are accepted by both buyer and seller and alone make the transaction possible.

3. Measurement in the Tree. The same conditions which brought about the adoption of log rules apply to the standing timber. When an owner sells trees to be cut and removed by the purchaser, an attempt to measure the contents of the standing trees is far less desirable than log measurement after cutting. A sale based on logs after cutting, however, postpones the settlement and requires supervision or inspection of the log scaling by the owner of the timber. When logging is to be postponed for a considerable period, a basis of sale is required for the standing timber. This is true when an owner plans the consolidation of his timber holdings by the purchase of many small tracts.

In the measurement of the standing trees, the products to be derived from the logs as measured by a log rule form the basis of the tree volumes. The tree volumes are equivalent to the sum of these standardized log volumes. This requires measuring heights to various points on the bole of the tree, and diameters at these respective points, which are necessarily observed outside the bark. However, diameter inside bark is the measure usually required, which necessitates a further approximation. On the other hand, if the tree is felled, the log contents inside the bark are easily found and totaled to determine the tree volume. If this information is tabulated prior to the measurement of the standing trees, the volumes thus obtained may be applied to trees of similar dimensions. Tabulated values obtained from measurement of felled trees take the form of standard volume tables. Their construction, based upon the laws of averages and of tree form, is an important part of forest mensuration.

When the form of trees or their merchantable contents is so variable as to make it impossible to standardize or average such tree volumes, it is still possible to tally the diameters of each individual log which a tree contains. Such a tally involves the deduction by approximation of the double width of bark, and judgment of both height and diameter by eye. Occasional checks may be made on down trees with instruments for measurement. By one or the other of these methods, the contents of standing trees may be obtained with some approach to accuracy.

4. The Measurement of Stands of Timber. Any attempt to obtain a record of dimensions of logs or bolts from ocular inspection of individual standing trees introduces a factor of personal error even among skilled observers. When tree volume tables are available that have been found to agree with the actual average volumes of the trees to be measured, standards of accuracy are possible which closely approach those set up for log measure.

If the volume of standing timber on a considerable area of land is desired, the expense of a complete tally of every tree is usually prohibitive. An error of 10 per cent in the estimate of tree volume may fall on either of the two parties to the trade; if low, on the owner; if high, on the buyer. But when a stand is measured, an expense that equals 10 per cent of its value on the stump is a direct loss of this amount, whichever party undertakes the estimate. He can recover this loss only if it results in a still greater saving, either by bettering his side of the bargain in purchase or by effecting economies in logging. Hence the cost of measuring standing timber must always be less than the probable economies resulting from this cost. Only in this manner can the cost of estimating add anything to the value of the timber. Its necessity as a basis of sales is obvious.

Standing timber is worth only a small fraction of its value as a final manufactured product. For example, a creosoted fence post may be worth $1, while this post still in the tree sells for 5 cents. Therefore, a correspondingly small sum should be spent in measuring standing timber.

Options in prospective sales are frequently for short terms only, requiring the estimation of large tracts in a comparatively short time. Correspondingly cheap and rapid methods are required for such work.

For these reasons, a standard of accuracy in measuring standing timber equal to that applied to logs is not to be expected. The cost of obtaining such measurement is reduced by substituting partial estimates or samples, as is done in determining the quantity and quality of unmined iron ore. These samples, in the form of plots or strips, are laid off in a manner calculated to attain the highest degree of accuracy possible for the average obtained by the sampling. The size or extent of the sample depends upon the cost limits imposed by timber value and by the time and funds available.

These practices form the art of timber estimating. When combined with the construction of maps of topography and forest cover, and with the inventory of such other forest resources as land, forage, water, game, and recreational utility, the undertaking constitutes the forest survey.

5. The Measurement of Growth. The purpose of measuring the growth of trees is to predict the quantity of wood products which may be cut from definite areas of forest land at the end of given periods of years, reckoning from the present. Such growth prediction aids an owner in estimating the profits to be made in holding his standing timber for cutting at a future date as against sale at the present time.

The growth of single trees is of interest chiefly in indicating the growth of stands of timber. Past growth of trees or stands is useful only as it reveals what may be expected in the future. The study of growth is therefore not a simple problem of counting and measuring the number and width of the annual rings on felled or standing trees, though this is the basis of some growth studies. Growth may also be studied by measuring the trees which make up the total volume of the stands and then determining the ages of the stands. For short periods, the difference in volume of the stand or of the trees at the beginning and end of the period is required. It is also necessary to know the number of trees of each size surviving in the final year. Growth rings may be counted or measured with great accuracy, but not in great numbers because of the cost.

The accuracy of growth predictions depends in turn on three qualifying or limiting determinations involving possible errors, *viz.*, first, volumes of standing trees; next, the existing volumes of stands of all degrees of stocking; and, finally, the prediction of what trees such stands will still contain many years hence. This threefold process involves a progressively increasing possibility of error which is similar to the increasing inaccuracy of the three successive steps in measurement of volume itself, based on the log, the tree, and finally the stand. The final step, that of prediction of growth on the entire area, is thus dependent entirely upon the accuracy of estimates or inventory of the present stand. No amount of care in growth measurement itself will decrease this source of error. As a practical science, growth predictions may keep pace with the accuracy of timber estimating but cannot rise above this limitation.

Just as tables of log or tree volume may be brought to a high degree of accuracy and reliability, so standard tables of growth and yield per acre may possess similar reliability. Yield tables may form a starting point for application to the real problem of measuring growth on large areas with less than full stocking. As in the case of timber estimates, the larger errors occur in the application to the average run of timber.

References

BRUCE, DONALD, and C. E. BEHRE: Forest Mensuration Today, *Jour. Forestry*, **23**, 282–289 (1925).

────── and F. X. SCHUMACHER: Forest Mensuration, 2d ed., McGraw-Hill Book Company, Inc., New York, 1942, 424 pp., 108 figs.

CARY, A.: A Manual for Northern Woodsmen, 4th ed., Harvard University Press, Cambridge, Mass., 1932, 366 pp.

CHAPMAN, H. H.: Forest Mensuration, 2d ed., John Wiley & Sons, Inc., New York, 1924, 557 pp.

────── and D. B. DEMERITT: Elements of Forest Mensuration, 2d ed., Williams Press, Inc., Albany, N.Y., 1936, 451 pp.

COOK, H. O.: Forest Mensuration Tables for the Measuring of Logs, Trees, and the Growth of Stands, Wright & Potter Printing Co., Boston, 1921, 69 pp.

GRAVES, H. S.: Forest Mensuration, John Wiley & Sons, Inc., New York, 1906, 458 pp.

MEYER, W. H.: Abstracts of Certain Phases of European Literature in the Field of Forest Mensuration, U.S. Forest Service, 1931, 406 pp.

MUNNS, E. N.: A Selected Bibliography of North American Forestry, *U.S. Dept. Agr. Misc. Pub.* 364, 1940, 636 pp.

──────: Chairman, Committee on Forest Technology, Society of American Foresters: Methods of Preparing Volume and Yield Tables, *Jour. Forestry,* **24,** 653–666 (1926).

──────, T. C. HOERNER, and V. A. CLEMENTS: Converting Factors and Tables of Equivalents Used in Forestry, *U.S. Dept. Agric. Misc. Pub.* 225, 1935, 59 pp.

SCHUMACHER, F. X.: New Concepts in Forest Mensuration, *Jour. Forestry,* **36,** 847–849 (1938).

U.S. Forest Service: Woodlot Forester's Tool Kit, North Central Region, Milwaukee, Wis., 1946, 70 pp.

WINKENWERDER, H., and E. T. CLARK: Handbook of Field and Office Problems in Forest Mensuration, 2d ed., John Wiley & Sons, Inc., New York, 1922, 133 pp.

CHAPTER 2

ELEMENTARY COMPUTATIONS

6. Introductory. The literal definition of mensuration as given by Webster is "1. Act, process, or art of measuring. 2. The branch of applied geometry concerned with finding the length of lines, areas of surfaces, and volumes of solids, from certain simple data of lines and angles." Forest mensuration has been defined as "a science dealing with the measurement of volume, growth, and development of individual trees and stands and the determination of various products obtainable from them." Forest mensuration is therefore a study not only of the static relationships of lines, areas, and volumes as applied to trees, but also of the dynamic relationships involving change of these measurements through growth and losses. The definitions imply directly that measurements must be taken and analyzed and that each student must therefore be equipped with sufficient knowledge of methods of basic arithmetical and mathematical calculations so that he can attain an accurate result, most easily.

It is not the purpose of this text to give an exhaustive treatment of arithmetical, statistical, and graphical methods, but it is essential that the basic concepts be developed at least in the simpler techniques. A sympathetic understanding of the nature of scientific data and an ability to perform various elementary mathematical computations is essential even for those who are giving the field of forest mensuration only a casual survey. The more detailed the survey becomes, the more essential is a good knowledge of these concepts. For the investigator, a thorough rigorous course in statistical and graphical techniques becomes a requirement. It is attempted here to give initially a coordinated picture of the most useful methods that are basic to a wide variety of problems and then in later chapters to treat specific topics in more detail and to develop special techniques useful for a restricted class of problems. It should be assumed not that the demonstrated techniques apply solely to the described problems but rather that they have a broad scope and are tools of more general application, ready for use according to the ingenuity of the mensurationist. This chapter therefore includes a number of definitions and descriptions of elementary techniques of graphics and of statistical analysis up through the computation of standard deviation. Later chapters apply the techniques and introduce new procedures. Chapter 21 explains certain statistical conceptions and contains a brief review of the more advanced statistical procedures yet is not intended to replace an adequate textbook devoted solely to the subject. The present chapter can therefore be considered as containing general reference material, which must be covered in advance of any computations and then referred to frequently in further work.

In every mensurational job, before a single measurement is taken, the investigator usually asks himself several fundamental questions. What are the objectives of the study? What measurements should be taken, and how should they be taken? In what form should the measurements be listed? What form should the calculations take to produce the best results with the minimum amount of work? How can the calculations be simplified and made more efficient and accurate? What arithmetical, mathematical, or statistical expressions give the best interpretation of the data? These questions arise immediately, and each job will require individual treatment. A common pattern, however, prevails throughout. The methods here given are fairly well standardized, but similar assurance cannot be given as to some of the techniques introduced later. Forest mensuration is not a matured science but is constantly growing, and new techniques are evolved each year, expanding upon older procedures or even replacing them. The rules of preliminary calculation change only slightly, however.

7. Accuracy of Measurement. The first rule of forest mensuration involves accuracy of measurement. By this is meant not the taking of a measurement to the utmost degree of refinement but the adoption of a relative standard of accuracy which satisfies the demands, eliminates as much as possible any elements of error of poor instrumentation, and permits the control of costs within prescribed economic limits. Errors are unavoidable, but they can be definitely restricted if a conscious effort is made.

The *random error* is not an error in the popular sense of the term, *viz.*, a mistake in instrumentation or computation, but a variation which will occur in any collection of measurements, owing to the natural and expected deviation of individual measurements from the average of the whole sample. The "errors" of about half of the individual measurements will be minus, those of the remaining plus, but collectively the measurements give a good expression of the average condition. In the case of the heights of trees in a pure even-aged stand, for example, the exact height values will vary from tree to tree, sometimes covering a range of 30 ft. or more. It lies within the nature of tree growth itself to produce this variation. It is a type of variation or "error" which forms the basis of statistical analysis.

If the instrument used were out of order or adjustment, a second kind of error would be added to the measurement, *viz.*, a *biased error*. Thus, if the heights of the trees were being measured by a transit, supposedly an accurate instrument, but the transit were out of adjustment, biased readings would be obtained, all tending in the same direction. Biased errors must be avoided as much as possible, since there is no way of eliminating them in further calculations. *Systematic errors* are synonymous with biased errors.

Compensating errors are of another class and are of the kind which may give a reading that is too large or too small, but such errors tend to average out if enough measurements are taken. Thus, if the tree heights are taken with an Abney hand level, an instrument of moderate precision, the height of one tree may be overestimated slightly, the height of the next tree underestimated. This type of error does not belong to the same vicious class as the biased error but is largely the result either of using instruments that have a lack of precision or of speeding up the work.

Accidental errors are commonly associated with the novice and if on a major order can be detected readily. Thus, in measuring tree heights where the trees range between definite limits, if a height is read that exceeds any yet recorded and is out of line with the average run of heights for trees of that diameter and species, it should be checked immediately, since the reading may be the result of careless work. This "over-all" check should be constantly applied to figures that normally should lie within a certain range of values with which the computer is familiar through his knowledge of the problem; *e.g.*, the misplacement of a decimal point in a timber estimate may introduce an error of 1,000 per cent, which should easily be detected. Minor accidental errors are not detectable and may take the form of biased, systematic, or compensating errors. Training in sound technique and systematization of measurement are the best ways to eliminate this class of error.

8. Significant Figures. Since accuracy of measurement is a matter of relative precision, depending upon the nature of the investigation, the question of *significant figures* always arises. In other words, how many digits should be carried, and how should they be recorded? Tree heights will seldom be measured to tenths of feet; they may be measured only to the nearest foot or may even be estimated only to the nearest 5 or 10 ft. and still be accurate enough for the purpose intended. A standard of accuracy less than the highest obtainable is not only permissible but compulsory in most problems, because of limitations of time and cost and the purposes to be served. The one requirement is consistency, or adherence to the standard set.

When a measurement is taken to definite limits of accuracy, however, it should be recorded appropriately. Thus a tree which is measured to the nearest foot, say 96 ft., should not be recorded as 96.0 ft., since the additional digit, zero, implies that it was measured to tenths of a foot; the 96-ft. reading means usually that the height could be between 95.6 and 96.5 ft. but was read to the nearest foot only. Likewise, if trees are measured by 5-ft. classes, this tree or any other tree between 92.6 and 97.5 ft. in height will be listed as being in the 95-ft. class, and the added accuracy of measurement will not be used. The last digit is therefore not accurate but signifies that the measurement has been rounded off to this nearest number. It is misinforming and inaccurate to add digits when they represent refinements of measurement which were not made. Zero is a digit and is treated the same as the others.

In many calculations, it is necessary in one way or another to compound measurements containing a different number of significant figures. Thus, if one figure of moderate accuracy, such as 25.3, having three significant digits with the last one a close approximation, is to be multiplied by a precise figure such as 15.2564, with six significant digits, giving 385.98692, the full result obtained is of doubtful value, since all digits of the product following the third are affected by the digit .3 and depart from exactness. A product (or quotient) is accurate to no more places than that of the value which has the least number of significant figures. Hence the above result should be stated as 386 (three significant figures), an answer which is obtained by multiplying 25.3 by 15.26 and dropping all digits beyond the third in the result. The rule is to include in the multiplier (or divisor) only one more significant figure than in the multiplicand (or dividend). It should be noted from the above example that significant figures have no relation to decimal points. They are counted from the left to the right until the last significant digit is reached.

9. Rounding Off. Rounding off is a device commonly employed to simplify records and calculations. The limit to rounding off is dictated by the desire for number of significant figures. Few rules can be stated without prior knowledge of the specific problem, but the method of rounding off can be systematized. Thus, in the case of tree diameters, it may be desired to carry a tally listing trees only to the nearest inch;

all trees above the .6 measurement in any diameter class are rounded off to the next higher class. The .5 point itself stands exactly halfway between the two class values but is usually rounded off down to the next lower class. In some instances, the practice has been adopted to compensate this slight error by making the previous figure an even number. For example, if a measurement of 24.5 in. has been taken, the figure preceding the .5 is already even, and the rounded-off value is called 24. In the case of 25.5 in., the previous figure is odd, and the value is rounded off upward to 26 in.

An average value obtained by computation is never rounded off, since an average is accurate to more places than the individual values comprising it. The greater the number of values, the more accurate are the decimal places. If an average is rounded off and expanded later to a sum by multiplying the average by the number of values, the error will be equal to the difference between the actual average value and its rounded-off value multiplied by the number of values.

10. Appearance of Data. Systematization of the recording of data is essential to speed, accuracy, and understanding. Each problem has its own characteristics and must be scheduled by itself. As a specific illustration, one could take the case of Table 1, listing the diameters of Douglas fir. The trees were listed as measured. From this list, only a superficial grasp is gained of the character of the forest as it relates to diameter. By inspection, an observer can get a good idea of the total range, 5 to 28 in., and a vague impression of the distribution of diameter classes. Only by long and tedious calculation can he obtain averages and other statistical values. The first simplifying step would be to assemble the measurements by inch classes. This could and should have been done in the first place when the stand was measured, thus saving much time and space in recording. When measurements can be tallied by classes, as is here the case, a convenient form of tallying is found in the dot-dash system, in which items are entered in the proper class first by dots, then by dashes, finally making a closed, crossed square when the count reaches 10, as follows:

Number of items............ 1   2   3   4   5   6   7   8   9   10

Score..................... .   . .   . | . | .   ⌐ .   ⌐ ⌐   ⌐ ⌐   ⌐ ⌐
                                                                         | | | ⊠

The three columns of Table 2 show the measurements arranged in this logical fashion. One immediately gets added impressions just from the appearance of the data. The trees have been grouped into diameter classes, showing effectively a characteristic distribution pattern. In addition, a more efficient form for the purposes of calculation has been obtained.

The construction of appropriate forms for registering data involves consideration of certain definite points. First of all, the form should be designed to facilitate the tallying of data in a regular, if not automatic, fashion and in this way to eliminate mistakes. Second, all columns should be clearly labeled; or, if sufficient room is not avail-

TABLE 1. DIAMETERS AT BREAST HEIGHT ON 1 ACRE IN A DOUGLAS FIR FOREST, 67 YEARS OLD OF SITE QUALITY III
(Diameters listed in the order taken)

16	18	18	14	10	21
10	10	14	8	18	15
9	15	12	12	9	16
22	14	16	13	13	17
24	16	14	13	15	21
8	22	15	12	18	27
6	8	18	9	11	19
12	19	13	16	18	11
7	7	15	11	17	20
17	20	19	10	20	12
16	11	9	17	23	17
17	14	15	15	15	11
13	20	9	15	11	18
20	10	13	8	19	18
12	15	14	10	10	13
13	26	11	10	21	15
15	21	14	16	20	12
16	19	17	14	18	11
13	14	6	15	11	12
23	14	16	16	16	12
13	20	14	13	12	19
11	24	14	9	17	11
17	9	20	9	10	19
21	19	10	13	20	8
14	14	19	18	16	13
8	10	14	15	10	17
21	17	15	22	22	16
9	12	19	11	12	10
14	12	13	15	17	14
6	17	16	17	16	21
18	17	15	15	12	14
9	7	8	13	19	23
18	23	17	12	17	17
26	28	13	15	11	11
7	18	11	10	16	12
14	25	13	16	5	16
8	14	12	9	15	

TABLE 2. DIAMETERS AT BREAST HEIGHT FOR A DOUGLAS FIR STAND TALLIED IN 1-IN. CLASSES

Diameter at breast height, in.	Tally	Number of trees
5		1
6		2
7		4
8		8
9		11
10		14
11		15
12		17
13		17
14		20
15		20
16		19
17		18
18		13
19		11
20		9
21		7
22		4
23		4
24		2
25		1
26		2
27		1
28		1
Total.......	....	221

able, notations at the bottom of the sheet or lists of instructions should be available so that no doubt exists as to the exact meaning of any column or row. Third, the form should anticipate that calculations will be made later and should therefore be arranged so that these computations will be facilitated. In some instances, computational columns may be included on the form. Fourth, the form should be as compact as possible.

Notations on the form should be clear and precise. Numerical values should be clearly written with a pencil that will not blur or blotch. Erasures are to be avoided, since they can cause trouble later. Many mistakes are made because values are illegible or are confused with some other value.

11. Proficiency in Calculation. Proficiency in calculation can be developed by any person, provided that he deliberately seeks the methods to gain efficiency. Output can be doubled and trebled and at the same time the work can be made far less arduous if a few simple rules are adopted.

1. Develop a calculation form, assuring a progressive calculation.
2. Use clear notation. Sloppy work should never be countenanced.
3. Make the form as compact as possible without crowding.
4. Reduce the work to automatic steps as much as possible, and arrange to do similar operations consecutively.
5. Plan the work to be a one-man job. An added man seldom if ever doubles output. For example, in the case of adding, the novice often has another man call off the values to him as he enters them into the adding machine. This is hardly worthwhile, except when values have to be selected from different places.
6. Arrange for systematic checks, if possible, and always be on the lookout for "bulls."
7. Use adding machines and calculators wherever possible, and slide rules for approximations and checks, and gain proficiency in the use of these tools.
8. If several forms have to be worked over simultaneously, arrange the forms compactly on the desk and in one plane to reduce eyestrain caused by frequent shifting of the eyes and refocusing to various planes.
9. In columnar tabulations, always have the digits directly in line, without any staggering, and carry all numbers to the same number of places in a column.
10. Whenever a result does not "look right," check immediately, and save the time of going back later over a series of calculations and of making corrections throughout. Many series of data follow

some general trend, departures from which indicate probable errors. "Impossible" results must be checked immediately. A thorough understanding of the limit within which values should fall constitutes a major "ocular" check in all calculations.

11. Remember that there is no perfect human calculator. Each man will make mistakes at one time or another, but some people admittedly are more inaccurate than others.

12. Population and Sample. The complete aggregate of individuals in one category, such as all Douglas fir trees of the same age and site class growing in fully stocked stands, composes the *population*, or *universe*. Very large populations which are studied in forest mensuration are practically *infinite*, *i.e.*, composed of practically unlimited numbers of individuals, as in the case of Douglas fir; only a relatively few are definitely limited or *finite*, as in the case of deodar in Lebanon, a species which formerly covered vast areas but now is found only in one small isolated stand. Since the usual case is one of the infinite population, a study of the characteristics of the population must necessarily be based upon measurements or observations of small portions, or *samples*. A single sample does not exactly define the characteristics of the population, but reasoning in the form of *statistical methods* permits valid inferences of these characteristics.

In the Douglas fir stand of Table 1, a sample of one tree could easily lead to a wrong estimation of the character of the stand. The chances are favorable that it may have a diameter which is within several inches of the true diameter; yet, by chance also, one of the very small or very large trees might be taken and give a completely erroneous impression. A sample of two trees will not be much better. The measurement of 200 trees as in the sample shown should impart a fair measure of confidence in the result. Yet there is the objection that the 200 trees measured are all located on one acre and that other acres in the same 67-year-old stand may be somewhat different. Hence, in order to get a still better estimate of the condition of the entire stand, a number of acre plots should be taken. Thus we have a scheme of sampling within sampling. The single trees of the plot can be considered *sub-samples*, while the plots themselves will be taken as the full *samples*, and the entire collection of plots will constitute the *sampling scheme* or sample of the entire area. Now, if the forest area is composed of two recognizably different age classes, site classes, or forest types, it must be anticipated that the conditions will vary greatly and that all sample plots taken in the area should not be thrown together. Instead the area should be first divided into its component parts and the sampling within each part kept separate. This divisioning is called *stratifying*, and the samples procured are *stratified samples*. Stratifying is recommended practice in accurate timber estimating. (For further discussion of sampling, see Sec. 167.)

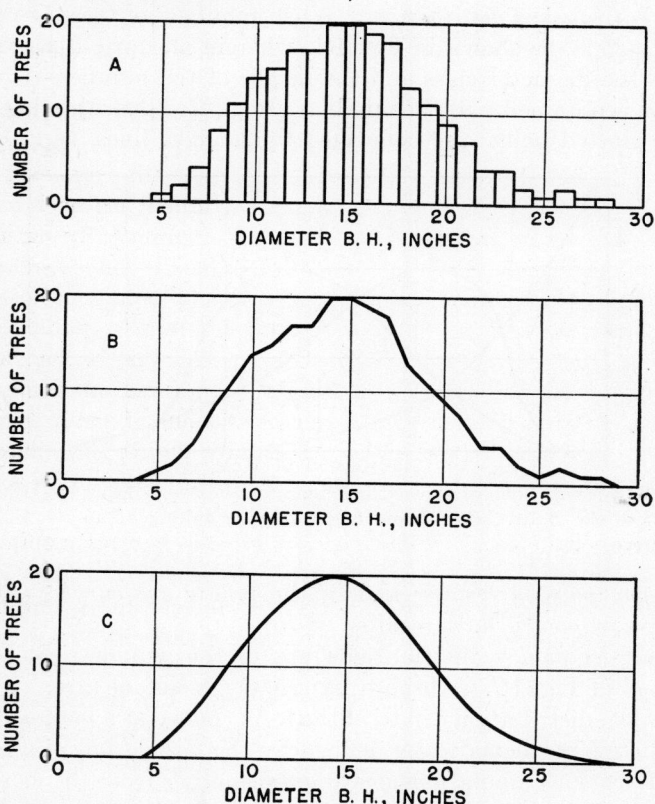

FIG. 1. Representations of the frequency distribution of tree diameters in a Douglas fir stand. *A*. Histogram. *B*. Frequency polygon. *C*. Frequency curve.

Since much of the basic statistical theory is founded upon the entire population, special adaptations have been developed for application to samples and to finite populations in order to give results which will best approximate the truth. A few of these special techniques will be introduced later at appropriate places, particularly those which are applicable to the field of mensuration and to the type of problem illustrated.

13. Frequency Distributions. The case of diameters in Douglas fir stand, as shown in Table 2, is one of numerous instances arising in forest mensuration in which a definite pattern of frequency of occurrence of units in each of a series of equal classes is obtained, on the basis of a simple characteristic, such as diameter or height, into which the range of distribution is divided. The general term used for the pattern is *frequency distribution*. When numbers of trees are plotted over diameters, as shown in Fig. 1A, wherein an upright bar has the width of the diameter class and the height of the numbers of trees, a *histogram* is obtained. If the numbers are plotted over the class center and the plotted points are connected by straight lines, as in Fig. 1B,

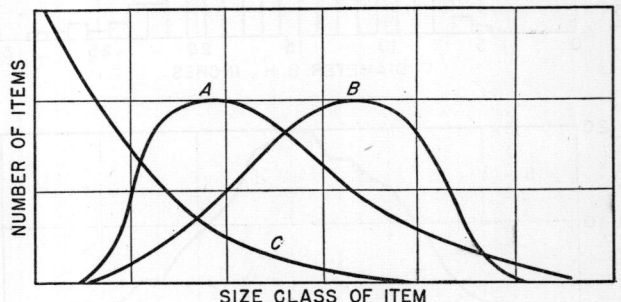

FIG. 2. Representations of skewed frequency curves. *A.* Skewed positively, or tailing off to the right. *B.* Skewed negatively, or tailing off to the left. *C.* J-shaped curve.

a *frequency polygon* results; and, if the points are curved out, as in Fig. 1C, a *frequency curve* results.

Not all frequency distributions are of the symmetrical order as illustrated in Fig. 1C. Absolute symmetry is in fact rare. The true symmetrical distribution can be of the form classed as a *normal distribution*, which is mathematically defined by the *normal law of error* (Sec. 168). Excess accumulations one way or another from the center lead to *skewness*. If this accumulation is to the left of center with a consequent long tailing off to the right, the skewness is termed *positive* (Fig. 2A); if to the right with a tailing off to the left, the skewness is termed *negative* (Fig. 2B). When distributions are arbitrarily cut off by, let us say, a minimum diameter limit in the case cited, a *curtailed distribution* is obtained. In the extreme case, a *J-shaped distribution* can be obtained (Fig. 2C). This would be typical, for instance, if a tally were made of diameters in an all-aged, or selection, forest.

The summarized diameters of Douglas fir trees by diameter classes in Table 2 have been plotted and curved out (see Fig. 1C). The plotting will appear to approach closely that of the symmetrical normal curve with slight positive skewness, or tailing out longer to the right than to the left. This is a most common form of warping of the normal curve. From Table 2 and Fig. 1C, several significant impressions, which were not available from Table 1, may be drawn without additional computation. In addition to the complete range of values (5 to 28 in.), one can now see that the classes of greatest frequency are the 14- and 15-in. classes and that, because of the fairly symmetrical distribution, the average diameter must lie close to this size class. On each side of the 14- and 15-in. classes, the numbers of trees fall off first

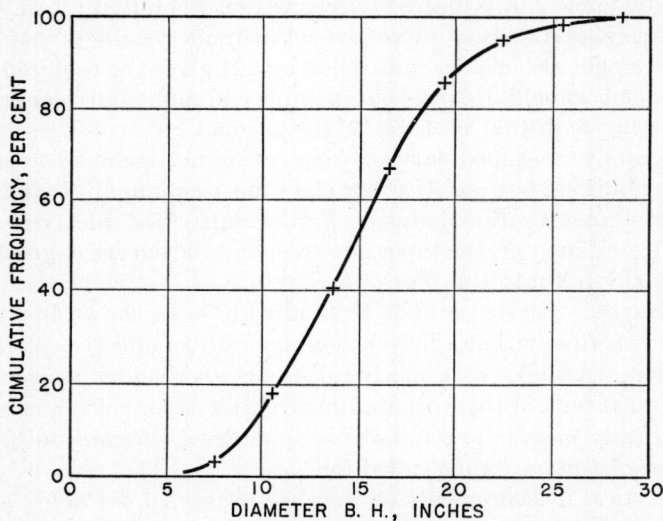

FIG. 3. Cumulative frequency distribution in per cent for tree diameters in a Douglas fir stand.

slowly, then rapidly, and then again slowly when small numbers are reached. An abstract definition of this character is not satisfying; some more precise measures should be available for adequate description. Such measures are in fact discussed in Secs. 15 and 16, covering measures of central tendency and measures of dispersion.

Frequency distributions can be recorded in actual numbers or can be converted to percentages. The latter device is frequently used since the plotting of numbers will then be on a common vertical scale for the various patterns treated.

There is occasional need for *cumulative distributions* in terms of either actual number or percentages. If, for example, the values of Table 4, which summarizes the Douglas fir trees by 3-in. classes, were totaled progressively from the smallest diameter class to the largest and then converted to percentages by dividing each total by 221 and if the percentages were then plotted over the upper class limit, an *S-shaped curve* of the type shown in Fig. 3 would be obtained. Cumulative distributions should be plotted at the upper limit of diameter classes to which they refer and not on the center of the diameter class, as was done in Fig. 1. One of the uses of a cumulative distribution can be illustrated immediately from this graph, as follows: The graph was actually constructed on the basis of 3-in. classes, but for some reason, assuming the absence of Table 2, it is desired to reconstruct the tally by 1-in. classes. The differences between successive class-limit readings from the S curve by single inch classes multiplied by 221 gives the required result. The student should follow the example through and compare the derived tally with that of Table 2, the original.

Incidentally, S-shaped curves, or *ogees*, are not confined to cumulative distributions but constitute a class most commonly found in the plotting of mensurational data. For example, the plotting of tree heights over diameter, the taper of a tree, and the curves of growth and yield usually fall into this class.

14. Graphs. Several graphs have already been shown in the foregoing. It is timely that a few elementary rules of effective plotting be mentioned. A choice of properly ruled paper is the first consideration. For the great bulk of the work, ordinary graph paper ruled horizontally and vertically in even graduations is applicable. *Rectangular*, or *cartesian, coordinates* are thus called for.

The vertical measurements on graphs are termed *ordinates*, and the horizontal measurements *abscissas*. The vertical scale of ordinates is always reserved for the *dependent* variable and is called the Y axis, and the horizontal scale of abscissas is reserved for the *independent* variable and is called the X axis. The dependent variable is the one of an associated pair whose value is considered to depend on that of the other variable in the pair. For example, with height and diameter of a tree, one usually thinks of height as depending upon or associated with diameter. Hence heights would be measured off on the scale of ordinates, and diameters (in this case the independent variable) on the scale of abscissas. In some cases, two or more independent variables can be used in association with one dependent variable; for example, the volume of a tree depends upon diameter and height. The graph would be prepared to show volume on the vertical scale and diameter on the horizontal scale. The second independent variable (height) would be introduced by the use of separate lines for each of several arbitrarily chosen height classes, such as 20, 30, 40 ft., etc. If it were possible to construct a three-dimensional chart, the system of lines for height would be replaced by a surface, thus permitting the reading of volume for any height as well as diameter.

The *scale*, or *graduation*, of an axis is chosen so that the full range of classes of the variables will be represented on the chart. The scales of the dependent and of the independent variable need not be the same, and such is rarely the case. It is desirable, but not necessary, to have the zero values of each variable appear on the chart, since they often give some degree of control when a curve is drawn through plotted points. A single point in a two-dimensional graph is located by a pair of values, the *coordinates*, of which the first is the abscissa and the second the ordinate. For example, if a tree of 15 in. has a height of 76 ft., its coordinates are 15, 76.

To a certain degree the size of the graph depends upon the range of the data, but this is only relatively the case, since the scale can be chosen at will. It is good practice to standardize on size, making fullest possible use of letter-sized sheets ($8\frac{1}{2}$ by 11 in.) or double letter-sized sheets (11 by 17 in.). These sizes can easily be inserted in the usual files and in manuscripts. Odd-sized pieces, small or large, are to be avoided. The most convenient type of rectangular graph paper to use is that with 20 subdivisions to the inch, of which the 10-lines are heavy, the 5-lines medium, and the unit lines light in weight of printing. Ten by ten graduations to the inch are suitable also for many kinds of work. Graph paper is available in many kinds of graduation systems, color of printing, and weight of paper.

The scales should be clearly labeled and the graduations carefully inscribed to prevent later confusion. Before a chart is set aside, it should be titled, dated, and signed. Plotted points without labeling and titling have no significance whatever, except in the mind of the preparer, who, if he does not date and sign the graph, may quickly forget its meaning.

Chart drawing is not confined to rectangular coordinates. In forest mensuration, much use is made of *semilogarithmic* paper, where one scale, that of the ordinates, is graduated logarithmically; of *logarithmic paper*, where both scales are graduated logarithmically; and of *probability paper*, where the scale of the abscissas is graduated according to the cumulative normal frequency distribution. Chart drawing also includes the drafting of figures even without the use of graph paper. Large books have been written on the topic of chart preparation, and there is no intention here to offer a complete manual. It is necessary at this time only to emphasize the nomenclature and the simple basic rules of clarity, neatness, and compactness.

15. Measures of Central Tendency. The vast majority of problems require the computation of a central value, usually some kind of *average* or *mean*, but these are not the only measures of central tendency that are available. The most useful of such measures in forest mensuration are

    a. The arithmetic mean.
    b. The weighted arithmetic mean.
    c. The geometric mean.
    d. The median.
    e. The mode.

Of these, the last two apply specifically to frequency distributions, but the first three are not so limited. Examples of the estimation of each will be drawn from Table 2, which happens to be a frequency distribution.

    a. Arithmetic mean. *Long Method.* If in the case of Table 2 the diameter is denoted by D and the number of trees within a diameter class by f, each D is multiplied by its corresponding f and the products totaled (Table 3). This sum is then divided by the total number of trees to arrive at the arithmetic mean. The formula in this case would be

$$\text{Average, or mean, diameter} = \frac{\text{sum }(fD)}{\text{sum }(f)} = \frac{3{,}258}{221} = 14.74 \text{ in.}$$

For the purpose of expressing the above equation concisely and of generalizing its form, the following notation will be adopted for the final formula:

$$M = \frac{S(fX)}{S(f)}$$

where M = mean value
       S = summation
       X = variable, which takes on different values
       f = frequency in any one class of X

The same result (14.74 in.) would have been obtained if the diameter values in Table 1 had been added and divided by their total number. Strictly speaking the above formula is for a weighted mean, but in the sense here used the term "weighted mean" will be reserved for a special meaning, later described.

    Short Method. Calculation efficiency can be increased by coding and grouping. Coding will be described first. It is obvious that, for some classes of data, the X value may be in terms of large figures, entailing much computation. Coding simplifies the work, particularly in the case of frequency distributions. An estimate is first made of the position of the average value, called an *assumed mean*, and a new value of 0 is then set at this point. Classes below the assumed mean are noted progressively $-1, -2, -3$, etc., and classes above the mean $+1$, $+2, +3$, etc. (see column 4, Table 3). The classes must of course be of even width. The classes have now been *coded*. The code values times the frequencies in the classes are computed and added, taking due regard for minus and plus signs (column 5, Table 3).

TABLE 3. CALCULATIONS OF A MEAN DIAMETER BY LONG AND SHORT METHOD

Diameter, in. D	Frequency f	Long method fD	Short method	
			Deviation from assumed mean $X_a - D = x_a$	fx_a
5	1	5	-10	-10
6	2	12	-9	-18
7	4	28	-8	-32
8	8	64	-7	-56
9	11	99	-6	-66
10	14	140	-5	-70
11	15	165	-4	-60
12	17	204	-3	-51
13	17	221	-2	-34
14	20	280	-1	-20
15	20	300	0	
16	19	304	$+1$	$+19$
17	18	306	$+2$	$+36$
18	13	234	$+3$	$+39$
19	11	209	$+4$	$+44$
20	9	180	$+5$	$+45$
21	7	147	$+6$	$+42$
22	4	88	$+7$	$+28$
23	4	92	$+8$	$+32$
24	2	48	$+9$	$+18$
25	1	25	$+10$	$+10$
26	2	52	$+11$	$+22$
27	1	27	$+12$	$+12$
28	1	28	$+13$	$+13$
Total.........	221	3,258		-417 $+360$ Net -57
	$M = \dfrac{3,258}{221} = 14.74$		$M = \dfrac{-57}{221} + 15 = 14.74$	

The mean is now computed as $-57/221 +$ assumed mean (15 in.) $= -0.24 + 15 = 14.74$ in. The general formula for this case, where a class holds only one unit (here 1 in.), is

$$M = \frac{S(fx_a)}{S(f)} + X_a$$

where $X_a =$ assumed mean
$x_a = X_a - X$
$S(f) =$ total number $= n$

The result is identical to that obtained by the long method, and the computation of products has been reduced to mental arithmetic. An adjustment in the above formula is needed where the deviations (x_a) from the assumed mean are not in the same units as for X. For example, if the data are recorded by 2-in. classes instead of 1-in. classes, and if the x_a are listed progressively from X_a by differences of 1, then the fraction in the formula must be adjusted for the class width, and the revised formula will be

$$M = \frac{S(fx_a)}{S(f)} C + X_a$$

where C is the number of units of X contained in one class of x_a. This formula is used in the next case, where the calculation is simplified still further.

TABLE 4. CALCULATION OF A MEAN DIAMETER AFTER GROUPING, SHORT METHOD

Diameter, in. D	Deviation from assumed mean $X_a - D = x_a$	Frequency f	fx_a	
6	−3	7	− 21	
9	−2	33	− 66	
12	−1	49	− 49	
15	0	59		
18	+1	42		+ 42
21	+2	20		+ 40
24	+3	7		+ 21
27	+4	4		+ 16
		221	−136	+119
			Net −17	

$$M = \frac{-17}{221} \times 3 + 15 = -0.23 + 15 = 14.77 \text{ in.}$$

Grouping of data into broad classes is often necessary for convenience of recording and calculating. Table 2 is now condensed into

ELEMENTARY COMPUTATIONS

3-in.-diameter classes, the assumed average diameter X_a is again taken as 15 in. and the x_a are recorded in successive differences of 1 from X_a. In this case each difference of 1 in x_a covers 3 in. of diameter.

A slight difference of 3 one-hundredths of an inch has been introduced, which for many purposes is insignificant. Grouping therefore simplifies calculations greatly and introduces very little error, if not made to cover too wide a range in each class.

Grouping has been used in the above without a preliminary discussion of the terminology involved when classes are used. *Class width* is the range covered by a single class. In Table 2, where diameters were set up by unit inch differences, it is correctly inferred that the value stands at the mid-point of the class, called *class mark* or *class center*, and that the class extends halfway on each side to the adjacent classes. Thus, for example, 15 covers a *class range* from a *lower class limit* of 14.6 to an *upper class limit* of 15.5 in. In the calculation of the preceding paragraph, 15 is the center of a 3-in. class and therefore covers from 13.6 to 16.5 in. Since the .5 overlaps from class to class, it is common practice to call the class width as ranging from the .6 below to the .5 above (Sec. 9).

TABLE 5. CALCULATION OF A MEAN DIAMETER, WEIGHTED BY BASAL AREA

Diameter, in. D	Basal area, sq. ft. BA	Frequency f	$f \times BA$
6	0.196	7	1.372
9	0.442	33	14.586
12	0.785	49	38.465
15	1.227	59	72.393
18	1.767	42	74.214
21	2.405	20	48.100
24	3.142	7	21.994
27	3.976	4	15.904
Total............	.....	221	287.028

Average basal area = 287.028/221 = 1.299 sq. ft. Diameter corresponding to average basal area of 1.299 sq. ft. = 15.42 in. (see Table 13).

b. Weighted arithmetic mean. The special use here given this term is a mean which has been weighted in some other way than by number of items alone. The most clear-cut example lies in the use that forest mensurationists make of an average diameter weighted by basal area. The diameter sought is equal to the diameter of the average cross-sectional area at breast height and not the average of the diameters directly. The first is always larger than the second. Theoretically this mean diameter weighted by basal area is closely related to the diameter of a tree of average cubic volume, since volume varies roughly as the square of the diameter. Deviations from an assumed mean can no longer be used, but the calculation must proceed as shown in Table 5. Basal areas for specific diameters read from Table 13.

This special case of a weighted mean diameter is the one most commonly used in forestry. It is here classified as a weighted mean, since the same result could have been obtained if the squares of diameters had been used instead of the basal areas and the mean determined by taking the square root of the average square.

c. Geometric mean. The geometric mean has some use in forest mensuration, although it has not always been recognized as such. For example, geometric means are those obtained when the logarithmic method of cubic-foot volume table construction is used (Sec. 95). It is obtained by taking the nth root of the product of n terms, a procedure which is not at all complicated when logarithms are used. The formula reads

$$G = \sqrt[n]{X_1 \cdot X_2 \cdot X_3 \cdots X_n}$$

where X_1, X_2, etc., are the specific values taken on by the variable X. If these X's are classed with frequencies of f_1, f_2, etc., respectively, the formula is changed to read

$$G = \sqrt[n]{X_1^{f_1} \cdot X_2^{f_2} \cdots X_n^{f_n}}$$

In this form the frequencies appear as exponents and not as multipliers as in the case of the arithmetic mean. The apparent difficulty of computing the fth power of an X value, then combining the various X^f's, and finally extracting the nth root is eliminated by conversion of the X's to logarithms and then applying the f's as multipliers, as shown in the sample calculation of the geometric mean given in Table 6.

The first two columns of Table 6 show the numbers of trees by 3-in. classes. The third column lists the logarithms of the diameters, and the fourth column contains the multiples of logarithms times frequencies. The sum of the fourth column divided by the total number of trees gives the logarithm corresponding to the average sought. By reference to a logarithm table, this average, the geometric mean, is 14.05 in. The geometric mean is smaller than the arithmetic mean, which in turn is smaller than the mean weighted by basal area. All three of these means have useful applications in forest mensuration.

ELEMENTARY COMPUTATIONS

TABLE 6. CALCULATION OF THE GEOMETRIC MEAN

Diameter, in. D	Frequency f	log of diameter	$f \times \log$
6	7	0.77815	5.44705
9	33	0.95424	31.48992
12	49	1.07918	52.87982
15	59	1.17609	69.38931
18	42	1.25527	52.72134
21	20	1.32222	26.44440
24	7	1.38021	9.66147
27	4	1.43136	5.72544
	221		253.75875

Average logarithm = 253.75875/221 = 1.14823. Diameter corresponding to average logarithm = 14.05 in.

d. The median. The median of a series of values which have been arranged, or *ranked*, in order of size is that value which stands exactly halfway through the series. If the number of items is odd, it is found by taking (total number -1)/2, then counting this number along the series from either end, and accepting the value immediately next to the last count as the median. If the total number of values is even, it would be obtained by taking (total number)/2, then counting along the series, and accepting as the median a value which was halfway between the last one counted and the one directly after it.

When the series is grouped into classes, as in Table 2 or 3, the tree in the middle of the series cannot be picked out directly but can be estimated by assuming that the frequency within the class is constant for every small interval of the class. This assumption, incidentally, is not quite correct, since the frequencies for small intervals of the class should increase in the direction of the mode (see next paragraph). An approximation formula is as follows:

$$\text{Median} = \text{lower class limit} + \frac{f_m - f_b}{f_c} C$$

where f_m = median number as determined above
f_b = cumulative frequency below the class in which the median lies
f_c = frequency in the median class
C = class width

In the example of Table 5, the median number is the 111th tree, and the median value is

$$13.5 + \frac{111 - 89}{59} \times 3 = 13.5 + 1.25 = 14.75.$$

In this case, therefore, the median and the average appear to be almost identical.

e. The mode. When a group of measurements is arranged by a series of classes and frequencies (Tables 2 and 3), in many instances one or a few classes show the highest frequencies. This class or these classes indicate the mode. The mode can be determined roughly by simply naming the class of highest frequency, or it can be somewhat more accurately placed by use of the following formula:

$$\text{Mode} = \text{lower class limit} + \frac{f_a}{f_a + f_b} C$$

where f_a = frequency in the class *above* the modal class
f_b = frequency in the class below the modal class
C = class width

This formula permits the location of the actual mode away from the center of the modal class, depending upon the frequencies on either side of the mode.

Many distribution series have only one mode and are hence called *unimodal*. Occasionally one may observe more than one mode. There may be, for example, two distinct humps in the curve, in which case the curve will be called *bimodal*. This case will arise, for example, if, in the case of the Douglas fir stand described previously, there were in addition an understory stand of young Douglas fir or western hemlock. Each age class or general story would be itself unimodal, but when one class is combined with the others a bimodal curve is formed. In many instances the bimodal curve arises exactly in this fashion, *viz.*, by joining two distinctly different distributions.

The calculation of the mode according to the formula given above in the case of the Douglas fir stand (Table 3) is as follows:

$$\text{Mode} = 13.5 + \frac{42}{42 + 49} \times 3 = 13.5 + 1.39 = 14.89 \text{ in.}$$

More accurate formulas are available for the computation of the mode, but these are applicable only in cases of mathematical frequency curve fitting, which is beyond the scope of this text.

16. Measures of Dispersion. The total number of observations and the measures of central tendency are fundamental to the description of a series or group of values. In many cases, this description is insufficient, and other measures must be used for a finer definition.

Measures of dispersion serve this purpose, since they define the extent to which the individuals in a set of measurements vary from its central tendency. The measures of dispersion most commonly used are

    a. Total range.
    b. Mean deviation.
    c. Standard deviation.
    d. Coefficient of variation.

a. Total range. The total range is a simple measure, being the total interval between the smallest and the largest values. It gives preliminary information on the scatter, or variability, of the values. In the case of Douglas fir, where the largest trees are in the 28-in. class and the smallest in the 5-in. class, the range is from 28 to 5 in., inclusive, or 24 in.

b. Mean deviation. The mean deviation is defined as being the average of the differences or deviations of the single observations from their mean, not considering the sign (+ or −) of the differences. It is not satisfactory from the statistical point of view but is often used in forest mensuration as spanning the gap between a crude measure of dispersion, such as the total range, and an accurate measure, such as the standard deviation. The calculations are performed in Table 7 for the Douglas fir data, grouped in 3-in. classes. This measure of

TABLE 7. CALCULATION OF THE MEAN DEVIATION

Diameter, in. D	Deviation from mean $14.77 - D$	Frequency f	$f(14.77 - D)$
6	− 8.77	7	− 61.39
9	− 5.77	33	−190.41
12	− 2.77	49	−135.73
15	+ .23	59	13.57
18	+ 3.23	42	135.66
21	+ 8.23	20	164.60
24	+11.23	7	78.61
27	+14.23	4	56.92
Total disregarding sign.	....	..	836.89

$$MD = \frac{836.89}{221} = 3.78 \text{ in.}$$

dispersion is useful in instances where quick results are desired. It leads to an approximation of the standard deviation, as will be explained later.

c. Standard deviation. This measure of dispersion is the accepted accurate expression of dispersion and is essential for further statistical developments. Most simply defined, the standard deviation is the square root of the average of the squared differences from the mean. Its squared value is denoted as the *variance*. In the case of the standard deviation, careful distinction must be drawn between the population and the sample. One formula applies for the population and a modified formula for the sample. Since the investigator deals principally with samples, the modified formula is usually used. For the population, the formula reads

$$\sigma = \sqrt{\frac{S(x^2)}{n}} \quad \text{or} \quad \sqrt{\frac{S(f_x{}^2)}{S(f)}}$$

where σ is the *standard deviation of the population*, x the deviate from the exact mean, f the frequency of the deviate, S the sum, and n the total number of deviates. For purposes of convenience in calculating, it is often desirable to work from an assumed mean (or even from 0 itself), in which case the formula reads

$$\sigma = \sqrt{\frac{S(X_a - X)^2}{n} - \left[\frac{S(X_a - X)}{n}\right]^2}$$

or

$$\sqrt{\frac{Sf(X_a - X)^2}{S(f)} - \left[\frac{Sf(X_a - X)}{S(f)}\right]^2}$$

where X is the full value of the observation and X_a the provisional or assumed mean. Although the second formula looks more formidable than the first, in actual practice it is much more easily handled.

The *standard deviation estimated from the sample* requires a slightly changed formula. The reason for the modified treatment lies in the argument that the measure one desires when dealing with a sample should be descriptive of the population. It has been proved mathematically that the best estimate of the standard deviation for the population by computation from a sample lies in substituting $n - 1$ in the denominator of the first formula given above; hence

$$s = \sqrt{\frac{S(x^2)}{n - 1}} \quad \text{or} \quad \sqrt{\frac{S(fx^2)}{S(f) - 1}}$$

where s is the best estimate of the standard deviation of the population, and the other symbols have the same meaning as before. It is seen, therefore, that a standard deviation computed without the corrected denominator is always smaller than one with it. In other words, the

ELEMENTARY COMPUTATIONS

standard deviation of a population is as a rule somewhat larger than that of the sample itself, with the difference increasing with the decrease in number of observations. The use of the factor $n-1$ has been described in some texts as being used as a factor of safety. Rather it is used because it has been mathematically proved to give the best estimate of the standard deviation of a population about which one is trying to draw inferences.

For convenience, it is again advantageous to work from an assumed mean, in which case the formula is revised to read

$$s = \sqrt{\frac{S(X_a - X)^2 - [S(X_a - X)^2]/n}{n-1}}$$

or

$$s = \sqrt{\frac{Sf(X_a - X)^2 - [Sf(X_a - X)^2]/S(f)}{S(f)-1}}$$

The calculation according to the last formula will be carried out in Table 8 for the data of Table 2, where the assumed mean will be taken as 15 in. (For further discussion of standard deviation and its application, see Secs. 172, 174, and 175.)

It should be noted that the calculation in Table 8 has the same first four columns as required for the calculation of the mean and that only one more column has been added. The signs ($+$ or $-$) disappear in the last column, since the sign of a squared value is always plus.

d. The coefficient of variation. It is sometimes desirable to express dispersion on a relative, or percentage, basis. Thus, for example, if two means are not the same and have differing standard deviations, it is not possible to state whether the degree of variation of the two is similar unless the deviations are transformed to a comparable basis. This is done through *coefficient of variation*, where c.v. $= s/M$, in which c.v. is the coefficient of variation, s is the standard deviation, and M is the arithmetic mean.

e. Relation between the measures of dispersion. If the frequency distribution of the observations is symmetrical in pattern and follows the so-called "normal law," then the several measures of dispersion are definitely related to one another. Using the standard deviation as the base, the theoretical relations are as follows:

Total range is approximately 6 times the standard deviation.

Mean deviation is 0.7979 times the standard deviation.

17. Standard Error of the Mean. When a sample is chosen from a population, its mean may not be that of the population. A second sample will have a slightly different mean. If a large number of sam-

FOREST MENSURATION

TABLE 8. CALCULATION OF THE STANDARD DEVIATION

Diameter, in. X	Deviation from assumed mean $X_a - X$	Frequency f	$f(X_a - X)$	$f(X_a - X)^2$
5	−10	1	−10	100
6	− 9	2	−18	162
7	− 8	4	−32	256
8	− 7	8	−56	392
9	− 6	11	−66	396
10	− 5	14	−70	350
11	− 4	15	−60	240
12	− 3	17	−51	153
13	− 2	17	−34	68
14	− 1	20	−20	20
15	0	20		
16	+ 1	19	+19	19
17	+ 2	18	+36	72
18	+ 3	13	+39	117
19	+ 4	11	+44	176
20	+ 5	9	+45	225
21	+ 6	7	+42	252
22	+ 7	4	+28	196
23	+ 8	4	+32	256
24	+ 9	2	+18	162
25	+10	1	+10	100
26	+11	2	+22	242
27	+12	1	+12	144
28	+13	1	+13	169
		221	Net −57	4,267

$$s = \sqrt{\frac{4{,}267 - [(-57)^2/221]}{220}} = \sqrt{\frac{4{,}267 - 14.70}{220}} = \sqrt{\frac{4{,}252.30}{220}}$$
$$= \sqrt{19.78} = \pm 4.45 \text{ in.}$$

ples were taken and if the means for each were computed and arranged in a frequency table, they would form a definite distribution from which a standard deviation applying to the means could be computed. This deviation is the *standard error of the mean* and shows the range within which the means vary. Fortunately it is not necessary to take many samples, since a satisfactory estimate of the standard error of the mean can be obtained directly from the mean of a single sample. The method is mathematically sound. The formula reads

$$s_M = \frac{s}{\sqrt{n}}$$

where s_M is the standard error of the mean, s the standard deviation as computed from the sample, and n the number of observations in the sample. In the case of the average diameter of Douglas fir, it is

$$s_M = \frac{\pm 4.45}{\sqrt{221}} = \pm 0.30 \text{ in.}$$

This expression has important applications, as will be shown later (Chap. 21).

References

ARKIN, HERBERT, and R. R. COLTON: An Outline of Statistical Methods, 3d ed., Barnes & Noble, Inc., New York, 1938, 228 + 47 pp.

BRINTON, W. C.: Graphic Presentation, Brinton Associates, New York, 1939, 512 pp.

BRUCE, DONALD: Some Possible Errors in the Use of Curves, *Jour. Agric. Res.*, **31**, 923–928 (1926).

CHAMPION, H. G.: Use of Statistical Methods, *Indian Forester*, **59**, 488–494 (1933).

KITTREDGE, J., JR.: Use of Statistical Methods in Forest Research, *Jour. Forestry*, **22**, 306–314 (1924).

MILLS, F. C.: Statistical Methods Applied to Economics and Business, Henry Holt and Company, Inc., New York, 1938, 756 pp.

RACILIS, A. P.: Use of Statistical Methods in Forest Research, *Makiling Echo*, **12**, 92–99, 131–141 (1933).

WAUGH, A. E.: Elements of Statistical Method, 2d ed., McGraw-Hill Book Company, Inc., New York, 1943, 532 pp.

WRIGHT, W. G.: Suggested Applications of Statistical Methods in Forestry Practice, *Jour. Forestry*, **22**, 372–385 (1924).

CHAPTER 3

CORD OR STACKED MEASURE

18. Utility of Cord Measure. When wood is utilized in bulk, as for fuel or paper pulp, much smaller and more irregular sticks are merchantable than when products such as boards must be sawed from them. The true basis of value for wood utilized in bulk is its solid or cubic contents. For fuel, weight of air-dry wood is the most reliable basis for determining the relative value of different species. For example, dead juniper fuelwood is sold frequently in the West by weight. Weight would be the best measure of value were it not for the large amount of water in green wood and the variation of this water content with species and degrees of seasoning. With these extremely variable factors always present, weight is not a reliable measure of value or price. Furthermore, wood must often be measured in the woods where it cannot be weighed.

On the other hand, volume of wood changes less than does weight with seasoning. The loss in cubic volume when wood is seasoned may be from 9 to 14 per cent. Most of this shrinkage is tangential or in circumference, producing radial cracks and checks; but the diameter changes only slightly. Since the value of small irregular pieces of wood cannot be determined either accurately or at a reasonable cost, actual cubic volume is not a convenient standard for these bulk forest products.

A cheaper method of volume determination for these products is therefore used in practice, which is to cut the sticks to a standard length and then pile them in stacks, or ricks. The length, height, and breadth of the ricks may be measured and the stacked cubic space computed regardless of how much solid wood is in the pile.

19. Measurement by Weight. Although weight has obvious deficiencies, it is being used more and more in commercial operations, where direct measurement of logs or stacked units is difficult. For example, in the relogging of cutover Douglas fir lands in the Pacific Northwest for pulpwood where many odd-sized pieces and shapes are handled, the weight of the rough wood on the truck or car has been substituted for volume. Since this raw material has a low value, it cannot bear the expense of more accurate measurement. The errors involved in weight measurement are therefore recognized and accepted, and an allowance is made for them. In a combined pulp and sawmill operation in the Southeast, the weights of truckloads of green logs or bolts are used as the basis of measurement, and factors have been worked out to convert the weight of raw green wood with bark to board-feet log scale (by the International rule for ¼-in. kerf) of the load in saw logs or to standard cords of peeled wood, if it is pulpwood.[1] This study shows that the volume-weight ratio for pulpwood varies significantly from one area to another, and, therefore, a single conversion factor cannot be used. This is to be expected. In addition, a variation within the same tract would also be expected, depending upon the season when the wood is cut and possibly upon the cutting method, such as a thinning, an improvement, or harvest cutting. These variations must be studied and the results applied in practice if measurement by weight is to be successfully used.

20. Units of Stacked or Cord Measure. A standard cord is a pile of stacked wood containing 128 cu. ft. of air and wood. The standard dimensions are 4 × 4 × 8 ft., or sticks 4 ft. long, piled 4 ft. high, in a rick 8 ft. long. End stakes are commonly used, giving a square-ended rick. The vertical face of a standard cord is 4 × 8 ft., or 32 sq. ft. in area. One cord foot contains ⅛ cord. With 4-ft. wood, its cross-sectional area is 4 sq. ft. In measuring stacked cordwood, a pole is used marked in feet and tenths. When the ends of the stack are not supported vertically, the average length is taken. Heights should be measured at intervals close enough to obtain an average height. The height should be taken to a point where spaces that are left unfilled on the top layer but would contain sticks in tiers below are balanced by the upper portion of sticks above this level.

On sloping ground either the length of the pile must be measured parallel to the slope and its height at right angles to this plane, or else the two dimensions must be measured horizontally and vertically, regardless of slope. Measurement of length along the slope and of height vertically gives more cubic volume than the stack actually contains (see Fig. 4).

Wood may be cut in lengths greater or less than 4 ft. A rick 4 ft. high and 8 ft. long, of bolts of a uniform length of less than 4 ft., may by local custom still be called a cord. This is a short, or running, cord. Firewood sawed from 4-ft. bolts into 4, 3, or 2 sticks may be sold as

[1] SCHUMACHER, F. X., Volume-Weight Ratios of Pine Logs in the Virginia–North Carolina Coastal Plain, *Jour. Forestry*, **44**, 583–586 (1946).

short cords measuring 1 ft., 16 in., or 2 ft. in breadth, respectively. Five-foot wood piled 4 × 8 ft. gives the long cord containing 160 stacked cubic feet. Bolts of different lengths for special products, such as staves, tool handles, or shingles, are also stacked for measurement and are frequently measured in short or long cords depending on their length.

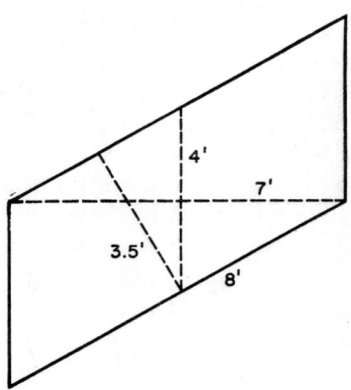

Fig. 4. Methods of measuring cordwood when stacked on sloping ground. The surface of the parallelogram formed by the stack can be found only by taking the two measurements at right angles to each other. When so measured, the pile is either 4 × 7 ft. or 3.5 × 8 ft., either of which gives 28 surface ft. or 112 ft. of cordwood, rather than 128 ft., obtained by measuring length of pile along the surface and height vertically. The pile contains but 87½ per cent of a full cord of 128 cu. ft.

In all cases, either the cord is based on surface measure taken as 32 sq. ft. regardless of length of sticks, or it is reduced to standard cords. The reduction is made by multiplying the area of the vertical face by the stick length and dividing by 128. The equivalent of any short or long cord in standard cords thus found serves as a basis for comparison when local custom permits the use of other than the standard cord. The volume of wood loosely thrown into a wooden box of certain size in some localities is called a cord.

The pen is used widely in the South as the unit of volume for pulpwood. This is an open square crib of bolts, with the sides only one layer in thickness. This method of piling appeals to the cutter who can easily pile up a 6-ft. pen by hand, wherever he has a pile of bolts. Depending upon the size of the bolts, 5 to 7 pens are commonly used to convert pens to standard cords.

21. Solid Cubic Contents of Stacked Wood. Every dealer in cordwood knows that the actual cubic volume of wood in a cord varies over a wide range. The shorter, straighter, smoother, and larger the sticks, the greater will be this solid volume. The wood content decreases along with the average of the sticks and also when their length, surface irregularities, and crooks or bends increase.

The solid content of a stacked cord in cubic feet is useful to convert cords to cubic feet and vice versa. To determine these contents, the round bolts may be measured for cubic volume by Smalian's or Huber's formula (Sec. 28), then split, piled, and measured in the stack. Their volume may also be found very accurately by submerging each piece or whole cords in a specially prepared tank in which the water displacement may be read in cubic feet and tenths. This type of apparatus is known as a *xylometer*.

A photographic method of measuring the solid content of stacked wood has been developed recently. The stacks or loads of bolts are photographed, and the air spaces along mechanically spaced lines are measured or scanned on the photograph. A special instrument has been devised to aid in this scanning.[1] The sum of air spaces times 100 divided by the total length of line scanned gives the percentage of air space which when applied to the standard cord of 128 ft. gives an estimate of the volume occupied by air. This gives a slightly more reliable determination than the supposedly accurate xylometer method.

The greatest actual solid content per cord of piled wood is about 105 cu. ft., or 82 per cent of the 128 cu. ft. of air space and wood. This is for straight bolts 8 in. and over. For 3-in. cordwood the solid volume may drop to 65 cu. ft. and for 1- to 2-in. wood to about 50 cu. ft.

Average solid contents of piles are shown below:

Type of wood	Solid cubic volume per cord, cu. ft.
Large smooth logs or bolts	96.0–102.4
Average split firewood	76.8–96.0
Round wood, tops, branches	64.0–83.2

Since the maximum solid cubic content of a standard cord is about 100 cu. ft., or 78.12 per cent of 128 cu. ft. of air and wood, this fact was used early in New Hampshire to prepare a local cubic log rule for measuring cordwood, known as the Humphrey caliper rule. This rule gives the contents of logs or bolts in cords and hundredths and is derived directly from the value in cubic feet of cylinders by dividing by 100. When this rule is used for pine logs, the log is measured in the middle by a special caliper on which the volumes are inscribed.

Some Canadian companies have adopted the unit of 100 cu. ft. of solid wood as a standard for converting stacked pulpwood to cubic feet. The actual solid cubic content of wood in a cord is expressed as a percentage of this standard, *viz.*, wood running 90 cu. ft. per cord is 90 per cent, or 0.9, of the standard yield per cord.

For stacked cords of given species, such as spruce, loblolly pine, or mixed hardwoods, experience based on frequent measurements of the

[1] KEEPERS, C. H., A New Method of Measuring the Actual Volume of Wood in Stacks, *Jour. Forestry*, **43**, 16–22 (1945).

ratio of solid to stacked contents may permit the adoption locally of an average or flat converting factor such as 70, 83, or even as high as 100 cu. ft. per cord. Several factors cause these variations (Sec. 22). Since they cannot be controlled, however, the stacked cord, which is the commercial unit for buying and selling, is wholly unsuited for scientific studies; but, for the measurement of the volume of wood in large quantities for manufacturing processes, a knowledge of the ratios of solid to stacked contents is indispensable.

22. The Causes of Variation in Solid Contents of Cords. A stick of round cordwood is roughly cylindrical but tapers. Perfectly straight

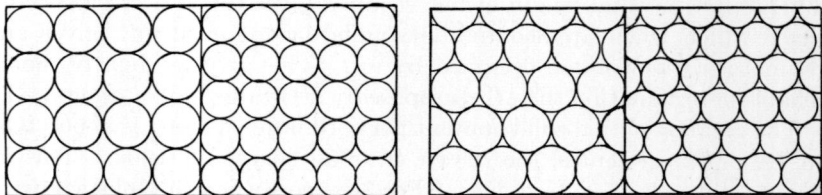

FIG. 5. Square versus hexagonal piling of cylinders. The solid content is increased by piling hexagonally, but for perfect cylinders it is not changed by increasing or decreasing the average diameters of the sticks.

smooth round sticks (cylinders) all of one size can be piled in alternate or six-sided formation (Fig. 5). These perfect bolts occupy 90 per cent of the 128 cu. ft. of space, or 115.2 cu. ft. per standard cord. This would be true for either small or large sticks. Losses from this theoretical maximum are caused by departures from perfect form due to crook, knots, or swellings, and to variable form or taper, including flaring butts. The method of piling is another important factor. If a hexagonal arrangement is not strictly adhered to, the cubic volume will be reduced. If the sticks are placed one over the other, as in the left-hand diagram of Fig. 5, the pile will contain only 100.5 cu. ft. per cord, or 78.5 per cent solid wood. This is only 87.2 per cent of a pile stacked hexagonally.

The effect of surface irregularities, such as knots, is relatively less for large than for small bolts. This effect varies directly with surface area (πDL), hence with circumference (πD) and with diameter (D). For example, if an 8-in. and a 4-in. bolt have one knot each protruding 1 in., the loss in stacked volume occasioned by the knot on the 8-in. bolt will be only twice that of the 4-in. bolt, although its cubic volume ($\frac{1}{4}\pi D^2 L$) is four times that of the latter. Hence, with equal total amounts of irregularity, the percentage of solid contents will increase as the average diameter of the sticks increases.

Loss due to crook increases roughly as the square of bolt length. A bolt 8 ft. long with a 2-ft. crook will not have that much crook in each piece when sawed in two. The maximum for each of the two pieces will be a 6-in. crook. If the crook is abrupt and centrally placed, both pieces may be nearly straight. A cord of 8-ft. wood showing a 20 per cent loss from crook may be expected to show but one-fourth of this loss, or 5 per cent, if bucked into 4-ft. sticks. The more crooked the wood, the greater the loss and the more marked the increase in solid contents when cut into short lengths. For either crooked or knotty wood, these maximum losses are reduced by careful piling.

Four-foot cordwood when bucked into stove lengths consequently shrinks in stacked space and increases in solid contents per standard stacked cord. The contents of standard cords of 128 cu. ft. will therefore depend, first, upon the character of the species, whether cylindrical and thin- or smooth-barked, as in paper birch or balsam, or tapering, knotty, and rough-barked, as in many other species; second, upon the average diameter of the sticks; third, upon their length; and fourth, upon their straightness or crookedness. Because of their irregular form and crook, stacked cords of hardwoods usually contain less wood than do stacked cords of conifers, but with notable exceptions, such as white birch. Because of small diameter and excessive crook, tops and branch wood contain less solid wood in a cord than does bole wood,

TABLE 9. INFLUENCE OF LENGTH OF STICK UPON THE SOLID CUBIC CONTENTS OF A CORD FOR CONIFERS*

Length of stick, ft.	Solid contents per cord, sticks over 5 ft. 5 in. in diameter at small end, cu. ft.	Per cent in terms of 4-ft. sticks	Solid contents per cord, sticks from 2.5 to 5.5 in. in diameter at small end, cu. ft.	Per cent in terms of 4-ft. sticks	Solid contents per cord, sticks 1 to 2.5 in. in diameter at small end, cu. ft.	Per cent in terms of 4-ft. sticks
1	91.80	+ 3.2	85.25	+ 3.4	65.69	+ 3.2
2	90.80	+ 2.2	84.35	+ 2.3	65.32	+ 2.7
4	88.92	0	82.42	0	63.62	0
6	86.45	− 2.8	80.00	− 3.0	61.60	− 3.2
8	83.75	− 5.8	77.20	− 6.3	59.40	− 6.6
10	81.00	− 8.9	74.30	− 9.9	56.90	−10.5
12	78.05	−12.2	71.20	−13.6	54.25	−14.7
14	74.85	−15.8	67.95	−17.5	51.50	−19.0

* ZON, RAPHAEL, *Forestry Quart.*, **1**, 132 (1903).

even when the latter is split. A proper price adjustment should of course be made for varying solid contents of different classes of wood.

The approximate allowances for losses or gains due to different lengths of sticks are shown in Tables 9 and 10. Longer sticks have greater taper, more crook, and hence more variables in addition to greater length, all of which reduce the solid cubic volume. These results were verified by tests on balsam fir in the Adirondack region of New York.

In the Lake states, the following test was made of peeled spruce pulpwood. Eight-foot sticks were first cut to 4-ft. bolts and then to 2-ft. bolts, and restacked and remeasured at each stage. The number of sticks per standard cord and the solid-wood volume varied with the length of the bolt as follows:

Length, in.	Number of bolts per cord	Solid-wood volume, cu. ft.
96	58	94.8
48	122	98.0
24	262	103.4

The effect of reducing the length of the bolt on the solid contents of the pile is clearly evident.

TABLE 10. RELATION OF VOLUME OF SOLID WOOD PER CORD TO STICK LENGTHS AND QUALITY*

Length of stick, ft.	Straight sticks		Crooked sticks		Knotty sticks	
	Volume, cu. ft.	Difference, %	Volume, cu. ft.	Difference, %	Volume, cu. ft.	Difference, %
1	99.81	+8.3	93.47	+14.1	89.60	+20.7
2	97.28	+5.5	89.60	+ 9.4	84.48	+13.8
3	94.72	+2.8	85.76	+ 4.7	79.36	+ 6.9
4	92.16	0.0	81.92	0.0	74.24	0.0
5	89.60	−2.8	78.08	− 4.7	69.12	− 6.9
6	87.04	−5.5	74.24	− 9.4	64.00	−13.8

* From GRAVES, H. S., Forest Mensuration, p. 104, John Wiley & Sons, Inc., New York, 1906, as cited from Dr. Müller's Lehrbuch der Holzmesskunde.

Also in the Lake states, a study was made of the variation of the solid-wood contents of stacks of black spruce bolts cut in 100-in. lengths from trees utilized to a 4-in. top. Solid-wood content was related to tree diameter and tree form as shown in the following tabulation:

Tree diameter, in.	Solid-wood content per standard cord, cu. ft.
6	93.7
8	95.6
12	99.0

Tree form	
Bole cylindrical, straight, smooth, no injuries.....................	100.4
Lower part of bole may be somewhat irregular and rough, but in most cases cylindrical, slight lean or crook admissible................	96.5
Bole frequently lopsided or fluted, often with churn butt and generally leaning, crooked, or containing heavy swell................	93.9

23. Deduction for Defect in Cordwood. On large pulpwood or wood-distillation operations where only sound wood fiber is acceptable to purchasers, unsound sticks must be deducted. Careless piling is another variable for which allowance must be made, since no purchaser would care to pay full price for a loosely piled rick.

Operators or purchasers usually specify the greatest amount of defect or rot they will accept in sticks of different diameters. If the defect is greater than this, the entire stick is rejected; if less, it is accepted and included in the sound scale.

TABLE 11. CULL TABLE FOR CORDWOOD

Diameter of stick, in.	Basal area per stick, sq. ft.	Number of sticks for 32 sq. ft. solid surface	Number of sticks for 32 sq. ft. stacked surface, ratio 85/128 = 0.664	Percentage of stacked cord in one stick
4	0.087	367.8	244	0.0041
6	0.196	163.2	108	0.0093
8	0.349	91.6	61	0.0164
10	0.545	58.7	39	0.0256
12	0.785	40.8	27	0.0370

For this purpose, attempts have been made to prepare cull tables giving the percentage of a standard 4-ft. cord to be deducted for sticks of each diameter. Obviously this deduction will depend on the character of the sticks and will vary with every factor that affects the ratio of solid to stacked contents (Tables 9 to 11). For the same reason which permits the wood to be purchased by stacked measure, averaging out the cubic contents, deduction per stick may also be averaged. In Table 11 the cull per stick is based on an average solid content of 85 cu. ft. per cord, in which case, for sticks of each indicated diameter, the number per cord required to give 85 cu. ft. is shown in column 4 and the deduction in percentage of one cord in column 5.

By tallying the number of cull sticks under their diameters and multiplying by the cull factors, the total cull is expressed in percentage of cord volume. The ratios used in column 5 can be averaged to apply to the entire range of diameters or can be applied separately to each diameter class as tests of cubic contents per cord indicate the factors best adapted to the material purchased. Table 11 is shown merely as an illustration of the method of constructing a cull table.

Cull specifications consider the maximum allowable defect in sticks of different diameters. Defect greater than that shown in Table 12 makes the entire stick a cull.

TABLE 12. CULL SPECIFICATIONS, SPRUCE AND BALSAM PULPWOOD

Diameter of bolt,* in.	Maximum allowable defect	
	Sap rot, in.	Heart rot
6	½	¼ of diameter
8	1¼	¼ of diameter
10	2	3/10 of diameter
12	2½	3/10 of diameter

* Bolts larger than 12 in. take 12-in. specifications.

Deduction for careless piling is determined by the number and size of spaces in the ricks that would normally hold one or more sticks. A cull table similar to the table for defective culls is used for poor piling. For example, the gross scale of a 4 × 4 × 8 ft. rick is 1 cord. If, by more careful piling, a stick 12 in. in diameter could have been included in the sum of unnecessary holes, the space of the pile would be reduced by the "stacked" volume of a 12-in. stick. From Table 11 this is 0.037 cord, and therefore the net scale would be 0.963 cord. This is a reduction of 3.7 per cent of the gross scale.

Because deductions for individual sticks appear to make so little difference, they may easily be overlooked. However, if a purchaser is buying a large quantity of cordwood each year, such small items may amount to a considerable loss. Careful measurement, including proper deductions for defects, and thorough inspection usually will save considerable in the cost of logging or in the value of stumpage on large operations.

24. The Volume of Bark. When cordwood is to be used for fuel or for wood distillation, the bark is usually left on and included in the measurement. The same is true, though less frequently, in the case of pulpwood. In the past, the bark of hemlock, chestnut, and certain oaks was peeled, stacked, and later used as a source of tannin. This bark was then stacked in cords for measurement. Substitutes are rapidly replacing these sources of tannin, however.

For pulpwood, excelsior, or kindred uses where only the wood fiber is to be used, the wood is most frequently peeled before delivery. If the bark is included, the price is adjusted to the net cubic contents of peeled wood. Bark makes up from 6 to 40 per cent of the total cubic volume of a tree, depending upon its thickness relative to the tree diameter. Bark percentage is greatest in young trees growing rapidly on exposed sites and least in old sheltered slow-growing trees. It varies with species as well as with conditions. Average bark volumes for various species run from 10 to 20 per cent of the total wood plus bark, steadily diminishing with increasing size of the tree because of slower growth and loss in shedding.

25. Board-foot Equivalents of Cordwood. Cordwood theoretically should not be expressed in board feet unless cut from round bolts large enough for lumber. Bolts 10 in. in diameter and larger will scale from 500 to 600 bd. ft. per cord, differing with the log rule used. No lumber could be sawed from small 3- to 5-in. bolts or from split cordwood, though board-foot converting factors are frequently employed when it is desired to summarize all volumes in a board-foot standard. Local custom determines the converting factor used. The U.S. Forest Service sometimes uses 333 bd. ft. per cord for firewood made from small crooked material such as dry western cedar. The factor of 5½ bd. ft. per cubic foot is recommended for general use to convert cords to board feet for statistical summaries of wood production. The factor of board feet per cord would then depend on the average cubic volume per cord.

References

BARRETT, L. I., J. H. BUELL, and J. F. RENSHAW: Some Converting Factors for Mixed Oak Cordwood in the Southern Appalachians, *Jour. Forestry*, **39**, 546–554 (1941).

BRYANT, O. F.: Pulpwood Measurements and Some Factors Involved in Chipping and Baling Pulpwood, *Pulp and Paper Mag. Canada*, **14**, 431–436 (1916).

CARY, A.: Fourth Memorandum on Pulpwood Volume Tables and Solid Contents of the Cord, *Pulpwood*, **3** (1), 5–8 (1930).

Cook, H. O.: Measurement of Fuel Wood, *Jour. Forestry*, **16**, 920–921 (1918).
Hawley, R. C.: Measuring Cordwood in Short Lengths, *Jour. Forestry*, **17**, 312–317 (1919).
Keepers, C. H.: A New Method of Measuring the Actual Volume of Wood in Stacks, *Jour. Forestry*, **43**, 16–22 (1945).
Kellogg, R. S.: What Is a Cord? *Jour. Forestry*, **23**, 608–610 (1925).
Lake States Forest Experiment Station: How Much Solid Wood in a Cord?, *Tech. Note* 74, 1934, 1 p.
———: Solid Contents of Standard Cords (Peeled Black Spruce Pulpwood), *Tech. Note* 66, 1933, 1 p.
Lodewick, J. E.: How Much Is a Cord of Pulpwood? *Pacific Pulp & Paper Indus.*, **9** (9), 5–10 (1935).
MacKinney, A. L., and L. E. Chaiken: Converting Factors for Loblolly Pine Pulpwood, *Appalachian Forest Expt. Sta. Tech. Note* 20, 1936, 9 pp.
Ontario Royal Commission on Forestry: *Report*, Chap. XVII, 1947.
Schnur, G. A.: Converting Factors for Some Stacked Cords, *Jour. Forestry*, **30**, 814–820 (1932).
Schumacher, F. X.: Stacked and Solid Volume of Southeastern Pulpwood, *Jour. Forestry*, **44**, 579–582 (1946).
———: Volume-Weight Ratios of Pine Logs in the Virginia–North Carolina Coastal Plain, *Jour. Forestry*, **44**, 583–586 (1946).
Soderston, H. R.: Some Mensuration Problems in Logging, *Jour. Forestry*, **27**, 234–240 (1920).
Veness, J. G.: Volume Tables in Cords and Conversion Factors for Pulpwood Operators, Canada Department of Interior, Forest Service, 1926, 6 pp.
Wright, W. G.: Solid Content of a Stacked Cord of Pulpwood, *Pulp and Paper Mag. Canada*, **22**, 846–847 (1924).
Zon, Raphael: Factors Influencing the Volume of Solid Wood in the Cord, *Forestry Quart.*, **1**, 125–133 (1903).

CHAPTER 4

CUBIC VOLUME OF LOGS

26. Logs and Bolts. In felling a tree, an undercut is first made on the side toward which the tree is to fall. This cut may be made with the saw and then chopped out. The saw is then started on the opposite side of the tree slightly above the lower side of the undercut. Wedges are usually needed to prevent the saw from pinching and to ensure that the tree falls in the desired direction. The limbs or branches are then removed from the bole for the length of its merchantable portion. The desired lengths for logs are measured on the bole allowing from 2 to 3 in. for trimming in the mill, after which the tree is crosscut or bucked. If these pieces are 8 ft. long or over, and suitable for lumber, they are called logs; if under 8 ft., bolts. Logs and bolts are, therefore, the round pieces with square-cut ends into which the bole is first divided in the process of manufacture of a tree into its final products.

The basal area is found from the diameter as follows:

B = basal area
D = diameter of circle
π = ratio of D to circumference or 3.1416

Then
$$B = \frac{\pi D^2}{4} = \frac{3.1416 D^2}{4} = 0.7854 D^2$$

Since the diameter D is in inches and the cross-sectional area B is desired in square feet, the formula is converted from square inches by dividing by 144, or,

$$B, \text{ in square feet} = \frac{0.7854 D^2}{144} = 0.06545 D^2$$

27. The Cylinder as a Basis for the Cubic Volume of Logs or Bolts. The cubic contents of logs is usually measured in cubic feet. This volume is equal to the average basal area multiplied by the length.

Table 13 gives the areas of circles in square feet for each tenth inch

TABLE 13. AREAS OF CIRCLES (OR TABLE OF BASAL AREAS) FOR DIAMETERS TO NEAREST $\frac{1}{10}$ IN.*

Diameter, in.	Area, sq. ft.	Diameter, in.	Area, sq. ft.	Diameter, in.	Area, sq. ft.	Diameter, in.	Area, sq. ft.	Diameter, in.	Area, sq. ft.
1.0	0.005	2.0	0.022	3.0	.049	4.0	.087	5.0	.136
.1	.007	.1	.024	.1	.052	.1	.092	.1	.142
.2	.008	.2	.026	.2	.056	.2	.096	.2	.147
.3	.009	.3	.029	.3	.059	.3	.101	.3	.153
.4	.011	.4	.031	.4	.063	.4	.106	.4	.159
1.5	0.012	2.5	0.034	3.5	.067	4.5	.110	5.5	.165
.6	.014	.6	.037	.6	.071	.6	.115	.6	.171
.7	.016	.7	.040	.7	.075	.7	.120	.7	.177
.8	.018	.8	.043	.8	.079	.8	.126	.8	.183
.9	.020	.9	.046	.9	.083	.9	.131	.9	.190
6.0	.196	7.0	.267	8.0	.349	9.0	.442	10.0	.545
.1	.203	.1	.275	.1	.358	.1	.452	.1	.556
.2	.210	.2	.283	.2	.367	.2	.462	.2	.567
.3	.216	.3	.291	.3	.376	.3	.472	.3	.579
.4	.223	.4	.299	.4	.385	.4	.482	.4	.590
6.5	.230	7.5	.307	8.5	.394	9.5	.492	10.5	.601
.6	.238	.6	.315	.6	.403	.6	.503	.6	.613
.7	.245	.7	.323	.7	.413	.7	.513	.7	.624
.8	.252	.8	.332	.8	.422	.8	.524	.8	.636
.9	.260	.9	.340	.9	.432	.9	.535	.9	.648
11.0	.660	12.0	.785	13.0	.922	14.0	1.069	15.0	1.227
.1	.672	.1	.799	.1	.936	.1	1.084	.1	1.244
.2	.684	.2	.812	.2	.950	.2	1.100	.2	1.260
.3	.696	.3	.825	.3	.965	.3	1.115	.3	1.277
.4	.709	.4	.839	.4	.979	.4	1.131	.4	1.294
11.5	.721	12.5	.852	13.5	.994	14.5	1.147	15.5	1.310
.6	.734	.6	.866	.6	1.009	.6	1.163	.6	1.327
.7	.747	.7	.880	.7	1.024	.7	1.179	.7	1.344
.8	.759	.8	.894	.8	1.039	.8	1.195	.8	1.362
.9	.772	.9	.908	.9	1.054	.9	1.211	.9	1.379
16.0	1.396	17.0	1.576	18.0	1.767	19.0	1.969	20.0	2.182
.1	1.414	.1	1.595	.1	1.787	.1	1.990	.1	2.204
.2	1.431	.2	1.614	.2	1.807	.2	2.011	.2	2.226
.3	1.449	.3	1.632	.3	1.827	.3	2.032	.3	2.248
.4	1.467	.4	1.651	.4	1.847	.4	2.053	.4	2.270
16.5	1.485	17.5	1.670	18.5	1.867	19.5	2.074	20.5	2.292
.6	1.503	.6	1.689	.6	1.887	.6	2.095	.6	2.315
.7	1.521	.7	1.709	.7	1.907	.7	2.117	.7	2.337
.8	1.539	.8	1.728	.8	1.928	.8	2.138	.8	2.360
.9	1.558	.9	1.748	.9	1.948	.9	2.160	.9	2.382

TABLE 13. AREAS OF CIRCLES (OR TABLE OF BASAL AREAS) FOR DIAMETERS TO NEAREST 1/10 IN.—(Continued)

Diameter, in.	Area, sq. ft.	Diameter, in.	Area, sq. ft.	Diameter, in.	Area, sq. ft.	Diameter, in.	Area, sq. ft.	Diameter, in.	Area, sq. ft.
21.0	2.405	22.0	2.640	23.0	2.885	24.0	3.142	25.0	3.409
.1	2.428	.1	2.664	.1	2.910	.1	3.168	.1	3.436
.2	2.451	.2	2.688	.2	2.936	.2	3.194	2	3.464
.3	2.474	.3	2.712	.3	2.961	.3	3.221	.3	3.491
.4	2.498	.4	2.737	.4	2.986	.4	3.247	.4	3.519
21.5	2.521	22.5	2.761	23.5	3.012	24.5	3.275	25.5	3.547
.6	2.545	.6	2.786	.6	3.038	.6	3.301	.6	3.574
.7	2.568	.7	2.810	.7	3.064	.7	3.328	.7	3.602
.8	2.592	.8	2.835	.8	3.089	.8	3.355	.8	3.631
.9	2.616	.9	2.860	.9	3.115	.9	3.382	.9	3.659
26.0	3.637	27.0	3.976	28.0	4.276	29.0	4.587	30.0	4.909
.1	3.715	.1	4.006	.1	4.307	.1	4.619		
2	3.744	.2	4.035	.2	4.337	.2	4.650		
3	3.773	.3	4.065	.3	4.368	.3	4.682		
.4	3.801	.4	4.095	.4	4.399	.4	4.714		
26.5	3.830	27.5	4.125	28.5	4.430	29.5	4.746		
.6	3.859	.6	4.155	.6	4.461	6	4.779		
.7	3.888	.7	4.185	.7	4.493	7	4.811		
.8	3.917	.8	4.215	.8	4.524	8	4.844		
.9	3.947	.9	4.246	.9	4.555	.9	4.876		
31.0	5.241	32.0	5.585	33.0	5.940	34.0	6.305	35.0	6.681
36.0	7.069	37.0	7.467	38.0	7.876	39.0	8.296	40.0	8.727
41.0	9.168	42.0	9.621	43.0	10.085	44.0	10.559	45.0	11.045
46.0	11.541	47.0	12.048	48.0	12.566	49.0	13.095	50.0	13.635
51.0	14.186	52.0	14.748	53.0	15.321	54.0	15.904	55.0	16.499
56.0	17.104	57.0	17.721	58.0	18.348	59.0	18.986	60.0	19.635

* Values rounded to the nearest 0.001 sq. ft.

of diameter. These tables also serve as the basis for cubic log rules such as the Humphrey rule (Sec. 21).

The volume of the cylinder in cubic feet is BL when L is the length in feet. A log is assumed to have a circular cross section at any point on its axis. Actually, logs are seldom perfect in this respect, and some are decidedly elliptical or even oval in cross section. Such irregularities are always reduced to terms of a circle by taking an average of two diameters, preferably the longest and the shortest, and assuming that this average is the diameter of the equivalent circle. The area of the circle determined by the above formula is slightly greater than that of the cross section from which the diameters are taken, though this discrepancy is accepted in practice.

The object of measurement is to obtain an average cross-sectional area which will be equal to the basal area of the cylinder of the same length, whose volume is then equal to that of the log. For this reason, when the diameter at the point measured is abnormal because of excessive flare of butt, crotches, limbs, knots, or swellings, it should be reduced to the normal diameter by measuring the bole at the two nearest points free from these abnormalities, above and below the desired point, and taking the average of these two measurements.

28. The Form and Cubic Volume of Logs. A tree diminishes in diameter from the ground to the tip; hence any log cut from it must have a tapering form. In fact, *taper* is the term applied to this loss in diameter. It is expressed in inches for given lengths, most commonly 16-ft. lengths. Because of swelling near the roots, butt logs have excessive taper in their first 2 or 3 ft. Taper is usually least in the logs between the butt log and the beginning of the live crown. From this point to the tip of the tree, it again increases.

The average taper of the merchantable portion of the bole, which would include saw-log sizes, is for many commercial species 2 in. in 16 ft. This taper rate may drop to 1 in. for southern pines or rise to 4 in. in hardwoods and certain West Coast conifers.

It is evident that the form of logs is not that of a cylinder, since they taper. Let it be assumed that a single cross-sectional area can be obtained which is the true average of all the cross sections of the log. Then the product of the average cross section and the length will give the cubic volume of the log regardless of its form, as is explained below.

Four forms of geometric solids are circular in cross section at right angles to their long axes, *viz.*, the cylinder, paraboloid, cone, and neiloid. All these except the cylinder taper to a point and may be compared with the entire bole of the tree, which, however, seldom approximates either the cone or the neiloid and never the cylinder. Truncated sections severed by crosscuts are termed *frustums* and may be compared with logs. The truncated paraboloid most nearly approaches the form of the average log. Butt logs may resemble the truncated neiloid owing to flaring butts, while small top logs may have the shape of a truncated cone (Fig. 6).

The volume of a complete paraboloid is one-half that of a cylinder possessing an equal base and height. The volume of a frustum of a paraboloid is equal to the product of its average cross-sectional area and its length. Logs are assumed to be frustums of paraboloids in measuring their cubic volume. When they resemble frustums of the cone or neiloid, an error is incurred in the above assumption. This error may be excessive for butt logs.

There are two standard methods of determining the average cross-sectional area of logs. The first is to determine the area at each end and obtain the average. The second is to measure the area at the middle of the log.

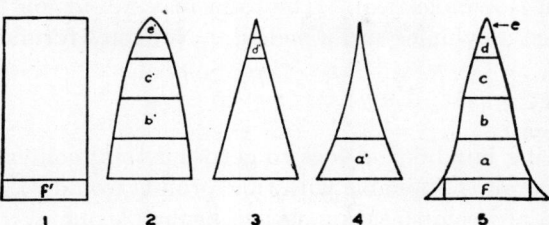

FIG. 6. Comparison of full and truncated cylinder, paraboloid, cone, and neiloid with form of tree and of logs.

In this figure the cylinder (1), paraboloid (2), cone (3), and neiloid (4) are shown. Number 5 is the average form of a tree. The successive logs, a to d, inclusive, resemble truncated portions of the respective geometric solids indicated by a' to d'. The stump, f, resembles a truncated neiloid, though measured as a cylinder, f', and the butt swell causes the first log also to resemble the similar portion of the neiloid figure. Logs b and c resemble truncated portions of the paraboloid, while the short top log, d, resembles a truncated cone. The tip, e, is usually taken as a paraboloid.

The formulas for cubic volume determination by these two methods are shown below and are known as

Smalian: $$V = \frac{(B + b)}{2} L$$

Huber: $$V = B_{\frac{1}{2}} L$$

in which V is volume in cubic feet; B, area in square feet at the large end; b, area in square feet at the small end; $B_{\frac{1}{2}}$, area in square feet at the middle; and L, length in feet. Either of these formulas gives accurately the volume of a frustum of a perfect paraboloid but is incorrect for frustums of cone or neiloid.

It has been pointed out that logs vary in their rate of taper. If the taper is such that the cross-sectional area at the middle of a log is greater than the average of the end areas, which rarely occurs, the log volume lies between that of the frustum of a paraboloid and that of a cylinder, approaching the latter with increasing cross-sectional area. When the cross-sectional area at the middle is less than the average of the end areas, which occurs frequently, and almost always in butt logs and top logs, the actual log volume approaches that of the frustum of a cone or even of a neiloid. In either case, it is obvious that its volume cannot be measured accurately by either of the above formulas. The so-called "Newton's formula," used by engineers in obtaining cubic yardage in cuts or fills, will give the true volume of any log whether resembling the frustums of a paraboloid, cone, or neiloid, provided that its form is symmetrical. The form may fall at any point in the series between a cylinder and a neiloid. Newton's formula is

$$V = \frac{(B + 4B_{1/2} + b)}{6} L$$

In measuring hundreds of logs to get their total contents, the close degree of accuracy obtainable with this formula would be unnecessary. These errors are compensating, hence negligible for average runs of logs, and are therefore ignored in scaling practice. If the log resembles a conic or neiloidic frustum, the middle-diameter measurement, or Huber formula, would give too little volume and the end, or Smalian formula, too much. The errors are just twice as large when the Smalian, or two-end measurements, are taken as when the middle diameter is used. In spite of the relatively greater accuracy of the Huber formula, the Smalian is the one most commonly used in mensuration. The reason for its preference is explained in Sec. 30.

29. Scientific Standards of Accuracy in Log Measurement. There are two standards for measuring logs, the scientific and the commercial. The scientific standard is required when the volumes are obtained for the construction of tables, either for logs or for entire trees, to be used as standards of measurement. In this case, average diameter is read to the nearest tenth inch from two readings, and lengths of log to the nearest tenth foot. Either the Smalian or the Huber formula may be employed to determine the total cubic contents, but the length of the logs can be made as short as required for the degree of accuracy sought; *viz.*, logs may be measured at intermediate points, in short sections, the sum of whose volumes give the total volume of the log. By either method, the shorter the length measured, the less the probable error in assuming its form to be that of a truncated paraboloid, hence the more accurate will be the result. This is shown by the following figures.

Proportion of height of frustum of paraboloid to height of completed paraboloid, %	*Proportion of volume of frustum of paraboloid to that of a cylinder of equal height, %*
100	50
25	87
12½	94

30. Commercial Methods of Log Measurement. The measurement of logs as a commercial operation is undertaken in order to determine their contents as a basis for sale or for payment of contractors or saw crews, or as a record of logging output. For these purposes, log rules are used (Sec. 32).

In commercial measurements, there is but one practice which will give the total cubic contents inside bark with reasonable accuracy; i.e., to measure the log with calipers outside bark at the middle point, deduct double bark thickness, and apply a log rule derived from the Huber formula. To accomplish this, the logs cannot be decked in rollways but must be scaled singly, either in the woods prior to skidding, at the yard, or as they enter the mill. Calipers on whose bar is printed the volumes of logs of different diameters inside bark and lengths are in use in this country. The Humphrey caliper cubic log rule is constructed on this basis.

Logs are frequently piled on rollways before it is practicable to measure them for volume. In theory, both ends of the logs could be taken as they lie in the pile, an average obtained, and the Smalian formula applied. This would require double work and would be difficult to perform accurately in large rollways. If the two diameters are averaged, the area thus obtained is always less than the actual average of the two end areas (See 15b). Under these conditions, the larger cross section is not given its proportional weight. This is an advantage in the case of butt logs.

To illustrate this point more clearly, assume that a log is 16 ft. long and that the small end measures 12 in. and the large end 14 in. The average of the two end diameters is 13 in. The area of a 13-in. circle is 0.922 sq. ft. The volume of the log on the basis of averaged end diameters would then be determined as 0.922 (16) = 14.752 cu. ft. If, however, the end areas are averaged from the 12- and 14-in. measurements, half the sum of 0.785 and 1.069 is 0.927 and the volume is 0.927 (16) = 14.832 cu. ft.

The most convenient commercial method of measuring cubic volume is one based on a diameter measurement at the small end of the log. If the taper were uniform on all logs, a rule could easily be worked out giving the full cubic contents of logs of given lengths and diameters at the small end. But if the log is treated as a cylinder whose basal area equals that of the small end, it will be underscaled by the amount of the cubic volume lying outside this cylinder. The percentage of error thus incurred increases with total taper of the log in inches, which in turn increases with the length of log. This percentage also increases with the ratio between total taper and the log diameter, being greatest for small logs. The error is constant only for given diameters and given total tapers. On actual logs, the taper for logs of the same diameter is not constant but varies with many factors. For this reason, there is no practical basis for a log rule in cubic contents whose area measurement is based upon the small end. Log rules constructed on this basis would therefore give inconsistent results as well as underscale the cubic contents of the logs. Such a log rule, if constructed, should attempt to show average log contents based on diameter, length, and a fixed average taper per 16-ft. length, disregarding variations in this rate of taper, instead of being based on a cylinder measured at the small end of the log.

Fig. 7. Cubic volume lost by using small end of log for scaling diameter.

31. Bark Measurements and Scaling Diameter. Unless the entire volume of logs with bark is accepted for bulk products, it is necessary to determine the average diameter inside bark as a basis for volume computation. It is a simple matter to exclude the bark from the measurement on the cut ends of logs; but when a caliper rule is used, the diameter must be taken at the middle, requiring a deduction for bark thickness. This measurement can be obtained with accuracy only by cutting through the bark on both sides of each log at the point of measurement. The caliper reading will then give the inside bark measurement directly. The diameter by calipers is usually obtained from a single reading and is therefore not quite so accurate a determination of average diameter as when two readings at right angles to each other are taken. In practical use of the caliper scale, the outside bark measurement is first taken, and then the double of *average* bark thickness is subtracted to obtain the desired scaling dimension. Where a uniform average bark thickness can be arbitrarily adopted for all sizes, the deduction can be set off directly upon the caliper rule.

For commercial work, the scaling diameters in inches are recorded. One common practice is to round off to the nearest inch. This is done by dropping all tenths from the 0.5 down and raising to the next inch when the reading is 0.6 or more. A 10-in. log would then measure anywhere between 9.6 and 10.5 in. In a large number of logs the errors thus incurred will compensate in the total scale.

In the measurement of diameters at the end, the scale stick is held to cover solid wood only. For accurate work, two diameters should be read, preferably the largest and smallest, although commonly in commercial work, the average diameter is selected by eye. Occasionally the smallest diameter is taken, which is again a device advantageous to the purchaser of logs. Another modification which is used to eliminate some of the underestimating inherent in one of the common log rules, *viz.* the Doyle rule (Sec. 52), is to measure the inside diameter of the log plus one bark thickness.

32. Log Rules Based on Cubic Contents of Logs. A log rule is a table giving the standard contents of logs of different diameters and lengths. The contents may be expressed in any unit of measure desired, such as cubic feet or board feet. The log contents indicated by the rule are inscribed on a *scale stick*, from which they are read directly for each log scaled. Log contents are then recorded in a *scale book*.

The U.S. Forest Service has authorized the optional use of a cubic log rule in the Northeast. It is a caliper rule applied at the middle of the log and may be read either with or without bark according as bark is merchantable. It is based upon the volume of cylinders and the Huber formula. The unit of measure is the cubic foot.

Metric log rules are used in continental Europe and the Philippine Islands. Their unit of measurement is the cubic meter, which contains 35.3156 cu. ft. Efforts have been made to substitute the cubic meter for the cubic foot in the United States with little success.

Two or three different cubic log rules have been used in the Adirondack region of New York. Their basis is the cubic content of a cylinder of a fixed diameter and length which is accepted as standard. One such standard, known as the *market*, is a cylinder corresponding to a log 13 ft. long and 19 in. in diameter, containing 25.6 cu. ft. The diameter measurement is taken at the small end. Logs of other sizes are measured as cylinders, and their volumes are expressed in terms of the number of markets they contain. This constitutes the log rule. Since taper is disregarded, the cubic scale is too low by an amount depending on the size of log and proportion of total taper to diameter.

In New Hampshire a different type of cubic measure was adopted by legislation. It consisted of an arbitrary "cubic foot," actually 1.4 cu. ft., called the "Blodgett foot." The standard consists of a block 1 ft. long and 16 in. in diameter. The Blodgett rule gives the contents of logs in terms of this standard. To determine the contents of any log in Blodgett feet, the formula used was

$$V = \frac{D^2}{16^2} L$$

where V indicates Blodgett feet; D, diameter at the middle of the log in inches; and L, length of the log in feet. Later attempts to modify this rule to apply at the small end seriously impaired its usefulness. Its real intent was to serve as a basis for conversion to board feet at a ratio of 10 bd. ft. to 1 Blodgett foot. For reasons described in Sec. 50, this rule never was satisfactory for purposes of conversion to board feet.

33. Cubic Volume of Squared Timbers. The cubic volume of squared timbers may be found by using the proportion between the circle and its inscribed square. The formula for the side of an inscribed square, in terms of the diameter of the log in inches, is

$$S = \sqrt{\frac{D^2}{2}}$$

in which S = side of the inscribed square
D = diameter of the circle

The desired side of the square for a 17-in. circle is 12 in. Each linear foot of a 17-in. log contains 1 cu. ft. of squared timber. A log rule may be made on this basis, by using the formula

$$V = \frac{D^2}{17^2} L$$

in which D = log diameter, in inches
L = length of log, in feet
V = volume of the squared timber, in cubic feet

The inscribed square of the log wastes 36.6 per cent of the contents of the cylinder whose base is the small end of the log. In addition, all taper is disregarded. Such log rules come far from expressing the cubic contents of logs. They are, instead, an attempt to give the merchantable contents only on the basis of such utilization as is practiced in certain parts of the tropics, where logs are hewed square before being dragged out of the jungle.

In Great Britain and India, the cubic volumes of logs are measured by a log rule known as the "quarter girth," or "Hoppus rule," in which, when girth (G) is measured in inches at the middle of the log and length (L) in feet,

$$\text{Volume in cubic feet} = \left(\frac{G}{4}\right)^2 \frac{L}{144}$$

This rule gives 78.4 per cent of the cubic volume of the cylinder, thus allowing a waste of 21.6 per cent. To obtain the full cubic volume of cylinders, the formula would be

$$\text{Volume} = \left(\frac{G}{4}\right)^2 \frac{L}{113}$$

References

ILLICK, J. S.: A Practical Xylometer, *Jour. Forestry*, **15**, 859–863 (1917).
RAPRAEGER, E. F.: The Cubic Foot as a National Log Scaling Standard, Northern Rocky Mountain Forest Experiment Station, 1940, 40 pp.
TAYLOR, R. W.: Red and White Fir Xylometer Test, *Forestry Quart.*, **12**, 24–25 (1914).
WAKELEY, P. C.: American Forestry and the Metric System, *Jour. Forestry*, **25**, 966–980 (1927).

CHAPTER 5

THE MEASUREMENT OF BOARD FEET IN THE LOG. LOG SCALING

34. The Necessity for Log Measurement. Sawed timber in any form, either as 1-in. boards or in thicknesses greater or less than 1 in., has a much higher value than an equivalent volume in cubic feet of unprocessed wood, such as firewood. Since logs suitable for sawing must meet minimum standards of diameter, length, quality, and even species, the value of the final product is reflected in the value of the raw material. For an example, an oak saw log will bring a higher price than another log of the same volume that is good only for firewood. The value of the log in the standing tree is the margin left after deducting the costs of logging and manufacture, plus a fair profit on these operations, from the value of the finished products. Since timber estimating is an essential step in appraising the value of standing timber, it becomes necessary to measure tree contents in terms of lumber or other final products (Sec. 4).

Boards are sawed not from trees but from logs of different lengths and diameters. The volume of lumber obtained from logs of identical diameters and lengths is not constant but varies with several factors dependent on sawing practice (Sec. 2).

If the owner of standing timber could wait until the trees were felled, taken to the mill, and sawed, and could then secure payment on the basis of sawed lumber by accurate measurement of the output of his particular logs, his compensation might be greater, but the cash settlement would be delayed. In most cases it is necessary to sell either the logs or the standing timber, leaving the purchaser free to saw in any manner and at any time that suits him (Sec. 1).

Determination of the volume of lumber before the log is sawed is practically universal procedure for these reasons. Log measure, therefore, and not actual sawed output, is the proper basis for timber estimating.

35. Log Rules as a Standard for Log Measure. A standard measure is a fixed quantity, accepted by buyers and sellers in determining the content of goods of the class for which the standard is designed.

The board foot is such a standard, and contains 144 cu. in. of sawed lumber, or the equivalent of a board 12 by 12 in. of surface and 1 in. thick. In practice this refers to green rough boards. Planing and shrinkage reduce the thickness by from $\frac{1}{16}$ to $\frac{3}{32}$ in. Trade practices, induced by the desire to reduce weight for shipment, have at times permitted even greater reductions up to $\frac{7}{32}$ in.

When the thickness is standardized at less than 1 in., as with box boards, or greater, as with dimension lumber, the quantity cannot be estimated solely from surface measure but must be computed on the basis of full board feet of standard thickness. For lumber $\frac{1}{2}$ in. or less in thickness, however, surface measure is often the basis of sale.

For a standard log rule the standard of measurement of contents cannot vary. Since the boards and dimension stock to be cut from logs will vary in thickness in an unpredictable way, an arbitrary standard of thickness must be agreed upon, ignoring any variations which may be obtained in sawing (Sec. 50).

A *board-foot log rule* is therefore a table sanctioned by custom or by law, giving the board-foot content of logs for boards of a standard thickness (usually 1 in.). Its values are given for each 1-in.-diameter class and for lengths usually differing by 2 ft. up to a given maximum length.

A *scale stick* is a rule on which inches are marked along one edge, and the log rule volumes are placed on the sides and remaining edge. Each row of figures gives the volume for a given log length and the range of diameters. Since a scale stick is 3 to 4 ft. long, it can be used to measure the total length of logs as well. After the length is obtained, the volume is read directly from the stick instead of from a separate table. Given the scaling diameter and length of a straight and sound log, the volume may be read for these dimensions in the office as well as in the woods.

Scaling is the measurement of the volume of logs, usually by means of a scale stick, with deductions from full or gross scale for defects which reduce the sound board-foot contents.

A *scale* of logs is their total net contents as scaled by a given log rule, applied according to the *scaling practice* agreed upon. These practices, which may differ with the same log rule in different or within the same localities, specify the methods for measuring diameter and length and deducting for defect.

36. Scaling Sound Straight Logs—Diameters. Most board-foot log rules are based on the assumption that a log is a cylinder. It is customary to measure the diameter at the small end of the log inside bark. Practice which includes one or both thicknesses of bark or requires the use of only the smallest diameter on a cross section is in error. Logs are not perfectly round, and a difference may exist of from one to several inches between the largest and smallest diameters, especially on large logs.

If the diameter measurement is taken to the nearest inch, it would include in 90 per cent of the cases a fractional inch. Past scaling practice, to favor the buyer, was based on the use of the smallest diameter and the rejection of all fractions above the nearest exact full inch. But since boards can be sawed from these fractional inches, they are not lost. This practice is therefore unfair to the sellers. The U.S. Forest Service requires the measurement of the average diameter and the rounding off to the nearest inch; *viz.*, a log 8.0 by 10.6 in. would be 9.3 in. and would be scaled as a 9-in. log. The average of one measuring 9.0 by 10.6 in. would be 9.8 in. and be scaled as a 10-in. log. On the other hand, under "scant" practice, they would be scaled as 8- and 9-in. logs. It is the U.S. Forest Service custom to drop diameters halfway between whole inches to the next lower inch. For example, a 10.5-in. log is scaled as a 10-in. log.

37. Scaling Sound Straight Logs—Lengths. Length is an important element in scaling practice. Logs taper from 1 to 4 in. in 16 ft., exclusive of the abnormal butt swell. In all but the International log rule (Sec. 51), the volume increases in exact proportion to the length of the log. For example, a 24-ft. log 12 in. in diameter with a taper of 4 in. would be scaled as having twice the volume of a 12-ft. log. But if this log were cut into two 12-ft. logs and each log scaled separately, the butt piece would measure 14 in. and the volume of the two short logs of any "cylindrical" log rule would be at least 15 per cent greater than that of the single long log. It is therefore best to adopt regulations limiting the length of the cylinder to be scaled as one log. Sixteen feet is a common limit for softwoods and 12 ft. for hardwoods. The limit thus set is called the *maximum scaling length*. The minimum length, on the other hand, is fixed by the length of the shortest boards which will commonly be merchantable as lumber in the region.

Logs longer than the maximum are divided by the scaler into two or more lengths. For softwoods, lengths are in 2-ft. intervals. If a log does not divide evenly into 2-ft. lengths, as, for instance, an 18-ft. log, the butt log is given the extra 2 ft. The 18-ft. log would then be scaled as a 10- and an 8-ft. piece, while a 20-ft. log would be scaled as two 10-ft. pieces. In a few regions, particularly in the Pacific Northwest and Alaska, long lengths, exceeding the usual nominal 16-ft. length and running from 20 up to 40 ft. in exceptional cases, are sometimes

LOG SCALING

acceptable and are scaled as a single log with a diameter equal to the top diameter of the long log. This practice leads to marked underscaling and high overrun (Sec. 38).

When long logs are scaled in two or more sections, in theory the scaler is supposed to determine the actual diameter inside the bark at these points. But in practice this cannot often be done, since the logs are piled in rollways. At best, a caliper measurement can be made with a guess at the bark thickness and a mental calculation to deduct twice this bark to arrive at the diameter. Scalers therefore estimate taper by measuring both ends of the log, or on the basis of experience, allow from 1 to 2 in. above the dimensions of the small end and scale the butt log accordingly. To obtain uniformity of practice, custom usually provides for a fixed amount of taper allowance in inches for logs of different lengths. This practice applies also to logs that must be divided into three or more sections. Although so divided, the sum of the scaled sections is found and entered as the total scale of the single long log and not as separate logs. The scale record requires a scale for each log or piece, the total entries thus agreeing with the total number of logs scaled.

Logs intended to saw out lumber of standard 2-ft. lengths must be cut slightly longer, in order to permit the boards to be trimmed to square ends in the mill without producing scant lengths. From 1 to 3 in. is allowed, depending upon the length and diameter of the log, as a trimming allowance. Larger logs require somewhat greater trimming allowance than small logs. This extra length is ignored in scaling. If it exceeds the permissible allowance, the log may by agreement be scaled 1 to 2 ft. longer as a penalty.

38. Overrun. The scale of a quantity of logs is the total net volume given by the log rule used as the standard of log measure. The actual sawed output in board feet may overrun or exceed the scale for several reasons. In rare instances it may underrun or fail to equal the scale.

Since the log rule is the standard, and not the sawed output, the *overrun*, as it is called, is always expressed as a percentage of this scale. For example, for logs scaling 10,000 bd. ft. and sawing out 12,500 bd. ft., the overrun is 25 per cent. Obviously a log rule that gives abnormally low values will result in a correspondingly high percentage of overrun, but one that coincides closely with sawing practice will have but a small percentage of overrun. On the other hand, any influence, such as thin saws or close utilization of slabs or sawing lumber thicker than 1 in., that increases the board-foot product of logs correspondingly increases the percentage of overrun.

The log rule volumes must be accepted as the standard of measurement of logs and of standing timber. Yet where a consistent overrun is commonly secured, both buyer and seller soon learn to reckon this overrun in price adjustments. Thus in effect they base the volumes and values of the logs and timber on the average volume of the sawed lumber. On the other hand, a log rule that is inconsistent, such as the Doyle rule, for different sizes of logs gives rise to different overruns for logs of different average diameter and different distribution of logs in the several diameter classes. Such a rule fails to meet the basic requirement of a standard of measure.

39. Deductions from Gross Scale—Sound Scale. Log rules cannot include deductions from the scale of sound straight logs without destroying their value as standards. When logs possess defects of such a character that the boards sawed from the defective parts will be unfit for lumber, the scaler must make the proper deduction from the standard scale of the log to allow for the loss. The *net scale* of a quantity of logs is termed the *sound scale* and is the scale of sound lumber by the given log rule, after deductions are made for defects.

No deductions are made for any defect which merely lowers the grade or value of the merchantable product but does not reduce the volume of sound lumber. However, when knots are so large and numerous that they render the lumber unmerchantable, the entire log may be culled or left in the woods.

The basis from which deductions are made is the full scale of the log, *viz.*, the volume of the cylinder with a diameter equal to that at the small end of the log (but see Sec. 51, International log rule). The center of the scaled cylinder must coincide with that of the butt end of the log and not be shifted sideways to avoid defects, such as fire scars. The lumber that may be sawed from the taper or part falling outside the circumference of this cylinder, as well as the lumber in the slab and edging collar inside the cylinder, is disregarded as if nonexistent, whether sound or defective. The lumber sawed from this part of the log affects only the overrun and does not influence either the log scale or deductions from the scale.

Defects are deducted as 1-in. boards. The part of the log beyond the length of the defect must itself have a length equal to that of the shortest merchantable boards. If less, the deduction must also include this sound part; *viz.*, if in a log 10 ft. long, with an 8-ft. minimum length for lumber, more than 2 ft. is deducted from the length, the entire log is rejected; but in a log 16 ft. long, an 8-ft. piece may be deducted with no further reduction in scale. In defects of unequal length, the average length may be taken.

For defects which appear at both ends of a log, the diameter or size of the defect, increased by 1 in., at the end where the defect is largest, is taken. This is usually the butt end of the log, since defects tend to follow the grain or rings. The diameter of the defect at the point or points measured is compared with the scaling diameter of the log. The U.S. Forest Service has adopted this rule as standard practice.

One of the log rules adopted by the Forest Service is the Scribner Decimal C, based on diagrams of circles on which 1-in. boards are drawn. It allows $\frac{1}{4}$ in. for saw kerf for each board, and widths of boards to the next lower full inch, fractional inches being dropped. The deduction for defects under this rule must consequently allow one-fifth waste for saw kerf (Sec. 50). The standard rule for deducting defects is therefore $\frac{4}{5}$ of the board rule, or

$$\frac{4}{5}\left[\frac{(W \times T \times L)}{12}\right] = \frac{(W \times T \times L)}{15},$$

where W is width of defect in inches, T is thickness in inches, and L is length in feet; *viz.*, a defect enclosed by a rectangle measuring 10 by 12 in. by 8 ft. long gives $(10 \times 12 \times 8)/15 = 64$ ft. deduction, which would be rounded off to 60 ft. (see Scribner Decimal C log rule, Appendix).

Deductions are not made from individual logs unless evidences of defect are visible either at the ends or on the surface of the log. For such species as cypress, incense cedar, or redwood, a percentage deduction for hidden defects due to rot may be allowed on the total scale.

40. Interior Defects—Deductions. Interior defects are those which cause waste in the interior of logs. These are caused by rot, ring shake, center splits or heart checks, and pitch seams. Rot which enters from decayed knots is irregular in shape, extending above and below the point of infection, the distance depending on the species of tree and the stage of infection. The scaler must see such logs sawed out in order to get a basis for deduction from this source. Studies made by forest pathologists lead to the formulation of rules for discounting for rot due to specific fungi. The presence of three or four rotten knots in different places on the surface, together with rot showing at each end, may cause the rejection of the entire log. In a log with defect to this extent, not enough lumber could be sawed to repay the cost of logging and manufacture.

To both dimensions of a rectangle enclosing an interior defect, 1 in. should be added to ensure sufficient margin for the production of sound lumber in sawing around the defect. In overlength logs, with center rot showing at both ends, the average diameter of the defect plus 1 in. is the basis for deduction from each of the two or more sections of the log (Fig. 8).

Center rot, circular in form, would by this rule be squared using the formula $[(D + 1)^2/15] L$ for total deduction; viz., rot measuring 12 in.

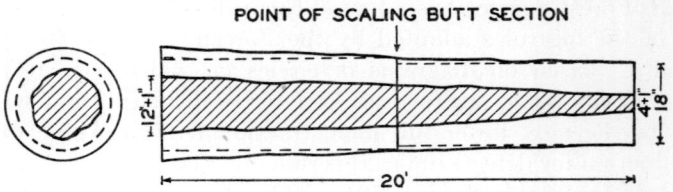

Fig. 8. Rot extending through a log. In this case the log exceeds a maximum scaling length of 16 ft., and the scale will be the net total of two 10-ft. pieces. The average of the diameters of the rot at each end, plus 1 in., or $\frac{13 \times 5}{2} = 9$ in. The deduction will be $\frac{(9 \times 9) 20}{15} = 108$ ft. or 11 Decimal. The scale for a log 20 in. × 10 ft. is 17; that for a log 18 in. × 10 ft. is 13. The total is 30; subtracting 11 gives 19, or 190 bd. ft.

in diameter and 8 ft. long is scaled as $(13^2/15)8$, or 90 bd. ft. This deduction is too liberal, the actual waste amounting to only about 83 per cent of the deduction, or $0.83 [(D + 1)^2/15] L$. The standard deduction therefore allows liberally for unseen irregularities in center rot. Several rules of thumb are in use for making deductions for cir-

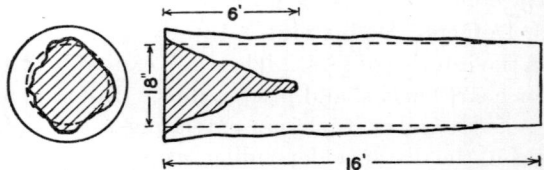

Fig. 9. Butt rot. Deduction from length of log, in this instance 6 ft., as the entire area of the *cylinder* at the butt is defective. With a smaller area, the method of enclosing the rot by a rectangular block, whose length equals that of the defect, would be used. The scale for a log 18 in. × 10 ft. is 13, or 130 bd. ft. The full scale for a log 18 in. × 16 ft. is 21, or 210 bd. ft.

cular center defects, but none gives consistent results and some are far from accurate.

Ground, or butt, rot is a common defect which seldom extends beyond 5 or 6 ft. into the butt log. Its probable length is determined from sawing tests, and deduction is made as above from the dimensions of the center scaled cylinder (Fig. 9).

Other forms of interior rot that appear as irregular patches and are usually due to infection from limbs are boxed out, allowing liberal margin for probable variation of shape within the log. In many instances, the scale of a badly defective log may be reduced by an arbitrary percentage.

Shake is caused by the separation of the growth rings due probably to swaying in the wind. These shakes cause the boards to split apart. Shake is treated as a circular defect. If it appears on both ends of the log, the deduction is the same as for center rot of like dimensions. When it so happens that the core within the ring is sound, the scale of a log of this diameter, measured at the small end, is treated as sound and the gross deduction reduced by this amount; *viz.*, a ring shake in a 16-ft. log measures 10 in. at the butt and the sound core is 7 in. at the small end. The initial deduction is $(11 \times 11 \times 16)/15 = 129$ bd. ft., scaled as 130. The scale of a 7-in. log 16 ft. long is 30 bd. ft. The net deduction is 100 bd. ft.

Pitch rings in certain western species are treated as defects and are deducted by the same methods used for ring shake.

Splits, heart checks, or pitch seams in the interior of logs are boxed out by a rectangle which completely encloses the defect. When the check or seam occurs in a log with twisted grain, the deduction must be large enough to enclose the entire defect, though it may be somewhat reduced to allow for the sawing of short boards of more than the minimum length from the sound portions of the rectangle at either end.

41. Side or Exterior Defects—Deductions. Side defects affect the outer surface of logs and penetrate more or less deeply into the cylinder. When the defect occurs at the butt of a tapering log, it is ignored unless it affects the volume of lumber in the scaled cylinder. This will not happen unless it goes deep enough to be evident after a slab 1 in. thick is removed from this cylinder in addition to the entire depth of the taper, which in large butt logs may be 2 to 4 in. The usual method of deduction for a catface or fire scar is by slabbing off the defective portion. The length of the slab is measured to the approximate point at which the defect leaves the 1-in. collar in the scaled cylinder. Its dimensions are taken as the average width of the boards which could be sawed from the slab within this cylinder, multiplied by the thickness of the slab after 1 in. is deducted for the collar. The rectangle formula is then applied; *viz.*, a fire scar extends 6 ft. up the cylinder and penetrates to a maximum depth of 9 in. on a log 25 in. in scaling diameter. The average width of board after allowance is made for the 1-in. collar is judged by inspection to be 17 in. The reduced depth within the

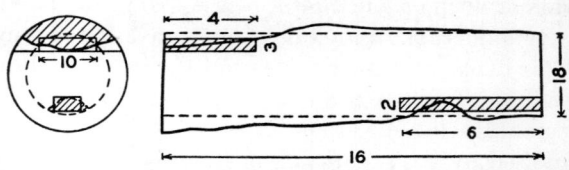

Fig. 10. A fire scar at the butt penetrates 4 in. into the cylinder. Deducting 1 in. for slab and taking the average width of boards lost as 10 in. and their length, inside the cylinder, as 4 ft., the deduction is $(3 \times 10/15)4 = 8$, called 1 in Decimal. A catface shown above penetrates 3 in. into the cylinder, 5 ft. from the end of the log. Deducting 1 in. for slab and taking the average width of boards lost as 6 in. and their length as 6 ft., the deduction is $(2 \times 6/15)6 = 4$, or 0 in Decimal. Since the gross scale of a log 18 in. $\times$ 16 ft. is 21 in Scribner Decimal C, the net scale for the log with the above two defects is $21 - 1 = 20$, or 200 bd. ft.

right cylinder after a 2-in. taper and 1 in. for slab are deducted is 6 in. The deduction scales $(17 \times 6 \times 6)/15 = 41$ bd. ft., scaled as 40 (Fig. 10).

Where the defect is deep and V-shaped, it can be enclosed in a sector of a circle. If the defect runs the entire length of the log, the deduction bears the same relation to the total volume as the sector bears to the circle. For example, a log 20 in. in diameter 16 ft. long and scaling 280 bd. ft. has a spiral lightning scar, which has been burned out, that runs its entire length and can be enclosed in a sector equaling one-fourth of the circumference. The deduction is one-fourth of 280, or 70 bd. ft. The net or sound scale is 210 bd. ft. (Fig. 11).

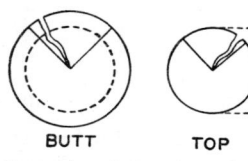

BUTT   TOP

Fig. 11. A frost crack or other deep defect, with a twist amounting to one-fourth the circumference of a 16-ft. log, is enclosed by a sector of equal dimensions extending through the length of the log. The full scale of an 18-in. log 16 ft. long is 21 Scribner Decimal C. One-fourth is 5. The net sound scale is 16, or 160 bd. ft.

Worm holes of a character to cause rejection of lumber are deducted by either the slab or sector method.

Surface checks, termed wind or sun checks, are cracks penetrating the surface of the log along the radii. The loss of boards thus caused is allowed for in scaling by dropping back one-half the depth of the checks to obtain the scaling diameter; viz., checks on a 24-in. log penetrate 6 in. on all sides, and the log is scaled as $24 - (3 + 3)$, or 18 in. in diameter.

Rotten sapwood is deducted by scaling only the sound core at its small end. These are the only two cases in which deductions are made by reducing the diameter of the log.

42. Crook or Sweep—Deductions. The extent of the actual loss of boards by reason of crook or sweep depends on the minimum length of a merchantable board and on the extent of the sweep and its character. The advocated method is to prolong a right cylinder from the butt end to the top, but throwing the crook into the top part of the log. If the log is unsound, the end that has the most sound material is used as the base. The extent of the probable loss that would result from sawing the log in this position is gauged by comparing the imaginary straight log with the crooked piece. The loss in sawing this crooked portion is approximated rather than calculated, as amounting to a given percentage of length which includes the crooked portion; *viz.*, a 16-ft. log

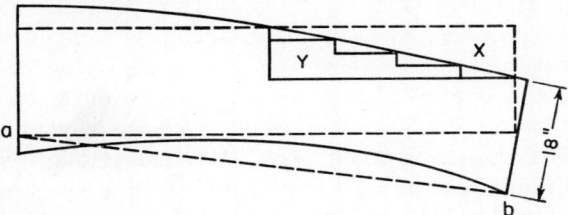

FIG. 12. Scale of a crooked log. (1) The half cylinder XY of the theoretical straight log contains $\frac{1}{4}$ of the total scale. The crook eliminates approximately one-half of XY, equal in volume to $\frac{1}{2} \times \frac{1}{4}$ or $\frac{1}{8}$ of the total scale. For a log 18 in. $\times$ 16 ft, scaling 21 Decimal the deduction is 3, giving a net scale of 18 Decimal. (2) Measuring the distance from the line ab to the surface at the middle of log gives 3 in. crook. The ratio of crook to diameter is $\frac{3}{18}$ or $\frac{1}{6}$, which gives a cull reduction factor greater than for methods 1 and 3. (3) Shortening the length of the log by $\frac{1}{8}$ reduces the scale correspondingly and gives the same result as method 1.

18 in. in diameter has a crook which deviates 9 in. from the projection of the straight portion. One-quarter of this log will suffer a loss in scale. After allowing for the sawing of 10-, 12-, and 14-in. boards from a part of this quarter, it is concluded that the net loss will be one-half of the affected quarter. The full scale of the log is 210 bd. ft. One-eighth of 210 is 28 bd. ft., which is deducted. The net scale is 182 bd. ft., rounded off to 180 (Fig. 12).

When boards are sawed the full length of the log and the sweep is distributed evenly on the carriage, the maximum sweep for the above log is 3 in. or less, as measured by stretching a line from the small end to the point of intersection of the margin of the scaled cylinder at the butt, thus allowing for taper. It has been found that losses bear a direct relation to the percentage of sweep based on diameter of log, being about one-third greater than this percentage. In the above log, the relation is $\frac{3}{18}$, or 17 per cent. Increasing this by one-third gives about 24 per cent, which for a log scaling 210 bd. ft. would call for a deduction of 48 bd. ft. To conform to the close-sawing practice of the first rule, this additional deduction of one-third can be omitted, in which case the loss is 36 bd. ft., rounded off to 40. The sawing and scaling of short boards justifies the smaller deduction. Losses by either method bear a direct relation to the percentage of sweep. A conservative rule allowing for short boards would therefore be to deduct from the scale the same percentage that the amount of sweep in inches, measured at the middle of the log, bears to the scaling diameter of the log.

43. Merchantable and Cull Logs. Logs are merchantable when the value of the products which they yield will return all costs of logging and manufacture and leave a margin sufficient to equal the price charged for stumpage multiplied by the net scale of the log. The sawmill operator who logs his own timber considers both the logging and milling costs and will take logs containing a large percentage of defect, provided that the quality and value of the remaining sound lumber will pay for handling the log. The logger, who may be paid by contract, is interested in the size and weight of the log. His costs are the same regardless of deductions from full scale due to defects. For this reason, when payment for logging is based on net sound scale, the defective portions are unprofitable for him to handle. When defects reach a large percentage, the log will not pay even to saw. Logs lying close to the track may be taken, while similar logs at the extreme limit of skidding would be left.

When logs are handled by contract or purchased on sound scale, an agreement is reached under which a fixed percentage of the total scale of any log must be sound if the log is to be scaled as merchantable. Such an arrangement is necessary to obtain a practical basis for administration. This percentage depends on the species, the more valuable requiring a lower percentage of sound scale; *viz.*, western white pine on sales in Idaho may have as much as 67 per cent deduction, while with ponderosa pine the limit is set at 50 per cent. Thirty-three per cent scale therefore makes a merchantable log for white pine and 50 per cent for ponderosa pine. This percentage refers to the total scale which the log would have if straight and sound, the deduction including crook, rotten sap, splits, or any other defects present.

Logs must conform to the minimum diameters and lengths which by agreement are considered merchantable and required to be logged. When smaller than these minima, the logs are classed as unmerchantable.

Unmerchantable logs are termed "cull" and so marked in the scale book. Where possible, trees or logs which would be culled when scaled are left in the woods. Hence a definite understanding is necessary as to the standards of scaling, in order to save the logger the expense of handling culls.

44. Sample Scaling. The principles governing the distribution of units of a population, as set forth in Sec. 12, apply to the volumes of logs; hence an appropriate system of sampling should give reliable information on the nature of the "population" of logs, particularly as to the total volume. Under certain conditions, sample scaling becomes a feasible practice, while under others it cannot be accepted. The guiding rules for feasibility are (1) the logs constitute a fairly homogeneous lot and not a mixed assortment of various species with widely different average log volume and stumpage price, particularly if it is desired to determine value and volume by species; (2) the logs can be concentrated in one place in sufficient numbers so as to gain a great saving in time of scaling; and (3) the total number of logs involved is large.

The problem is to determine, for any given total, the number of logs which must be scaled in order to confine the maximum sampling error in total scale within limits acceptable to both seller and purchaser. The maximum error is commonly taken as twice the standard error and is a limit which can be exceeded in either a plus or minus direction only once in 22 cases. Thus, if the agreed maximum error is 5 per cent, the corresponding standard error is $\pm 2\frac{1}{2}$ per cent. It has been previously shown (Sec. 17) that the standard error of the mean (s_M) is related to the standard deviation (s) in the following way:

$$s_M = \frac{s}{\sqrt{N}}$$

The equation can be transformed to

$$N = \frac{s^2}{s_M^2}$$

and in this form is a convenient expression for determining the number required in a sample, assuming the standard deviation is known and having an accepted limit of error of the final scale. It should be obvious that, in using this formula, both s and s_M should be in the form of percentages. The first task, therefore, is to gain a knowledge of the volume and variation of volume in the species to be scaled. This is done from records of past operations, and the variation is expressed in terms of standard deviation (Sec. 16d) and then in terms of standard deviation/mean or coefficient of variation (Sec. 16e). Assuming for purposes of illustration that the coefficient of variation for the timber in question is found to be 40 per cent and the maximum error of 5 per cent (corresponding to a standard error of 2½ percent) has been agreed upon, the number of logs to be scaled will be found as

$$N = \frac{0.40^2}{0.025^2} = \frac{0.1600}{0.000625} = 256$$

Thus 256 logs will have to be scaled, without regard to the total number of logs involved. If now it has been estimated previously that the sale area contained 2,500 logs, only one out of every 10 logs, well distributed throughout the entire lot of logs, will have to be scaled in order to get a total volume within the accepted limits of accuracy. In getting the scale, there should also be a total count of logs, since the originally estimated 2,500 may not be correct. The average volume per scaled log multiplied by the counted number of logs will then give the required total estimate. Tables can be set up, or alinement charts drawn (Chap. 11), which will give quickly the ratio of logs to be scaled for any sized lot, to any degree of accuracy, but all must be within one standard deviation class. This has been done for a number of species in the United States, and sample scaling has become standard practice in some forest regions.

Sample scaling has its limitations, however. If the timber is of such a character as to make the standard deviation unreliable, it invalidates the method of sample scaling. Factors which are of this nature are

1. Wide variation in percentage of cull.
2. Variation in species, particularly if the species are of different growth habit.
3. Wide variation in value by species and by size. A standard deviation determined on the volume basis will be much too small, if expected to apply to a value basis.
4. Variation in length of log. Within the same species, long logs show a greater standard deviation than short logs.

A second major restriction is that the use of sample scaling should lead to economies in time and cost which are sufficiently worth while to more than offset the maximum losses that might occur. If it is not possible to accumulate logs in the woods or on the landings so that a scaler may come in for one or two days a week, it will be necessary to have him on the job constantly. If this is the case, then he should be kept occupied, if not at scaling, at some other job or jobs. In many instances, the logs flow into the landing constantly and are loaded and transported out immediately. Sometimes even in this case it is possible to apply sample scaling to a string of loaded cars or trucks at a point where they are assembled.

A third restriction lies in the fact that the logger or purchaser may not easily be persuaded to accept the risk of taking a loss, which, although it happens only very infrequently, may hit him. On the other hand, he stands an equal chance of realizing a gain. Over many sales, the errors are compensating and he would not suffer, but usually he is interested only in a single sale and is apt to take the blackest point of view.

A fourth restriction is that the method has no use in small sales. The saving in time is inconsequential, and the ratio of tally so high that the lot of logs might just as well be scaled completely and accurately.

45. Scaling for Cubic Contents—Quebec. The taper of logs, accentuated in butt logs, imposes difficulties when the logs are to be scaled in rollways. The Province of Quebec established the cubic-foot standard on all logs in 1936 but in 1945 permitted logs intended for manufacture into lumber to be scaled by a new log rule, the Roy rule, in board feet. The scaling for cubic feet is based on end diameters, but these are averaged for butts and tops by one of two systems. For logs 10 ft. or less in length, the scaler selects one side of the rollway and measures the diameters of butts and top ends indiscriminately. For logs 11 ft. and over, the log is measured in two half-log sections, the smaller half based on top diameter and the larger, or butt, section based on butt diameter. With flaring butts, judgment is used in arriving at a fair diameter. Inches are rounded off at the $\frac{1}{2}$-in. point. For saw logs, lengths over 16 ft. are scaled by taking an average diameter of butt and top. Inches are rounded off at the $\frac{3}{4}$-in. point.

46. Scaling Records and Check Scaling. Records of the scale are kept in scale books in which are entered the number of each log its length and net scale, and the deduction made from full scale. The record is kept separate for different species whenever the price paid for each differs.

The log number is chalked on the scaling end of each log, to permit a check on scaling. Check scaling may consist of a rescale of a number of logs, then comparing their total scale with the original. This is done in case of complaint as to the scale and is the basis for adjustments.

For checking and correcting the methods of the scaler, individual logs should be rescaled and the causes of any differences in scale determined. It does more harm than good to criticize a tendency to over-

70 FOREST MENSURATION

or underscale, since this comment is practically always productive of a worse error in the opposite direction.

Check scales should come within 1 per cent on sound logs; on logs up to 10 per cent defective, within 2 per cent; on logs 11 to 20 per cent defective, within 3 per cent; on logs over 20 per cent defective, within 5 per cent.

Below is a sample of two pages from a scale book:

Purchaser, **James Brown**
Date of Sale, **6/31/32** End Mark, **Ⓑ**
Species **White Pine**

Where Scaled, **Landing No. 7**
Compartment, **4**; Sec. **17**; T. **20**; R. **4**; Date, **7/30, 1932**

Log No.	Length	Ft. B.M.	Log No.	Length	Ft. B.M.	Log No.	Length	Ft. B.M.	Log No.	Length	Ft. B.M.	Log No.	Length	Ft. B.M.	Remarks
301	16	4	321	18	40	341	18	10	361	16	18	381	18	Cull	
2	16	14	22	16	12	42	18	6	62	16	4	82	20	Cull	
3	14	10	23	16	15	43	16	4	63	18	6	83	20	Ⓔ 16	
4	18	20	24	18	Cull	44	18	24	64	16	4	84	18	9	
5	24	21	25	16	8	45	18	9	65	20	10	85	16	9	
6	20	Ⓔ 15	26	16	3	46	14	9	66	16	7	86	16	16	
7	16	8	27	16	15	47	12	Ⓔ 12	67	22	16	87	16	21	
8	16	3	28	18	10	48	12	Cull	68	16	6	88	16	7	
9	16	10	29	22	Ⓔ 17	49	18	20	69	18	Ⓔ 7	89	16	14	
10	12	Ⓔ 8	30	24	30	50	16	16	70	18	9	90	16	7	
11	18	19	31	20	18	51	16	10	71	16	5	91	16	2	
12	18	7	32	20	7	52	22	16	72	12	5	92	18	20	
13	16	15	33	20	19	53	24	7	73	14	7	93	14	Ⓔ 10	
14	16	10	34	14	8	54	14	7	74	14	10	94	14	Ⓔ 17	
15	22	26	35	16	Ⓔ 9	55	20	17	75	24	16	95	22	27	
16	14	Ⓔ 20	36	16	Ⓔ 15	56	20	18	76	24	19	96	18	18	
17	18	Cull	37	16	Cull	57	20	14	77	22	22	97	22	16	
18	16	10	38	16	11	58	16	3	78	24	28	98	16	14	
19	14	11	39	18	12	59	16	16	79	20	10	99	14	19	
20	14	10	40	22	15	60		16	11	80	20	9	00	14	10	

Totals: 2410, 2620, 2290, 2180, 2520

— Total this page 120,20
— Brought forward 175,340
— Total since last report 187,360
— Reported to 7/15/32
— Total to 7/30/32 740,200
92,560

Scaled by **Wm. Curry**

References

Association of Forest Engineers, Quebec: Status of the Board Foot, Cubic Foot and Cord Units of Wood Measurement, *Jour. Forestry*, **26**, 912–928 (1928).

Belyea, H. C. and T. R. Sheldon: Some Anomalies in the Board Foot Measurement of Logs, *Jour. Forestry*, **36**, 963–969 (1938).

Berry, J. B.: Measuring Woodland Products, *Ga. Agric. Col. Ext. Bul.* 142, 1918, 16 pp.

Blythe, R. H., Jr.: Saving by Sampling, *Timberman*, **46** (5), 42–45, 80 (1945).

Bradner, M., and P. Neff: Log Scale versus Lumber Tally: A Discussion of Overrun and the Factors Affecting It, *Timberman*, **27** (9), 46–48, 50–52 (1926).

Briegleb, P. A.: Sample Scaling for West Coast Logs, *Timberman*, **45** (4), 98–102 (1944).

Brown, R. M.: Defect Deduction for the One-quarter Inch International Log Rule, *Jour. Forestry*, **33**, 760–762 (1935).

Clark, F. G.: The Application of an Old Rule for Scaling Interior Defects, *Jour. Forestry*, **28**, 94–95 (1930).

Fogh, I. F.: Sampling Methods in Log Scaling, *Forestry Chron.*, **19**, 127–138 (1943).

GEVORKIANTZ, S. R., and H. E. OCHSNER: A Method of Sample Scaling, *Jour. Forestry*, **41**, 436–439 (1943).
GRAVES, H. S.: Methods of Scaling Logs, *Forestry Quart.*, **3**, 245–254 (1905).
HUBERT, E. E.: Manual of Wood Rot for Cruisers and Scalers of the Inland Empire, *Timberman*, **28** (3), 42–46; (4), 43–48; (5), 48–52; (6), 48–53 (1927).
KNOUF, C. E.: Basic Principles in Scaling Pine, *Timberman*, **29** (1), 64–68 (1927).
———: Early History and Development of Log Scaling Practice, *West Coast Lumberman*, **36** (427), 41, 54; (428), 40, 64–65 (1919).
———: Log Scaling in the Inland Empire, *Timberman*, **24** (3), 42–51; (4), 42–46 (1923).
———: Trade Course in Log Scaling for Idaho Woods, *Idaho Vocat. Ed. Bul.*, **5** (5), 108 pp. (1922).
KOEHLER, A.: A New Hypothesis As to the Cause of Shakes and Rift Cracks in Green Timber, *Jour. Forestry*, **31**, 551–556 (1933).
LEXEN, B. R.: Application of Sampling to Log Scaling, *Jour. Forestry*, **39**, 624–631 (1941).
———: Better Measurements for Small Logs, *Lumber Trade Jour.*, **98** (9), 34–36 (1930).
———: Short Cutting the Scaling Job, *Timberman*, **44** (9), 20, 22, 23, 26, 28, 30 (1943).
LOCKARD, C. R., and R. D. CARPENTER: Interim Sawlog Grades for Southern Hardwoods, Southern Forest Experiment Station, January, 1948 (Multigraphed), 9 pp.
MCINTYRE, R. W., B. BRERETON, C. E. KNOUF, E. FRITZ, and E. I. KARR: Log Scaling, *Timberman*, 1936, 157 pp.
Michigan State College, Agricultural Experiment Station: Michigan Log Marks, *Memoir Bul.* 4, 1941, 89 pp.
ROY, H.: Log Scaling in Quebec, *Jour. Forestry*, **36**, 969–975 (1938).
SAMMI, J. C.: A Rule of Thumb for Log Scaling, *Jour. Forestry*, **34**, 181–182 (1936).
———: A Simple Log Rule, *Jour. Forestry*, **35**, 307–308 (1937).
TIEMANN, H. D.: The Log Scale in Theory and Practice, *Soc. Amer. Foresters Proc.*, **5**, 18–58 (1910).
———: Methods of Making Discounts for Defects in Scaling Logs, *Forestry Quart.*, **3**, 354–357 (1905).
U.S. Forest Service: Instructions for the Scaling and Measurement of National Forest Timber, rev. ed., 1938, 103 pp.
WOOLSEY, T. S.: Checking Check Scalers, *Soc. Amer. Foresters Proc.*, **11**, 245–248 (1916).
———: Scaling Government Timber, *Forestry Quart.*, **5**, 166–173 (1907).
ZACH, L. W.: The Application of Sample Log Scaling in Region One, *Northern Rocky Mountain Forest and Range Expt. Sta. Res. Note* 31, 1943, 10 pp.
ZILLGIT, W., and S. R. GEVORKIANTZ: Estimating Cull in Northern Hardwoods, *Lake States Forest Expt. Sta. Paper* 3, 1946, 7 pp.

CHAPTER 6

CONSTRUCTION OF LOG RULES

47. Factors Causing Variation in Sawed Output of Logs. Log rules arose from the necessity of obtaining a standard by which the contents of logs in terms of sawed lumber could be measured while still in the log (Sec. 2). It would have been of great benefit to lumbermen and forest owners if a single standard log rule had been devised, which could measure the contents of logs of all sizes consistently.

No rule can give the actual sawed contents of all logs, since this depends on the method of sawing. A log of a given diameter and length, if sound and straight, may yield different amounts of lumber from the following causes:

1. Thickness of saw. The waste from saw kerf increases with thicker saws; hence a $\frac{1}{8}$-in. band saw produces more lumber than a $\frac{1}{4}$-in. circular saw.

2. Thickness of lumber. The thicker the lumber sawed, the fewer will be the number of saw cuts, and the greater the quantity of lumber produced.

3. Utilization of slabs. Large slabs may be remanufactured into lumber to prevent unnecessary waste.

4. Width of lumber. A narrow minimum width permits the sawing of a few more boards from what would otherwise be slabs.

5. Minimum length of boards. Short boards may be sawed from slabs and put through the edger and trimmer. This output comes from the taper of the log and from sound portions of defective logs.

6. Method of sawing the log. A log may be sawed through without squaring it, by the method of slash sawing, but this is done only when the resulting boards are intended for close utilization of small dimension in a cut-up plant. When the resulting wedge-shaped or tapering boards must be squared in the edger, more lumber is produced by squaring the log and slabbing from four sides. The gain may be from 6 to 22 per cent in a 20-ft. log, and proportionately greater with smaller diameters.

The six factors just mentioned all apply to lumber when measured at its actual thickness and converted to board feet by the factor of 1 bd. ft. equals 144 cu. in. Although the nominal thickness of boards is 1 in., the lumber trade has accepted as a standard for dressed and dried lumber a thickness of $25/32$ in. (Sec. 35). Even after allowing for losses in thickness by shrinkage in drying and by planing, it is possible that a good well-adjusted band saw can be set for $15/16$ in. and even for $7/8$ in. and still produce finished lumber of the required thickness. Thus, in large logs, for every fifteen, or seven, cuts, respectively, there is an additional cut, which would not be obtained if the saw had to be set at the full inch. On the other hand, circular saws cutting hardwoods are often set to cut boards $1\frac{1}{4}$ in. thick to meet the standards.

If the log is measured by a standard log rule, these six variable factors will be ignored by the rule and will thus affect the overrun.

48. Standardizing by Means of Diagrams. Diagram Rules. The early inventors of log rules, in determining what standard contents should be adopted for a log of a given size, had to decide arbitrarily how the log would be sawed. For instance, the Scribner rule is based on a saw kerf of $\frac{1}{4}$ in., boards 1 in. thick, probably not less than 8 in. wide and the full length of the log. The log lengths given were from 10 up to 24 ft. and the diameters from 12 to 44 in.

Scribner eliminated the variables due to different methods of squaring and slabbing by drawing diagrams of circles of different diameters and plotting on each area the ends or cross sections of the boards which might be sawed from it. Dropping fractions of inches, he computed the total cross-sectional area in the boards in square inches and reduced them to board feet by the divisor 12. He then got the standard log contents by multiplying this result by the length of the log in feet. The facts that his diagrams were in some cases eccentric and that the resulting values for successive inch classes increased in an irregular manner were ignored. The diagram was the standard. The original values still hold good regardless of later efforts to introduce modifications or improvements in the rule.

By such a method, waste is automatically deducted, and the rule will give no overrun if the log is sawed according to the diagram and taper is ignored or wasted. Logs of all sizes are represented. If all circles have been treated alike, the overrun on all logs will be practically the same for given conditions of utilization. A diagram rule is usually a good rule.

The Scribner log rule has been extended downward to cover logs 6 in. in diameter and upward from the original 44 in. to 120 in. for Pacific Coast timber. Other good log rules constructed by diagrams are the

Spaulding rule of California and the Maine rule. The latter is more consistent than the Scribner.

The process of adding long columns of figures required in totaling a scale book gave rise in the Lake states to the practice of rounding off the last figure in the scale of a log to the nearest 10 and dropping off the cipher. This modification is called the "Scribner Decimal rule." In extending the values of the decimal rule downward below 12 in., three separate sets of values resulted, one of which, termed the "Scribner Decimal C log rule," became the standard in the Lake states and was adopted by the U.S. Forest Service throughout the United States as the result of operations on the Minnesota National Forest in 1902.

49. Constructing Log Rules by Formula—The Board Foot-Cubic Foot Ratio. All log rules regard logs as cylinders. In the International rule, the maximum length of the cylinder is 4 ft.; in other rules it equals the log length up to the maximum given in the rule. The volume in board feet of a log of any length may therefore be found by multiplying the board-foot contents of a 1-ft. cylinder by the length of the cylinder in feet.

The board-foot volume of a cylinder is related to the number of square inches in the area of the circle of the given diameter. Since the cross-sectional area of 1 bd. ft. contains 12 sq. in., one could obtain board-foot volume of a cylinder 1 ft. long by dividing the cross-sectional area in inches by 12, provided that there were no waste in saw kerf and slabs. Formula rules attempt to make adjustments for these losses in order to obtain the net board-foot contents of the cylinders.

The area of a circle in square inches is $\pi D^2/4$ or $0.7854 D^2$. If there were no waste from kerf and slabs, the equivalent board-foot content of the 1-ft. log would be $0.7854 D^2/12$ or $0.06545 D^2$. The Constantine log rule, used for veneer logs, is actually based on this formula and is expressed as $0.06545 D^2 L$, where L is the length of the log in feet. The Brereton log rule is identical with the Constantine rule and is used in the export trade.

Other log rules make use of the board foot–cubic foot ratios applied to the contents of the 1-ft. cylinder. These ratios express the number of board feet that can be sawed from each cubic foot of log contents. Since percentage of waste due to slabbing decreases with increase in size of the log (Sec. 50), this ratio is not constant but increases with the diameter of the log. Two sets of ratios are available, one based upon actual sawed-lumber content compared with total cubic content, and the second based upon estimated (scaled) board-foot content compared with total cubic content. An example of the first is given in Table 14, where actual sawed contents are estimated from a special log rule devised by H. D. Tiemann. It is seen that the ratio increases from 1.27 to a maximum of 8.22 bd. ft. for each cubic foot. An example of

TABLE 14. RELATION OF CUBIC- AND BOARD-FOOT CONTENTS OF 16-FT. LOGS WITH A TAPER OF 1 IN. IN 8 FT., BASED ON TIEMANN'S LOG RULE, $\frac{3}{16}$-IN. SAW KERF

Diameter inside bark at middle of log, in.	Cubic contents, cu. ft.	Sawed contents, Tiemann log rule, bd. ft.	Ratio bd. ft. to 1 cu. ft.	Volume utilized, %
3	0.79	1	1.27	10.5
4	1.40	4	2.85	23.8
5	2.18	9	4.13	34.4
6	3.14	15	4.77	39.5
7	4.28	23	5.37	44.8
8	5.59	32	5.71	47.7
9	7.07	43	6.08	50.7
10	8.73	55	6.30	52.5
11	10.56	69	6.53	54.4
12	12.57	84	6.68	55.7
13	14.75	101	6.85	57.0
14	17.10	119	6.96	57.9
15	19.63	139	7.08	59.0
16	22.34	160	7.16	59.7
17	25.22	183	7.26	60.5
18	28.27	207	7.32	61.0
19	31.50	233	7.39	61.6
25	54.54	419	7.68	64.0
31	83.86	659	7.86	65.5
37	119.47	954	7.99	66.5
43	161.35	1,301	8.06	67.2
49	209.52	1,703	8.13	67.7
55	263.98	2,159	8.18	68.2
61	323.96	2,669	8.22	68.5

the second could be made by comparing the values of one of the log rules, such as the Scribner rule, with the total cubic contents of the log. In this type of case, it is found that the ratio is inaccurate, since the overrun inherent in the log rule is neglected. It is important to note that the board foot–cubic foot ratio varies with diameter and that therefore no one ratio should ever be chosen to apply to all classes. To do so would mean a large overestimate for logs smaller than the one to which the ratio applied and a large underestimate for logs larger than this size. Nevertheless, several log rules were thus constructed. The New Hampshire rule (Sec. 32) accepts the ratio of 6/1 and applies it to cubic-foot content. The Vermont rule accepts 6.8/1. In the first, only a 9-in. log would be scaled accurately and in the second only a 13-in. log.

50. Waste from Saw Kerf and Slabs. The increasing ratio of board feet to cubic feet is due to decreasing waste with increase in the size of log. Waste is separated into that portion lost in saw kerf and that lost in slabs and edgings. The waste from saw kerf is a fixed percentage of the volume of the cylinder, and that from slabs and edgings is related to the circumference or diameter.

Let k = width of saw kerf in inches

t = thickness of board in inches

Then $[k/(t + k)] \times 100$ gives the percentage of total volume wasted in saw kerf. For example, with a saw kerf of $\frac{1}{4}$ in. and boards 1 in. thick, the sawdust is $[0.25/(1 + 0.25)] \times 100 = 20$ per cent. Calling this loss b and introducing it into the board-foot formula of the previous section, one could express a revised formula as $(1-b)0.06545D^2L$, provided that waste from slabs and edgings is omitted.

TABLE 15. COMPARISON OF WASTE IN SAWING LOGS OF DIFFERENT SIZES

	Diameter at small end of log, in.		
	10	20	40
Waste due to slabbing and edging, %	27.75	14.44	7.27
Comparative ratios in cubic content	1	4	16
Comparative loss from slabbing and edging (approximate only)	4	2	1
Waste from kerf in residual area, % of net volume sawed	20	20	20
Equivalent of waste from kerf in residual area in terms of total content of log, %	14.45	17.11	18.54
Total waste, %	42.20	31.55	25.81
Residual part used, %	57.80	68.45	74.19
Board foot–cubic foot ratio	6.9	8.2	8.9

Waste from slabs and edgings has been found to be related to circumferences or πD and, therefore, since π is a constant, to diameter alone. The waste may thus be figured as equivalent to a ring or collar whose thickness is the same for logs of all diameters except those which are too small to produce the minimum width boards. The average normal thickness of this ring has been found to be between $\frac{3}{4}$ and 1 in.,

CONSTRUCTION OF LOG RULES

which when doubled would reduce the effective diameter of the log by 1.5 to 2 in. It follows that the *percentage* of waste from slabs and edgings will decrease with increasing diameter of log. The percentage of waste from this source, as actually determined from logs, is shown in Table 15 for logs 10, 20, and 40 in. in size. Since this form of waste can be expressed as a reduction in diameter in inches, the formula given above can be still further revised to

$$\text{Volume in board feet} = (1 - b)0.06545(D - a)^2 L$$

where a is double the width of the collar. For example, with a 12-in. log 16 ft. long, a saw kerf of $\frac{1}{4}$ in. and a $\frac{3}{4}$-in. collar, equivalent to a deduction of 1.5 in. from diameter, the formula gives

$$\text{Volume in board feet} = (1 - 0.20)0.06545(12 - 1.5)^2 16$$
$$= 0.05232(110.25)16 = 92$$

51. Log Rules Based on Formulas—International Log Rule. For a formula to produce a log rule whose values for logs of all sizes give consistent percentages of overrun, three conditions are necessary: a correct formula must be used;[1] the deduction for slabs and for saw kerf must be in proper quantities and relations to each other; and the taper in long logs must be controlled or accounted for.

The International log rule fulfills these three conditions. It was constructed by Judson F. Clark in 1906, originally for a saw kerf of $\frac{1}{8}$ in. plus $\frac{1}{16}$ in. allowance for shrinkage, or $\frac{3}{16}$ in. total deduction, giving a waste allowance of

$$\frac{\frac{3}{16}}{1 + \frac{3}{16}} = \frac{3}{19} = 0.158$$

Then
$$1 - b = 1 - 0.158 = 0.842$$

The taper is controlled by scaling the logs in separate sections or cylinders 4 ft. long, giving $L = 4$, and obtaining the total scale by adding the computed contents of these cylinders. Then

$$0.842(0.06545 D^2)4 = 0.22 D^2$$

---

[1] By the first of two correct formulas, the deduction for waste in slabs is made in the form of a plank of given thickness, whose width equals the diameter of the log (International rule); viz., $[(1 - b)0.06545 D^2 - (aD/12)]L$. By the second, it is made by decreasing the diameter of the log by the double width of a ring representing this waste in slabs; viz., $(1 - b)0.06545(D - a)^2 L$.

The slabs are next deducted as a *plank* (first formula) 2.12 in. wide, which reduced to board feet becomes, for a 4-ft. log,

$$\frac{2.12D}{12} \times 4 = 0.71D$$

The total net board-foot content for each 4-ft. length is therefore $0.22D^2 - 0.71D$.

A taper, or increase of ½ in. for each 4 ft., is allowed, and a log of 8 or 12 ft. is measured by adding the contents of two or of three 4-ft. lengths, respectively, each larger by ½ in. than the initial piece. It is the only log rule which builds up the contents of a longer cylinder from the sum of shorter lengths. All other log rules regard the log as a single cylinder, and scaling practice alone limits the length to be scaled as such.

In 1920, Clark was induced by H. H. Chapman to issue a new log rule, derived from the original ⅛-in. basis but applicable to ¼-in. kerf. The older rule is not applicable to mills using the larger saw kerf, and on the basis of the converting factor of 0.905 based on comparative waste from saw kerf alone it would give an underrun of 9.5 per cent, which thus discredits the ⅛-in. rule for adoption as a standard for commercial scaling practice. The purchaser of logs, being the risk taker, cannot afford to incur the certainty of an underrun.

Clark's ¼-in. rule, after at least three unauthorized modifications had been published, was finally in 1941 accepted as standard in the form in which the author issued it. It has been adopted by statute in New York, Michigan, and Connecticut and is almost universally used by foresters and timberland owners throughout the South and elsewhere as the basis for measuring growth of stands in board feet, this despite the retention of other rules, notably the Doyle, in scaling practice.[1] It was adopted by the Forest Service in 1939 for scaling over 2 billion bd. ft. of hurricane timber in New England with an underrun of 1.98 per cent for pine; and by Regulation S 17, 1939, it was made optional throughout the United States. With this exception, the Scribner Decimal C rule is the Forest Service standard.

The International ⅛-in. rule is still employed in some quarters for scientific measurements but has no advantage for this purpose over the ¼-in. rule.

52. Log Rules Based on Formulas—The Doyle Rule. The Doyle rule reads "Deduct 4 inches from the diameter of the log for slabbing,

---

[1] CHAPMAN, H. H., The International Log Rule for ¼-inch Kerf—Can It Replace the Doyle Rule? *Jour. Forestry*, **40**, 224–234 (1942).

square one-quarter of the remainder and multiply by length of log in feet to obtain the board-foot content of the log." (Sec. 51, second formula.) In algebraic form it is Volume = $[(D-4)/4]^2L$ or $0.0625(D-4)^2L$ and is therefore directly comparable with the basic formula $(1-b)0.06545(D-a)^2L$. The deductions for slabbing and for saw kerf are both incorrect, however, leading to inconsistent overruns, heaviest in small logs and decreasing as diameter increases, resulting finally in an underrun on large logs.

Since $1\frac{1}{2}$ in. is the normal deduction for slabs, 4 in., taken actually in the form of a ring, is more than twice too much, and the resultant percentage deducted from this source accounts for the progressively higher overruns with decreasing diameter. On the other hand, the saw-kerf deduction is too small, as will be shown later, giving the element of underrun, which increases rapidly in relative weight with increasing diameter, as the effect of slab deduction becomes less. In the construction of the Doyle rule, a $\frac{5}{16}$-in. saw was intended, the deduction for kerf to be 25 per cent, or $(1-b)$ to be equal to 0.75. But, in applying it, the author of the rule fell into an error which gave him a kerf allowance of only 4.5 instead of 25 per cent. This is seen by comparing the coefficients 0.0625 of the Doyle rule stated above and the expression $(1-b)0.06545$ of the correct allowance. The fraction $(0.06545-0.0625)/0.06545$ should give the value for b. Evaluated, this is equal to $0.00295/0.06545 = 0.045$ or 4.5 per cent deduction for kerf, or less than one-fifth of the intended deduction. Thus

$$(1-b) = 1 - 0.045,$$

which gives $95.5(0.06545D^2)L$ instead of $(1-b) = 0.75$, giving $0.75(0.06545D^2)L$. Yet, since the rule sounded right, it became firmly entrenched in many portions of the United States and Canada, in spite of its vicious inaccuracy. The net resultant of the two errors is to underscale small logs, because of the large slab deduction, and to overscale large logs, because of insufficient kerf deduction, for at this diameter the overscale from too small a saw-kerf deduction overtakes the underscale from slab deduction. The rule favors the buyer of small logs as against the owner and producer of second-growth timber and therefore tends to discourage reforestation and care of immature stands. Its continued existence and use in regions of second growth is due to the force of custom, ignorance of the defects of the rule, and the desire of log purchasers to perpetuate a concealed advantage.

With the increasing utilization of small timber for saw logs, this rule underwent a curious modification. Since the scale by the Doyle rule is zero for 4-in. tops and negligible under 8 in., logs of sizes from 7 in. down are scaled as containing 1 bd. ft. per foot of length, regardless of diameter. This scale reverses the trend and produces an underrun for logs of less than 6 in. in scaling diameter, when compared with the International ¼-in. rule!

Estimates of standing timber furnished by consulting firms in the South are often expressed, by diameter classes, in terms of the International ¼-in. rule, and paralleled by conversion of each diameter breast high class into terms of the Doyle rule by the use of separate converting factors, adopted to tree sizes. Where hardwoods or other

TABLE 16. OVERRUN, DOYLE RULE, ONTARIO, 12-FT. LOGS

Diameter of log at small end, in.	Scale by Doyle rule, bd. ft.	Actual output of inch lumber, bd. ft.	Overrun percentage
6	3	14	366
8	12	30	150
10	27	50	85
12	48	76	58
14	75	108	44
16	108	144	33
18	147	186	26
20	192	234	22

logs are cut from old-growth timber of large sizes, the Doyle rule may approximate the actual cut. A similar result may be obtained when tree-length logs are scaled as one piece based on middle diameter inside the bark.[1] Such exceptions do not justify universal application of the rule.

This rule should not be used, especially when small timber is being scaled. In this case, it is both very low and extremely inconsistent. Unfortunately the rule was long ago adopted as the standard scale by certain Southern states when most of the cut was in old-growth timber. Growth in board-foot volume, when predicted on the basis of this log rule, has the appearance of being excessively rapid on timber already 6 to 8 in. in diameter at the small end of the logs but up to this size is negligible. The rule would also favor those who must pay severance taxes based on the log scale, to the loss of the public.

Above 28 in. the Scribner log rule gives lower values than the Doyle

[1] MOORE, E. B., A Defense of the Doyle Log Rule in New Jersey, *Jour. Forestry*, **41**, 577–580 (1943).

rule. A combination rule in which the low values of each are used (Doyle to 28 in. and Scribner above 28 in.) is called the "Doyle-Scribner." This hybrid rule has had a wide use in southern pine. A second hybrid rule, the Scribner-Doyle, adopted by the state of Louisiana, gives the high values of each rule. It has been ignored in practice.

53. A Universal Log Rule for Board Feet. Without doubt, the International log rule for $\frac{1}{4}$-in. kerf might well be adopted as the universal standard for log measure, but this outcome has so far been retarded by the force of custom in the use of other log rules. Its use will tend to increase as the old-growth timber disappears.

The Scribner Decimal C log rule is practically universal in states where the national forests control most of the timber, and it may eventually drive out competing rules throughout the West. In Canada, provincial log rules, only one of which is accurate (that of British Columbia, a formula rule), will probably be continued. The difficulty elsewhere may eventually be partly solved by the substitution of cubic measure, cords, and piece products.

54. Log Rules Based on Mill Tallies. The manufacture of white pine into round-edged lumber for boxboards and factory uses by slash sawing gave rise to a log rule in Massachusetts in which these boards are scaled or measured at the average width of the narrowest face, instead of being trimmed with parallel edges and scaled accordingly. This rule was constructed by tallying the contents of such boards for logs of different diameters and lengths as actually sawed. It is known as the "Massachusetts log rule" and gives much higher values than any other log rule, especially for small logs.

It is sometimes the purpose of a log rule, based on mill tallies, to bridge the gap between the sawed output and the standing timber and give a basis for tree volume tables which measure the sawed output, rather than to set up an intermediate standard by introducing log scaling between the tree and the product. For such a purpose, it is useful in regions of small custom mills where payment for logs may be made on the basis of sawed lumber and timber is purchased on a similar basis.

The Massachusetts log rule fits such a case. Another example is the practice of sawing hardwood lumber into $2\frac{1}{4}$-in. plank, with 1-in. boards from the slabs. A log rule, such as the International, based on a standard 1-in. thickness, underruns the sawed output in such a case. The owners of timber are benefited in dealing with purchasers by using a tree volume table based on a log rule which is constructed from mill tallies of lumber sawed to these dimensions. Such a log rule might never be used to scale logs, since the basis of purchase would be either the sawed contents or the standing timber. This log rule would be needed only for the average length of log used in estimating the timber.

References

ASHE, W. W.: Loblolly or North Carolina Pine, *N.C. Geol. Survey Bul.* 24, 1915, 98 pp.

BENTLEY, JOHN, JR.: A Comparative Study of Two Log Rules as Applied to Timber in Central New York, *Forestry Quart.*, **12**, 390–394 (1914).

BERRY, SWIFT: Volume of Western Yellow Pine Logs from an Actual Mill Tally, *Jour. Forestry*, **15**, 615–618 (1917).

BRERETON, B.: Percentage Comparisons between Scribner, Spaulding, Doyle, and British Columbia "Allowance" Log Scales and Brereton Scale, *Timberman*, **38** (5), 24–26 (1937).

BRUCE, DONALD: A Formula for the Scribner Rule, *Jour. Forestry*, **23**, 432–433 (1925).

BUELL, C. R.: Relation between Various Scales in Redwood Logging Operations, *Timberman*, **30** (8), 156–160 (1929).

CHAPMAN, H. H.: The International Log Rule for ¼-inch Kerf—Can It Replace the Doyle Rule?, *Jour. Forestry*, **40**, 224–234 (1942).

———: The Measurement of Logs and the Construction of Log Rules, *Southern Lumberman*, **1369**, 114–117 (1921).

CLARK, JUDSON, F.: Measurement of Sawlogs, *Forestry Quart.*, **4**, 79–93 (1906).

COOK, H. O.: Handbook on Forest Mensuration on the White Pine in Massachusetts, Boston, 1908, 50 pp.

DANIELS, A. L.: The Measurement of Sawlogs, *Vt. Agric. Expt. Sta. Bul.* 102, 1903, 8 pp.

———: The Measurement of Sawlogs and Round Timber, *Forestry Quart.*, **3**, 339–345 (1905).

FISHER, R. T.: Utilization and Round Edge Lumber, *Soc. Amer. Foresters Proc.*, **11**, 386–393 (1916).

FROTHINGHAM, E. H.: The Northern Hardwood Forest: Its Composition, Growth and Management, *U.S. Dept. Agric. Bul.* 285, 1915, pp. 63–65.

GRAVES, H. S.: Recent Log Rules, *Forestry Quart.*, **7**, 144–146 (1909).

HAWES, A. F.: Hemlock in Vermont; Comparative Study of Log Rules, *Vt. Forest Service Pub.* 9, 1912, 32 pp.

HERRICK, A. M.: A Defense of the Doyle Rule, *Jour. Forestry*, **38**, 563–567 (1940).

———: Grade Yields and Overrun, *Purdue Univ. Agric. Expt. Sta. Bul.* 516, 1946, 59 pp.

McKENZIE, H. E.: A Discussion of Log Rules: Their Limitations and Suggestions for Correction, *Calif. State Bd. Forestry Bul.* 5, 1915, 56 pp.

———: A Mill Scale Study of Western Yellow Pine, *Calif. State Bd. Forestry Bul.* 6, 1915, 171 pp.

McNAUGHTON, N. R.: A Comparison of the Doyle and Scribner Rules with Actual Mill Cut for Second Growth White Pine in Pennsylvania, *Forestry Quart.*, **12**, 27–30 (1914).

MOORE, E. B.: A Defense of the Doyle Log Rule in New Jersey, *Jour. Forestry*, **41,** 577–580 (1943).
RAPRAEGER, E. F.: Influence of Ponderosa Pine Log Size and Quality on Overrun, Lumber Grades and Conversion Values, *West Coast Lumberman*, **59** (8), 12–14, 20 (1932).
REYNOLDS, R. R.: Factors for Converting Log and Tree Volumes or Values from One Common Scale to Another, *Southern Forest Expt. Sta. Occas. Paper* 68, 1937, 4 pp.
SCHUMACHER, F. X., and W. C. JONES, JR.: Empirical Log Rules and the Allocation of Sawing Time to Log Size, *Jour. Forestry*, **38,** 889–896 (1940).
——— and H. E. YOUNG: Empirical Log Rules According to Species, Groups, and Lumber Grades, *Jour. Forestry*, **41,** 511–518 (1943).
SIMMONS, F. C.: International Log Rule for Long Logs, *Jour. Forestry*, **42,** 136–138 (1944).
SPAULDING, N. W.: Table for Measurement of Logs, San Francisco, 1909, 20 pp.
STETSON, I. G.: A Comparison of Maine and Blodgett Log Rules, *Forestry Quart.*, **8,** 427–432 (1910).
ZIEGLER, E. A.: The Standardizing of Log Measures, *Soc. Amer. Foresters Proc.*, **4,** 172–184 (1909).

CHAPTER 7

PIECE PRODUCTS. MEASUREMENT

55. Classes of Piece Products and Standards of Measurement. In addition to stacked cubic measure and log measure, wood products take the form of either round or shaped pieces intended for special uses. These classes include piling, poles, posts, mine timbers and props, and some smaller products such as mine lagging, converter poles, hop poles, mine sprags, and stakes. Railroad crossties are a special product requiring shaping. Ties are either hewn flat on two sides or sawed on four sides.

Smaller manufactured or shaped products such as staves are manufactured from bolts, which are counted for measurement.

Round products are measured by the piece or by the cubic foot, hewn and squared ties by the piece or by the board foot when sawed. All piece products are classed, first by species, and second by dimensions, into definite grades which are the basis of price. Practically all these products are intended for use exposed to the weather, and all but the most durable species are customarily treated with creosote or some other preservative, especially those parts which are in contact with the ground. The improvement in the technique of wood preservation has permitted the use of a wide range of species which are of low durability in their native state.

Each class of product is bought by grades covered by specifications, setting forth the dimensions and defects which qualify the grade of the piece. Specifications may show the following factors:

1. Standard lengths, with variations permitted.
2. Diameters or circumference for different lengths, tops, and butts.
3. Species accepted in different grades.
4. Amount and character of defect allowed by grades, such as crook or sweep, taper, rot, splits, knots, catface, or even coarse grain.
5. Factors considered as sufficient to render the piece unfit for the use intended.

The standards thus set up are enforced by buyer's inspection, which may be rigid or lax. The best practice is to define uniform standards as to defects that are permissible or will cause rejection and then to enforce such standards uniformly at all times, governing the supply through price adjustments rather than by fluctuating standards of inspection.

Round products and crossties are peeled when cut. Before cutting and preparing any form of round products, copies of the specifications should be obtained from the purchaser and carefully studied.

56. Round Products. *Poles.* The most common species for untreated poles is either eastern or western cedar. For treated poles, southern yellow pine and, lately, lodgepole pine are used extensively.

Poles are classed according to length: from 16 up to 22 ft., in 2-ft. lengths; and from 25 up to 90 ft., in 5-ft. lengths. They are further classed according to minimum *circumference* at the top into seven classes running from 27 to 15 in., for each of which class minimum circumference at 6 ft. from the butt is specified; and into three additional classes, from 18 to 12 in., with no butt requirements.

The standard dimensions for poles of different species are published by the American Standards Association (latest, Mar. 17, 1945). These standards do not preclude the acceptance of poles having greater circumference at the points of measurement than those shown above.

Piling. The qualities desired in piling are straightness, small uniform taper, and soundness. Piling is commonly creosoted and sometimes sheathed as a protection against the teredo. The minimum top diameter (not circumference) is stated, usually decreasing with longer piles. Both maximum and minimum diameter of butt are prescribed, measured at 3 ft. from butt.

Specifications for timber piles are published in the *Proceedings of the American Society for Testing Materials,* 1937 Supplement to Book of A.S.T.M. Standards, pages 194–198, for untreated round timber suitable for timber piles.

Posts and Small Poles. Round products over 10 and under 20 ft. are sometimes classed as small poles. From 10 ft. down to $6\frac{1}{2}$ ft. they are called posts. Western red cedar posts run up to 20 ft. Lengths of fence posts are $6\frac{1}{2}$, 7, and 8 ft. Diameters for round posts are from 3 to 8 in. For split posts, the circumference is measured at the small end. Scant measure is allowed as follows:

Diameter, in.	Circumference, in.
4	$11\frac{3}{4}$
5	$14\frac{1}{2}$
6	$17\frac{1}{4}$
7	21
8	24

Split posts must be triangular, quadrangular, or half round. One-way sweep may be permitted, especially in western juniper.

The creosoting of posts and other forms of preservation have, as for other round products, permitted the utilization of nondurable species and prolonged the supply.

57. Mine Timbers. The consumption of wood in coal and other mines calls for large quantities of material, most of it in the round. This falls into four general classes: timber or framing called "stulls," props, lagging, and mine ties. The length of mine stulls and of mine props is determined by the depth of the veins, and specifications differ for each region. Stulls, also called "drift timbers" or "gangway timbers," include mine caps and crossbars and are hewn or sawed on two or more faces. They constitute the main timber supports and strength is the chief qualification. Maximum sizes may be up to 15 in. and 27 ft.

Bark is not removed from mine props or posts unless undergoing preservative treatment. Most of the mine props in the past were used for temporary workings, decayed rapidly, and were abandoned.

Mine props may be round or split. The split props either must in some cases have a cross section of the same area in square inches as a round prop, or in other cases must be large enough to hold an inscribed circle of the diameter of the round prop. Lengths vary from $3\frac{1}{2}$ to 16 ft. and diameters from 3 to 13 in. Four-inch props are common. The diameter is sometimes measured in the middle of the stick instead of at the small end. For props 10 ft. and over in length it may be specified that crook shall not exceed 3 in.

Mine props are frequently bought by green weight. Cubic contents is commonly accepted as a basis of price adjustment between small and large sizes. Prices may be quoted by the piece, according to sizes. Chute timber is another form resembling props in specifications.

Mine ties may run from 5 to 7 ft. in length, with some shorter dimensions for special cases. In thickness and breadth of face, they range from 4×5 in. to 6×7 in. in various grades differing by regions and mines. They may be hewn on two sides, or squared. Mine sprags are small short pieces used in braking mine cars.

58. Crossties—Specifications. Hewn, or pole, ties are cut from trees measuring from 11 to 17 in. diameter breast high, yielding one or more ties. Trees of larger diameter if made into ties are sawed rather than hewn. It seldom pays to cut a tree which will yield less than two ties, one of which should be of a better grade.

The Railway Engineering Association's specifications are given below. These are merely advisory, and many railroads do not adhere to them. Under these rules, the grades are numbered from the lowest to the highest, running from 0 to 6. Lengths may be 8 ft., 8 ft. 6 in., or 9 ft., and the dimensions are as shown in Table 17.

The top surface of the tie is taken as the narrower of the two sides, or the one with the narrower or no heartwood if both surfaces are of the same width (as in sawed ties). Ties may exceed these minimum dimensions by not over 1 in. in thickness, 3 in. in width, or 2 in. in length.

TABLE 17. STANDARD SIZES FOR CROSSTIES BY GRADES

Grade	Sawed or hewn top, bottom, and sides		Sawed or hewn top and bottom	
	Thickness, in.	Width on top, in.	Thickness, in.	Width on top, in.
0	*	*	5	5
1	*	*	6	6
2	6	7	6	7
3	6	8	6 or 7	7 or 8
4	7	8	7	8
5	7	9	7	9
6	7	10	7	10

* None accepted.

The dimensions given are measured through both sections lying between 20 and 40 in. from the center of the tie. A tie must be sufficiently straight in a horizontal plane for its axis to fall entirely within the tie. In the vertical plane, the axis of the tie must fall not more than 2 in. from the top or bottom. Ties should not show over ½ in. difference in thickness of the sides or ends. Specifications, covering defects, should be obtained from the purchasing corporations or agents.

59. Crossties—Tie "Log Rules." The estimation of the number of crossties which a tract of timber will yield requires, for accuracy, a tie "log rule." Ties must be estimated in the grades which will be produced. Most estimating has been based on rough ocular methods or tie counts, but the results of such efforts seldom give a satisfactory degree of accuracy. Stumpage value of the different grades of ties differs by greater percentages than does the price of the ties after manufacture. The yield of different grades from bolts of different diameters is quite variable, since it depends on the alternate choice of subdividing the bolt into two or more ties of different grades and sizes. Hewn ties can utilize "pole" timber only up to a certain maximum diameter approximating the maximum size of the largest tie accepted. On hewn-tie sales, therefore, the larger trees may be left standing. When ties are sawed, bolts of all sizes will be subdivided into squared ties, with additional products in the form of boards from the slabs.

A tie log rule, for ties of given standard length, such as $8\frac{1}{2}$ ft., is the first need in estimating ties. Such a log rule based on the five standard tie grades has been prepared by the Lake States Forest Experiment Station, St. Paul, Minn.

References

BÈHRE, C. E.: Chart for Application of Percentile Taper Curves to Trees of Any Size, *Jour. Forestry*, **24**, 272–274 (1926).

BONNEY, M. C.: Use of the Cull Caliper in Scaling White Oak Stare Bolts, *Jour. Forestry*, **38**, 351–355 (1940).

GIBBONS, W. H.: How to Figure Cedar Bolts in the Terms of Feet Board Measure, *West Coast Lumberman*, **37** (438), 21 (1920).

GIRARD, J. W., and U. S. SWARTZ: A Volume Table for Hewed Railroad Ties, *Jour. Forestry*, **17**, 839–842 (1919).

HAWES, E. T.: A Method of Determining Southern Pine Pole Classes from D. B. H., *Jour. Forestry*, **45**, 204–205 (1947).

HUNT, G. M.: Calculating the Volume of Poles and Piles, *Amer. Wood Preservers Assoc. Proc.*, **14**, 245–249 (1917).

Lake States Forest Experiment Station: Yields from Tie Bolts, *Tech. Note* 194, 1943.

MEYER, H. A.: Bark Volume Determination in Trees, *Jour. Forestry*, **44**, 1067–1070 (1946).

SCHNUR, G. A., and A. C. MCINTYRE: The Measurement of Mine Props, Linear Foot, Top Diameter, Weight and Volume Tables, *Pa. Agric. Expt. Sta. and School Agric. Bul.* 269, 1931, 24 pp.

STEVENS, R. D.: Stave Volume and Defect in Old Growth White Oak, *Univ. Ark. Col. of Agric. Bul.* 362, 1938, 26 pp.

U.S. Railroad Administration: Specifications for Crossties, *Jour. Forestry*, **16**, 837–839 (1918).

UPSON, A. T.: Volume Table for Lodgepole Pine, *Forestry Quart.*, **12**, 319–329 (1914).

CHAPTER 8

DETERMINATION OF VOLUME OF FELLED TREES FOR THE CONSTRUCTION OF VOLUME TABLES

60. Standard Tree Measurements. Felled trees are measured primarily for the purpose of constructing tree volume tables for standing timber. Although it may seem advisable to take only such measurements as are necessary for the specific table required, such practice is undesirable, because standards for utilization change, and for a different table the field work would then have to be repeated. If, instead, a complete set of tree-diameter measurements is taken and the tree records filed, this work will never require repetition but would serve all subsequent needs, including the study of tree forms.

Tree measurements are of two classes:

1. Those required to determine the tree class or group. The tree classes, which form the basis of the averages of tree volumes, are three in number: d.b.h., total height, and form class.

a. D.B.H. This term is an abbreviation for diameter breast high, or the diameter of the tree outside bark at $4\frac{1}{2}$ ft. above average ground level. In continental Europe, this point is taken at $4\frac{1}{4}$ ft. On sloping ground, the height is measured from the average of upper and lower points. The upper point is the standard in Great Britain and Canada. Trees whose merchantable diameters do not range above 36 in. are usually divided into 1-in. classes. Where the range is greater, as in West Coast forests, 2-in. d.b.h. classes are used.

Although the volume given in a tree volume table may not include the bark, the d.b.h. classes given in the table are always based on measurement outside the bark.

D.b.h. is the most important measurement of either felled or standing trees, and errors made in determining it are correspondingly serious.

On felled trees measured for a volume table, the $4\frac{1}{2}$-ft. point, if not taken before the tree is felled, is found on the butt log by subtracting the stump height from $4\frac{1}{2}$ ft. and measuring off this distance on the log. Care must be taken that on slanting cuts the measurement of stump and log coincide as they would on a standing tree.

b. Height. For second-growth timber and for species with regular form such as most conifers and some hardwoods, total height is preferred as a basis for tree classes. The height classes may cover 5, 10, or 20 ft., depending on the total height attained by the species.

c. Form class. This is discussed in Sec. 61. Trees are grouped into shape or form classes based on a numerical index of form which requires the measurement of the diameter of the stem at some fixed point above d.b.h.

2. Measurements required to determine the dimensions of the logs or bolts. Standard practice requires that diameter measurements be taken at fixed intervals along the entire bole of the tree, always including the unmerchantable top. In no other way can a permanent tree record be assured, which may be used for any purpose that may arise because of changing standards of utilization or new products.

The sum of the volumes of the logs so measured gives the volume of the tree.

61. Tree Form as a Variable in Tree Volumes—Form Classes. Although trees may have the same d.b.h. and height, they will vary in volume. This is due to variation in the taper or form of the boles. The form varies from a solid somewhat like a cone to one like a paraboloid. The more "cylindrical" the tree, the greater will be its volume for a given d.b.h. and height.

These variations in form may be represented by the ratio of an upper diameter to d.b.h. One of the best diameters to use in this ratio is the diameter at one-half the height above breast height. The diameter outside bark at this point when divided by d.b.h. outside bark (o.b.) gives a ratio known as the *form quotient*, which is always less than 1.000. The form quotient for a tree with a d.b.h. of 17.1 in. and an upper diameter of 10.5 in. o.b. at one-half height above b.h. is 10.5/17.1, or 0.614. A more accurate index of the form of the solid wood bole, not including bark, is the ratio of the same two diameters inside bark. Inside bark (i.b.) measurements should always be used for form quotients, since the thickness of the bark varies measurably.

Form quotients are grouped into *form classes*. The class interval is usually 0.05, as follows: 0.50, 0.55, 0.60, etc., with limits ranging from 0.476 to 0.525, 0.526 to 0.575, 0.576 to 0.625, etc., respectively.

An average form class may apply fairly consistently for a given species throughout its botanical range but will not apply equally well to associated species, owing largely to specific differences in the average thickness of bark and average stump taper which affects d.b.h. For longleaf pine the average form class is 0.70, while for shortleaf pine it is 0.75. Within a given species, form class is lowest for open-grown long-crowned trees and increases with density of stand and relative shortness of the live crown. Since these differences average out on large areas, standard volume tables usually ignore form-class variation. For restricted areas and for stands which have greater, or less, density than the average, corrections must be made by sampling and then applied in modifying a local table which has been derived from the standard table.

62. Total versus Merchantable Height. In all cases total tree height including stump should be measured to the tip of the tree, preferably with a tape, and recorded to the nearest ½ ft. Broken tops may often be reconstructed; if not, the trees should be rejected. For spreading crowns, the crown level at the highest point should be taken.

For large old-growth timber which is to be manufactured into lumber (Sec. 151), it is more convenient to use the merchantable height as a basis in estimating; hence it is commonly required that volumes for such timber be classified on this basis rather than on that of total height. This requires that the taper measurement which falls approximately at the end of the last merchantable cut be the point to which the standard height is measured for the height class. In order to obtain a satisfactory standard, the following difficulties must be overcome:

1. The top diameter of the merchantable part of the bole is not constant but varies with d.b.h. and character of the crown.

2. On logging operations, logs are cut to varying lengths, to utilize the bole efficiently, approximately in multiples of 2 ft. for conifers and 1 ft. for hardwoods. From 2 to 6 in. over the log length is allowed for trimming.

3. A timber cruiser does not want to tally merchantable lengths in sizes differing by less than 8 ft. and prefers to confine the classes to 16 ft.

Aside from irregular trimming lengths, log lengths also vary in practice for trees of the same d.b.h., height, and form. Because of the effect of taper on sawed volumes, this would introduce an unnecessary fourth variable into tree volume tables. It is sufficient to choose a standard average length of log suitable for timber estimating. This standard may be 12 ft. for hardwoods and is practically always 16 ft. for conifers, with half-log divisions when required.

When 4-ft. wood is the basis, total rather than merchantable heights should always be adopted. The last 4-ft. bolt which falls within the merchantable limit at its small end is included in the merchantable portion. The standard top diameter is 3 in. outside bark. Branch wood in 4-ft. bolts is usually calipered in the middle of the stick.

63. Rules for Measuring Standard Merchantable Heights. Where total height is the standard, no difficulties arise. The merchantable length is measured nearest the last actual cut (see 3 below), unless there is inexcusable waste to be avoided. It may be taken to a fixed diameter only when utilization permits such a standard.

Where merchantable height is the standard, the following rules should be observed in measurements:

1. Adopt a standard stump height and lay off lengths for taper measurement from this point. Since actual stump height varies, it should never be used. The stump height of large trees is greater than that of small trees because of greater butt swell. On the other hand, where close utilization is practiced, the stump height is lowered. Standard stump heights commonly adopted are

½ ft. for small second growth under 12 in. d.b.h. to be cut for cordwood or pulpwood.
1 ft. for second growth 12 to 24 in. d.b.h. to be cut for saw logs.
1½ ft. for timber over 24 in. d.b.h.

2. Beginning at the standard stump height, measure the bole in equal standard lengths. The general practice is to use 4-ft. lengths for small trees to be used for cordwood or pulpwood, 8-ft. lengths for trees up to 80 ft. in merchantable length to be cut for saw logs, and 16-ft. lengths for larger trees. On each 16-ft. section a *trimming allowance* of 0.3 ft. should be added, or 0.15 ft. for 8-ft. sections.

3. Since the merchantable length is to be based on actual logging practice with a variable top diameter, take the diameter which falls at the nearest fixed interval, either above or below, the top of the last log which will be cut, as giving the total merchantable length to be used in the table; *viz.*, the last actual cut on a tree is at 61 ft. above the standard stump height point. It must be measured at 64 ft. and is called a 4-log tree. If this tree is cut at 60 or 59 ft., it is measured at 56 ft. and called a 3½-log tree. It is incorrect to reject all merchantable lengths lying above the last standard taper point, since this would give a biased negative error in volumes.

The sum of the lengths of these logs or bolts, disregarding trimming allowances, gives the standard total merchantable height expressed in log lengths. The tree is then classified in the group in which this height will fall; *viz.*, if the classes are 8 ft., or half a 16-ft. log, the 4-log class averages 64 ft. but embraces trees from 61 to 68 ft. in actual net merchantable length. If a bole contains a bolt 5 ft. or more long beyond the 64-ft. point, it falls in the 72-ft. or 4½-log class. If the bolt is 4 ft. or less in length, the tree stays in the 4-log class.

64. Additional Measurements for Butt Swell. Butt swell usually extends to a point above the stump cut. For this reason the cubic volume of butt sections when determined by the Smalian formula may, for flaring butts, give errors as high as plus 10 per cent (Sec. 28). For complete permanent records of tree form and for accurate determination of cubic volume, taper measurements are taken at 1, 2, and 3 ft. above the ground. Combined with d.b.h., these tapers permit accurate calculation of butt log cubic volume when needed.

65. Diameter outside and inside Bark—Thickness of Bark. The diameter outside bark is termed d.o.b. For most purposes, diameter inside bark, or d.i.b., is needed instead. Bark thickness is apt to be measured inaccurately unless its purpose is realized, which is to obtain the true difference between the d.i.b. of the tree and the d.o.b. as determined from placing the caliper arms parallel on opposite sides of the bole. On rough-barked trees or those having deep fissures, this measurement coincides not with any average thickness of bark as such but with average maximum thickness measured on the ridges. Hence the height of ridges must be measured.

Tree measurements are taken on a logging operation after the trees are felled and before the logs are skidded away or the wood stacked. The method of obtaining d.i.b. depends on whether the bole is sawed into logs or bolts at the exact point indicated for standard tapers as for cordwood or into varying lengths for individual logs. In the first case, the d.i.b. is measured directly with a ruler or scale stick graduated to inches and tenths. Two diameters at right angles are taken and averaged. The width of bark, usually on two opposite sides, or even on four sides, is then taken. The single average width of bark obtained from these measurements is recorded. This measurement is then doubled and added to the d.i.b. to give the d.o.b.

In the second case where the lengths as actually sawed do not coincide with the standard length, d.o.b. is the first, or standard, measurement. Bark thickness must be determined, doubled, and subtracted from d.o.b. to get d.i.b. Bark is almost never removed for the sole purpose of obtaining direct d.i.b. data.

A common method is to chop out a notch, extending slightly into the wood and having a flat face perpendicular to the bole. The zero end of the rule is then placed on the cambium, or juncture of bark and wood, and the thickness of bark is measured to the nearest twentieth inch, or 0.05. In measuring bark thickness on cuts, one is apt to get too small a result. This is due, first, to the loss of loose scales on the surface; second, to the tendency to average the thickness of fissures and ridges instead of observing the maximum requirements of a caliper measurement; and third, to compression of the bark on the cut when severed with the ax. These errors may amount to a reduction of between 0.05 and 0.10 in. in double bark thickness, giving a corresponding excess in d.i.b.

66. The Swedish Bark Gauge. The bark on practically all species of trees is penetrated more easily than the sapwood. Based on this fact, the Swedish bark gauge is constructed so as to penetrate only the bark. The width of bark is then measured on a scale on the shank of the instrument. This scale should be graduated in inches and twentieths. In use, the shaft is forced into the bark. When the cambium is reached, its progress is stopped because part of the cutting edge is blunt or flattened. The flange is then pressed down into contact with the bark to indicate the required reading of bark thickness. The flanges of the instrument are apt to be curved on too small a radius for many of our American tree sizes, thus giving a reading which is slightly high. This defect should be remedied easily in manufacture[1] (Fig. 13).

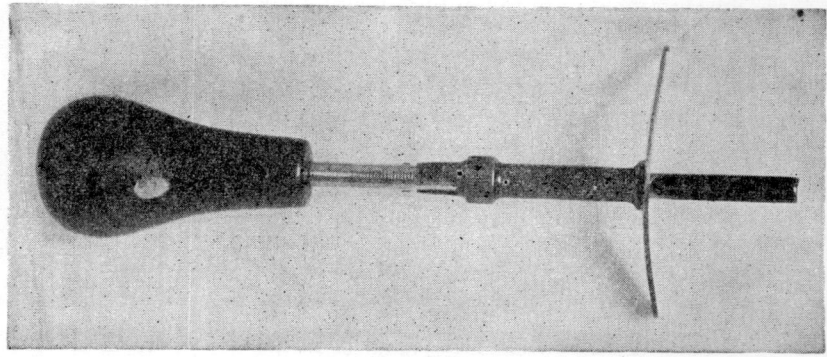

Fig. 13. Swedish bark gauge.

67. Procedure for Measuring Felled Trees

Equipment:

100-ft. tape, calipers,[2] ax, graduated rule or bark gauge, pole 8.15 ft.

[1] The Swedish bark gauge is obtainable in the United States. Inch and twentieth graduations should be specified in ordering.

[2] Tree calipers and their use are described in Sec. 110.

long graduated at 4.5 ft. and at 2-ft. intervals, record blanks. If not on a logging operation and when trees must be felled specially for measurement: crosscut saw, axes, wedges, and kerosene.

Order of work:

A crew of three men is most efficient, though two men can do the work when necessary.

1. D.b.h. Locate d.b.h. and also 1-, 2-, and 3-ft. points from ground Caliper, take bark thickness, and record d.o.b. and d.i.b.

2. Height. Measure with tape; locate standard stump height, measure total height including stump, locate one-half height above breast height, and notch the point for caliper and bark measurement.

3. Diameters o.b. and i.b. at each fixed interval from standard stump height to tip. Locate with pole and notch for bark measurement. Caliper and measure bark at located points.

The average diameters at any point on felled trees or logs, including d.b.h., may be determined to the best advantage with tree calipers. To obtain the two measurements at right angles as a basis for recording the true average to the nearest tenth inch, each measurement is taken where possible at an angle of 45 degrees with the perpendicular. It is usually impossible to slide the caliper arms onto the log till the crossbar touches. For this reason, special care must be exercised to keep the calipers in perfect adjustment.

4. For complete tree records, taper measurements must be continued at regular standard intervals until the length of the remaining tip is less than this interval. Where the measurements are intended for construction of a volume table for merchantable volume only, two additional steps may be taken to distinguish the tapers in the crown from those below, as follows:

  a. A heavy vertical line is drawn on the record blank to the right of the diameter recorded for the merchantable height which falls nearest, either above or below, the top of the last log.

  b. The total utilized length is measured and recorded by tape as a check.

5. Additional data, optional, to complete the tree record.

  a. Average crown width as judged from the prostrate crown.

  b. Crown length in feet to the point of origin of lowest living branches constituting part of the functioning crown; small epicormic shoots are ignored.

  c. Branch wood, when merchantable, is kept separate from bole wood.

96   FOREST MENSURATION

6. The age of the stump section may be recorded from a ring count. This count takes little extra time and indicates the relation between age, size, and form of the tree.

7. Systematic growth data may be taken on the stump and at upper sections. This subject is dealt with in Chap. 26.

8. Each tree is recorded on a separate sheet.

68. The Cubic Content of Felled Trees. In computing the cubic-foot volume of a felled tree from the recorded dimensions on the form or record sheet (see page 97), Smalian's formula is nearly always used. Volumes are therefore computed for the sections lying between each two consecutive taper measurements on the bole. The single average width of bark has been recorded, doubled, and deducted from d.o.b. to get d.i.b. The cubic volume can then be found with bark or without bark, as desired.

A form used by the U.S. Forest Service, as shown, gives on one side the form for recording diameters and on the reverse side a form for growth data. Form 558a, also issued by the Forest Service, is commonly used for plotting diameters directly in the field (Sec. 69).

The cubic volume of each log, either with or without bark, has been obtained by Smalian's formula as above, but it remains to compute the volume in the tip of the tree to get the total volume of the bole. This is done by treating the tip as a paraboloid, involving the formula $BL/2$, where B is the cross-sectional area at the base of the tip and L is the length of the tip. The stump is regarded as a cylinder with volume BL, in which B is the area of the top of the stump and L is the stump height in feet. Using one term for each log or section, the total tree volume is then

$$V = B_0 L_0 + \frac{B_0 + B_1}{2} L_1 + \frac{B_1 + B_2}{2} L_2 + \frac{B_2 + B_3}{2} L_3 + \cdots + \frac{B_n}{2} L_n$$

where B_0, B_1, B_2, etc., = cross-sectional areas at the stump and each successive log in square feet

L_0, L_1 L_2, etc. = stump height and length of each successive log in feet

L_n = length of the tip

When a tree is measured at fixed intervals, making $L_1 = L_2 = L_3$, etc., the formula may be consolidated as

$$V = B_0 L_0 + \left(\frac{B_1 + B_n}{2} + B_1 + B_2 + B_3 + \cdots\right) L + \frac{B_n}{2} L_n$$

If only the merchantable content of the tree is desired, the term at each

VOLUME OF FELLED TREES

Locality _____ T. _____, R. _____, _____ M.

Species _____, Tree No. _____, D. B. H. _____, Tot. height _____

Date _____, 19   Clear length _____, Used length _____, Mer. length _____

Section.	Age.	Length.	Diam. inside bark.	Width of bark.	Diam. outside bark.				Remarks.
	Years.	Feet.	Inches.	Inches.	Inches.				
1 Stump									
2									
3									
4									
5									
6									
7									
8									
9									
10									
11									
12									

Remarks: _____
_____
_____

(Name) _____

REVERSE SIDE

Tree class ...., Quality ...., Crown length ...., Crown width ....

Cross Section.	Distance on average radius from heart to each 10th ring—Inches. (Count in, measure out.)															
	1	2	3	4	5	6	7	8	9	10	11	12	13	14	15	16
1 Stump.																
2																
3																
4																
5																
6																
7																
8																
9																
10																
11																
12																

Cross Section.	Distance on average radius from heart to each 10th ring—inches. (Count in, measure out.)															
	17	18	19	20	21	22	23	24	25	26	27	28	29	30		
1 Stump.																
2																
3																
4																

Remarks: ....

Form and condition: ....

end of the right-hand side of the equation is omitted, plus any volume lying above the diameter limit of merchantable wood below the tip. Branch sections, measured at the middle point, are computed by Huber's formula.

Because of the flare of the butt section, it may be necessary to subdivide this section when the standard length has been chosen as 8 ft. or more. The b.h. point falls 3 ft. above a 1½-ft. stump, or 3½ ft. above a 1-ft. stump. Since most of the butt flare is below the b.h. point, it is customary in the above computations to divide an 8-ft. butt log into two sections at the b.h. point. The two sections are therefore either 3 and 5 ft. long, or 3½ and 4½ ft., respectively. Most of the error of applying the Smalian formula is then eliminated.

69. Graphic Determination of the Cubic Volume of Felled Trees. Since a tree tapers from the ground to the tip, the diameter may be said to depend on the height above ground. Using a system of rectangular coordinates, the diameters at different points on the bole can be plotted as the dependent variable over height above ground as the independent variable. For example, the following taper measurements on a longleaf pine tree were taken at 8.15-ft. intervals above a 1-ft. stump.

VOLUME OF FELLED TREES

TABLE 18. LONGLEAF PINE TAPERS

Section	Length, ft.	D.i.b., in.	Width of bark, in.	D.o.b., in.
At stump height	1.0	11.4	0.8	13.0
At 2 ft.	1.0	11.2	0.7	12.6
At 3 ft.	1.0	11.0	0.6	12.2
At 4½ ft.	1.5	10.9	0.55	12.0
1	4.65*	10.4	0.4	11.2
2	8.15	9.5	0.25	10.0
3	8.15	8.9	0.25	9.4
4	8.15	8.4	0.2	8.8
5	8.15	7.6	0.2	8.0
6	8.15	6.6	0.2	7.0
7	8.15	4.1	0.2	4.5
Tip	10.4			
Total............	68.45			

* Length of first section = 1.0 + 1.0 + 1.5 + 4.65 = 8.15.

If the diameters are plotted over the heights as in Fig. 14, the resulting graph shows at once the trend of the taper. The difference between the upper and lower curves gives a picture of the bark thickness at different points on the tree. This graph can be used to determine the

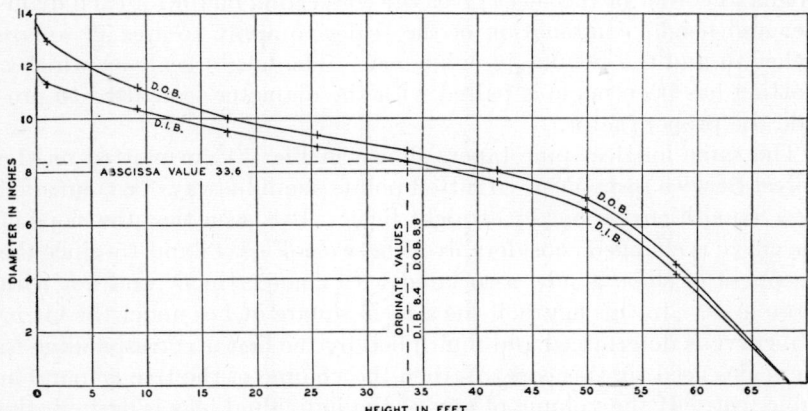

FIG. 14. Tapers or diameters outside bark (d.o.b.) and inside bark (d.i.b.) for one longleaf pine. The dimensions at the height of 33.6 ft. are plotted where the abscissa value of height for 33.6 intersects respectively the ordinate values for 8.4 and 8.8 in. The remaining dimensions are plotted in the same manner, and the profiles of the tree outside and inside bark are drawn through the consecutive points thus plotted.

diameters at points not measured on the bole. It can be read with considerable accuracy, either i.b. or o.b., at any point on the stem.

The volume of the stem in terms of contents of any given length of log cannot be read directly from the graph but must be computed by finding the areas corresponding to diameters at regular sections and using Smalian's formula.

The ordinate, or diameter scale, for a graph of this sort may be modified so that tree volume may be obtained more directly. L. H. Reineke of the U.S. Forest Service has devised a form (558a) in which the ordinate scale is on the basis of cross-sectional area and not of diameter. The graduations of this scale are given directly in terms of equivalent diameters for ease in plotting. The scale graduation in inches corresponds to $B = 0.7854D^2/144 = 0.005454D^2$; but to make the chart more useful by covering compactly a wide range of diameter classes, the ordinate scale is marked off in four diameter scales, on the basis of multiples of 0.5, 2.0, 12.5, and 50 times the initial cross-sectional area graduation. Thus one scale applies to diameters or tapers running from 0 to 6 in., the second scale to 12 in., the third to 30 in., and the fourth to 60 in. The abscissa scale, or height scale, is likewise marked off with four systems of height, *viz.*, 5, 10, 20, and 40 ft. per inch of scale.

Figure 15 is the completed chart. The small square in the lower left-hand corner of the sheet gives the converting factor for each diameter and height combination of the scales to apply to area in square inches to find the volume in cubic feet. The height scale on which a plotting has been made is paired with the diameter scale used to provide the proper factor.

The same longleaf pine tapers shown in Fig. 14 are plotted on the Forest Service form 558a. Plotted points should always be connected by a smooth curve, not by straight lines. It is seen that the slope of the curve is changed considerably as between Figs. 14 and 15, since the steepness of slope is now associated with changes in D^2 and not with those in D. In this figure, if the area in square inches under the i.b. or o.b. curve is determined and multiplied by the factor corresponding to the scales used (in this case, 5), then the volume of the tree is found in cubic feet. If the volume of any of the individual logs is desired, the area under the portion of the curve including the log is found and multiplied by the factor.

The most convenient method of finding the area is by use of the polar planimeter. This instrument is designed to give areas of plane surfaces directly in square inches. The weighted arm (see Fig. 16) is usually placed outside the area to be measured. The tracer point is

VOLUME OF FELLED TREES

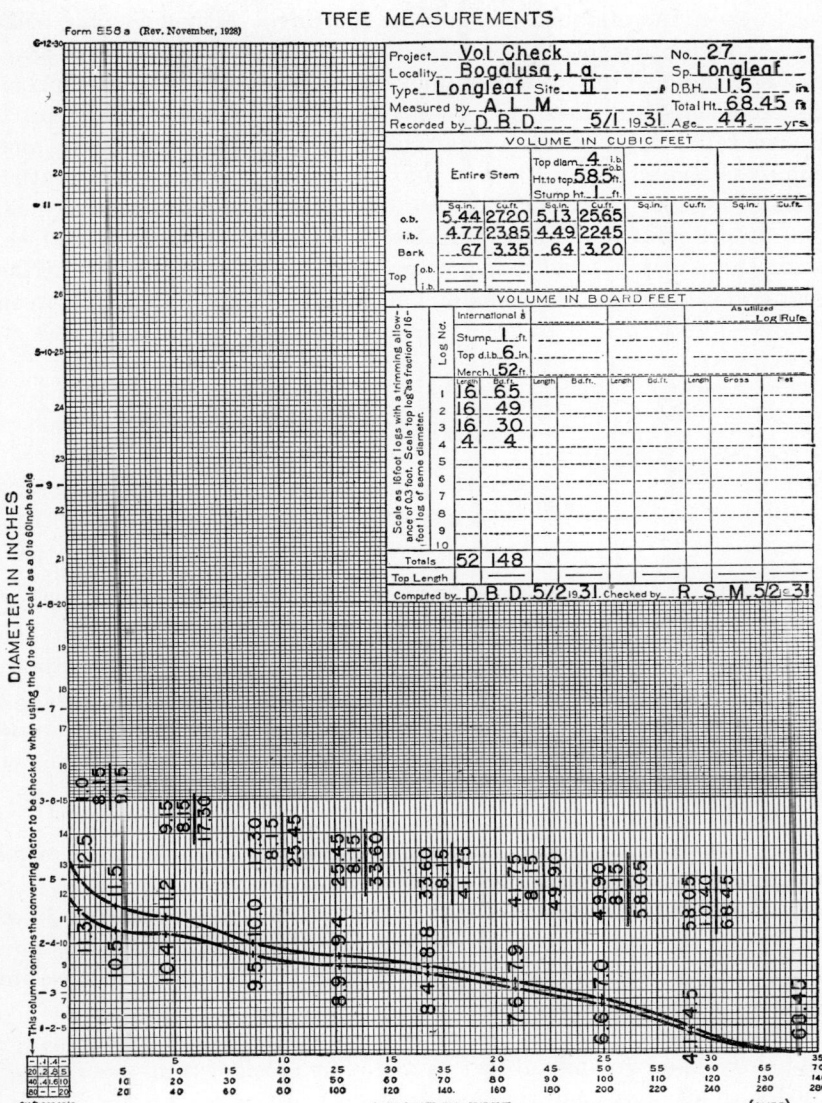

Fig. 15. Longleaf pine plotted on form 558*a* showing dimensions and the record of diameters and height at different points on the bole.

set at a definite starting point on the plotting. Although some technicians prefer to set all scales to read zero at this time, it is better practice not to do so but to make and record a reading of the small horizontal scale, the wheel, and the vernier as they stand. The small horizontal scale is graduated in tens of inches; the wheel makes one complete revolution for 10 sq. in. and is graduated in inches and tenths; and the vernier permits a fine reading to the hundredth square inch. The tracer point is moved *clockwise* around the plotting, or surface, until the starting point is again reached. A second reading of the scale. wheel, and vernier is made and noted. The difference between

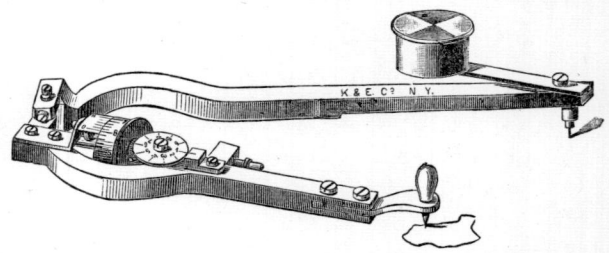

FIG. 16. Polar planimeter for measuring areas in square inches.

the two readings gives the area of the surface in square inches. To detect errors and ensure accuracy, the plotting is again traced without resetting the scales, and a third reading is made. The second reading subtracted from the third again gives the area. The two determinations of area should not differ from one another by more than 1 per cent. The average of the two is then used as the correct area.

The advantages of the tree measurement sheet are as follows:

1. Computation of tree volume, or any partial tree volume, is rapid and accurate.

2. *Trees may be measured in the field exactly as bucked* and bark thickness directly measured on log ends. Log dimensions of any length may be read from the curve on the sheet. This tends to expedite the field work.

3. The sheets serve as a permanent record of tree form of each individual tree, and volume tables may therefore be made from them at any time without additional tree measurements in the field.

4. Diameters may be read from the curves at any point. Volume tables may be made from these measurements for any standards of utilization desired.

5. By use of special overlays, board-foot as well as cubic-foot volumes may be computed rapidly.

The chief disadvantages may be overcome if proper care is taken in the field measurements, plotting, and measurement of area by the planimeter. They are:

1. Tapers must be taken to points relatively near the tip of the tree; otherwise the plotting of this last portion will be left to guesswork. This may result in an error of several cubic feet in extreme cases.

2. Care must be observed in plotting, or the several scales will become confusing. Hence the scales not used should be crossed out.

70. The Board-foot Volume of Felled Trees. a. Mill tally volume tables. Board-foot volumes of individual trees may be obtained by marking the individual logs of each tree with an identifying number, following them through the mill, and totaling the sawed output. As was shown in Secs. 34 and 35, this total will vary with several factors concerning the methods and equipment used in manufacturing, the lengths of the logs cut, the top diameter of the last log, and the thickness of the lumber. Tree volumes derived in this manner are applicable only to the species and areas where the above factors are roughly constant. The tables are useful only to the extent that they reflect the general trend of output for the region, species, and product. The variable elements are not controlled or standardized in a volume table constructed directly from sawed output.

One example of such a tree volume table is that constructed in Massachusetts for round-edged boards to be used for box material. Owing to the short lengths used in the box industry, it is advantageous not to run the boards through the edger directly after sawing on the head rig, as is the usual practice. For this use, the board-foot content of each board is taken multiplying length by width at the middle of the board on its narrower face. The volume overruns that of square-edged lumber by an increasing percentage with decrease in size of log.

In Connecticut a tree volume table for hardwoods was based on the average sawed product of logs of different sizes, in which considerable dimension lumber and sawed ties were produced. Log lengths were those actually cut.

b. Log rule volume tables. Except where the volumes of the logs contained in the tree are computed and averaged directly from the output of the mill, the board-foot content of felled trees is determined directly from the scaled contents, as taken from a standard log rule, for the sections into which the tree is divided.

For most conifers, standing timber is estimated in 16.3-ft. log lengths and half logs. The length of logs measured in the tree must then conform to this standard, which becomes the basis for board-foot measurement of trees for a volume table. The logs must be measured by a standard scaling practice, based on rounding off fractional inches, rather than dropping all fractions, and taking the average and not the smallest scaling diameter (Sec. 36). In trees containing the odd half log, *viz.*, 1½, 2½ logs, the scale consists of the volumes of all full-length logs plus one-half the scaled volume of a 16-ft. log with a diameter equal to that of the half log, except for the International rule where a separate scale is given for an 8-ft. log. The sum of the scaled log contents gives the tree volume.

Hardwoods may be scaled in similar standard lengths, which would be 16 ft. unless the average log length is consistently 14 or 12 ft., in which case the volumes would overrun the 16-ft. volumes. A more common practice, short of direct tally of tree output in the mill, is to adopt a suitable standard log length and then add the contents of the top section by scaling it in proportion of its length to that of the standard, *viz.*, a 6-ft. top portion is $6/16$ of a standard 16-ft. log with the same top diameter, except again for International rule.

Standard volume tables for board feet are constructed by scaling each log as if it were sound, ignoring defects. Allowances for defect can then be made in each tree or stand according to its condition.

On this basis, tree volumes in board feet can be measured from the records of diameters taken at fixed intervals on the bole, or from the tree forms recorded on form 558a.

For piece products, as, for instance, sawed crossties, hewn ties, shingle bolts, stave bolts, or similar products, the tree volume depends on the length, quality, and cross-sectional area of the bolts contained in the tree which will yield these products, and the volume, or number per bolt. The content is best determined from the bolt itself, before removal from the site of the felled tree. Where more than one piece per bolt is obtained, as in sawing crossties or splitting staves or shingles, the intermediate step is in effect a "bolt" table of contents for bolts of different diameters and the required fixed lengths, from which total tree content is derived.

References

BEHRE, C. E.: Computation of Total Cubic Contents of Trees, *Jour. Forestry*, **22**, 62–63 (October, 1924).

CHAPMAN, R. A.: Errors Involved in Determination of Tree Volumes by the Planimeter Method, *Jour. Forestry*, **36**, 50–52 (1938).

REINEKE, L. H.: The Determination of Tree Volume by Planimeter, *Jour. Forestry*, **24**, 183–189 (1926).

SATTERLY, J.: The Hatchet Planimeter, *Univ. Toronto Studies, Math. Ser.*, 2, 1921, 24 pp.

CHAPTER 9

LOCAL TREE VOLUME TABLES— CONSTRUCTION—LIMITATIONS

71. The Volume Table Problem. Trees with identical breast high diameters and total heights, even for the same species, do not necessarily have the same volume. A single universal volume table that would apply to all conditions and species is therefore not possible. The basic factors that cause variation in volume may be grouped as follows:
1. Tree form.
2. Unit of product.
3. Closeness of utilization.

The effect on volume of 2 and 3 is man-caused, but 1 is the result of natural forces. Cubic measure gives the closest approximation to the true total cubic volume of the tree. But for practical purposes it is necessary to know the net volume of usable forest products, thus deducting or ignoring all parts of the tree that must be wasted. In parts of Europe and in Asia nothing is wasted; twigs and small branches are bound into bundles of fagots, bark is saved, and stumps are excavated for fuel. Even then the more valuable parts of the tree must be measured separately if any useful data are to be obtained for the construction of a table giving the net volume of the standing trees. Tables giving the total volume of trees including stump and top, with or without bark, are of little practical use. In most tables the volume of branch wood, which is as valuable for fuel as the top of the bole, is omitted.

Since bark is not removed from fuelwood, its volume is included in tables of cubic contents of trees used only for this purpose but is deducted from trees used for lumber or other sawed or peeled products. Bark may even be measured separately, as was the case with hemlock when bark alone was utilized for tannin and the wood was wasted.

Even for total cubic contents of trees, some limitation to the diameter and length of the smallest stick is needed. The limit most widely adopted is 3 in. o.b. with a 4-ft. minimum length of stick, which permits inclusion of most of the larger branch wood. The height of stump to be excluded is dependent on average practices, which under modern conditions for pulpwood, cordwood, and small trees approaches 6 in. or less, especially when power saws are used.

Because the main purpose of volume tables is to permit the measurement of the volume of the tree by a unit of volume that will indicate its stumpage value as closely as possible, the expression of this volume in cubic content, even after conversion to its equivalent in standard cords, serves only for smaller trees and parts of larger trees which cannot be utilized for lumber, crossties, poles, piling, or other products of greater value than bulk wood. That part of the total stand and of each tree which will be devoted to these higher uses calls for a volume table expressed in board feet, standard ties, or other units of product. Since ties can be measured by board-foot content and for poles and piling only diameter and total length are required, the board-foot unit is the principal unit of volume used in addition to the "merchantable" cubic feet. In this case, and in regions where the entire merchantable volume is used, with each part of the tree devoted to its best use, volume tables may show first the board-foot content without bark to the merchantable limit for saw logs, and then the volume of cordwood in the top to a 3 in. diameter; or it may include acid wood, or highway and fence posts, when the volume of these additional products must be determined.

72. Construction of a Local Volume Table Based on D.B.H. Classes. When felled trees are measured in the same locality and site, a volume table can be constructed directly from these tree measurements, based solely on d.b.h. classes and applicable only to the average heights of the different diameter classes on the site in question. Such a table can be based on the measurement of 25 to 30 trees, covering the range of diameters present. The process of construction is as follows:

1. Sort the tree records, which are on separate sheets, into their respective 1-in. diameter classes, and get the average diameter to the nearest $\frac{1}{10}$ in. for each class.

2. Total the volumes of the trees in each d.b.h. class, and divide by the number of trees in the class to get the average class volume.

3. Average the heights of the trees in each d.b.h. class in order to indicate the heights (and site) to which the table applies.

4. By plotting a curve of volume over d.b.h., obtain the curved average volume corresponding to each full inch of diameter. This requires a knowledge of the proper graphical methods for fitting a curve. These methods are used not only for the construction of local

volume tables but also for many kinds of tables, such as tables for growth and yield, in fact, for any purpose which involves the determination of curved averages over a range of another variable, such as age, size, and form class.

73. The Graphic Plotting of Averages

1. Cross-section paper is chosen with rectangular coordinates graduated 10 (or 20) squares to the linear inch. The horizontal axis, or scale of abscissas, is marked off to indicate d.b.h. in inches. The vertical axis, or scale of ordinates, is marked off for volume in cubic feet. In this case, volume depends on diameter; hence volume is the dependent variable and diameter is the independent variable (Sec. 14).

2. The average volume of the trees in each diameter class is plotted over the respective exact average diameter; *viz.*, if the trees of the 10-in. class average 10.1 in. in diameter, their average volume is plotted over 10.1 and not over 10. The purpose of plotting volumes over diameters in this manner is to discover the trend or law which may govern their relationship. Volume is generally known to increase with d.b.h., but other factors, such as form and height, also have an effect. When only a few trees are measured, these other influences may obscure the true nature of the d.b.h.-volume relationship. Since volume is proportional to basal area or the square of the diameter and since the scale of abscissas is based on diameter unsquared, it must be expected that the average volumes will fall along a curve which rises sharply and concavely upward, resembling one wing of a parabola and not a straight line. When a large number of trees are measured, it may even be possible to detect true trend without resort to graphic representation.

The merchantable cubic volume of 26 red and scarlet oaks, excluding a 1-ft. stump and the top above a 2-in.-diameter limit, are plotted in Fig. 17. Opposite each average is entered the number of trees in the diameter class, which gives the *weight* of the plotted point. Although these average volumes increase with diameter, the trend is not perfectly smooth. This irregularity is due to the fact that, for the average of trees of a given diameter class, the average height and average form quotient may be low or high as compared with those of other diameters. With a greater number of trees in each class, these irregularities tend to disappear, as is seen for the 9- and 10-in. classes. With a large number of trees in each class, the averages would follow a smooth curve, thus illustrating the underlying law of these averages.

3. Draw a tentative curve, weighting its position according to the relative number of trees in each average but preserving for the entire

curve the shape indicated by the general relationship of volume to diameter.

4. There is now some doubt as to whether this curve is too high or too low or has the proper trend. To test the curve, the total sum of deviations, plus and minus, should be checked. The deviation of a plotted value from the curve is the difference in volume between the actual plotted average and the reading on the curve directly below or above it. This difference is multiplied by the weight of the point, or the number of trees in the average. The algebraic sum of the weighted

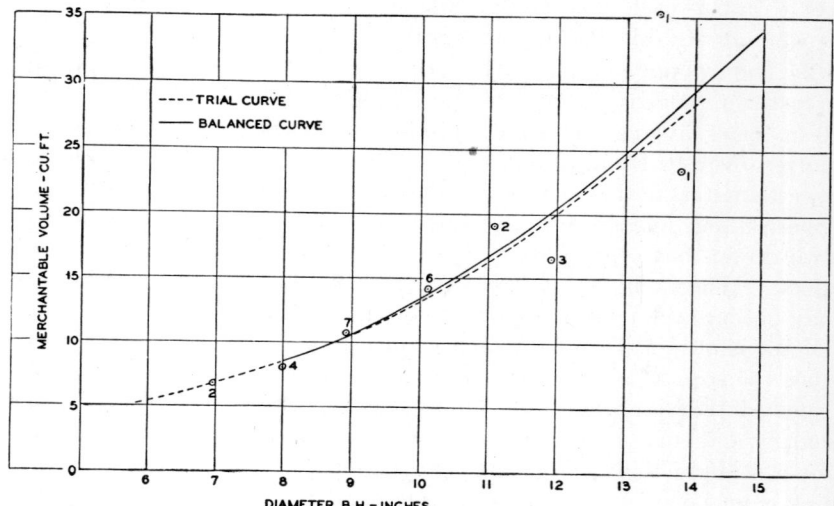

Fig. 17. Trial and final curves. Volume on diameter b.h., red and scarlet oaks, 26 trees, East Lyme, Conn. Merchantable volume of stem wood, with bark, above a 1-ft. stump to a 2-in. top outside bark.

deviations indicates whether the curve as a whole is too low or too high. But this in itself is not sufficient, for if one half of the curve is too high and the other half too low, the deviations may still balance. If, however, the curve is so drawn that the sum of the deviations *regardless* of sign is reduced to the lowest possible figure, the result is satisfactory.

5. To test the trend, or tilt, of the curve in order to reduce the sum of the deviations to the lowest possible value, divide the total number of trees by two on the basis of diameter. Repeat the calculations separately for each half of the curve. In this example, the 10-in. trees are divided equally between the two halves to make the trees in each

LOCAL TREE VOLUME TABLES

7-in. class		8-in. class		9-in. class		10-in. class		11-in. class		12-in. class		13-in. class		14-in. class	
D.b.h.	Cu. ft.	D.b.h.	Cu. ft.	D.b.h.	Cu. ft.	D.b.h.	Cu. ft.	D.b.h.	Cu. ft.	D.b.h.	Cu. ft.	D.b.h.	Cu. ft.	D.b.h.	Cu. ft.
6.9	6.60	8.2	8.37	8.8	9.75	10.2	13.10	10.9	19.08	11.9	14.13	13.5	35.64	13.8	23.50
7.0	6.91	8.2	10.00	8.7	10.84	9.8	12.88	11.2	10.10	12.2	16.60				
		7.8	6.64	8.9	11.12	10.3	12.41			11.7	19.12				
		7.7	7.07	8.7	9.55	10.2	16.58								
				8.6	9.64	10.4	17.05								
				9.2	11.98	9.8	12.70								
				9.2	11.94										
av. 6.95	av. 6.75	av. 8.0	av. 8.02	av. 8.85	av. 10.69	av. 10.1	av. 14.12	av. 11.05	av. 19.09	av. 11.95	av. 16.62				

110   FOREST MENSURATION

half equal in number. If plus and minus sums are equal, that half of the curve is properly located. If they are unequal, the position of the curve must be changed until a balance is obtained while the general trend and continuity are still preserved as a complete curve. The amount of the shift is gauged by dividing the net total, *i.e.*, the aggregate deviation, by the number of trees.

The diameters and cubic-foot volumes of the 26 oak trees, sorted by 1-in. classes, were as shown on page 109. Here the average diameters are rounded off to the nearest 0.05 in.

The average points are plotted in Fig. 17, and the first, or trial, curve is drawn as a dotted line and tested for balance, as follows:

Diameter, in.	Lower half of curve, deviation, cu. ft.	Number of trees	Total deviation, cu. ft.	
6.95	0.0	2		
8.00	−0.6	4	−2.4	
8.85	+0.3	7		+2.1
Total...............	.....	13	−2.4	+2.1
Net deviation...........	.....	..	−0.3	
Average per tree.........	.....	..	−0.023	

Diameter, in.	Upper half of curve, deviation, cu. ft.	Number of trees	Total deviation, cu. ft.	
10.10	+0.8	6		+ 4.8
11.05	+2.5	2		+ 5.0
11.95	−3.0	3	− 9.0	
13.50	+8.4	1		+ 8.4
13.80	−4.0	1	− 4.0	
Totals...............	.....	13	−13.0	+17.2
Net deviation...........	.....	..	+4.2	
Average per tree.........	.....	..	+0.323	

It is evident that the lower half of the curve is approximately correct, but the upper half must be raised by .323 cu. ft. This is approximately two-thirds of one small square on the original 10 × 10 coordinate paper used in plotting.

The new curve as shown by the solid line in Figure 17 gives the following deviations for the upper half:

LOCAL TREE VOLUME TABLES

Diameter, in.	Deviation, cu. ft.	Number of trees	Total deviation, cu. ft.	
10.10	+0.3	6		+ 1.8
11.05	+2.4	2		+ 4.8
11.95	−3.5	3	−10.5	
13.50	+8.6	1		+ 8.6
13.80	−5.0	1	− 5.0	
Totals.....................	.....	13	−15.5	+15.2
Net deviation................	.....	..	−0.3	
Average per tree..............	.....	..	−0.023	

The curve as a whole is now slightly above its true position by 0.023 cu. ft. per tree, an error too small to plot accurately.

74. Reading and Rechecking the Curve

1. When the curve is adjusted as well as possible by this simple check, the values are read for each exact full inch to make the local volume table. Curves enable one to read values at any point, and those now desired are for exact inches of diameter. These volumes are tabulated as follows:

D.b.h., in.	Volume, cu. ft.
7	6.8
8	8.5
9	10.8
10	13.5
11	16.7
12	20.5
13	24.8
14	29.3

This table may be extended by prolonging the curve at either end, as shown, to include

D.b.h., in.	Volume, cu. ft.
6	5.4
15	34.3

This volume curve actually starts at zero, or 6 in. to the left of the smallest diameter read.

2. The final check of the curved values is to compare the total actual volume of all trees with the total volume estimated from the curves or table. The check is similar to that made in balancing the curve, except that total volumes are now read instead of deviations from the

curve. As before, the readings of volume should be for the actual average diameters of the trees measured, as 10.1 in. and not 10 in. Each volume is multiplied by the number of trees in the class, and a total is obtained. The difference between the total of estimated volumes and that of actual volumes should be less than 1 per cent. In the foregoing example, the total of curved volumes is 353.4 cu. ft. and that of actual volume is 352.3 cu. ft. The curve checks within +0.31 per cent.

75. Importance of the Sample, or Measured Values. This check shows the total volume as read from the table to be accurate within

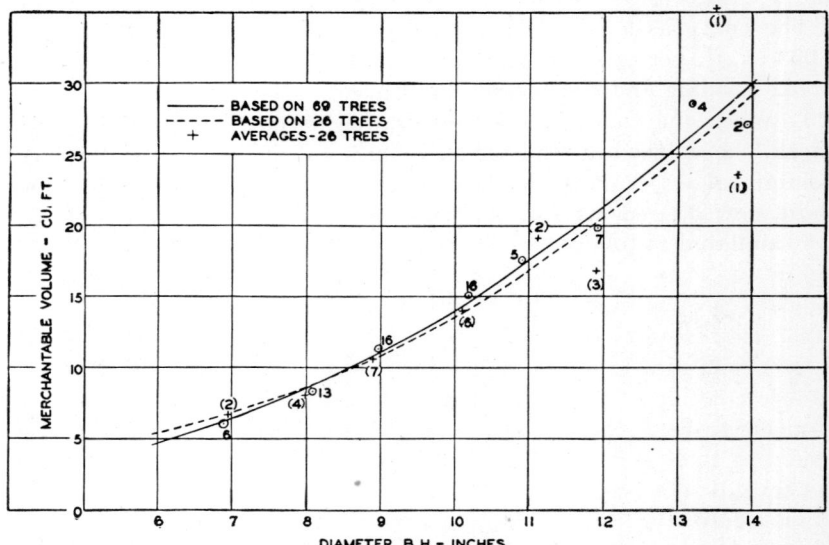

Fig. 18. Comparison of curve of cubic volume based on d.b.h. for 69 red and scarlet oaks with original curve for 26 of these same trees. East Lyme, Conn. Merchantable volume of stem wood above a 1-ft. stump to a 2-in. top outside bark.

0.31 per cent, but only for the actual trees from which the table was constructed. In other words, the curve is balanced for the "sample," or basic data at hand. But are these sample trees representative of the actual average volumes, based on d.b.h., of all the oaks on the given site? Does the curve represent the true average relationship between increasing diameter and corresponding increase in cubic volume?

One method of testing this assumption is to obtain additional sample measurements. The results of such a test are shown in Fig. 18, where

LOCAL TREE VOLUME TABLES 113

43 new trees are added to the original 26 trees, thus totaling 69 trees for the site. The solid line is the balanced curve for the 69 volumes and checks within 0.3 per cent of the measured volumes of the 69 trees. The resulting volume table is as follows:

D.b.h., in.	Volume, cu. ft.
7	6.5
8	8.6
9	11.1
10	14.0
11	17.5
12	21.3
13	25.6
14	30.0

By comparing this table with the original, it is seen that the volumes are higher in the enlarged sample. The original 26 trees give volumes averaging 4.1 per cent too low, according to the new curve. The original curve is shown in Fig. 18 by the dotted line.

It is necessary to remember that in all such statistical calculations the results obtained from the use of averages cannot exceed the accuracy of the sample or basis obtained from the field measurements.

76. The Board-foot Unit—Log Rules. In timber estimating for board feet, no attempt is made to determine the actual volume of lumber that may be *sawed* from the stand. The emphasis lies on estimating the volume that would be obtained by *scaling* the logs according to an accepted log rule (Chap. 5). Hence a separate standard table is required for each log rule used in timber estimating in the region concerned. Fortunately, the use of the International log rule for ¼-in. kerf (Sec. 51) is steadily increasing, both as an official standard for individual states and by the U.S. Forest Service, both for scaling logs and to a still greater extent for estimating standing timber by private owners of forest land. Whatever rule is used in scaling in the region, standard tables should always be constructed first for International ¼-in. rule and then for other rules only if wide common use makes them indispensable. Of the many existing rules, the only ones recommended for volume tables are the Scribner Decimal C rule (Sec. 48), which is used throughout the pine regions of the Lake states and on the national forests, and the Doyle rule (Sec. 52), which despite its glaring and inexcusable inaccuracy is used throughout the hardwood region and for all species in the South. Converting factors are sometimes given, by which the volume of stands in International rule may be changed to volume in terms of another rule, but this procedure is not satisfactory. As long as log rules remain in common use, standard volume tables are required in terms of each.

77. Merchantable Diameter Limits. How should the average minimum merchantable diameter limit for saw logs in the tree be determined? It should not be a theoretical limit, but one which will be taken in average actual practice. This fact rules out fixed diameter limits, even in regions and for species closely utilized for lumber, down to a 6-in. or even a 5-in. top. For even here it is found that the larger the tree and rougher the top, the larger will be the upper diameter of the top log. Volume tables are applied to standing trees only, whose height and merchantable length will be measured by eye or instrument from the ground. The cruiser's job is to estimate the tree to the point where the last cut will be made. He can readily train himself to do this. Hence no table is directly serviceable for its intended purpose unless the average top diameters used in its construction coincide with those of the top logs cut on logging operations. This rule requires that the merchantable contents for the volume table be taken from felled trees already sawed or marked into logs. When this practice is followed, the average top diameter is found to be a more or less consistent ratio of d.b.h., usually about 50 per cent, except in very close utilization. Hence the actual top diameter increases with d.b.h. This percentage is more uniform for conifers than for hardwoods. Since the bole of many hardwoods divides into large branches, the top diameter of hardwood saw logs is both greater and more variable than for conifers.

78. Log Lengths. In actual logging operations, the saw logs vary in length. For volume table construction it is necessary to adopt a uniform standard length. With conifers the logs may run from 8 to 16 ft. in 2-ft. graduations. Longer logs are scaled as two or more short logs. There are, however, usually more 16-ft. logs than any other length, which justifies the adoption of the 16-ft. length as the standard. Half-log graduations, *viz.*, 2, 2½, 3 logs, are recognized in volume tables to facilitate interpolation. This standard length is almost universally accepted in timber cruising and hence is the adopted practice for constructing volume tables.

For hardwoods, the logs are often cut in odd-foot lengths, averaging on the whole less than 16 ft. The variation in length is greater than for conifers. Under these circumstances, volume tables may be constructed on the basis of totaling the scaled volumes of the logs actually cut, rather than those of fixed standard lengths.

79. Tree Heights and Standard Volume Tables. Trees on one and the same area may vary considerably in total or merchantable height within each diameter class, but the height curve based on diameter can be used to represent the local condition in the construction and application of the volume table. On another area, a similar variation will be found, but the location and shape of the height curve may be quite different. Such a difference is commonly associated with change in site quality. Average height associated with age is in fact commonly used to indicate quality of site (Sec. 218). A volume table based on the first set of conditions and giving values by diameter only and not by height will not apply in the second case and is therefore strictly local in application. Such a table is therefore called a *local* volume table.

Although it is possible to show the average heights on which the local volume table is based and then make a suitable correction in applying the table to timber of different average height, this is not standard practice. Although local volume tables are more widely used than any other form of volume table, they are not the accepted basis for constructing volume tables. Instead a two-way table is made, in which volumes are classified by both diameter and height. This form is called the *standard* volume table. A table of this kind can give reliable estimates of volume throughout the range of the species, requiring only minor adjustments in special cases where a consistent departure in form can be detected (Sec. 81). An example of such a table is given in Table 19.

Derivation of a local volume table from a standard table simply requires the construction of a curve of height on d.b.h. for the site or stand to be estimated. By interpolation in the standard table, the volume is found for the actual feet of curved height for each d.b.h. class. Once constructed, standard volume tables thus serve as the basis for local tables, fitting all sites and stands having the same average volumes for given d.b.h. and height classes; and this includes all large areas and the majority of single stands.

The degree of accuracy of a volume table applied in estimating depends of course upon the difference of the actual average volume of the estimated trees from the theoretical average volume as obtained from the table. The likelihood of error increases under the following conditions:

1. When the volume of individual species is taken from a table for groups of species, such as hardwoods in general, or from a "universal" table of all species.

TABLE 19. SECOND-GROWTH RED OAK (*Quercus borealis* Mich.)
Conn., Md., N.Y., Ohio, Va., and W. Va.
J. H. Buell, 1928, merchantable stem with bark, cubic feet 4.0 in. top d. o. b.

Diameter breast high	Total Height in Feet									Basis: No. of trees
	30	40	50	60	70	80	90	100	110	
Inches	Volume in Cubic Feet									
4	.22	.57	.90	1.30						5
5	1.11	1.59	2.17	2.78						12
6	2.15	2.86	3.64	4.48						6
7	3.39	4.33	5.32	6.48	7.65					16
8	4.78	5.95	7.25	8.75	10.30	11.8				29
9	6.28	7.70	9.45	11.05	13.0	15.1	17.6			40
10	7.9	9.8	11.8	14.0	16.3	18.9	22.0			34
11	9.8	12.0	14.3	17.0	20.0	23.0	26.7			25
12	11.8	14.3	17.1	20.4	23.8	27.4	31.6	36.5		31
13		16.8	20.2	23.9	27.9	32.3	37.0	42.9		21
14		19.7	23.7	27.8	32.5	37.4	43.2	50.2		22
15		22.8	27.3	32.2	37.3	43.0	49.9	57.2		14
16			31.0	36.5	42.6	49.2	56.5	65.2		15
17			35.1	41.4	48.2	55.5	63.9	73.0		9
18			39.4	46.4	53.8	62.0	71.0	82.0		7
19			43.8	51.6	60.0	68.8	78.8	90.5	102	7
20			48.5	57.0	66.5	76.0	87.4	100.5	112	4
21			53.4	63.0	72.8	83.5	95.5	110.5	124	3
22						91	105	121	136	—
23						100	115	132	148	—
24						108	125	144	160	—
25						117	135	156	174	1
26							145	168	188	—
27							156	180	205	—
28							168	195	232	1
29							180	218	262	—
Basis: No. of trees	3	20	69	103	50	39	16	2	—	302

Prepared by the Central States, Appalachian, and Allegheny Forest Experiment Stations.
Top diameter outside bark 4.0 in.
Stump height 1.0 ft.
Volume measured by planimeter.
Prepared by alinement chart method.
Aggregate deviation: table 0.66 per cent low.
Average percentage deviation, 297 trees 5.0″+, ±8.14 per cent.
Block indicates extent of basic data.

2. With diminishing size of the area to be cruised, unless site classes are distinguished and separate local tables are applied to each site.

3. With diminishing number of trees, unless the same standard for statistical accuracy is applied to control the standard error of the mean (Sec. 17) by increasing the proportion of the trees measured out of the total number.

4. When board-foot volumes are estimated, since the standard deviation of board-foot volumes in a volume table is always greater than that of cubic-foot volume tables.

Such possible errors must be controlled by refinements of application which are obvious from the above; *viz.*, corrections for differences in form of species (Sec. 81); increase in the scope of the sample; and checks to detect average variation from the table for trees of identical d.b.h. and height (Sec. 86).

80. Total versus Merchantable Heights. In constructing standard volume tables, a choice must be made between the use of total height in feet and that of merchantable length in number of saw logs. Total height is always preferred for cubic volume of conifers and young or small trees of any species. In cruising, one never measures or estimates the height of all the trees but only of sufficient numbers to obtain a reliable curve of height on diameter upon which the local volume table will be based.

For board-foot volume of old and large timber, and especially for hardwoods, total height is much less satisfactory and is more difficult to measure accurately and rapidly. Merchantable height is not a consistent ratio of total height, but the actual merchantable top limit can easily be determined. Under these conditions, standard tables are usually expressed in terms of different merchantable lengths in logs for each d.b.h. class. For conifers the unit is the full 16-ft. log or the half log, such as 2, $2\frac{1}{2}$, 3 logs. For hardwoods, divisions of 4 ft. in total merchantable length are preferred.

In applying tables based on diameter and log classes, owing to small size of area, high value of individual trees, and different values of species, it may be necessary to round off and tally the merchantable length of each tree under the nearest height class given in the table. For large areas and numbers of trees and less individual variation in volume and value, a local volume table may be prepared from the standard table based on average log length by diameter class. This is done by plotting a curve of merchantable height in logs on d.b.h. and interpolating in the standard table to the nearest $\frac{1}{10}$ log to obtain the average volume for each d.b.h. class.

81. Tree Form and Bark Thickness—Form Class. The construction of a single standard volume table for all species and sites would be possible if the only variables were diameter and height. But other variables exist, consisting of varying standards of utilization and resulting variability of top diameters; differences in log rules for board feet (Sec. 76); and differences in the average form within a d.b.h.-height class for separate species. Though it is possible to construct a single standard volume table to average top diameter limits of utilization, it will be composed of a sample in which many species are represented, having different actual average volumes within a given d.b.h.-height class, and it will therefore be accurate or within accepted standard of error only for the few species that conform to the average.

Can correction factors for each species be determined and applied to the universal volume table? In a set of standard volume tables for Connecticut hardwoods based on 3,896 trees and 17 hardwood species, the following general tables were found necessary:

1. Cubic feet, entire tree with branches, to 3-in. limit o.b.
2. Board feet, International ¼-in. log rule, by 5-ft. total height classes.
3. Board feet, International ¼-in. log rule, by 4-ft. merchantable height classes.
4. A special table similar to 3 for oaks because of their importance and specific difference from the table for 3.
5. Cubic feet, plus cubic volume of highway posts, for the residual volume not utilized in saw logs, based on 5-ft. height classes.
6. Identical with 5 but expressed in standard 4 × 4 × 8 ft. cords, with bark.
7. Number, volume, and size of highway posts per tree based on d.b.h. only, for oak, then other species grouped.

The aggregate differences (which resulted in a separate table for oak) were now determined for each of the above tables 1, 2, 3, and 5 for each species. The total range of differences by species was:

Table	Per cent
1	+11.93 to − 6.54
2	+18.71 to −21.29
3	+21.10 to − 7.77
5	+13.75 to −25.86

The errors differed both for species and for the separate tables. They are of a magnitude which rules out the use of the general table without correction for any one species or product.

Variation in form of the merchantable bole covers the range between the truncated paraboloid and the truncated cone (Sec. 28). Species with full or paraboloidal boles yield greater volume than trees with more conical boles. Beech and basswood volumes, for instance, in the foregoing four tables are greater than the average by 10 to 20 per cent, while black birch volume consistently underruns the average.

These differences do not depend solely on the form of the bole itself. A second cause is variation in bark thickness. No diameter other than d.b.h. outside bark is generally acceptable as the standard diameter of classification, except for such species as swamp cypress, gums, and tropical species that have buttressed trunks, in which case the measurement of diameter at breast height is impractical, and a diameter must be measured at a point higher from the stump.

These differences in volume due to differing average form have been averaged out in the construction of standard volume tables for each separate species.

A second method has recently been developed (Sec. 101) by which form class, independent of species, is made the basis of a series of standard volume tables, each based on a different ratio between diameter inside bark at the top of the first 16-ft. log and d.b.h. outside bark. The ratios, and consequent tables, for use are readily determined in the field. These are known as the Girard form class tables.

A third method adopted in Scandinavia requires the construction, for each separate species, of standard volume tables for each difference of 0.05 units in breast high form quotients (Sec. 61). Where only two or three coniferous species exist, this refinement is possible. It is impractical for conditions in North America, where over 150 commercial species are found, for each of which one of the Girard form class tables will apply.

References

BEDELL, G. H. D.: Form Class of White Pine and Jack Pine as Affected by Diameter, Height and Age, *Canadian Forest Serv. Silvicultural Res. Note* 89, 1948, 11 pp.

BLADES, C. J.: Rules of Thumb for Converting Tree Volumes from One Log Rule to Another, *Jour. Forestry*, **36**, 612–614 (1938).

BRUCE, DONALD: The Problem of the Regional Volume Table, *Jour. Forestry*, **19**, 722–735 (1921).

MASON, F. R.: Rules of Thumb for Volume Determination, *Forestry Quart.*, **13**, 333–337 (1915).

SIMMONS, F. C.: Simplifying Tree Measurement Sale Procedure, *Jour. Forestry*, **40**, 963–966 (1942).

CHAPTER 10

THE CONSTRUCTION OF STANDARD VOLUME TABLES FOR CUBIC FEET

82. Standard Volume Tables. With standard volume tables of the required specifications, the only additional requirement is that a curve of height on d.b.h. be prepared for each area or locality (Sec. 86). A local table may then be derived from the standard table. This is usually sufficient, but in case the form of the timber differs from the average form included in the standard table, a further step may be necessary involving a percentage correction.

Standard volume tables may be on the basis of total height or of merchantable height (Sec. 80). If the former, the height curve is based on the total heights of trees. If the latter, particularly in the case of hardwood stands or old-growth timber, it is customary in estimating to tally the heights of the trees in terms of number of logs contained as well as d.b.h. and thus dispense with the height curve and the local volume table. The merchantable lengths of trees of the same d.b.h. may vary as much as two to three 16-ft. logs, especially in old-growth timber of large diameter.

83. Standard Volume Tables for Cubic Feet—Basis. The measurement of 300 to 2,000 sample trees is necessary in order to control sampling errors (Sec. 17) for the preparation of a standard volume table, depending on the size of the region for which the table is being constructed. The measurements in the field are the same as those described in Sec. 60. For the local table the heights of the individual sample trees were recorded only for the purpose of determining their average height as an indication of the stands to which the table would apply. In the standard table, however, the height will be used as one of the bases of tree classification. The final volume table will show the average volume of trees of different diameters at breast height and either total or merchantable height classes, as desired.

As in the case of the local volume table, the sample trees may vary in volume for trees of the same d.b.h. and height, because of variations in diameter and height from that representing the class center and because of differences in form or taper of the individual trees.

STANDARD VOLUME TABLES FOR CUBIC FEET   121

The factor causing greatest variation in volume within a given class is form.

Many methods have been evolved for the preparation of standard volume tables, most of which can be classified by one of the following four groups:
1. Graphical analysis with the use of harmonized curves (Sec. 84).
2. Volume form factors (Sec. 88).
3. Calculation by statistical methods (Sec. 89).
4. Alinement charts (Chap. 11).

The four general methods will be described in order. Each has its place depending upon the accuracy desired, the facilities available for calculation, and the technical ability of the operator. Numerous details of methods outlined are applicable not only to volume table construction but also to other classes of mensurational data.

84. Volume Tables Constructed from Harmonized Curves. To obtain the average curved volume for a given d.b.h. and height class, graphic analysis may be used. Instead of two variables, as in the local volume table, there are now three. Both d.b.h. and height classes are predetermined or set at fixed intervals of value; hence they become the independent variables. Volume has at the beginning of the analysis an unknown relation to d.b.h. and height and therefore becomes the dependent variable.

Since on a two-dimensional chart the ordinates represent the dependent variable (volume) and the abscissas the independent variable (either diameter *or* height, but not both), a decision must be made on which to choose for purposes of plotting. It is customary to choose diameter for this scale. Any single curve will therefore show only the relationship between volume and diameter. But volumes must be shown also by height class. This is done by accepting specific height classes, such as 40, 50, and 60 ft., and drawing a separate volume-on-diameter curve for each class. This is shown in Fig. 19. A single volume-on-diameter curve bends upward, since volume depends approximately on the square of the diameter. The spacings between curves for consecutive height classes in each diameter class tend to remain constant, because volume varies approximately as height (unsquared). In the case of the cylinder, the equation of which is $V = 0.005454D^2L$, the curves are smooth and true.

In the case of natural phenomena, such as the volumes of trees to be used in constructing a standard volume table, the points plotted from volume averages will rarely fall into smooth curves. Also the spacing between the points on the same abscissa and between the tentative curves drawn through these points will rarely be regular. The smoothing and spacing of volume curves will always require the use of individual judgment to bring them into harmonious relations with each other and with the underlying laws of average volumes. In this way,

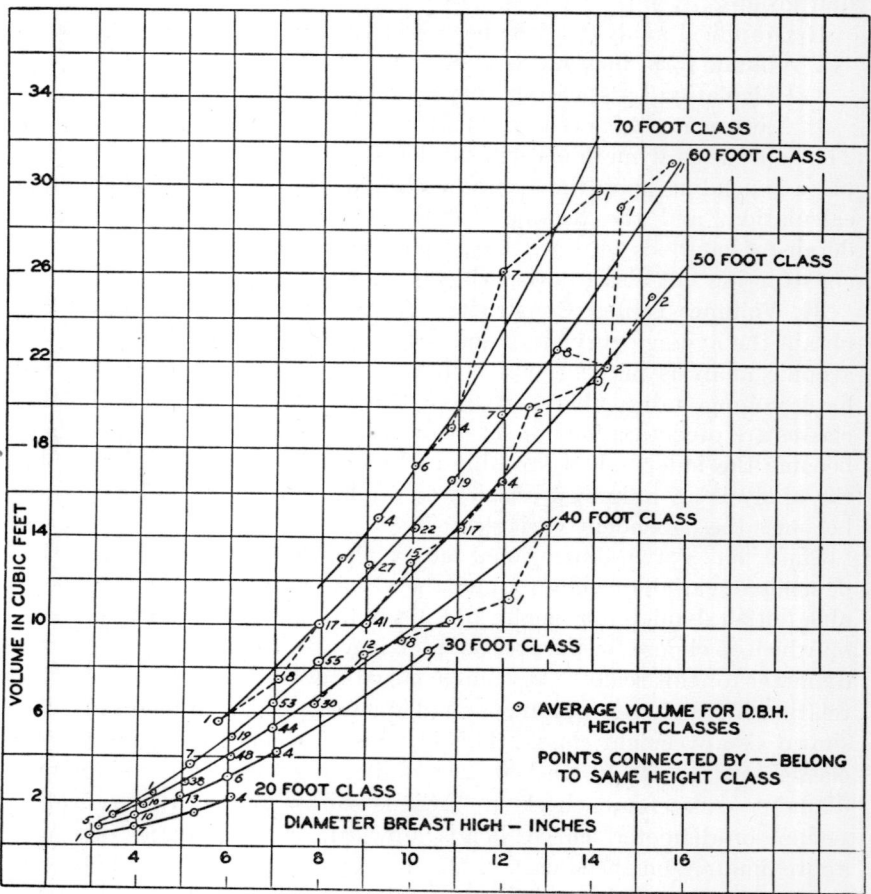

FIG. 19. Cubic volumes of red spruce plotted on d.b.h. by 10-ft. height classes.

a small sample gives results approaching a much larger one in which no effort is made to adjust the volumes. Such a process is called *harmonizing* and is described below.

In constructing a standard volume table, the first step is to sort the basic tree data into d.b.h.-height classes. When the diameter range is below the 40-in. class, it is customary to use 1-in. classes. Heights may be in 5- or 10-ft. classes, usually not over seven or eight in number. A convenient method of listing the data for plotting is shown in Table 20, which gives the average volumes by d.b.h. and height class for 607 red spruce trees in New England.

The next step is to lay off on a sheet of coordinate paper the proper range of volume and diameter graduations. The average volumes are plotted as ordinates and the average d.b.h.'s as the abscissas, as shown in Fig. 19, and the plotted points for each height class are connected by a lightly drawn line. Each series of connected points forms the basis of a curve, which is now drawn to smooth out the trend. Each curve is balanced by positive and negative deviations, as described in the case of local volume tables (Sec. 73). From these curves, the average curved volume for trees of the exact full inch of diameter can then be read.

Since the curve for each height class has been drawn without consideration of the position and shape of the neighboring curves, the series of curves will show slightly different trends and will have uneven spacings. The reason for this irregularity lies in large part in the variation of average height within each general height class and the lack of uniform differences in height from one diameter class to the next. For example, the average heights in the 40-ft. class vary from 38.0 to 43.2 ft.; in the 50-ft. class from 46.0 to 55.0, etc. In a single diameter class, the average heights do not progress by exact 10-ft. intervals; *viz.*, in the 9-in. class they are 43.2, 50.2, 59.7, and 67.3.

In the adjustment of the curves for height irregularity, one should not go back to the basic data of Table 20 and replot by height classes, since thereby the advantage obtained by curving out irregularities due to diameter will be lost. It is necessary to read volumes for the curves of Fig. 19 for the even-inch class and to replot these volumes on actual average height for the class, as is done in Fig. 20.

In transferring the volumes from the curves based on d.b.h. to a chart set up for volume on height, a strip of paper may be used. A zero point is marked on this strip near one end, and the strip is laid vertically along the ordinate representing a given d.b.h. class with the zero mark at zero volume. The intersections of the volume curves on the ordinate are then marked and labeled to avoid error. These markings are then transferred to the second graph, which has an abscissa scale for height, by shifting the strip stepwise to the right to correspond with the increasing height class. Each point must be plotted over the actual average height in the class and not over the class center; *viz.*, the curve volume for 11-in. trees which average

TABLE 20.—AVERAGE VALUES FOR PLOTTING STANDARD VOLUME TABLES. RED SPRUCE IN NEW ENGLAND*

D.b.h. class, in.	Height class, ft.	Number of trees	Average d.b.h., in.	Average height, ft.	Average volume, cu. ft.
3	20	1	3.0	24.5	.57
	30	5	3.2	30.9	.92
	40	1	3.5	38.0	1.38
4	20	7	4.1	20.0	.91
	30	10	4.0	32.3	1.40
	40	10	4.2	37.8	1.83
	50	1	4.4	46.0	2.30
5	20	9	5.3	20.6	1.51
	30	13	5.0	33.0	2.29
	40	38	5.1	40.2	2.87
	50	7	5.2	48.8	3.69
6	20	4	6.1	23.2	2.21
	30	6	6.0	32.3	3.09
	40	48	6.1	40.8	4.02
	50	19	6.1	48.9	4.96
	60	1	5.8	60.0	5.69
7	30	4	7.1	34.0	4.26
	40	44	7.0	42.0	5.39
	50	53	7.0	50.2	6.48
	60	8	7.1	57.7	7.56
8	40	30	7.9	41.0	6.40
	50	55	8.0	50.8	8.34
	60	17	8.0	59.5	10.09
	70	1	8.5	66.0	13.11
9	40	12	9.0	43.2	8.65
	50	41	9.0	50.2	10.17
	60	27	9.1	59.7	12.74
	70	4	9.3	67.3	15.00
10	30	1	10.4	34.0	8.88
	40	8	9.8	41.9	9.33
	50	15	10.0	51.1	12.91
	60	22	10.1	58.5	14.51
	70	6	10.1	69.4	17.31

STANDARD VOLUME TABLES FOR CUBIC FEET

TABLE 20. AVERAGE VALUES FOR PLOTTING STANDARD VOLUME TABLES. RED SPRUCE IN NEW ENGLAND*—(*Continued*)

D.b.h. class, in.	Height class, ft.	Number of trees	Average d.b.h., in.	Average height, ft.	Average volume, cu. ft.
11	40	1	10.9	38.0	10.28
	50	17	11.1	50.1	14.55
	60	19	10.9	59.0	16.69
	70	4	10.9	67.0	19.05
12	40	1	12.2	40.0	11.29
	50	4	12.0	50.5	16.70
	60	7	12.0	58.8	19.72
	70	7	12.0	70.1	26.16
13	40	1	13.0	38.0	14.66
	50	2	12.6	53.8	20.08
	60	8	13.2	58.6	22.71
14	50	1	14.1	53.0	21.30
	60	2	14.3	57.0	21.90
	70	1	14.1	66.0	29.80
15	50	2	15.3	55.0	25.03
	60	1	14.7	61.5	29.17
16	60	1	15.7	60.0	31.18
Total.....		607			

* Data collected by Northeastern Forest Experiment Station.

38.0 ft. instead of 40 ft. in height must be plotted over 38.0 ft. The plotted points are connected with lightly drawn lines.

Curves drawn through these averages of volumes based on height for each d.b.h. class would theoretically be straight lines (since volumes increase directly in proportion to height) were it not for the remaining variable of form or taper. Only in the case where the average form quotients of all d.b.h. classes falling within a given height trend line were identical would the curve be straight. Actually these average form quotients may vary systematically with d.b.h. and height, in addition to being somewhat inconsistent between adjacent classes. Hence the plotted averages may still produce an irregular line when connected. Should the average form quotient increase with diameter, the resultant curve would trend upward from the straight line, and vice versa.

126   FOREST MENSURATION

In Fig. 20, the lines connecting the averages in the lower three diameter classes fall in a continuous straight line. In other diameter classes there is a distinct slight upward curving. This means that the taller trees have higher average form quotient, thus giving these trees more volume.

In those diameter classes that are based on a large number of trees,

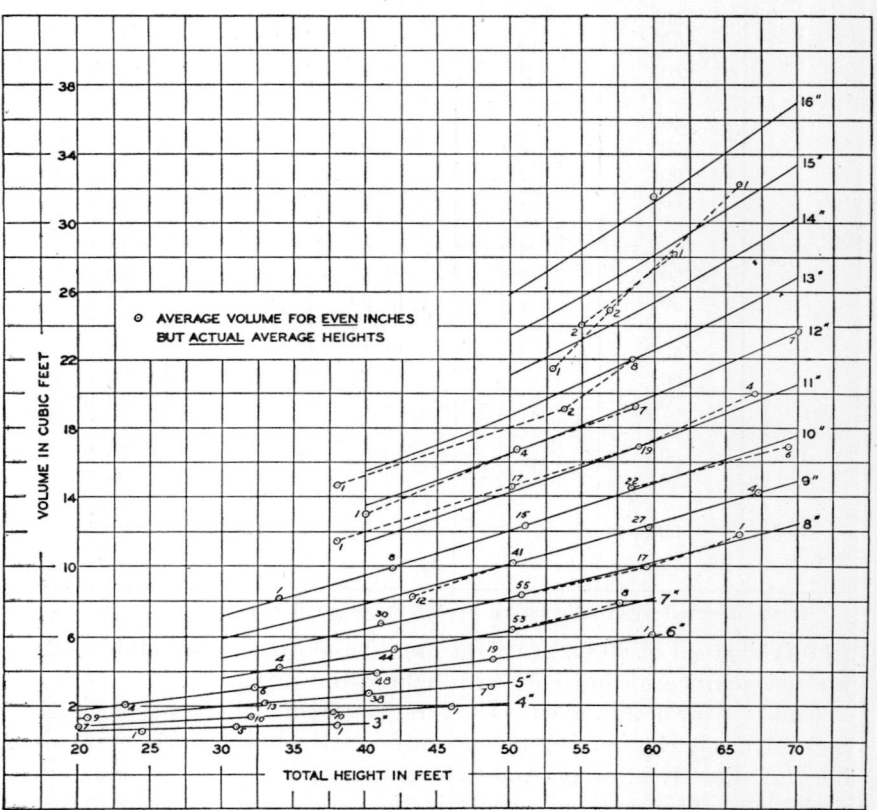

FIG. 20. Cubic volume of red spruce plotted on height, by 1-in. d.b.h. classes.

the fit of the curves is relatively good. In the larger diameters, however, for which there are few sample trees, the curves deviate widely from the general trend of the remaining curves. It would normally be expected that the average of a large number of trees in these larger diameters would show the same general trend as those of the smaller diameters. Such being the case, curves are fitted to these values following the same general trend of spacing and slope as in the smaller diameter classes without attempting to fit the curve to the points as plotted. Unless this were done, the curves for the 14- and 15-in. class would cross in Fig. 20. Such a situation would not be expected if more trees were available, since 15-in. trees, on the average, have more volume than 14-in. trees of the same height. For this reason, the 14- and 15-in. curves are drawn deliberately to fall nearly parallel with those in the smaller diameter classes but sloping upward more steeply because of the effect of diameter on volume. On the same basis, the location of the individual curves at their extremities is determined by the trend of volumes above or below in classes having a larger number of sample trees. It is always true that the extreme heights in the diameter classes will be represented by fewer trees than those near the average. Thus their average volume is more likely to vary than those derived from a large number of trees.

Since some personal judgment was used in fitting both the first and second sets of curves, the second set, based on height, may now be read off by the strip method and replotted on d.b.h. The result is shown in Fig. 21. This serves as a check on the harmonizing done in Fig. 20 and eliminates minor irregularities not smoothed out in curving over height.

The spacing should now be nearly even between the curves, and their trends are harmonized. The average volumes, by tree classes, obtained in this manner are probably as accurate as would be found by accepting uncurved averages from many times the number of measured trees, thus effecting a corresponding saving in cost.

85. Checking the Accuracy of the Curves. As in the case of the local volume table, this standard volume table should now be checked to determine the variation of the actual volumes of the trees from the curved volume equivalents. This process is much simplified if the spacing between the 10-ft. curves is divided into 10 equal parts and curves drawn through these divisions. These individual curves represent the 1-ft. intervals of height. Without such curves, it will be necessary to interpolate arithmetically for most of the values, since it is only by accident that individual trees have a height that falls exactly at the 10-ft. mark.

By reading from such curves the volume of each individual tree in the sample and comparing it with the actual volume, it is possible to find the aggregate difference and mean deviation (Sec. 16), both to be expressed in percentage. This process is very time-consuming and in this particular case has not been done.

The finished volume table may now be read from the final set of curves (Fig. 21) for each even inch of diameter and 10-ft. height curve. The number of trees and their distribution by diameter class are usually

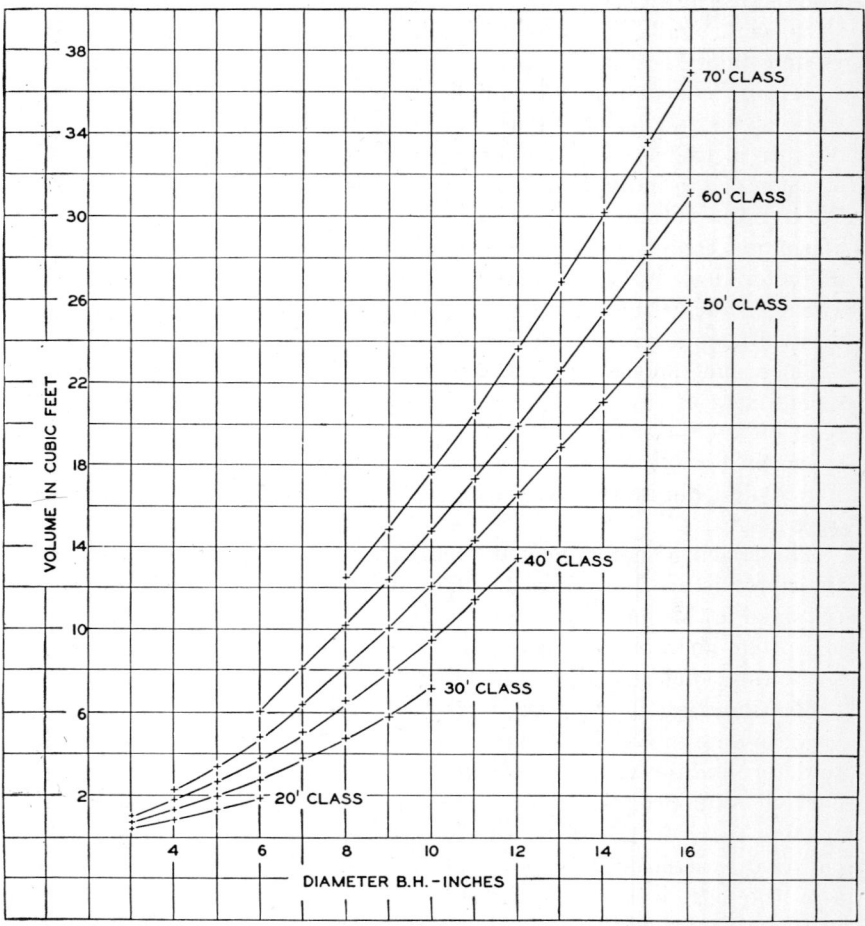

Fig. 21. Harmonized curves of cubic volume for red spruce replotted on d.b.h. by 10-ft. height classes.

given, as well as the utilization limits and the locality of the timber where the sample was obtained.

86. Derivation of Local from Standard Volume Tables. The steps necessary in the derivation of a local table from the standard table are:

1. The measurement of heights of standing timber for separate diameter classes.

2. Averaging these heights for each diameter class by means of a curve.

3. Obtaining the volumes corresponding to these average heights by interpolation in the standard volume table.

For example, each of the 69 oak trees measured for cubic volume (Sec. 75) was also measured for total height. Two other areas within ¼ mile and 1 mile, respectively, of this tract gave average heights differing from those on the first tract. Furthermore, the trees from the new tracts covered a much wider range of diameters. The three

TABLE 21. HEIGHT OF RED AND SCARLET OAKS ON THREE SITES AS READ FROM CURVES

D.b.h., in.	Tract 1, rocky dry site, height, ft.	Tract 2, medium site, height, ft.	Tract 3, good site, height, ft.
7	49	46	53
8	53	50	57
9	56	55	60
10	59	58	63
11	62	61	67
12	64	64	70
13	65	67	73
14	66	70	76
15	..	73	78
16	..	75	81
17	..	77	84
18	..	79	86
19	..	81	88
20	..	83	90
21	..	85	92
22	..	86	93

height curves are summarized in Table 21. These curves differ not only in total height but also in trend. Curve 1 is from a 70-year-old even-aged stand. In curve 2, although the larger trees were about 70 years old, smaller and younger trees lowered the average height of the 7- to 11-in. classes. The average heights of curve 3 are uniformly greater. Since the entire range of height and diameter is covered in the standard table, no difficulty is experienced in obtaining the volumes for any height curve, whatever its location and slope.

Height curves for different stands may thus have quite different trends. The tendency is, of course, for the curve to rise with increasing diameter in each case, except where the larger trees are growing relatively isolated in open parts of the stand. If the stand is even-aged, the difference in height between small and large diameters will not be great and the curve is comparatively flat. If the stand is many-aged, height increases more rapidly with increasing diameter, since in this case the smaller trees are also the younger trees and have not had time to attain the total height that small but old trees have in an even-aged stand.

The process of interpolation to obtain the local volume table from the standard table is as follows, after the average height for each diameter class has been determined:

Tract 2, 12-in. trees averaging 64 ft. in height. Standard volume table (Table 19)

12-in. 70-ft. trees contain	23.8 cu. ft.
12-in. 60-ft. trees contain	20.4 cu. ft.
Difference for 10 ft. of height	3.4 cu. ft.
4 ft. is 0.4 of a 10-ft. height class; hence volume difference is $0.4 \times 3.4 =$	1.36 cu. ft.

Adding this to volume of a 12-in. 60-ft. tree gives 21.76 cu. ft = volume of a 12-in. 64-ft. tree.

TABLE 22. LOCAL VOLUME FOR RED AND SCARLET OAK DERIVED FROM STANDARD VOLUME, TABLE 19

D.b.h., in.	Tract 1	Tract 2 volume per tree, cu. ft.	Tract 3
7	5.22	4.93	5.67
8	7.70	7.25	8.30
9	10.41	10.25	11.05
10	13.8	13.6	14.7
11	17.6	17.3	19.1
12	21.8	21.8	23.8
13	25.9	26.7	29.2
14	30.6	32.5	35.4
15	.....	39.0	41.9
16	.....	45.9	49.9
17	.....	53.3	58.9
18	.....	61.2	67.4
19	.....	69.8	76.8
20	.....	79.4	87.4
21	.....	89.5	98.5
22	.....	97.4	109.8

The resulting local volume tables for the three tracts are shown in Table 22.

87. Cordwood Volume Tables. If it is desired to estimate timber in standard cords, a cordwood tree-volume table is required. Trees are utilized for pulp, fuelwood, or other purpose to a size in the top commensurate with logging and manufacturing costs. Thus a cordwood table must be available for the top diameter limit to which the trees are merchantable.

Cordwood volume tables are usually prepared from existing cubic-foot tables by conversion. This means that knowledge is required of the average number of solid cubic feet per cord for trees of different diameters at breast height. Then, knowing the cubic-foot volume of a tree, its cord volume is found by dividing the tree volume by the solid-wood volume of a cord for the respective diameter. For example, if the solid-wood content of a cord composed of 8-in. trees is 95.6 (Sec. 21) and an 8-in. tree has a cubic volume of 8 ft., its cord volume is 8/95.6, or 0.084 cords. Converting factors for loblolly pine, cut in 5-ft. lengths, are given in Table 23 for trees of each d.b.h. class and are considered applicable to all heights within the class indicated.

In the absence of such a table, it is customary either to apply a fixed converting factor or to increase the factor with increasing d.b.h., as is the tendency shown in Table 23. The basis for determining the converting factor to apply by d.b.h. classes is the increasing size of the average stick in trees of successively larger d.b.h. classes.

The accuracy of a local cordwood volume table should be checked by the method, formerly described, of comparing the aggregate volume of felled trees with that estimated from the table (Sec. 85). This check might indicate a change in the solid cubic volume or the converting factor upon which the table is based.

88. Cubic Form Factors—Basis. In Chap. 8 it was shown that the bole of a tree resembles the form of a paraboloid, whose volume is 50 per cent that of a cylinder of equal diameter and length. The volume of a cone is 33.3 per cent that of the cylinder. Cubic volumes of cylinders are found easily, being $0.00545 D^2 L$, where D is diameter in inches and L is height in feet. If the cubic volume of trees bore a fixed relation to cylinders, this factor would be constant, and volume tables for cubic contents would be obtained with ease. In a general way, the form of the tree determines the value of this factor, which is therefore called the *form factor*.

Total height of the tree from ground to tip is taken as the height of the cylinder in determining form factors for cubic volume. The point at which diameter should be measured has been placed at b.h., since the taper of the stump below this point is too variable to permit consistent comparisons. Thus these dimensions of the cylinder and of the tree are compared not at the base but at a point 4½ ft. from the base. In a tree 18 ft. tall, this would be at 25 per cent of its height; for one 45 ft. tall, at 10 per cent; and for a 90-ft. tree, at 5 per cent of

TABLE 23. CONVERTING FACTORS, CUBIC FEET TO CORDS, BASED ON D.B.H.
(Loblolly pine, 5-ft. sticks, Eastern Coastal Plains)

D.b.h., in.	Rough wood			Peeled wood		
	Wood content of a standard cord of 128 cu. ft.		Converting factor, solid to stacked, cu. ft.	Wood content of a standard cord of 128 cu. ft.		Converting factor, solid to stacked, cu. ft.
	Cu. ft.	%		Cu. ft.	%	
5	75.9	59.3	1.686			
6	77.4	60.5	1.653			
7	78.8	61.6	1.623	95.3	74.5	1.342
8	80.2	62.7	1.595	96.8	75.7	1.321
9	81.5	63.6	1.572	98.0	76.6	1.305
10	82.6	64.5	1.550	99.0	77.4	1.292
11	83.7	65.3	1.531	99.8	78.1	1.280
12	84.6	66.1	1.513	100.6	78.6	1.272
13	85.5	66.7	1.499	101.3	79.1	1.264
14	86.3	67.3	1.486	101.7	79.5	1.258

total height. The nearer the tip, proportionately, that the d.b.h. falls, the relatively smaller will be the tree diameter with reference to its total form. The diameter at 4½ ft. determines the standard volume of the cylinder with which the tree is compared, and the volume of this assumed cylinder therefore diminishes as the "base," or diameter, is taken at greater percentages of its total height. For trees of identical form, therefore, the shorter trees will have the larger factor. The factor derived in this way is termed the *breast-high form factor*. Breast-high form factors are usually based on volume and diameter inside bark, since for most purposes bark is not used, and the form of the tree inside bark is more consistent. Bark at breast height varies in thickness more than at upper points.

Form factors may also be computed on the basis of comparison of form outside bark with the cylinder whose base is d.b.h. outside bark. The form factors for these measurements are apt to be lower than for those inside bark owing to relatively greater thickness of bark at b.h. than at upper diameters.

Standard volume tables can be constructed on the basis of breast-high form factors, but this method is now obsolete.

89. The Construction of Standard Volume Tables by the Method of Least Squares. A basic equation for tree volume is

$$V = 0.005454 D^2 H F,$$

where V is cubic foot volume, D is d.b.h. in inches, H is height in feet, and F is form factor. This expression could be further generalized to read $V = aD^b H^c$, if one felt that volume did not vary exactly as the square of the diameter and directly as height and preferred to compute the constant a and the two exponents b and c directly from the data. This equation can be transformed to the logarithmic expression

$$\log V = \log a + b \log D + c \log H$$

This form of the equation has been used successfully for the construction of standard cubic-foot volume tables and is now frequently used. Tests with many species have demonstrated that the cubic-foot volumes of trees obey this rule, particularly if the diameter range is not too great. This logarithmic equation is well suited to mathematical solution by the *least square method*, which is the fundamental method of fitting in statistical procedure. This method gives a fit in which the sum of the squares of the residuals are a minimum and the difference between the total actual and total estimated values, or the aggregate error, is zero.

Each tree is assumed to follow the volume equation. For example, a tree with a diameter of 15.0 in., a total height of 70 ft., and a measured volume of 38 cu. ft. could be expressed as

$$\log 38 = \log a + b \log 15.0 + c \log 70$$

For a single tree and equation, the coefficients remain unknown, but for three tree measurements they can be evaluated algebraically by simultaneous solution. In forest mensuration, it is necessary to measure large numbers of trees to get reliable results, in which case a modified form of simultaneous solution is required. This modified form is the least squares solution, which is based upon setting up a series of *normal equations*, equal in number to the number of unknown coefficients, in this case three. The first equation involves the sums of the three variables; the second, the sums of the products of the three variables times the first independent variable; and the third, the sums of the products of the three variables times the second independent variable as follows:

(1) Sum (log V) $\quad$ = log a(Sum of items) + b Sum (log D)
$\qquad$ + c Sum (log H)
(2) Sum (log D log V) = log a(Sum log D) + b Sum (log D log D)
$\qquad$ + c Sum (log D log H)
(3) Sum (log H log V) = log a(Sum log H) + b Sum (log H log D)
$\qquad$ + c Sum (log H log H)

To obtain the several sums, it is necessary to set up a calculation form of the following nature:

D	H	V	log D	log H	log V	(log D)^2	(log D log H)	(log D log V)	(log H)^2	(log H log V)
11.8	69	19.84	1.0719	1.7782	1.2975	1.1490	1.9061	1.3908	3.1620	2.3072
11.7	55.5	18.46	1.0682	1.7443	1.2662	1.1411	1.8953	1.3526	3.0426	2.2086
12.5	57.5	20.85	1.0969	1.7597	1.3191	1.2032	1.9302	1.4469	3.0965	2.3212

In the above form, the actual d.b.h., height, and volume of each tree is entered in the three columns to the left. The logarithms of each of the three values are then listed in the following three columns. In the seventh to ninth columns, each logarithm is multiplied by the logarithm of diameter. In the last two columns, the logarithms of height and volume are multiplied by the logarithm of height.

Each column from the fourth on is totaled and the sums introduced into the three normal equations. For illustrative purposes, the results of this preliminary calculation in a volume study of Connecticut hardwoods are taken, as follows:

$$\begin{aligned}
\text{Sum of items} = \text{number of trees} &= 3{,}890 \\
\text{Sum (log } D) &= 3{,}771.5547 \\
\text{Sum (log } H) &= 6{,}831.7229 \\
\text{Sum (log } V) &= 4{,}117.6418 \\
\text{Sum (log } D)^2 &= 3{,}746.9203 \\
\text{Sum (log } D \text{ log } H) &= 6{,}661.6616 \\
\text{Sum (log } D \text{ log } V) &= 4{,}204.8830 \\
\text{Sum (log } H)^2 &= 12{,}023.4701 \\
\text{Sum (log } H \text{ log } V) &= 7{,}329.7834
\end{aligned}$$

For convenience in the simultaneous solution, the left-hand member of each of the three foregoing normal equations is placed to the right end and given the minus sign, and the whole equation is set equal to zero. Substituting the determined sums into the equations, one gets

STANDARD VOLUME TABLES FOR CUBIC FEET 135

(1) $3,890 \log a + 3,771.5547b + 6,831.7229c - 4,117.6418 = 0$
(2) $3,771.5547 \log a + 3,746.9203b + 6,661.6616c - 4,204.8830 = 0$
(3) $6,831.7229 \log a + 6,661.6616b + 12,023.4701c - 7,329.7834 = 0$

The normal equations are now ready for algebraic solution, of which several methods are available. Of these, the Doolittle solution is probably the easiest to grasp and will be followed through. It should be noted at this point that absolute arithmetical accuracy is required throughout the calculations from the very beginning, since even minor errors may build up a cumulative effect and lead to a final error completely out of proportion to the original error. Extra decimal places are carried along to the end, when the coefficients finally determined will be rounded off to values that are easily handled.

The first step in the solution is to divide each normal equation by the first value in the equation. Thus the first equation is divided through by 3,890 the second by 3,771.5547, and the third by 6,831.7229, giving

(4) $ 1 \log a + 0.96955b + 1.75623c - 1.05852 = 0$
(5) $ 1 \log a + 0.99347b + 1.76629c - 1.11489 = 0$
(6) $ 1 \log a + 0.97511b + 1.75995c - 1.07290 = 0$

The term $\log a$ can now be eliminated by subtracting the fifth equation from the fourth, and the sixth from the fifth, giving

(7) $ 0 - 0.02392b - 0.01006c + 0.05637 = 0$
(8) $ 0 + 0.01836b + 0.00634c - 0.04199 = 0$

The first step is now repeated by dividing equation (7) through by its first value -0.02392 and equation (8) by its first value 0.01836, giving

(9) $ 1 b + 0.42057c - 2.35661 = 0$
(10) $ 1 b + 0.34532c - 2.28704 = 0$

the term b is eliminated by subtracting equation (10) from (9), giving

$$0 + 0.07525c - 0.06957 = 0$$
$$c = \frac{0.06957}{0.07525} = 0.924518$$

Coefficient b is obtained by substituting the now known value for c in either equation (9) or (10), thus

$$1 b + 0.42057 \times 0.924518 - 2.35661 = 0$$

or

$$1 b + 0.34532 \times 0.924518 - 2.28704 - 0$$

The first gives a value for b of 1.96779 and the second a value of 1.96760. The difference is due to insufficient decimal places, but since the final coefficients of a and b will be rounded off to two or three places, this is not worth worrying about.

The coefficient log a is obtained by substituting the known values for c and b into either equations (4), (5), or (6). Slightly different results will be obtained because of insufficient decimal places. One solution gives log $a = 2.47302$. This logarithm is equivalent to

$$\begin{array}{r} 10.00000 - 10 \\ -2.47302 \\ \hline 7.52698 - 10 \end{array}$$

and the antilogarithm, or coefficient a, is then 0.003365.

Inserting the three coefficients into the original volume equation, one gets

$$V = 0.00336 D^{1.968} H^{0.925}$$

For ready reference and use, the above equation can be put into alinement chart form, as described in Chap. 11.

Calculations are speeded up considerably without significant loss in accuracy if the original data are first assembled by d.b.h.-height classes and average diameter, height, and volume computed for each class. This was actually done in the case of the foregoing Connecticut hardwood volume study. Instead of handling 3,890 lines in the initial calculation form, only 142 lines were necessary as a result of the grouping. In this case, care must be observed to weight each logarithm or product of two logarithms by the number of trees in the class.

This method of transforming the absolute values of diameter, height, and volume to logarithms involves one systematic error, which arises from the fact that the logarithmic volume is balanced according to logarithms and not absolute values. In fact the method makes use of geometric averages, rather than arithmetic averages (Sec. 15). It is well known that the average of two or more logarithms does not lead to the same value as the average of the corresponding absolute values. For example, if the two extreme values of 10 and 100 are taken, the arithmetic average is 55, but the average obtained through logarithms is $(1.0000 + 2.0000)/2 = 1.5000$, a logarithm which corresponds to the value 31.62. In other words, although the aggregate error of volumes in terms of logarithms is precisely 0 by the least squares method, in terms of cubic feet the actual volumes will always be higher than those estimated from the completed equation. The difference is commonly of the order of 1 to 2 per cent. It has been proved that the error in per cent is independent of volume, and therefore the aggregate percentage error in terms of volume (not logarithms) can be used as a correction factor in drawing up the final table. A mathematical correction is also available,[1] based on the standard error of the single observation (in terms of logarithms) according to the following factors:

Standard error in logarithms	Increase the absolute values by the following percentages
0.03	0.24
0.04	0.43
0.05	0.67
0.06	0.96
0.07	1.31
0.08	1.71
0.09	2.17
0.10	2.68
0.11	3.26
0.12	3.89

For this purpose, the standard error of the single observation in terms of logarithms may be computed by taking a random or mechanical sample of 100 to 200 trees and computing the standard error according to the usual rules (Sec. 16c). This procedure is strictly valid only if the trees have been handled separately through out the computations, and it does not lend itself accurately in the event that class means were used.

The logarithmic expression of cubic-foot volume lends itself to ready application in the case of the local volume table, where height is not included as a variable. Here the equation is

$$\text{Volume} = aD^b$$

or

$$\log \text{volume} = \log a + b \log D$$

The coefficients a and b in the foregoing simple expression do not have the identical values as when height is included as a variable but must be recomputed on the basis of volume and diameter alone. On double logarithmic paper, this equation forms a straight line. In other words, if actual volumes are plotted over diameter on double logarithmic paper and if the plotted points fall in a straight line, one has direct proof that

[1] MEYER, H. A., A Correction for Systematic Error Occurring in the Application of the Logarithmic Volume Equation, *Penn. State Forest School Res. Paper* 7, 1941.

the equation is valid. In practice this fact is used. The points are plotted and the volume line drawn in ocularly by use of a straightedge without recourse to calculation. The result is a close approximation but not identical to the line determined by the least squares method. This graphical procedure has been found appropriate with many species, provided that the diameter range is not too great. It constitutes a quick but reliable method of constructing a local volume table for cubic feet. On the other hand, the logarithmic expression is not valid for board-foot volumes, which appear to obey another rule (Sec. 103).

References

BARROWS, W. B.: Reading and Replotting Curves by the Strip Method, *Soc. Amer. Foresters Proc.*, **10**, 65–67 (1915).

BRUCE, DONALD: The Height and Diameter Basis for Volume Tables, *Jour. Forestry*, **18**, 549–560 (1920).

CLARK, J. F.: On the Form of the Bole of the Balsam Fir, *Forestry Quart.*, **1**, 56–61 (1902).

CLAUGHTON-WALLIN, H.: The Jonson "Absolute Form Quotient" as an Expression of Taper, *Jour. Forestry*, **18**, 346–357 (1920).

DWIGHT, T. W.: Refinements in Plotting and Harmonizing Free Hand Curves, *Forestry Chron.*, **13**, 357–370 (1937).

FERNOW, B. E.: New Method of Measuring Volumes of Conifers, *Forestry Quart.*, **5**, 29–35 (1907).

JONSON, TOR: Form Class Tables (Massentabellar för Träduppskattning). Review, *Forestry Quart.*, **11**, 399–403 (1913).

MEYER, H. A.: The Standard Error of Estimate of Tree Volume from Logarithmic Volume Equation, *Jour. Forestry*, **36**, 340–342 (1938).

SCHUMACHER, F. X., and F. S. HALL: Logarithmic Expression of Timber-tree Volume, *Jour. Agric. Res.*, **47**, 719–734 (1939).

WINTERS, R. K., and P. B. WHEELER: The Suitability of Reineke's Planimeter Method for Volume Determinations of Delta Hardwood Species, *Jour. Forestry*, **30**, 429–434 (1932).

CHAPTER 11

THE CONSTRUCTION OF ALINEMENT CHARTS FOR STANDARD VOLUME TABLES FOR CUBIC FEET

90. Alinement Charts. Alinement charts possess definite advantages in the solving of some mensurational problems, including those of standard volume tables. Because of the complicated nature of these charts, it is considered desirable to outline the fundamental principles of construction before applying them directly to the case of volume tables.

An alinement chart may be defined as "a graphical-mechanical representation of the relationship between two or more variables." Such charts have long been used in engineering and other professions for this purpose. The best known example of an alinement chart is the slide rule in its various forms. On slide rules the relationship between variables may be found either by direct graduations in the case of two variables or by the sliding of graduated scales in the case of more than two variables.

It should be borne in mind that alinement charts may not be used to solve every kind of problem. An equation involving two or more variables that may be expressed in determinate form, however, may be expressed in an alinement chart. The preparation of tree volume tables in this form requires some knowledge regarding the more simple types of charts; hence these simpler charts will be discussed first.

The following two sections contain a brief description of the two fundamental forms of alinement charts, *viz.*, the addition chart and the multiplication chart. Only one method of construction, a graphical method, is shown. For a more thorough treatment of this important mensurational technique, the reader's attention is called to the references. The graphical method is not the only one that can be used. Mathematical methods[1] are available for locating the position of axes and for determining scale intervals.

91. Addition Charts. It has been stated that a mathematical equation can be expressed in alinement-chart form. One of the

[1] BRUCE, D., and L. H. REINEKE, Correlation Alinement Charts in Forest Research, *U.S. Dept. Agric. Tech. Bul.* 210, 1931, 87 pp.

simplest equations is that of the addition of two quantities, in the form $x + y = z$, in which x and y are any two quantities whose sum is z. Any desired value may be substituted for x and y, and the value of z depends on these substituted x and y values; z is thus the dependent variable and x and y are the independent variables.

The expression of such an equation in alinement-chart form requires only the knowledge of the range of x and y values that are to be added. That is, do the values of x range from 0 to 20 or 5 to 50, and do the values of y range from 5 to 50 or from 9 to 46? The method is the same in any case, but it is simpler if this range of values is known prior to chart construction. For purposes of illustration, assume that x varies from 0 to 10 and y from 0 to 20 and that it is desirable to have both scales of approximately the same total length on the alinement chart. The process of chart construction is as follows:

1. Lay off near the bottom of a sheet of cross-section paper a scale OP (see Fig. 22) with units of equal length covering the range 0 to 30 (the sum of the minimum values of x and y to the sum of their maximum values).

2. Erect a vertical axis for x near the left side of the chart and locate arbitrarily on it the minimum and maximum values, 0 to 10. Do similarly for y near the right side of the chart.

3. Above 0 on OP find location A on a horizontal with 0 of the x axis. Similarly point B is above 10 on OP and opposite 10 on the x axis. Connect the two points A and B by a straight line. This is the *graduating line* or *curve* from which all intermediate graduations for the x axis are obtained by reading from a given x value on scale OP vertically to line AB and then horizontally to the x axis, as is illustrated for the value of 6 in Fig. 22.

4. The graduating curve for the y axis line AC is constructed similarly and the y axis graduated.

5. The position of the z axis is found by the *method of intersections*. Two, or preferably three, values of z are chosen, in this case 10, 18, and 26, and two pairs of values of x and y for each z value are computed as follows:

x	y	z
6	4	10
3	7	10
10	8	18
5	13	18
10	16	26
6	20	26

CONSTRUCTION OF ALINEMENT CHARTS 141

With a straightedge, span a pair of x and y values, and draw a light line in the center of the chart. Span the second pair, leading to the same z value, and draw a light line intersecting the first line. Label the intersection with the corresponding z value. Repeat the process

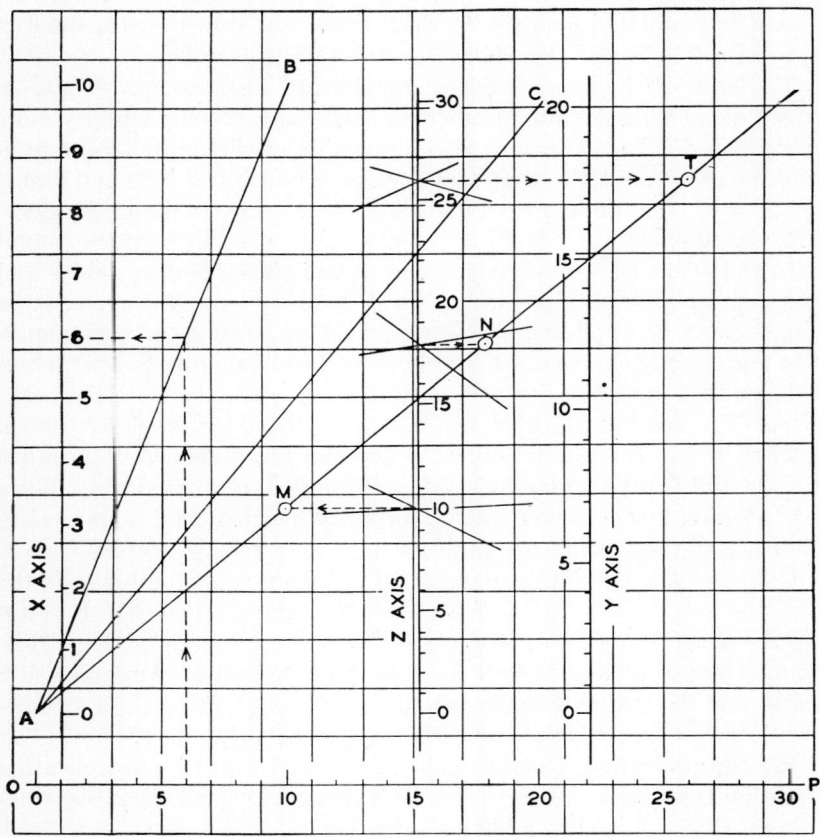

FIG. 22. Addition chart with different scales for independent variables.

for the other two z values. Finally draw an axis through the three intersections. This is the z axis.

6. The z axis is divided into units by means of a graduating curve (line MNT) in a manner similar to the graduation of the x and y axes. Point M is opposite the intersection $z = 10$ and above 10 on OP. N is opposite the intersection 18 and over 18 on OP. Similarly with T.

7. The chart is now ready for use. The value of z for any combination of x and y within the limits of the scale can be read by spanning

the combination with a straightedge and reading the intersection on the z axis.

The two cases dealt with the addition of two variables. In the case of an equation involving subtraction, for example, $x - y = z$, the form can be changed to read $x + (-y) = z$, where the $(-y)$ means that the y scale is graduated opposite to the x scale, *viz.*, downward instead of upward. The remainder of the construction is as before.

The case of the general linear or addition equation, $ax + by = z$, where a and b are definite numbers or coefficients, is only slightly more complicated. In this case, the computation of the desired z intersections takes account also that each x value is multiplied by a and each y value by b before finding z. Following this the construction is identical with the above.

In the event that an equation has three independent variables and one dependent variable, such as $aw + bx + cy = z$, the above techniques must be applied in two steps, working through a *turning axis*. The equation is changed arbitrarily to resemble an equation with two independent variables instead of three, for example, $aw + bx = T$, where T takes the place of $z - cy$. The chart for $aw + bx = T$ is constructed in the usual manner. The axis for T is located by intersection and is the turning axis. It need not be graduated fully, but a few selected intersections are retained. When this first step is completed, a new equation is handled where $T = z - cy$ or $T + cy = z$. The former T axis is retained, an axis for y is erected, and, between the two, the axis for z is located by intersection and graduated. In reading the chart, values for x and y are spanned by a straightedge. The point where the edge crosses the axis T is held without reading a T value, and the straightedge is shifted to go through the desired y value. The intersection with the z axis then gives the value for z.

92. Multiplication Charts. The simplest form of multiplication equation is expressed as $xy = z$. By the use of logarithms, this equation may be changed to addition form, thus

$$\log x + \log y = \log z$$

The process of constructing an alinement chart for this equation is exactly the same as for simple addition, except that the axes are graduated logarithmically but labeled with the number whose logarithm is scaled off on the axis.

In constructing a chart for multiplication, an axis length may first be assumed. For purposes of illustration, suppose the axis length is 8 in. The graduations for 1 and 10 may then be determined in inches by

CONSTRUCTION OF ALINEMENT CHARTS 143

multiplying the logarithms of the numbers by 8. Thus the distance of each successive graduation would be as shown below:

Graduation	Logarithm	Distance from bottom of axis, in.
1	0.0000	0.0000
2	0.3010	2.4080
3	0.4771	3.8168
4	0.6021	4.8168
5	0.6990	5.5920
6	0.7782	6.2256
7	0.8451	6.7608
8	0.9031	7.2248
9	0.9542	7.6336
10	1.0000	8.0000

Now if two axes are set up as x and y and these graduations are carefully measured on them, they will form the basis of a simple addition chart for the addition of the logarithms whose numbers are marked on the axes. The remainder of the chart construction is the same as before, i.e., locate the z axis by intersection of paired x and y values and by repeated intersection locate the graduations on this axis.

A simpler method of accomplishing the same result is to use logarithmic cross-section paper from the beginning. Any form will answer the purpose so long as it has at least one logarithmic cycle printed in one direction. In Fig. 23, semilogarithmic paper is used. On this paper let the abscissa at the base be the scale OR for purposes of graduating. Since the x and y-values will both vary from 1 to 10, draw a line as AB, with A above the minimum scale value 1, and B above the maximum scale value 10. Now set up the x and y axes at any convenient distance apart, and place the graduations, using line AB as a graduating curve.

Next locate the z axis by assuming three or more z values such as 6, 18, and 60; then compute pairs of x and y values to obtain these z values. In this case the following pairs were used for intersections:

x	y	z
3	2	6
2	3	6
6	3	18
3	6	18
10	6	60
6	10	60

144   FOREST MENSURATION

From the known positions of 6, 18, and 60 on the z axis, points M, N, and T may now be plotted above the OP scale and the z axis graduating curve (line AC) drawn through them.

If desired, smaller graduations than units may be plotted on the

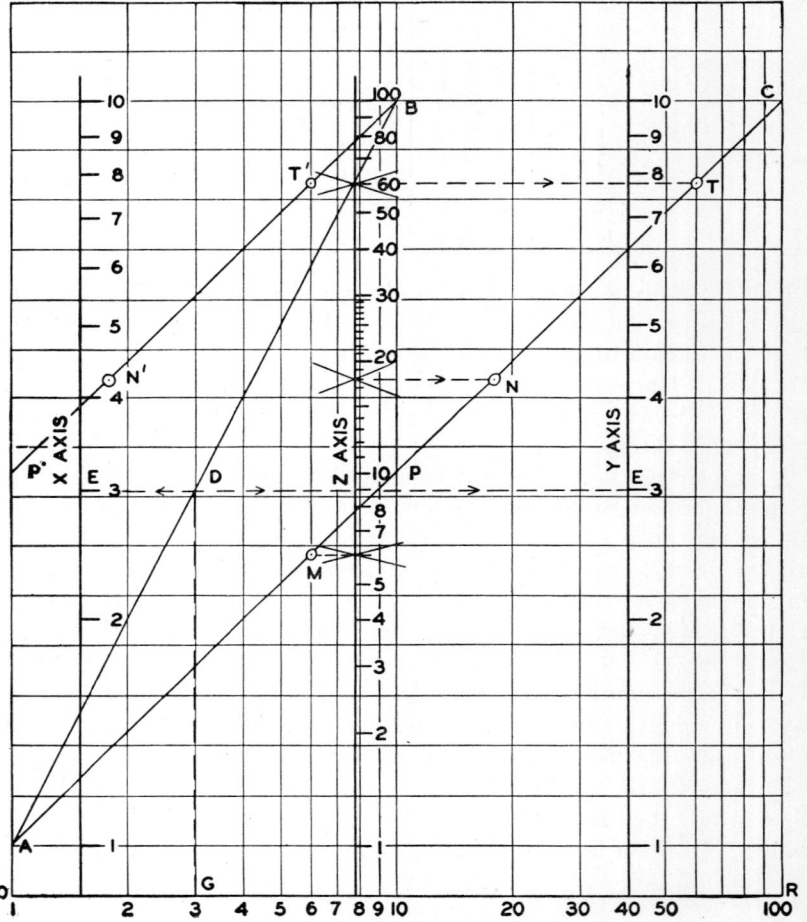

Fig. 23. Multiplication alinement chart.

axes by use of the graduating curves. This makes it possible to multiply such values as 1.1 by 3.2 or 11 by 32 with slide-rule accuracy.

93. Alinement Chart Volume Tables for Cubic Feet—Forest Service Method. The most immediate application of multiplication alinement charts comes in the preparation of tree volume tables, which may be derived from an adaptation of an alinement chart for cylinders (Fig. 24). The formula for cylinder volumes in cubic feet is $V = BL$ or, in terms of diameter, $V = D^2(0.005454L)$. Expressed logarithmically, this becomes

$$\log V = 2 \log D + \log (0.005454L)$$

Now, if the same rules are followed as in the case of the multiplication chart (Sec. 92), a cylinder chart can be prepared. This is not done immediately, however, since trees never have cylindrical taper. Their volume is better expressed by the formula

$$V = 0.005454 D^2 HF$$

where the additional F is the breast-high form factor. If all trees had the same form factor, an alinement chart could easily be prepared, giving the volume of trees accurately. Such is not the case, however, since form factors of trees usually vary from about 0.300 to about 0.500, averaging in the neighborhood of 0.400, depending upon species, crown class, site, and other factors. If an average form factor of 0.400 is chosen as a first approximation and introduced in the above equation, one gets

$$V = 0.005454 D^2 H (0.400) = 0.0021816 D^2 H$$

or

$$\log V = 2 \log D + \log 0.0021816 H$$

Figure 24 is a "base chart" prepared by the U.S. Forest Service, according to this modified cylinder formula. The method of construction is the same as that for the usual multiplication chart. In this base chart, paper used is 10×20 in. in size, and the diameter graduations run from 1 to 60 in. while the height graduations are from 10 to 200 ft. The paper is semilogarithmic, *i.e.*, graduated logarithmically in one direction (the abscissa in this case) and arithmetically in the other. Narrow blank spaces beside each of the axes are provided for regraduating the axes to conform to the tree volumes after the comparisons have been made.

The d.b.h., height, and volume graduating curves are indicated on the chart to simplify the process of axis regraduation later. The volume graduating curves for full cylinders are drawn, with dotted lines, and labeled V_c. V_{mc} is the "modified cylinder" graduating curve or cylinder times 0.4. The diameter and height graduating curves are labeled D_c and H_c, respectively.

The next step in table construction is to sort the trees in the sample

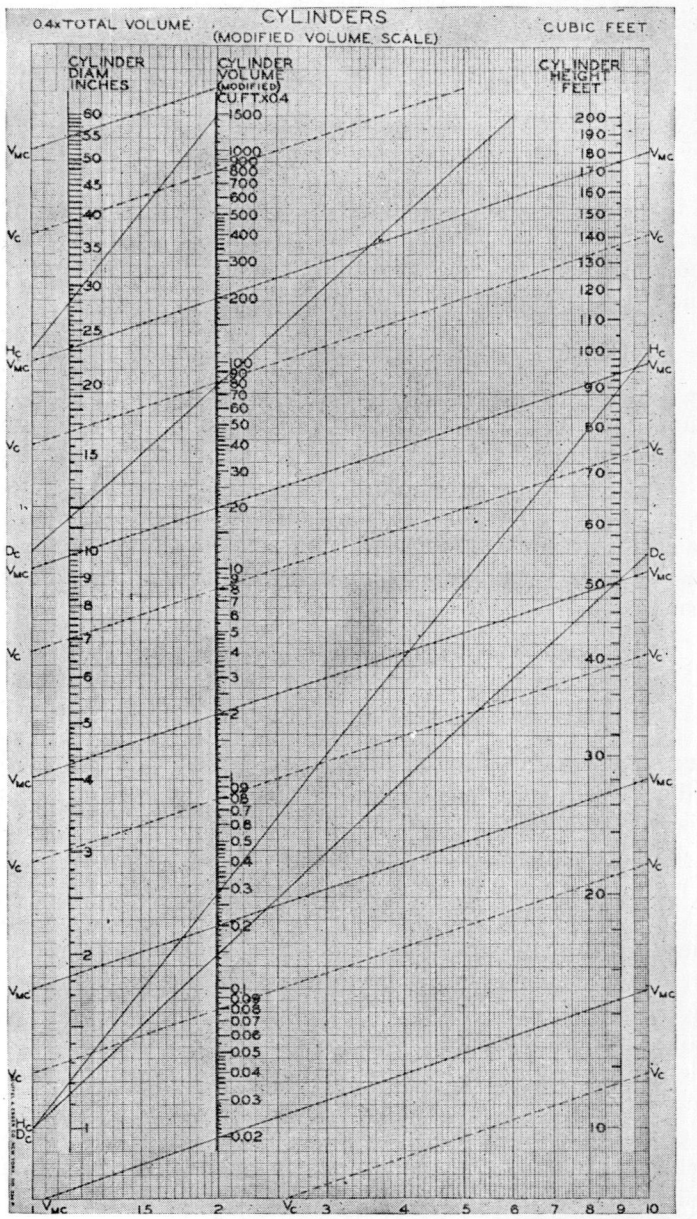

Fig. 24. Modified cylinder alinement chart used by the U.S. Forest Service.

into d.b.h.-height classes, the same as was done for the other volume-table construction. In this case, however, each individual tree must be listed as in columns 1 to 4 of Table 24.

TABLE 24. TABULATION OF INDIVIDUAL TREES FOR VOLUME-TABLE CONSTRUCTION BY ALINEMENT-CHART METHOD

Numbered trees	D.b.h., in.	Total height, ft.	Actual tree volume, cu. ft.	First estimate volume, cu. ft.	Second estimate volume, cu. ft.	Third estimate volume, cu. ft.	Fourth estimate volume, cu. ft.	Deviation last estimate, %
(1)	(2)	(3)	(4)	(5)	(6)	(7)	(8)	(9)
12-in., 60-ft. class:								
62	11.8	60	19.84	18.00	20.00	19.90	20.00	0.8
60	11.7	55.5	18.46	16.38	18.50	18.00	18.50	0.2
59	12.5	57.5	20.85	19.50	21.80	21.00	21.05	1.0
72	12.3	60.5	20.12	20.00	22.00	21.65	22.00	8.5
77	11.7	61	18.92	18.00	20.00	19.85	20.00	5.4
78	11.8	61	19.03	18.25	20.50	20.00	20.50	7.3
82	12.4	58	20.82	19.15	21.80	20.70	21.00	0.9
Total........	84.2	413.5	138.04	.....	144.60	141.10	143.05	24.1
Average.....	12.03	49.1	19.72	.....	20.66	20.16	20.43	3.4
12-in., 70-ft. class:								
109	11.6	70	24.82	20.35	22.50	23.00	23.30	6.4
108	11.8	70	26.28	21.00	23.50	23.60	24.00	9.1
116	12.3	73	28.84	24.00	26.00	26.70	27.20	6.0
117	11.6	73	25.08	21.45	23.80	24.20	24.50	2.4
115	11.8	69	23.88	20.90	23.00	23.20	23.60	1.1
121	12.3	69	23.29	22.50	25.00	24.85	25.40	8.3
128	12.5	67	30.94	22.40	25.00	24.85	25.20	23.0
Total........	83.9	491	183.13	.....	168.80	170.10	173.20	56.3
Average.....	12.0	70.1	26.16	.....	24.11	24.30	24.74	8.04

At least four columns should be provided for to the right of the column headed "actual tree volume" for entering volumes to be read later from the chart. The trees listed above are those included in two d.b.h.-height classes of an extensive sample of red spruce.

94. The First Chart Correction. It is now possible to compare the actual tree volume with the volume of the modified cylinder by reference to the alinement chart. The actual d.b.h. and total height of a tree are found on the respective scale; and, by use of a straightedge, the corresponding chart volume is found. The volume of the modified cylinder, corresponding to the 11.8-in. tree with a height of 60 ft., is found to be 18.00 and is recorded in column 5 of Table 24 as the "first estimate volume." This procedure is followed for each tree in the sample, and the first estimate volumes are averaged by d.b.h.-height classes. It is not to be expected that the chart will give a correct reading the first time, since the tree cannot be expected to have a form factor of 0.4. On the basis of the first estimates, however, the volume axis can be regraduated so that a closer reading will be obtained. To accomplish this, the class averages of the actual volumes are plotted as ordinates on double logarithmic paper over the corresponding class averages of first estimate volumes as the abscissas. The reason that the estimated volume is placed on the abscissa scale is that the alinement chart scale of modified cylinder volume is to be regraduated, and hence for this purpose the estimated volume must be considered the base or independent value, while the actual volume is temporarily the dependent variable. Each individual tree could be plotted, but that would lead to a multiplicity of points which would carry no specific advantage in later steps. Another method of plotting which has minor advantages involves the sorting of the individual trees by estimated volume classes, starting with 1- to 2- cu. ft. classes for small trees but increasing to 5- and 10-ft. classes or more with increasing size.

The points plotted in Fig. 25 were obtained by the latter process from first estimates made on 607 red spruce trees that were felled and measured for volume in cubic feet. A line is fitted to the plotted points, in this case a straight line. Had this line proceeded at 45 degrees through points 1,1; 10,10; and 100,100, one could logically assume that the chart graduations were correct. In practice the curve will rarely, if ever, be of this nature, since it presupposes that trees of all diameter classes are of form factor 0.40. The fitted line of Fig. 25 passes above the indicated points; in other words, actual volume is always greater on the average than estimated volume, and hence average form factor is greater than the assumed 0.40. Also the line does not have an exact 45-degree tilt, indicating that average form factor varies steadily from one diameter class to the next. Here the tilt is slightly less than 45 degrees, which means that form factor decreases slightly with size of tree. The shape of the fitted line need not be straight but is often curved downward at the upper end. This would normally be expected in trees taken from even-aged stands. Under these conditions, the largest trees are probably dominant in the stand

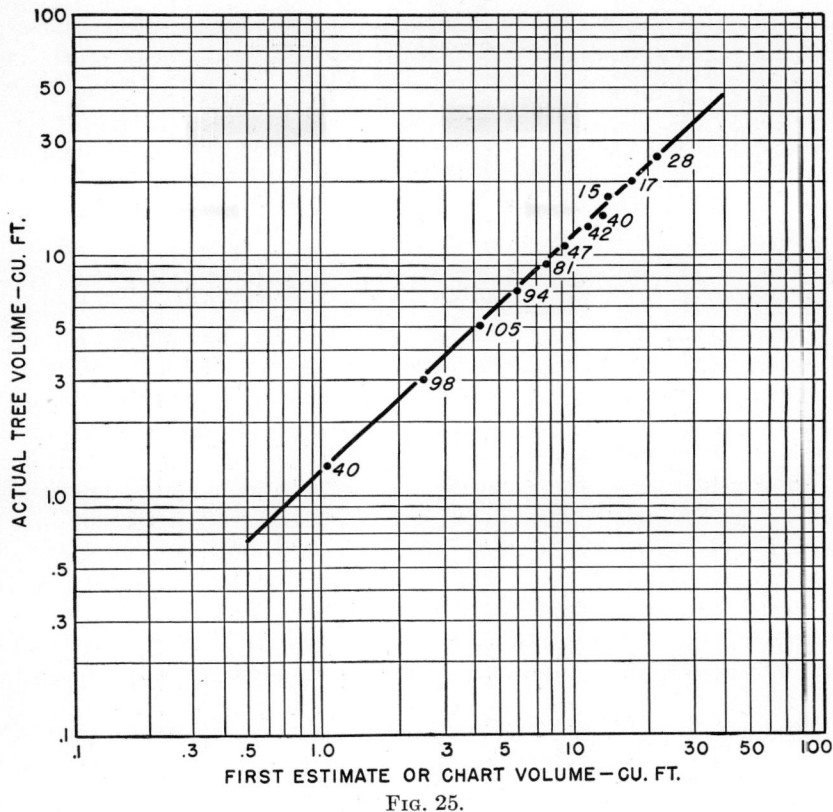

Fig. 25.

and have full crowns and fairly rapid taper. The trees of fuller form would be likely to be trees of poorer crown class and smaller size.

Figure 25 indicates that on the average trees of 1 cu. ft. in actual volume have been estimated as having slightly less than 0.8 cu. ft., trees of 10 cu. ft. actual volume as having about 8.7 cu. ft., and trees of 30 cu. ft. as having 27 cu. ft. If the chart graduations were renamed, viz., if 0.8 on the chart was called 1, 8.7 called 10, and 27 called 30, then the volumes would be read correctly. This is actually what is done. A number of points are read and plotted as the basis for a graduating curve, as in Fig. 26, by the line labeled V_1. This curve is now used to regraduate the volume axis on the blank space on the left side of the volume axis.

Frequently this process completes the work of chart construction, but occasionally the diameter or height axes may have wrong spacing

150   FOREST MENSURATION

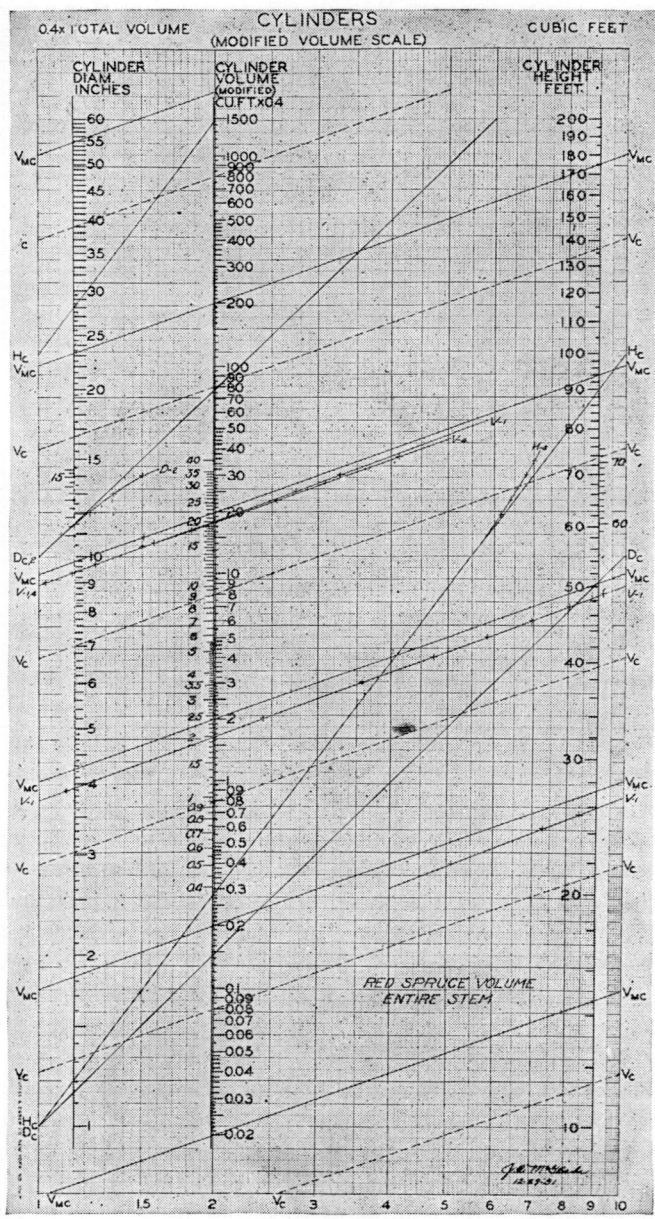

Fig. 26. Alinement chart volumes for red spruce derived from cylinder volume x 0.4.

and need correction. In order to determine whether this needs to be done, a second estimate is always read from the chart, using the regraduated volume axis.

95. The Second Chart Correction. The second estimate volumes are read off and tallied opposite the proper tree in column 6 of Table 24 above. They are first totaled and averaged by d.b.h. classes (Tables 25 and 26). After this operation, the trees are re-sorted on the basis of

TABLE 25. GROUPING OF TREES BY D.B.H. COMBINING ALL HEIGHTS

D.b.h. class, in.	Height class, ft.	Number of trees	Total d.b.h., in.	Total actual volume, cu. ft.	Total second estimate volume, cu. ft.
(1)	(2)	(3)	(4)	(5)	(6)
5	20	9	47.7	13.57	14.29
5	30	13	64.4	29.74	28.18
5	40	38	195.1	109.10	106.70
5	50	7	36.4	25.88	24.46
Total.....	..	67	343.6	178.29	173.59
Average..	..	..	5.1	2.66	2.59

TABLE 26. AVERAGE ACTUAL AND SECOND ESTIMATE VOLUMES BY D.B.H. CLASSES

D.b.h. class, in.	Average d.b.h., in.	Average actual volume, cu. ft.	Average second estimate volume, cu. ft.	Number of trees	New graduation distance, in.
(1)	(2)	(3)	(4)	(5)	(6)
3	3.24	0.93	0.89	7	3.32
4	4.10	1.47	1.43	28	4.13
5	5.10	2.66	2.59	67	5.18
6	6.10	4.11	4.02	78	6.17
7	7.00	6.04	5.94	109	7.02
8	8.00	8.11	8.16	103	7.98
9	9.00	11.01	10.90	84	9.03
10	10.00	13.47	13.71	52	9.87
11	11.00	15.88	16.40	41	10.88
12	12.00	21.01	20.97	19	12.00
13	13.00	21.50	22.92	11	12.52
14	14.20	23.73	27.85	4	13.00
15	15.10	26.41	30.70	3	13.92
16	15.70	31.18	35.00	1	14.75

152   FOREST MENSURATION

height class, and the volumes, both actual and estimated, are totaled and averaged by these new classes (Tables 27 and 28). Table 25 is the summary of the 5-in. class of the 607 red spruce trees, given for purposes of illustration, and Table 26 is a tabulation of all diameter class averages.

Now in order to check the diameter axis, a straightedge is placed on

TABLE 27. GROUPING BY HEIGHT COMBINING ALL D.B.H.'s

Height class, ft.	Diameter class, in.	Number of trees	Height, ft.	Actual volume, cu. ft.	Second estimate volume, cu. ft.
(1)	(2)	(3)	(4)	(5)	(6)
20	3	1	24.5	0.57	0.60
20	4	7	140.0	6.39	6.27
20	5	9	187.5	13.57	14.29
20	6	4	93.0	8.84	9.39
Total.....	..	21	445.0	29.37	30.55
Average..	..	..	21.2	1.40	1.45

TABLE 28. AVERAGE ACTUAL AND SECOND ESTIMATE VOLUMES BY HEIGHT CLASSES

Height class, ft.	Average height, ft.	Average actual volume, cu. ft.	Average second estimate volume, cu. ft.	Number of trees	New graduation distances, ft.
(1)	(2)	(3)	(4)	(5)	(6)
20	21.20	1.40	1.45	21	20.6
30	32.60	2.88	2.36	39	33.0
40	41.00	4.97	5.04	194	40.7
50	50.30	8.97	9.07	217	50.3
60	.59.20	14.54	14.45	113	59.8
70	72.90	22.21	19.16	23	77.3

the average class d.b.h. (column 2), intersecting the volume axis at the point indicated by the average second estimate (column 4). Holding the straightedge in this position gives an intersection on the height axis. Hold the height axis intersection with a needle or fine pencil and turn on this point as a pivot until the straightedge intersects the volume axis at the value indicated by the average actual tree volume in column 3. A diameter reading may now be made on the diameter axis and tallied in column 6 headed "New graduation distance." This is repeated for each diameter class.

Arithmetic cross-section paper is now prepared in a scale large enough to carry the range of diameters in the sample (Fig. 27). The

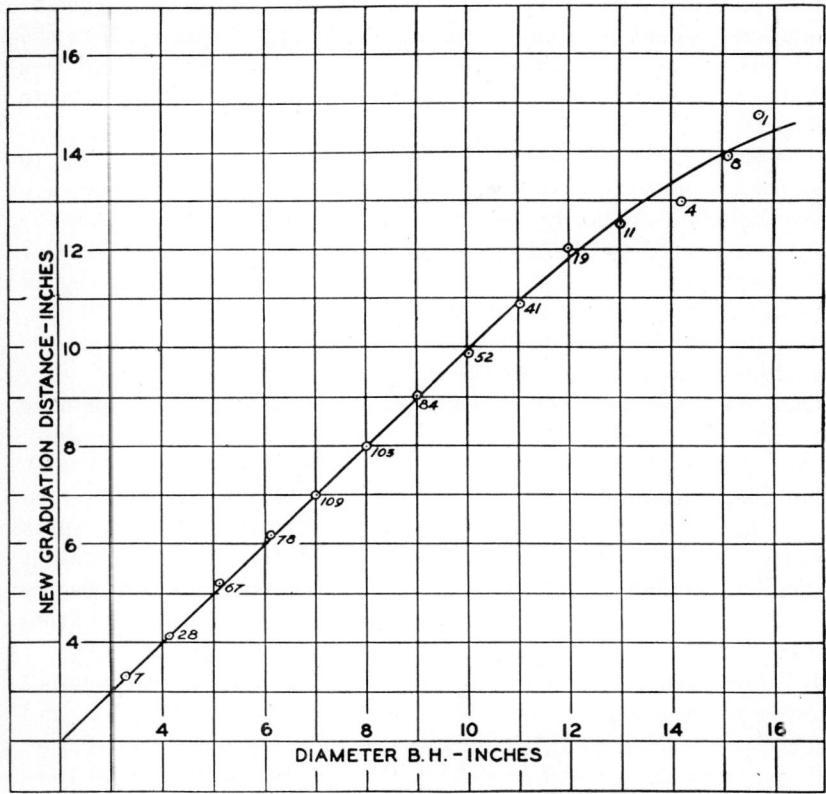

Fig. 27. Basis for regraduating d.b.h. axis.

abscissa stands for actual d.b.h. and the ordinate for new graduation distance, both on the same scale. The values in column 6 are plotted on those of column 2, and a smooth curve is drawn through the point.

The d.b.h. axis may now be regraduated by reading from the curve the new graduation distance corresponding to each actual tree d.b.h. Sufficient values can be read to plot a graduating curve on the chart, from which to regraduate the axis in the usual manner. The line labeled D_2 in Fig. 26 is this new graduating curve. It should be noted that, in this particular case, the curve of Fig. 27 is a straight 45-degree line from 3 in. to about 10.5 in. Above 10.5 in. it falls off. This means that the d.b.h. axis does not require regraduation except from 10.5 in. up.

96. The Third Chart Correction. It is next necessary to check the height axis in a similar manner. The trees are grouped by height classes, regardless of d.b.h., as shown in Table 27, which covers the 20-ft. class. Totals and averages are taken for height, actual tree volume, and second estimated volume. Table 28 gives the complete summary of averages for the 607 red spruce trees.

The new graduation distance (column 6) is found in a similar way to that for the diameter axis; *i.e.*, lay a straightedge from the average height value on the height axis and intersecting the central axis at the volume indicated by the average second estimate (column 4). Now hold the intersection on the d.b.h. axis and pivot the straightedge to intersect the volume axis at the actual average tree volume (column 3). Read the new graduation distance from the height axis, where the straightedge intersects it, and record the reading in column 6.

On arithmetic cross-section paper plot the new graduation distance for height on actual height, precisely as was done for the d.b.h. data. If the resulting curve is not a 45-degree straight line passing through equal abscissa and ordinate values, the chart height axis is corrected exactly as was done with d.b.h. Figure 28 shows the graph of heights based on these values. Note that in this chart the curve is a straight 45-degree line up to 60 ft. This means that regraduation is necessary only above 60 ft. Line M_2 of Fig. 26 is the corresponding graduating curve.

It is now necessary to make a third estimate reading of tree volumes in order to determine the effect of the regraduation of the d.b.h. and height axes. Since the d.b.h. axis was regraduated only from 10.5 in. and up and the height axis from 60 ft. and up, third estimates for trees less than this d.b.h. and height will give the same volume as the second estimates and hence need not be repeated. The third estimate is tallied in column 7 of Table 24 and totals and averages found as with the first estimate.

97. Completion of Chart and Test for Aggregate Difference. The third estimate averages may now be compared with actual tree averages (column 4, Table 24) by plotting on double logarithmic paper as was done with the first estimate (Fig. 25). If the diameter and height axes have been greatly changed, it may be necessary to regraduate the volume axis a second time. In the case of the red spruce, this was done, since the plottings did not fall in a 45-degree straight line pass-

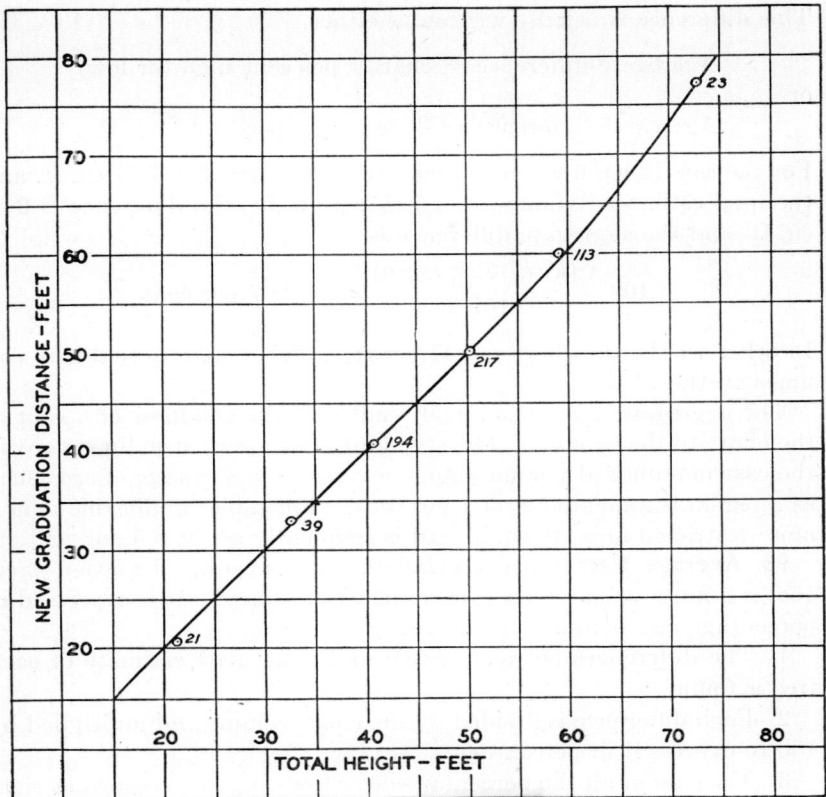

Fig. 28. Basis for regraduating height axis.

ing through equal abscissas and ordinates. Since the volume axis was again regraduated, it is necessary that a fourth estimate be made and a second check be followed through for the d.b.h. and height axes as before. This cycle is repeated until no further correction is indicated on any of the axes. The fourth and sixth estimates correspond to the second; the third, fifth and seventh to the first. In the case of the spruce, stability was reached at the fourth estimate.

From the final chart estimate, the aggregate difference is determined, i.e., the difference between actual tree volume of the whole sample and the last estimated volumes is found and expressed as a percentage of the actual volume, as follows:

$$100 \frac{\text{sum actual minus sum estimated}}{\text{sum actual}} = \text{aggregate difference in per cent}$$

This difference is usually written as either

Aggregate difference = chart x per cent high (or low)

or

Aggregate difference = + (or −) x per cent

For the case in hand where actual tree volume was 5,157.21 cu. ft. and the final estimated volume was 5,181.61 cu. ft., the difference is 24.4 cu. ft. and the aggregate difference is

$$100 \frac{5,181.61 - 5,157.21}{5,157.21} = +0.47 \text{ per cent}$$

In other words, the chart is 0.47 per cent higher than actual tree volumes on the whole.

The aggregate difference merely indicates the goodness of the fit of the chart to the sample. The standard of accuracy usually sought in the case of standard volume tables covering a large range of area such as a region is a maximum of 1 per cent. For tables applicable over a more restricted area, the standard is frequently set at 0.5 per cent.

98. Average Percentage Deviation. The average deviation (Sec. 16c) is a more valuable measure of chart accuracy. It is expressed in percentage and is found as follows:

1. The differences between actual and final chart estimate of each tree is found.
2. Each difference is divided by the chart volume and multiplied by 100 to express it in percentage.
3. The sum of all the percentage deviations, without regard to sign, divided by the number of trees in the whole sample gives the average percentage deviation. Good cubic-foot tables frequently have an average deviation of ±8 per cent or less.

In the example, the sum of the individual percentage deviations regardless of sign is 3,818.1, and this sum divided by 607, the number of trees, gives an average deviation of 6.29 per cent. Since in a normal distribution the standard deviation is 1.253 times the average deviation (Sec. 16), this fact is made use of to get a short-cut approximation of the standard deviation. In this case, $1.253 \times 6.29 = \pm 7.88$ per cent. A more accurate determination of standard deviation could be made by the method of squaring the residuals in the conventional manner (Sec. 16d). The value of 7.88 per cent represents the expected variation of a single observation from the tabular value and is commonly referred to as the *standard error of the single observation* or the *standard error of estimate*. This must not be confused with the standard error of the mean (Sec. 17), which can now be computed but has little significance in this case.

The volume table may now be read off in the conventional form as shown in Table 29. If the table is to be published, it is desirable to give also the alinement chart, since in this form the interpolation of volumes is simplified in the preparation of local volume tables.

The base chart founded upon the modified cylinder, in the above case cylinder volumes times 0.4, is only one of several possibilities for the purpose of cubic-foot volume tables. The volumes of paraboloids, or even cones, may be and have been used with equal success. No new techniques are involved, however, and the procedure is exactly the same as described above.

TABLE 29. VOLUME TABLE FOR RED SPRUCE IN THE NORTHEAST[*]
(McGlade and Demeritt)

D.b.h., in.	Total height, ft.						Basis, number of trees
	20	30	40	50	60	70	
	Volume, cu. ft.						
3	0.50	0.73	0.94	1.20			7
4	0.85	1.30	1.70	2.10	2.48		28
5	1.35	1.97	2.62	3.28	3.88		67
6	1.94	2.89	3.80	4.70	5.49		78
7	2.63	3.90	5.04	6.38	7.50	8.95	109
8		4.97	6.60	8.18	9.65	11.50	103
9		6.25	8.20	10.10	12.00	13.20	84
10		7.60	10.00	12.40	14.70	17.60	52
11		9.20	12.00	14.80	17.70	21.00	41
12			13.70	16.90	20.20	24.70	19
13			15.60	19.40	23.60	28.00	11
14			17.50	22.00	26.50	31.70	4
15				24.00	29.00	34.00	3
16				25.00	30.10	36.20	1
Basis, number of trees	21	39	194	217	113	23	607

[*] Data collected in New England by Northeastern Forest Experiment Station, Amherst, Mass. Heavy line indicates extent of original data. Volumes are for entire stem above a 1-ft. stump. Table prepared by alinement-chart method. Aggregate difference, table 0.47 per cent high. Average deviation ±6.29 per cent.

99. Cubic Volume Tables for Partial Utilization of the Stem. Frequently volume tables are required that show only the volume in the stem to a specified top diameter such as 6 or 4 in. The top diameter limit requirement depends upon the basis for which the timber is to be sold. For example, if the trees are to be sold for fuelwood and it is expected to utilize sticks as small as 4 in., this size will limit the merchantable volume of wood in the trees. Under these conditions, a volume table would be required that showed the volume to a 4-in. diameter limit and could be computed inside or outside bark as desired. If the products or material to be manufactured from the trees cannot be utilized economically in sizes smaller than 8 or 10 in., as in the case of veneer blocks, then this size becomes the one that limits the top diameter.

In other cases it may be possible to utilize small bolts or logs at the manufacturing plant, but because of large limbs or excessive taper and crooks in large trees it becomes impractical to cut logs from the tops. It will be found that in large timber the economical top diameter limit for many products increases with d.b.h. Thus for trees 8 in. at b.h., it may be economical to cut to a 6-in. top, while for trees 12 in. at b.h. it is not practical to cut to less than 8 in. In such cases, the top diameters of trees are averaged in the field data, and the volume table based thereon is said to be on a sliding, or variable, top diameter basis. The average top diameters should always be shown in the final volume table by d.b.h. class, if this variation exists.

The preparation of cubic-foot volume tables for partial utilization of the stem thus requires the computation of the volumes of individual trees to either a fixed or a sliding top diameter as desired in the final table. Any volume above the smallest merchantable point is not included. The process of constructing the table, if the method of harmonized curves is used (Sec. 84), is the same as before.

The height may be in terms of either total or merchantable height. In the case of merchantable height, the trees are grouped in classes to the nearest full or half-log length, *i.e.*, an actual merchantable length of 50 ft. will be called 3 logs, as will also a length of 46 ft., etc. As usual, all samples are first classified by d.b.h.-height classes and averaged before the harmonized curves are prepared.

The alinement-chart principle can also be used for partial stem volumes by methods similar to those outlined above for entire stem volume. The same cylinder charts could be used as before, but it is more practical to prepare a base chart for paraboloids of the same top diameter as is being used in the volume table. The preparation of this base chart requires the derivation of a formula for a frustum of a paraboloid in terms of the total height of the full paraboloid, when total tree height is the basis for the finished volume table.

The volume of a full paraboloid is

$$V = \frac{BH}{2} \qquad (1)$$

where V is cubic foot volume, B is basal area at the stump or base, and H is the height of the paraboloid minus a 1-ft. allowance for stump. The volume in the unmerchantable tip is

$$v = \frac{bh}{2} \qquad (2)$$

where v is the volume in the tip, b the basal area of the tip, and h is its length. Subtracting equation (2) from equation (1), we get

$$V - v = \frac{BH}{2} - \frac{bh}{2} \qquad (3)$$

Since a property of the paraboloid is that cross-sectional area varies directly with distance from the tip, in other words, $b/B = h/H$ or $h = bH/B$, equation (3) can be simplified by substitution to read

$$V - v = \frac{H(B^2 - b^2)}{2B} \qquad (4)$$

Expressed logarithmically for purposes of alinement-chart construction, equation (4) is

$$\log(V - v) = \log\frac{H}{2} + \log\frac{B^2 - b^2}{B}$$

The usual procedure for constructing a base chart is followed for any desired top diameter d, which would give a constant value to b and b^2.

References

BRUCE, DONALD: Alinement Charts in Forest Mensuration, *Jour. Forestry*, **17**, 773–801 (1919).

—— and L. H. REINEKE: Correlation Alinement Charts in Forest Research, *U.S. Dept. Agric. Tech. Bul.* 210, 1931, 87 pp.

CUMMINGS, L. J.: A Cubic-foot Volume Alignment Chart for Western Larch, *Jour. Forestry*, **35**, 415–417 (1937).

REINEKE, L. H., and DONALD BRUCE: An Alinement Chart Method for Preparing Forest-tree Volume Tables, *U.S. Dept. Agric. Tech. Bul.* 304, 1932, 28 pp.

CHAPTER 12

THE CONSTRUCTION OF STANDARD VOLUME TABLES FOR BOARD FEET

100. Board-foot Volume Tables—Basis. Standard volume tables in board feet are required for estimating timber that is to be sold as logs for manufacture into lumber. As in the case of other products, the volume table used in estimating must be made in accordance with the specifications under which the timber is to be cut and utilized. The table must be based on the same log rule that will be used in scaling these logs (Sec. 76). Otherwise, the contents of the trees by the volume table will not be comparable with the scale of the logs after the trees are felled. For example, sample trees are scaled by the International log rule for $\frac{1}{4}$-in. kerf and then made into a volume table. A tract containing small timber is estimated for board feet by this table. When the trees are actually sold, the logs are scaled by the Doyle rule. Other things being equal, the tract would scale in logs much less than indicated by the timber estimate.

In constructing standard volume tables for board feet, trees are measured in the woods, customarily in 8.15-ft. or 16.3-ft. lengths, and their tapers, d.b.h., stump height, bark thickness, total height, etc., tallied (Sec. 67). If merchantable heights are to be the standard basis for height classification, the trees must be classified and their volumes averaged in 16-ft. or 8-ft. merchantable height classes as desired.

The board-foot volume of each felled tree measured is determined by using the desired log rule. The tree tapers are taken at 8.15-ft. or at 16.3-ft. intervals and are tallied on a sheet in the conventional form. The log volumes are determined by scaling each log as 8 or 16 ft. long, using their top diameters inside bark and disregarding the 0.15- or 0.3-ft. trimming allowance.

As in the case of cubic-foot volume tables, the board-foot volume of from 300 to 2,000 trees give the basis for a board-foot table. The number of trees required depends on the extent of the region to which the table is expected to apply and on the range in sizes of the timber encountered in the field.

To construct a standard board-foot table by harmonized curves, the procedure is the same as that for cubic feet (Sec. 84).

101. Standard and Local Volume Tables Based on Butt Log Form Quotients—Girard Method. The purpose of volume tables is to supply an accurate average merchantable volume that will be cut from the tree in the ordinary processes of utilization. When timber is utilized for lumber, and logs are cut for this purpose from trees of different d.b.h. classes, the top diameter of the last merchantable log does not conform to a fixed standard of utilization such as 6, 8, or 12 in. Instead, the merchantable length varies from 60 per cent of total height for young, small, smooth trees to as low as 40 per cent for old timber with large limbs. It is affected by closeness of utilization, as for pulpwood, and by the market for knotty lumber. For the purpose of estimating the volume of saw logs for the production of lumber, the point on the bole which marks the limit of utilization is easily determined by eye. Tables based on a given standard of utilization and consequent upper diameter are inaccurate when smaller top diameters and a greater proportion of the bole are to be taken. The error in applying a table meant for a 3-log tree of 12-in. top to a "3-log" tree whose third log has only a 9-in. top and which should be a 2-log tree, according to table standards, is a serious overestimate of the actual tree volume; for in the table each of the three logs would have a larger diameter than that of the corresponding log in the tree tallied, a fact easily shown by diagram. The appraised tree is actually a 2-log tree plus the contents of an extra, or third, log with a 9-in. top. Conversely, when a table specifying tops as small as 6 or 5 in. is used for cruising to a 10- to 12-in. top, the estimate may be as low as 50 per cent of the actual volume.

It has been found that the tapers of trees per log for all species above the first, or butt, log approach uniformity within the same merchantable height and diameter classes. The form classes on which tree tapers are then based are determined by the ratio (decimal) which the d.i.b. at the top of the first 16-ft. log bears to the d.b.h. outside bark, provided only that the species does not normally have excessive butt swell extending above b.h. In the case of cypress, tupelo gum, and swamp ash, it is necessary to measure the lower diameter at a point 18 in. above the butt swell (called diameter above bottleneck, or d.b.n.), regardless of how close this point comes to the 16-ft. point above the stump, which 16-ft. point is the scaling diameter of the first log. When stumps are cut high, the 16-ft. point is correspondingly higher on the bole.

Form classes of the first log run from 0.65 to 0.90. A difference of 0.01 in form class is equal to 3 per cent difference in board-foot volume by the International ¼-in. rule. For this reason, standard volume tables for administrative use have been issued by the Forest Service for each form class from 0.65 to 0.90, and for three log rules, the International ¼-in., Scribner Decimal C, and Doyle, covering diameters from 10 to 40 in. and half-log length intervals from 1 to 6 logs, to the nearest board foot of contents. In constructing the standard tables for each form class, a set of standard tapers for the second and each succeeding log was adopted, as shown in Table 30. These tapers are held to apply to all trees when the d.b.h. and the number of 16-ft. logs are used as the basis for the tapers, but, as stated, they differ with each d.b.h. and each total merchantable length in logs. They therefore constitute the basis for a set of universal form class volume tables. Species which have identical form class, such as 0.80, are assumed to have the same tapers for each of the separate classes of merchantable length, whether 2, 3, 4, etc., logs. The basis of these tables is over 22,000 trees measured throughout the South and in Pennsylvania, and they are applicable throughout the United States.

For deductions to exclude defects, Table 31 is given, showing the distribution of total volumes of the tree within successive logs and indicating the amount by which the tree volume should be reduced for each defective log or portion of a log in estimating.

Since form is standardized for each d.b.h. and merchantable height class, the upper diameters of trees of any d.b.h. and form class, for each log, are found by subtracting the taper given in Table 30 from the d.b.h. inside bark, successively for each log, scaling the logs to 0.1 in. of diameter, and totaling the contents for the volume table. It will be noticed that the taper of the first log, which includes double width of bark, exceeds the d.i.b. taper of the second and succeeding logs except for the top log, or logs, which show increased tapers, thus conforming to the general shape of the paraboloid. Although averages of the form classes of common eastern species apply rather consistently, the cruiser should not depend on them but should determine on the ground the form class applicable to the timber to be estimated. For instance, southern pines, with average form class of about 78, may vary from 65 for small, branchy, old field pines to 83 for older trees growing in dense stands. The form class, and not the species, determines the table to use. It may happen, however, that average form departs from that derived by the average tapers given in Table 30. To test this possibility, about 50 trees should be measured for volume, distributed by

STANDARD VOLUME TABLES FOR BOARD FEET 163

		6-log tree				
	Fourth log	Fifth log	Sixth log			
	2.1	3.2	4.4			
	2.1	3.2	4.5			
	2.1	3.2	4.6			
	2.2	3.3	4.7			
	2.2	3.3	4.9			
	2.3	3.4	5.1			
	2.4	3.5	5.3			
	1.4					
	1.4					
	1.4					
	1.4					
	1.5					
	1.5					
	1.5					

[A second page with a "Conversion of Slope to Horiz. Dist." table is overlaid on the page, obscuring much of the original table.]

Conversion of Slope to Horiz. Dist.

Slope Distance in Links

	10	20	30	40	50	60	70	80	90	100
5	10	20	30	40	50	60	70	80	90	100
10	10	20	30	40	50	60	70	80	90	100
15	10	20	30	40	50	60	70	80	90	100
20	10	20	30	40	50	60	70	80	90	100
25	10	19	29	39	49	59	69	80	90	100
30	9	19	29	38	49	59	69	79	89	100
32	9	19	28	37	48	57	68	78	89	99
34	9	19	27	37	47	56	67	78	88	99
36	9	19	27	36	46	56	66	77	86	98
38	9	18	27	36	45	55	64	75	84	96
40	9	17	26	36	45	54	63	73	82	94
42	9	17	26	35	44	53	62	72	81	91
44	9	17	26	35	43	53	61	71	80	90
46	9	17	25	34	43	52	61	70	79	89
48	8	17	25	34	42	51	61	69	79	88
50	8	16	25	33	42	51	60	68	78	87
52	8	16	24	33	41	50	59	68	77	86
54	8	16	24	32	41	49	58	67	76	84
56	8	16	24	32	41	49	57	66	75	83
58	8	16	24	31	40	48	57	65	74	82
60	8	15	23	31	39	47	56	64	73	81
62	8	15	23	31	39	47	55	63	72	80
64	7	15	22	31	38	46	54	63	71	79
66	7	15	22	30	38	46	53	62	70	77
68					37	45	53	61	69	77
70						44	52	60	68	76
72							52	60	67	75
74									67	74

D.b.h., in.	2-log tree Second log									
10										
12										
14										
16										
18										
20										
22										
24										
26										
28										
40									2.3	

diameters and merchantable heights. For each aggregate difference of 3 per cent found by the check, this table should be shifted to that 0.01 class higher or lower than that for the determined form class, *viz.*, if the form class is 0.80 but the volume check gives a plus error of 3 per cent, that for 0.81 is used instead.

Tests made on 3,937 trees gave the following percentages of error for combined species by regions:

Region	Number of trees	Error, %
South	1,739	Negligible
Appalachian	1,203	−1.3
Northeast	493	+1.8
West	441	−0.2
Total	3,927	−0.83

Average form classes by species for the important softwoods and hardwoods fall between 0.76 and 0.84. If some of the hardwoods of the Mississippi delta of form class 0.76 to 0.77 are eliminated, also birch in the North with form class 0.84, southern old-growth pine with 0.84, and hemlock in the Lake states with 0.77, the remaining species fall within the narrower range of 0.78 to 0.82. For the same species, grow-

TABLE 31. AVERAGE DISTRIBUTION OF TREE VOLUME BY LOGS ACCORDING TO LOG POSITION

Usable length (number of 16-ft. logs)	Per cent of total tree volume in each log, by position					
	First	Second	Third	Fourth	Fifth	Sixth
1	100					
2	58	42				
3	42	33	25			
4	34	29	22	15		
5	29	25	21	15	10	
6	24	23	20	16	11	6

ing in different regions, the maximum difference between regions is 0.03 for red gum and 0.02 for any of the other 20 listed species.

102. Determination of Butt Log Form Class. This method of cruising is described by the Forest Service.[1] Proficiency in ocular

[1] MESAVAGE, C., and J. W. GIRARD, Tables for Estimating Board-foot Volume of Timber, U.S. Forest Service, 1946, 94 pp.

determination of form class is even more readily attained than that in judging diameters or log lengths. It depends, at first, on checking the form class after ocular estimate by actual measurement of d.b.h. and d.o.b. at 16 ft. from stump, deducting double bark thickness from the latter. To obtain the upper measurement a 16-ft. pole with crosspiece marked in inches colored alternately black and white, with the base of the pole held at stump height, provides a quick and convenient method. In judging upper diameter, sights should be taken on lines parallel to each edge of the tree by moving the eye and not from a fixed point. Bark allowance is learned from log scaling or volume measurements for the species and first log and varies with diameter. A substitute for the pole is climbing trees with irons or a light collapsible window cleaner's ladder and using calipers.

In order to train the eye, the form class itself should be estimated directly, without regard to the two diameter measurements, and then calculated by dividing the diameter inside bark at 16 ft. from stump by the d.b.h. outside bark, each to the nearest $\frac{1}{10}$ in., which gives the percentage form class figure. Proficiency can be obtained in a matter of days by men accustomed to ocular estimating of d.b.h. and height.

103. Alinement Chart Volume Tables for Board Feet. Base alinement charts have been constructed by the Forest Service for use in constructing board-foot volume tables. The procedure in volume-table construction with such a chart available is no different from the technique described for cubic feet (Chap. 11). Because the formula for board-foot volume of a geometric solid is very complex, it is difficult to express it in alinement-chart form. A fairly close approximation can be prepared, however. The process of preparation is so tedious and difficult, and its description is so cumbersome that it is omitted from this text.

The geometric solids used to construct board-foot alinement charts are paraboloids when total height is the basis, but cone frustums when merchantable height is used. The use of these solids makes the computation of board-foot volume easy, since paraboloids plot as straight lines on the Forest Service tree measurement form 558a and cones plot as straight lines on ordinary graph paper. Figure 29 shows one of the several base charts that are available. It deals with modified paraboloid volumes. It will be noted that the diameter and height axes are curved and that the graduating curves are no longer straight lines as in the case of the modified cylinder chart.

The procedure in regraduating this type of base alinement chart is no different from that already described for cubic feet.

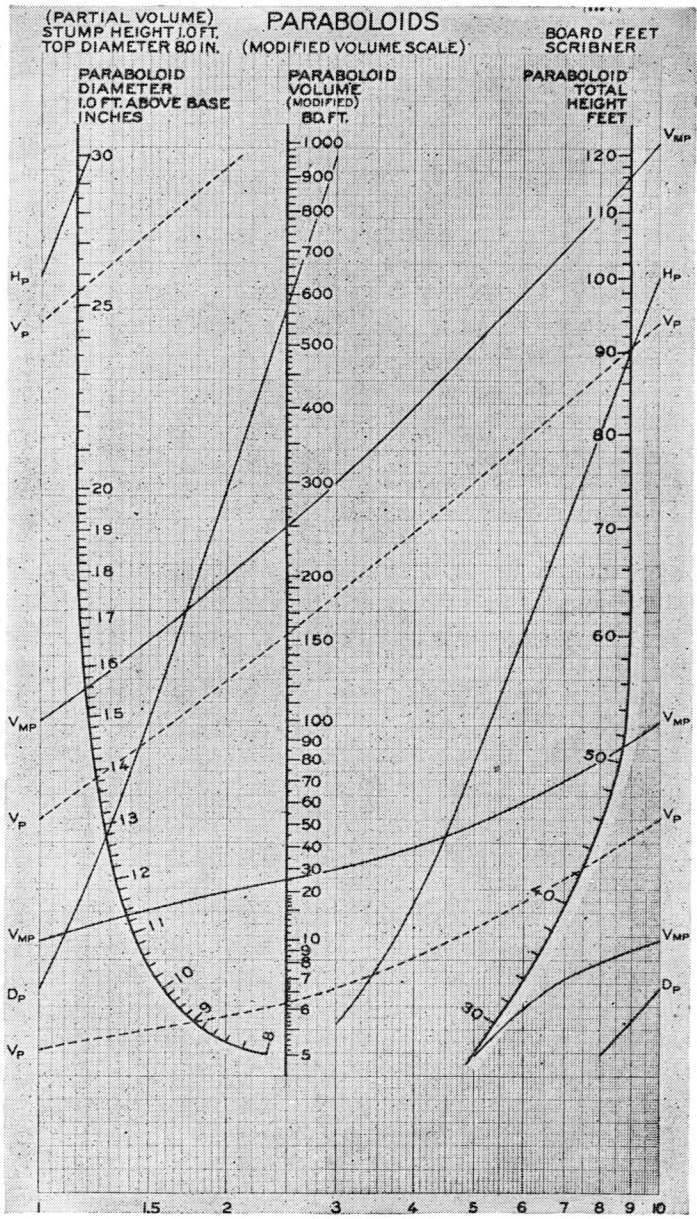

Fig. 29. Alinement chart volume table for board feet, Scribner log rule, for paraboloids. (*Courtesy U.S. Forest Service.*)

104. Data Which Should Accompany Volume Tables. For all standard volume tables, in addition to the volume for trees of each d.b.h. and height class, there should be shown:

1. Enclosed by heavy lines, the classes within which trees were actually found and measured (see Table 29).
2. The basis, or number of trees measured for each diameter class, on the right, and for each height class, below.
3. The region where the data were collected, and the class or age of the timber, as old growth or second growth, with age limits, if possible.
4. The unit of volume and the portion of the tree included, as total volume peeled or with bark, or merchantable volume; the standard height of stump; and the diameter limit or limits in the top.
5. Method of computing the tree volumes and of constructing the table.
6. Name of author and year.
7. Aggregate deviation of original data from volume of same trees as measured from the table, and average deviation or standard error of individual trees.

105. Adaptation and Conversion of Volume Tables. It is not possible to convert tables giving the contents by one log rule into those for another rule of a dissimilar character. If, however, the overrun of each rule is consistent for logs of all sizes, this can be done by finding the proper ratio. Log rules differing only in width of saw kerf can readily be converted from one to the other, and volume tables made on such log rules share this quality; *viz.*, to convert the International ⅛-in. rule to the ¼-in. rule, multiply by 0.905.

Volume tables expressed in terms of merchantable height may be converted to total height, and vice versa, if no change in top diameter utilized is involved. In either case, it will be necessary to measure the length of the unutilized top above the diameter limit. For conversion from merchantable to total height, these top lengths are averaged on the basis of merchantable heights and added to them. The volumes are then revealed as for trees of given odd total heights for each merchantable height class. The volumes of trees of the indicated height for each separate diameter class can then be plotted based on height, and from a curve drawn through these points the volume is read for the exact height class desired.

106. Checking the Accuracy of Standard Volume Tables for Board Feet. Volume tables cannot be applied without error to trees whose average form quotient and resultant form class is higher or lower than that of the trees used in the table, the error increasing with the amount of divergence in form. These deviations in average form find their expression in cubic- or board-foot volumes.

In checking such errors, the actual volumes of the trees measured for this purpose are totaled and compared with the volumes of the trees of the same diameter and height as taken from the volume table, preferably interpolated to $\frac{1}{10}$ in. in each case. The error is expressed in per cent of the table and may be used to modify the calculated total. The number of trees required to make these checks is a small fraction of that needed for the construction of a standard or even of a good local table.

At this point it is possible to utilize the aggregate difference and average deviation (Sec. 16c), if given in the volume table to be used.

To check the applicability of a table by use of the aggregate and average deviations, 15 to 20 sample trees should be felled, and tapers and other measurements obtained in the same way as for volume-table construction. Their volumes should be computed in exactly the same manner and to the same limits of utilization as used in the volume table itself. The d.b.h. and heights of the test trees should cover the range of diameters and heights in the stand to be estimated. The procedure in checking may best be explained by using an example.

Suppose 20 red spruce trees have been felled and their volumes computed to the same utilization limits as in the volume table to be checked (Table 29). Now list the d.b.h., height, and actual volume of the test trees, as in Table 32, and from the volume table interpolate their volume to the nearest hundredth cubic foot. List the tabular volumes as in column 4 of the table.

Compute the aggregate difference of the trees in the test sample by finding the difference between the sum of the actual and tabular volumes. Express this sum as a percentage of the total tabular volume. This percentage should be less than the value determined by the formula $2(a.d.)/\sqrt{n}$, where a.d. is the average, or mean, deviation given in the volume table being checked and n is the number of test trees. In the example (Table 32), the aggregate difference is actually too small to be tested further, but the calculation will be followed through for purposes of illustration.

$$\text{Aggregate difference of test trees from table} = \frac{287.31 - 287.33}{287.31}$$
$$\times 100 = -0.07 \text{ per cent.}$$

There are 20 trees in the test, and since the average deviation of the table is ± 6.29, the above formula gives

STANDARD VOLUME TABLES FOR BOARD FEET

$$\frac{2(6.29)}{\sqrt{20}} = \frac{12.58}{4.47} = \pm 2.81 \text{ per cent approximately.}$$

The aggregate difference of the test sample, −0.07 per cent, is therefore well below the value, 2.81 per cent, obtained by the formula.

TABLE 32. METHOD OF APPLYING CORRECTION TO A STANDARD VOLUME TABLE FOR LOCAL APPLICATION

D.b.h., in.	Total height, ft.	Volume, cu. ft.		Difference, cu. ft.	Deviation, %
		Actual	Tabular		
(1)	(2)	(3)	(4)	(5)	(6)
4.5	27	1.54	1.47	+0.07	+ 4.8
4.0	35	1.50	1.50	−0.10	− 6.7
6.3	38	3.88	3.97	−0.09	− 2.3
5.7	50	4.55	4.27	+0.28	+ 6.6
7.2	58	7.72	7.70	+0.02	+ 0.3
8.3	49	8.20	8.59	−0.39	− 4.5
8.3	54	8.23	9.40	−1.17	−12.4
9.6	56	13.07	12.76	+0.31	+ 2.4
9.5	44	10.83	9.96	+0.87	+ 8.7
9.0	49	8.93	9.91	−0.98	− 9.9
10.1	54	13.66	13.58	+0.08	+ 0.6
10.8	53	14.96	15.15	−0.19	− 1.3
10.9	64	19.02	18.70	+0.32	+ 1.7
11.5	46	14.74	14.65	+0.09	+ 0.6
12.7	55	21.40	20.61	+0.79	+ 3.8
13.0	61	25.04	24.04	+1.00	+ 4.2
14.0	56	22.92	24.70	−1.78	− 7.2
14.1	66	29.80	29.86	−0.06	− 0.2
15.7	60	31.18	29.77	+1.41	+ 4.7
15.5	55	26.26	26.72	−0.46	− 1.7
Total.....	..	286.33	287.31	+0.02	84.6

The average deviation of the test sample should next be found by taking the difference between each test tree volume and its table equivalent and expressing it as a percentage of the latter (column 6, Table 32). The sum of these percentages *regardless of sign* divided by the number of test trees gives the average deviation of the test trees. In the illustration this is 84.6/20 = 4.23 per cent. If this value is equal to or less than the average deviation of the table (as it is), then the table may be assumed to be applicable without correction to the stand in question.

If the checks show that either the average deviation or the aggregate difference exceeds the limit of acceptability, the table should be corrected. A good method of correction is to apply the same type of graph as used after making the first estimate when constructing the alinement chart volume table (Sec. 94). The test tree volume is plotted on tabular volume on double logarithmic paper, and a smooth curve is plotted through the points. The volume table may then be corrected by reading from the ordinate the curved test tree volume in terms of tabular volume. Figure 30 shows such a plotting for 14 red

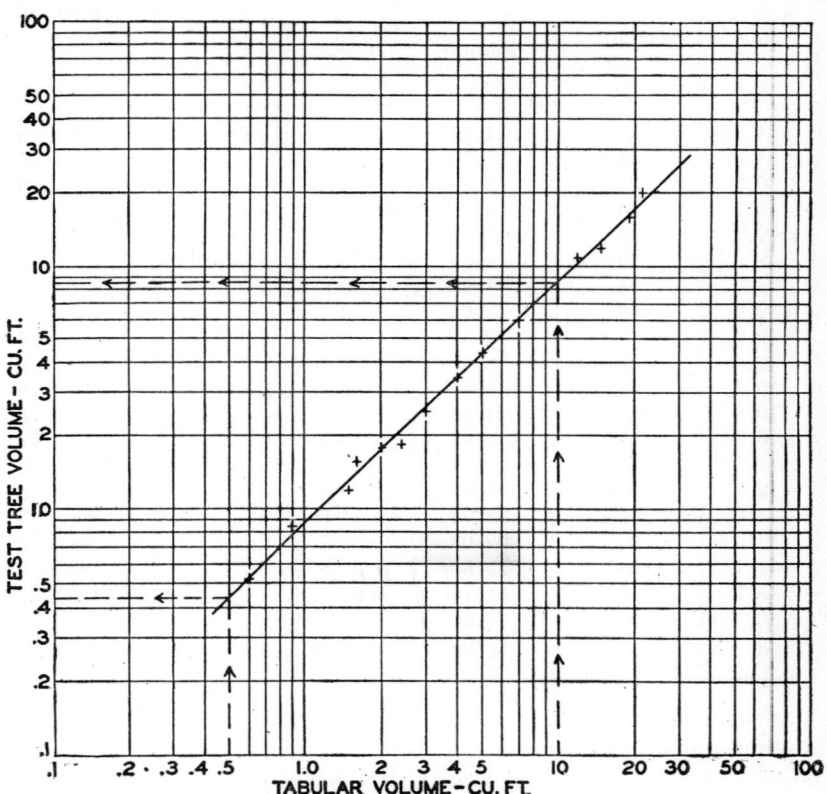

Fig. 30. Method of correcting a volume table.

spruce trees measured in a different stand from that used in the example above. Table 29 shows that the volume of a 3-in. tree 30 ft. tall is 0.50 cu. ft. To correct this value to apply to the test trees, go to 0.50 on the abscissa scale (Fig. 30), read from the curve 0.43 on the ordinate, and in the table replace the value of 0.50 by 0.43. If this is done for each volume in the table, the new table will fit the test sample.

Figure 30 also furnishes a second and quicker form of correction. The fitted straight line runs parallel to a 45-degree line running through the coordinates 1,1 and 10,10. This means that each actual volume is less than the tabular by the same percentage. For example, for the tabular volume 10, the actual volume is 8.6, or 14 per cent less. Likewise, for the tabular volume of 0.5, the actual volume is 0.43, or again 14 per cent less. In other words, the volume table can be corrected by applying a flat correction percentage of 14 per cent throughout, thus eliminating the reading of each individual value in the table.

This method of volume-table adjustment has been found useful to correct not only for varying form but also for variations in utilization practice, such as the scaling of long logs, instead of 16.3-ft. logs, differences in top utilization, and even to a certain extent allowing for breakage and average cull.

If the volume table were originally in alinement-chart form, the work of checking would be simplified, since interpolation would be unnecessary. In this case, the alinement chart could be modified or corrected on the volume axis in the same way as prior to the correction for second estimate (Sec. 94).

107. The Volume-diameter Ratio Method. When the volumes of logs or trees are plotted over diameter, a rising curve is formed suggestive of an exponential equation or one arm of a parabola. In the case of cubic-foot volumes, the exponential curve was commonly found to satisfy the relationship (Sec. 84). It fails, however, to satisfy the board-foot relationship, but the parabola in the form

$$V = c + aD + bD^2$$

is a definite possibility. In this equation, when D (diameter) is zero, V (volume) equals c. Since it is reasonable to assume that V should be zero when D is zero, one can conclude that c in the foregoing equation should also be equal to zero, thus shortening the form to

$$V = aD + bD^2.$$

Dividing this equation through by D, one gets $V/D = a + bD$, which plots as a straight line when V/D is plotted over D. In other words, if the V/D ratios are computed for a set of volume data and plotted over their respective diameters, and if the plottings are best fitted by a straight line, the proof is available that the curve of volume on diameter is a parabola. Tests of this relationship have been made with a large number of species[1] and the principle used for the construction of local and standard board-foot volume tables for Connecticut hardwoods.[2] A high degree of conformity was shown between theory and actuality.

Two general applications will be developed, viz., (a) for the local volume tables, and (b) for the standard volume table.

a. The local volume table. The tree volumes are sorted by 1- or 2-in.-diameter classes, and an average V/D ratio computed for each

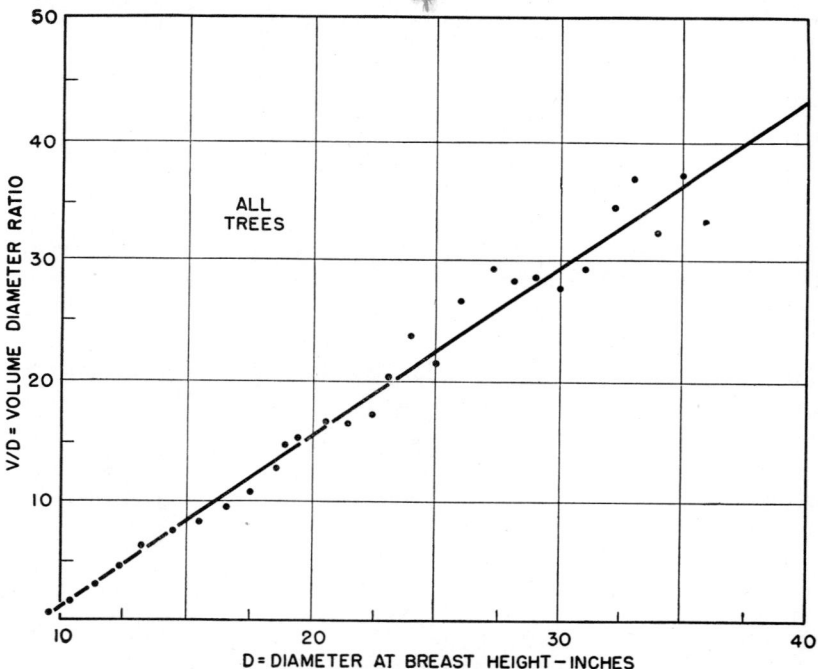

FIG. 31. Volume-diameter ratios for Connecticut hardwoods, plotted over d.b.h.

class. These are plotted over diameter. Figure 31 shows these ratios for 2,360 hardwoods of various species in Connecticut, scaled by International rule for ¼-in. kerf. A straight line obviously fits the points, and the equation $V/D = a + bD$ must hold. It remains to compute a and b.

[1] MEYER, WALTER H., Volume-diameter Ratios as a General Basis for Board-foot Volume Tables, *Jour. Forestry*, **43**, 49–53 (1945).

[2] MEYER, WALTER H., and RAYMOND KIENHOLZ, Volume Tables for Connecticut Hardwoods, *Yale Forest School Bul.* 54, 1944, 58 pp.

The coefficient b is the slope of the fitted line and can be found by taking two diameters such as 15 and 30 in., reading the curved V/D for each diameter, and then using the formula

$$b = \frac{V/D \text{ at } 30 - V/D \text{ at } 15}{30 - 15} = \frac{29.2 - 8.1}{15} = 1.41$$

The coefficient a represents the value of V/D when D equals zero and is called the *intercept*. It can be evaluated by extending the chart to include $D = 0$. It is often more convenient to compute it through the following expression:

$$a = \frac{V}{D} \text{ at chosen diameter} - b \times \text{chosen diameter}$$

Taking the diameter as 30 in. with its ratio of 29.4, this becomes

$$a = 29.2 - 1.41 \times 30 = -13.1$$

The completed equation for the local board-foot volume for Connecticut hardwoods is therefore

$$\frac{V}{D} = -13.1 + 1.41D$$

The volume for any d.b.h. is obtained in two steps: first, by inserting into the equation a given D and obtaining the corresponding V/D; second, by multiplying this ratio by D to get the volume; *viz.*, for a d.b.h. of 20 in., $V/D = -13.1 + 1.41 \times 20 = 15.1$ and

$$V = 15.1 \times 20 = 302 \text{ bd. ft.}$$

b. The standard volume table. When height is recognized as an independent variable in addition to d.b.h., the treatment is more involved. It consists basically in drawing up a volume-diameter ratio expression for each height class and then harmonizing the several expressions obtained.

The volume data are first sorted by d.b.h. and height class, as was done with the usual standard volume table (Sec. 84), and a volume-diameter ratio is computed for each class. These ratios are plotted over diameter, and the points within one height class are connected by a lightly drawn line. Figure 32 shows some of the plottings obtained in a study of the volume of eastern hemlock, scaled by the International log rule for $\frac{1}{4}$-in. kerf. A straight line fits each series of points within a height class. In successive height classes, the lines occupy higher levels and have as a rule progressively increasing slopes. Both the slopes and the spacing between lines must be harmonized.

The harmonizing of slopes is taken care of first. This is done by plotting the slope coefficients over height, as shown in Fig. 33. Again a straight line is formed, showing that slope is linearly related to

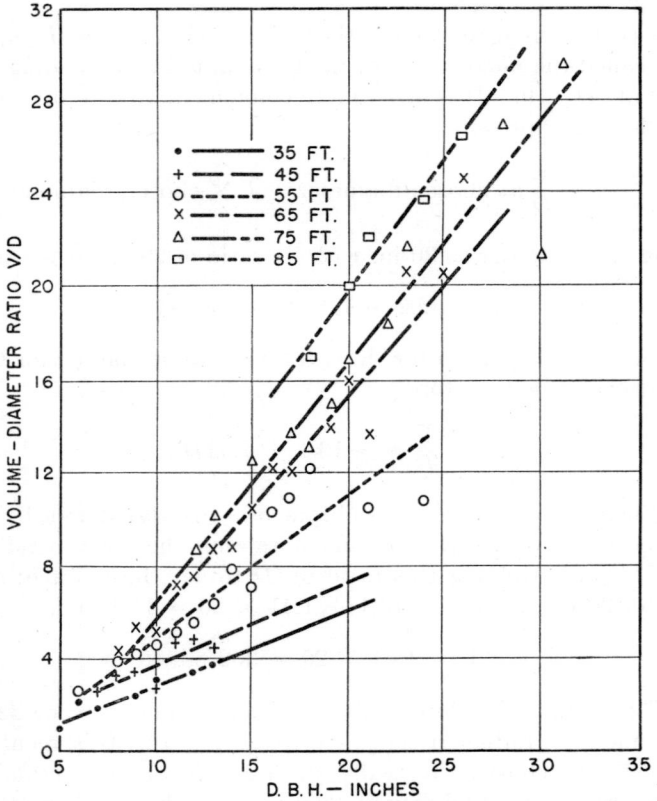

FIG. 32. Volume-diameter ratios for selected height class for eastern hemlock, plotted over d.b.h.

height and can be expressed by the equation $b = a' + b'H$, in which a' and b' can be determined, just as a and b were before. In this case, the following is obtained:

$$b = -0.444 + 0.0208H$$

The harmonizing of spacing takes advantage of the slope equation just obtained, by computing an *adjusted* V/D for a standard diameter class and plotting these adjusted ratios over height. The standard diameter class is one which is common to all height classes, in this case 15 in. The adjusted V/D ratio for 15 in. is obtained by the expression

$$\text{Adjusted } \frac{V}{D} = \text{average } \frac{V}{D} \text{ for the height class}$$
$$- b(\text{average } D \text{ for the height class} - 15)$$

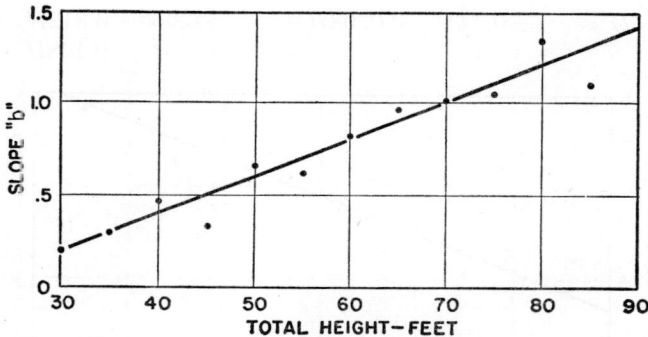

FIG. 33. Slope coefficients b for trend of volume-diameter ratio on d.b.h. for eastern hemlock, plotted over total height.

The average V/D for the 60-ft. height class is 8.48, and the average d.b.h. is 13.7 in., as determined from the original data. Inserting these values into the equation,

$$\text{Adjusted } \frac{V}{D} = 8.48 - b(13.7 - 15)$$

But b must be found from its equation

$$b = -0.444 + 0.0208H = -0.444 + 0.0208 \times 60 = +0.804$$

Then

$$\text{Adjusted } \frac{V}{D} = 8.48 - 0.804(-1.3) = 9.52$$

This is done for all height classes. The adjusted ratios for each height class are plotted over height as in Fig. 34, and when curved they give the spacing of the curves for the standard diameter of 15 in. The points can be smoothed out by a straight line and an equation of the form adjusted $V/D = a'' + b''H$ computed. In this case, it is adjusted $V/D = -2.82 + 0.204H$.

Returning to the basic equation $V/D = a + bD$, which applies separately to each height class and in which b was shown to be related linearly to height, one can now show that a is also linearly related to height. The equation is transformed to the form $a = V/D - bD$. Substituting in this expression the equations previously derived for the adjusted V/D and for b when $D = 15$, one gets

$$\begin{aligned}
\text{Adjusted } \frac{V}{D} & H = -2.82 + 0.204H \\
-bD = -(-0.444 + 0.0208H)15 &= +6.66 - 0.312H \\
= a &= +3.84 - 0.108H
\end{aligned}$$

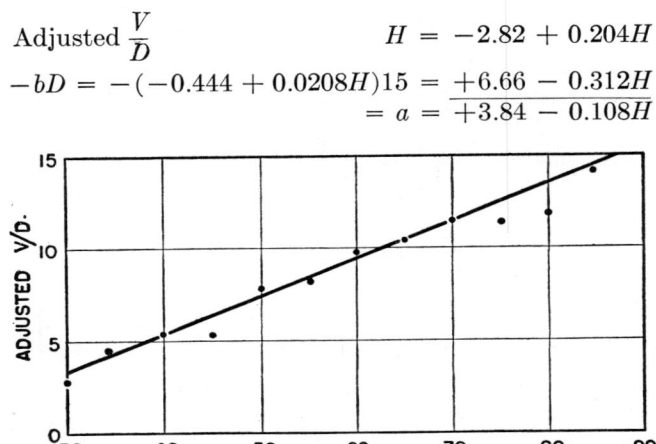

FIG. 34. Adjusted volume-diameter ratios (basis = 15 in.) for eastern hemlock, plotted over total height.

All the information is now in hand for the computation of the standard volume table itself:
1. The general expression $V/D = a + bD$.
2. The slope expression $b = -0.444 + 0.0208H$.
3. The intercept expression $a = +3.84 - 0.108H$.

The first calculation is to substitute a given height value in the last two expressions and get computed values for b and a. These computed coefficients are then inserted in the first equation, following which various D's are assumed to get the V/D for each diameter class. Finally, each V/D is multiplied by its respective D to get the board-foot volume. For a 13-in. tree, 70 ft. tall, the calculation is

$$\begin{aligned}
b &= -0.444 + 0.0208 \times 70 = +1.012 \\
a &= 3.84 - 0.108 \times 70 = -3.72 \\
\frac{V}{D} &= -3.72 + 1.012 \times 13 = 9.436
\end{aligned}$$

and finally
$$V = 9.436 \times 13 = 113 \text{ bd ft.}$$

If the linearity of the plottings for V/D on diameter by height class, for slope coefficient b on height, and for adjusted V/D or intercept a on height is valid, as tested by the above graphical methods, then a substitute method is available, through which a single equation can be derived by the least-squares method. This equation reads

$$\frac{V}{D} = a + bD + cDH + dH$$

wherein a and b are no longer the previous a and b but are arrived at independently in the solution. This solution has the advantage that the subjectiveness of the graphical method is eliminated and that correlation coefficients and standard errors of estimate can be determined directly. In a test of 359 loblolly pine stem analyses, the following were the four normal equations required for the solution:

a	+	bD	+	cDH	+	dH	=	V/D
359		6,725		83,570		4,394		8,583
6,725		130,143		1,639,424		83,570		168,323
83,570		1,639,424		21,464,282		1,082,986		2,186,687
4,394		83,570		1,082,986		56,294		110,289

In the preparation of these numerical values, height (H) has been coded as follows: 1-log trees given the value of 3, 1½-log trees that of 5, 2-log trees that of 7, and so forth. By following the methods of Sec. 89, the following specific equation results:

$$\frac{V}{D} = -0.654 + 0.332D + 0.085DH - 0.126H$$

with a correlation coefficient of 0.933 and a standard error of estimate of $2.77 V/D$ units.

References

BAILEY, I. W., and P. C. HEALD: Graded Volume Tables for Vermont Hardwoods, *Forestry Quart.*, **12**, 5–26 (1914).

BARRETT, L. I.: Recent Volume Tables for Some Southern Appalachian Species, *Appelachian Forest Expt. Sta. Tech. Note* 19, 2 pp. + 46 tables.

BEHRE, C. E.: Factors Involved in the Application of Form-class Volume Tables, *Jour. Agric. Res.*, **51**, 669–713 (1935).

BELYEA, H. C., and O. M. PORTER: Foresters' Tables for New York State, *N.Y. State Col. Forestry Bul.* 14, 1923, 83 pp.

BLYTHE, R. H., JR.: A Simplified Method of Constructing Merchantable Board-foot Volume Tables, *Jour. Agric. Res.*, **55**, 159–173 (1937).

BRANIFF, E. A.: Grades and Amount of Lumber Sawed from Yellow Poplar, Yellow Birch, Sugar Maple, and Beech, *U.S. Dept. Agric. Forest Serv. Bul.* 73, 1906, 30 pp.

BRUCE, DONALD: Further Notes on Frustum Form Factor Volume Tables, *Soc. Amer. Foresters Proc.*, **10**, 315–321 (1915).
———: The Height and Diameter Basis for Volume Tables, *Jour. Forestry*, **18**, 549–557 (1920).
———: A Proposed Standardization of the Checking of Volume Tables, *Jour. Forestry*, **18**, 544–548 (1920).
———: The Use of Frustum Form Factors in Constructing Volume Tables, *Soc. Amer. Foresters Proc.*, **8**, 278–288 (1913).
——— and F. X. SCHUMACHER: Revised Volume Tables for Second Growth Redwood, *Jour. Forestry*, **23**, 148–155 (1925).
Canadian Forest Service, Department of Interior: Form Class Volume Tables for Balsam, Jack Pine, Red Pine, Lodgepole Pine, White Pine, Black, White, Red Spruce, Ottawa, 1930, 200 pp.
CHANDLER, B. A.: A Study of the Frustum Form Factors of Hard Maple and Yellow Birch, *Vt. Agric. Expt. Sta. Bul.* 210, 1918, 38 pp.
CHAPMAN, H. H.: The Factor of Top Diameters in Construction and Application of Volume Tables Based on Log Lengths, *Soc. Amer. Foresters Proc.*, **11**, 221–225 (1916).
CLARK, J. F.: Volume Tables and the Bases on Which They May Be Built, *Forestry Quart.*, **1**, 6–11 (1902).
DEMERITT, D. B., and A. C. MCINTYRE: A Simple Method of Constructing Tree Volume Tables, *Jour. Agric. Res.*, **44**, 529–539 (1932).
DWIGHT, T. W.: The Percent Deviation Method of Constructing Volume Tables, *Forestry Chron.*, **13**, 409–416 (1937).
———: Refinements of Plotting and Harmonizing Freehand Curves, *Forestry Chron.*, **13**, 357–370 (1937).
GIRARD, J. W.: Volume Tables for Mississippi Bottomland Hardwoods and Southern Pines, *Jour. Forestry*, **31**, 34–41 (1933).
GRAVES, H. S.: Volume Tables, *Forestry Quart.*, **3**, 227–244 (1905).
HALLIN, W. E.: Volume and Taper Tables for Old-growth Redwood, California Forest and Range Experiment Station, 1941, 79 pp.
HAWLEY, R. C., and R. G. WHEATON: Studies of Connecticut Hardwoods; the Form of Hardwoods and Volume Tables on a Form Quotient Basis, *Yale Forest School Bul.* 17, 1926, 41 pp.
HORNIBROOK, E. M.: Scribner Volume Tables for Cutover Stands of Ponderosa Pine in Arizona, *Jour. Agric. Res.*, **52**, 961–974 (1936).
KORSTIAN, CLARENCE F.: The Use of Frustum Form Factors in Constructing Volume Tables for Western Yellow Pine in the Southwest, *Soc. Amer. Foresters Proc.*, **10**, 301–314 (1915).
LENTZ, G. H.: Top Diameter Utilization Limits for Delta Hardwoods, *Jour. Forestry*, **31**, 547–550 (1933).
MACKINNEY, A. L., and L. E. CHAIKEN: Volume Tables for Second Growth Loblolly Pine in the Middle Atlantic States, *Appalachian Forest Expt. Sta. Tech. Note* 21, 1936, 16 pp.
MAUGHAN, W. A.: Cubic Volume Table for Eastern Red Cedar, *Jour. Forestry*, **34**, 777–778 (1936).
MESAVAGE, C., and J. W. GIRARD: Tables for Estimating Board-foot Volume of Timber, U.S. Forest Service, 1946, 94 pp.
MEYER, W. H.: A Method of Volume-diameter Ratios for Board-foot Volume Tables, *Jour. Forestry*, **42**, 185–189 (1944).

—— and R. KIENHOLZ: Volume Tables for Connecticut Hardwoods, *Yale Forest School Bul.* 54, 1944, 58 pp.

Minnesota Agricultural Experiment Station: Tables for Determining Contents of Standing Timber in Minnesota, Michigan, and Wisconsin, *Tech. Bul.* 39, 1926, 99 pp.

MULLOY, G. A., and H. W. BEALL: A Comparison of Several Methods of Compiling Volume Tables, *Jour. Forestry*, **35**, 932–941 (1937).

MUNGER, T. T.: The Problem of Making Volume Tables for Use on the National Forests, *Jour. Forestry*, **15**, 574–586 (1917).

MUNNS, E. N., and R. M. BROWN: Volume Tables for the Important Timber Trees of the United States: Part I, Western Species, 159 pp.; Part II, Eastern Conifers, 146 pp.; Part III, Eastern Hardwoods, 104 pp., U.S. Department of Agriculture, 1925.

——, T. G. HOERNER, and V. A. CLEMENTS: Converting Factors and Tables of Equivalents Used in Forestry, *U.S. Dept. Agric. Misc. Pub.* 225, 1948, 48 pp.

REINEKE, L. H.: Discussion of Demeritt and McIntyre's Method of Constructing Tree Volume Tables, *Jour. Forestry*, **33**, 412–418 (1935).

REYNOLDS, R. R.: Volume Tables and Fixed Top Diameters, *Jour. Forestry*, **32**, 29–31 (1934).

ROTHACHER, J. S.: Percentage Distribution of Tree Volume by Logs, *Jour. Forestry*, **46**, 115–118 (1948).

WINTERS, R. K.: Use of Fixed Top Diameters in Volume Table Construction for Bottomland Hardwood Species, *Jour. Forestry*, **31**, 427–429 (1933).

CHAPTER 13

THE CONSTRUCTION AND APPLICATION OF TAPER TABLES

108. Application of Taper Tables for Different Form Classes to Trees of Different Species. The form of a tree is a natural adaptation to meet the mechanical stresses caused by wind and gravity and to resist breakage. Tree forms are created by the formation of wood in the annual rings and are altered by the laying on of relatively greater or lesser amounts of wood at different points on the bole and limbs. The form of the bole is in part a response to wind pressure exerted on the surface of the crown, bole, limbs, and leaves, and in part a provision for securing equal cross-sectional area of sapwood at all points below the green crown. If the crown surface extends to a low point on the tree, more wood, relatively, will be laid on below this point than above it. This results in a bole approaching the form of a paraboloid. The form class of the paraboloid is 0.5. A tree with a small top requires more nearly the same amount of sapwood at all points below this crown and will fall between the paraboloid and the cylinder, with a form class of about 0.7 or more. If a long-crowned tree loses its lower crown, a change in form in this direction will be induced.

Stump taper, or butt swell, of trees is in large part an adaptation against the pressure of wind in bending the tree at its base. When trees are freed from the surrounding stand by logging and are left as seed trees, the rate of growth at the stump is stimulated more than that at higher points on the bole, but the bole itself tends to approach, rather than depart from, the average of the species, as has been shown for red spruce and ponderosa pine. On leaning trees most of the growth is placed on the side toward which the tree leans, even to producing "flatiron" cross sections, as in redwoods. On limbs, the growth is largely on the lower side.

Based on these premises, it has been held that within a single form class the relative diameters at differing relative heights are consistent and that therefore, for any given form class, tree forms and tree volumes once found will apply generally to all species. Butt swell and

TAPER TABLES

bark thickness at the base, however, have been found to be complicating factors. Since form quotient is based on the ratio

$$(d \text{ at } \tfrac{1}{2}H)/\text{d.b.h.}$$

variable butt swell and bark thickness between species as it affects d.b.h. outside bark would cause variation in form factor, even though the trees had the same shape in the parts of the bole above their influence. The d.b.h. of the tree is increased beyond the normal size that would be in harmony with the form of the remainder of the bole. Since d.b.h. outside bark must continue to be used as the standard measurement of size (Sec. 60), the use of taper tables based on normal forms and common to several species will always require corrections for bark thickness and butt swell for specific application.

The measurement of butt swell at b.h. is made possible through the plotting of taper measurements of individual trees. The trend of the natural curve of the bole is plotted and extended to the ground. The normal diameter in inches at b.h. may be read from this curve and the difference from the actual d.b.h. determined. These differences can then be averaged by d.b.h. class and put into a table. Table 33 is of

TABLE 33. RELATION OF NORMAL DIAMETER TO ACTUAL D.B.H., CONNECTICUT HARDWOODS

D.b.h. outside bark, in.	Normal d.i.b., in.	Sum of butt swell and bark thickness, in.	Double bark thickness, in.	Butt swell, in.
2	1.9	0.1	0.1	0.0
4	3.6	0.4	0.3	0.1
6	5.4	0.6	0.5	0.1
8	7.0	1.0	0.8	0.2
10	8.7	1.3	1.0	0.3
12	10.4	1.6	1.1	0.5
14	12.2	1.8	1.2	0.6
16	14.0	2.0	1.3	0.7
18	15.8	2.2	1.3	0.9
20	17.6	2.4	1.4	1.0
22	19.4	2.6	1.5	1.1
24	21.3	2.7	1.6	1.1
26	23.1	2.9	1.6	1.3
28	24.9	3.1	1.7	1.4

this kind, applying to many species of hardwoods growing in the region surrounding New Haven, Conn. When the concept of normal d.b.h. inside bark and diameter at one-half-height of the tree above b.h. is introduced into the form-class formula, it has been shown that surprising agreement between the volumes of different species is obtained, provided that the comparison is made within the single form-class unit.

109. Construction of Tables for Percentile Tapers. If the necessity for correction by reduction to normal diameter is accepted and there exist no differences or departures from normal form sufficiently serious to reduce the accuracy of resultant volume below a required standard, tables for taper for each diameter, height, and form class would give the basis for volume tables for all species and all products. The tapers would be needed at any point on the bole required by the length of the products used, *viz.*, 8½ or 9 ft. for ties, and 12 or 16 ft., with trimming allowance, for saw logs. Such a set of tables would be quite bulky, but if form is similar for each form class, as is contended, diameter can be expressed in per cent of d.b.h. for each point on the bole expressed in per cent of total height. These are called *percentile tapers*. Several formulas have been recommended as a general expression of this curve, among which the hyperbolic equation[1] $y = x/(a + bx)$ appears to be the simplest and most generally applicable. In this equation, y is the ratio of diameter at any point of a tree to *normal* d.b.h., and x is the percentage of height from the tip to 4.5 ft. above ground, at which the diameter in question is located.

Percentile taper tables for average hardwoods in Connecticut are shown in Table 34. To apply such a table, the form class or classes are first determined by measurements on down trees or trees felled for the purpose. The number of trees measured can be much smaller than is required for the construction of a standard volume table. The normal d.i.b. is found for each felled tree, the d.i.b. at one-half-height above b.h., and finally the form class as a ratio of these two. The various estimates may be averaged by d.b.h. or species as a whole as is most suitable. The normal diameter is now taken as 100 per cent, and the diameters at each tenth of the stem length above b.h. are found by multiplying this diameter by the percentages of Table 34 in the proper form-class column.

These diameters are plotted on graph paper, using actual inches for diameter and feet for height instead of the percentages, and the points are connected. The diameters at any desired points can then be interpolated graphically. From these diameters and lengths it is possible to obtain volume tables for ties, mine timbers of different grades and sizes, saw logs, etc.

[1] BEHRE, C. E., Form-class Taper Curves and Volume Tables and Their Application, *Jour. Agric. Res.*, **35**, 673–744 (1927).

When an average form class is used as the basis for such a derived table, it is assumed that the yield of piece products, such as ties, from trees which will exceed this form will compensate for loss of products in trees of smaller form quotient. Where products range in grade down to fairly small dimensions, there is no serious error in this assumption, but for larger products the losses from trees or bolts falling below the minimum dimensions and thus rejected would exceed the gains in larger bolts on the average. In such cases, a straight average of the contents of all trees of given dimension classes in a volume table (Sec. 83), regardless of form, would come nearer the actual output.

When desired taper tables are lacking for the estimation of the volume of piece products such as mine timbers, posts, or crossties, approximate curves of tree form can be constructed, based on the form quotient for the species for each d.b.h. and required average height (Sec. 61).

TABLE 34. PERCENTILE TAPER TABLE FOR DIFFERENT FORM CLASSES. AVERAGE HARDWOODS, CONNECTICUT

Percentage of height from tip to b.h.	Form class								
	0.40	0.45	0.50	0.55	0.60	0.65	0.70	0.75	0.80	0.85
	Percentage of d.i.b. at b.h.								
10	7.3	8.4	9.7	10.7	11.9	13.0	14.0	15.7	17.7	20.4
20	14.4	16.8	19.3	21.6	23.9	26.2	28.7	32.2	36.2	41.6
30	22.0	25.6	29.1	32.6	36.0	39.5	43.7	48.6	54.4	62.2
40	30.2	34.5	38.9	43.7	48.2	52.9	58.0	63.5	69.7	76.6
50	40.0	45.0	50.0	55.0	60.0	65.0	70.0	75.0	80.0	85.0
60	52.5	57.8	62.4	66.9	71.1	74.7	77.9	81.6	85.1	89.0
70	65.4	69.8	73.6	76.7	79.8	82.1	84.2	86.7	89.2	91.8
80	77.9	80.8	83.2	85.1	86.8	88.2	89.6	91.3	92.8	94.7
90	89.7	90.9	91.8	92.9	93.5	94.3	94.9	95.6	96.4	97.3
100	100.0	100.0	100.0	100.0	100.0	100.0	100.0	100.0	100.0	100.0
Basis in number of trees..........	6	19	70	147	254	355	249	103	20	5

References

BAKER, F. S.: The Construction of Taper Curves, *Jour. Agric. Res.*, **30**, 609–624 (1925).

———: Taper Curves in Relation to Linear Products, *Soc. Amer. Foresters Proc.*, **9**, 380–387 (1914).

BARROWS, W. B.: The Construction of a Set of Taper Curves, *Soc. Amer. Foresters Proc.*, **10**, 32–40 (1915).

BEHRE, C. E.: Chart for Application of Percentile Taper Curves to Trees of Any Size Class, *Jour. Forestry*, **24**, 272–274 (1926).
———: Factors Involved in the Application of Form-class Volume Tables, *Jour. Agric. Res.*, **51**, 669–713 (1935).
———: Form-class Taper Curves and Volume Tables and Their Application, *Jour. Agric. Res.*, **35**, 673–744 (1927).
———: Preliminary Notes on Studies of Tree Form, *Jour. Forestry*, **21**, 507–511 (1923).
———: Is Taper Based on Form Quotient Independent of Species and Size? *Jour. Forestry*, **22**, 282–290 (1924).
BOWER, RAY F.: The Mathematical Expression of Tree Taper in Volume Table Construction, *Jour. Forestry*, **33**, 426–431 (1935).
FROTHINGHAM, S. H.: The Eastern Hemlock, *U.S. Dept. Agric. Bul.* 152, 1915, 43 pp.
JONSON, TOR: Methods and Aids in Tree Form Investigations and in the Calculation of Volume, Yield of Various Products, and Growth, *Internatl. Cong. Plant Sci. Proc.*, **1**, 729–739 (1926).
REINEKE, L. H.: A Test of Taper Tables, *Jour. Forestry*, **23**, 945–947 (1925).
SCHUMACHER, F. X.: A Method of Measuring Form Quotient of Standing Trees, *Jour. Forestry*, **24**, 552–554 (1926).
WICKENDEN, H. R.: The Jonson Absolute Form Quotient: How It Is Used in Timber Estimating, *Jour. Forestry*, **19**, 584–593 (1921).
———: New Devices for Solving Some Problems in Timber Mensurations, *Jour. Forestry*, **21**, 260–265 (1923).
WRIGHT, W. G.: Investigation of Taper as a Factor in Measurement of Standing Timber, *Jour. Forestry*, **21**, 569–581 (1923).
———: Taper as a Factor in the Measurement of Standing Timber, *Canada Dept. Int. Bul.* 79, 1927, 132 pp.

CHAPTER 14

THE MEASUREMENT OF DIAMETERS OF STANDING TREES

110. Tree Diameters—Tree Calipers. When dependence is placed on standard or local tree volume tables already constructed from felled trees, the task of determining the volume of standing trees becomes one of recording the diameters and heights of the trees to be measured, in predetermined classes, and multiplying the number of trees in each class by the volume from the table, with such deductions as are required in allowance for defects and cull.

Before foresters introduced the tree volume table, trees and stands were either estimated directly, by ocular judgment, or measured by estimating the number of logs and their average contents, expressed as log run (Sec. 145). These methods were not dependent on d.b.h., while heights were recorded as number of 16-ft. logs in the average tree, *viz.*, as 3-log or 4-log timber. These cruisers thus used the log rule, not the tree volume table, to measure the volume of standing trees.

It is not necessary to *record* the d.b.h. of standing trees to the nearest 0.1 in., except on permanent sample plots, where individual records are to be kept of tree growth. On temporary, *i.e.*, once measured, plots used in the preparation of yield tables or for check cruising, two diameter measurements must be taken at right angles, to 0.1 in., and the average calculated, even if it is recorded only in the diameter class in which the average falls. In ordinary timber cruising, tree diameters can be measured once and taken as they come. In this event, the 0.1 graduation indicates the class within which the tree falls, as 9.6 to 10.5 in. for the 10-in. class in 1-in. class graduations.

For either purpose, tree calipers can be employed which should be read to the nearest tenth inch. Standard calipers consist of a graduated bar with one fixed arm at the zero point set at right angles to the bar, and a second sliding arm, which is adjustable by a screw. When pressed against a tree, the sliding arm forms a right angle with the bar and is parallel with the fixed arm (Fig. 35).

When out of adjustment, the spread between the tips of the arms is greater than the diameter indicated on the bar. The reading will then be less than the true diameter. Conversely, if the caliper arms are not held firmly against the tree, the spread at the base of the arms may exceed the required distance, resulting in a reading greater than the true diameter. Calipers may be used conveniently for timber under 36 in. in diameter, but for greater sizes they are clumsy and impractical. Calipers are indispensable in training the eye to estimate diameters accurately (Sec. 152).

The average diameters at any point on felled trees or logs may be determined to the best advantage with the tree calipers. To obtain the two measurements at right angles as a basis for recording the true average to the nearest tenth inch, each measurement is taken at an

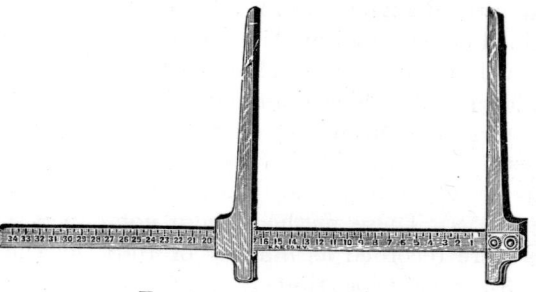

FIG. 35. Tree calipers.

angle of 45 degrees with the perpendicular. Since it is usually impossible to slide the caliper arms far enough on the log so that the crossbar touches the log, special care must be exercised to keep them in perfect adjustment.

In measuring standing trees for timber estimating, the height of 4½ ft. should first be marked on one's clothing in some convenient manner. The tree diameter is always measured at this height. Failure to observe strictly the height of 4½ ft. may result in high measurements through tilting up the points of the caliper arms. A far more common fault is lowering the calipers through lack of attention. Either practice results in appreciable errors in volume (Sec. 113) due to the difference in taper and the exaggeration of the error in diameter when converted to its D^2 equivalent in volume tables.

111. The Diameter Tape. Steel tapes graduated to read 1 in. for each 3.14159 in. of tape length may be used to read tree diameters. The graduations are based on the assumption that the measurement required is the diameter of a circle. Tapes are graduated in inches and tenths and may be obtained in lengths from 6 ft. up (Fig. 36).

Readings taken by the diameter tape are not subject to the variations incurred in using the tree calipers. These variations are caused by eccentric shapes of boles which give different diameters at different points on the same cross section. Whatever the shape, the circumference reading should not vary. For this reason, the diameter tape is often used instead of calipers for scientific records on permanent sample plots to be remeasured at stated intervals of time. It also finds a use in timber estimating for checking the diameter of trees too large to be measured conveniently by calipers.

Fig. 36. Diameter tape for measuring circumference of trees in terms of d.b h. On the reverse side, 1 in. in tree diameter measures 3.1416 standard inches on the tape. thus converting circumference into diameter.

The diameter tape cannot be used to measure circumference on logs and felled trees except at the exposed ends of the logs. The tree calipers are preferable for this purpose.

112. The Biltmore Stick. The Biltmore stick is a straight stick graduated in such a manner that the diameters of standing trees may be read from it when held horizontally against the trunk at b.h. and tangent to the bole of the tree. It is more convenient than the calipers but less accurate. It is widely substituted for calipers in timber cruising but is never used in scientific measurements.

In Fig. 37, the line EC represents the position of the stick, with the end C held on the line of sight AB' between the eye and the outer edge

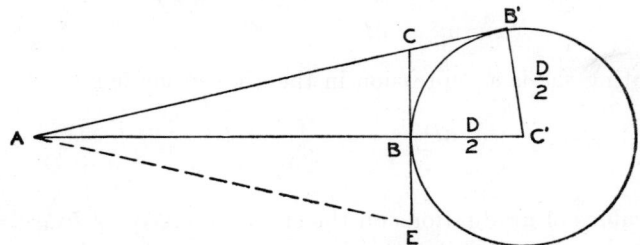

Fig. 37. Theory of the Biltmore stick.

of the bole. The radius of the tree, $B'C'$, to the same edge is not parallel to BC. It is farther off; therefore, the distance $2BC$ which should be graduated on the stick to correspond to the tree diameter, is less than $2B'C'$. The larger the tree the greater is the angle $AB'C'$ and the greater the difference between $B'C'$ and the corresponding distance on the stick. This difference becomes so great that for 65-in. trees a stick less than 36 in. long is required for diameter measurement.

The proper stick graduations for each diameter depend also on the length of the arm reach, represented in the figure by AB.

Let $AB = a$, or arm reach in inches

$2B'C' = D$, or diameter of any tree

Then

$$\text{Scale graduation} = \sqrt{\frac{aD^2}{a+D}}$$

The derivation of this formula is as follows:
In the right triangles ABC and $AB'C'$, the following proportion holds:

$$\frac{AB}{BC} = \frac{AB'}{B'C'} \quad \text{or} \quad \frac{AB \times B'C'}{AB'} = BC$$

Since $AB = a$ and $B'C' = D/2$, this is simplified to

$$BC = \frac{a(D/2)}{AB'} \quad \text{or} \quad 2BC = CE = \frac{aD}{AB'}$$

Now in the right triangle $AB'C'$

$$AB'^2 = AC'^2 - B'C'^2$$
$$= \left(a + \frac{D}{2}\right)^2 - \left(\frac{D}{2}\right)^2$$
$$= a^2 + aD + \left(\frac{D}{2}\right)^2 - \left(\frac{D}{2}\right)^2$$
$$= a^2 + aD = a(a+D)$$

Substituting the last expression in the above, one has

$$CE = \frac{aD}{\sqrt{a(a+D)}} = \sqrt{\frac{a^2D^2}{a(a+D)}} = \sqrt{\frac{aD^2}{a+D}}$$

The values of graduations on the stick are obtained by substituting desired values of a and D in this formula. Scale values for D are given in Table 35 for arm reaches of 23 to 27 in.

In determining the length of arm reach, the arm should be held straight in front of the body, without stretching or throwing the shoulder forward. The stick should have a beveled edge to bring the graduated scale flush with the surface of the tree at the point of contact.

TABLE 35. FIGURES TO BE USED IN GRADUATING A BILTMORE STICK*

Diameter of tree, in.	Distance from eye to tree, in.				
	23	24	25	26	27
	Distance to be marked on stick, in.				
3	2.82	2.83	2.83	2.84	2.85
5	4.53	4.55	4.57	4.58	4.59
7	6.13	6.16	6.19	6.21	6.24
9	7.63	7.68	7.72	7.76	7.79
11	9.05	9.11	9.17	9.22	9.27
13	10.39	10.47	10.54	10.61	10.68
15	11.67	11.77	11.86	11.94	12.03
17	12.89	13.01	13.12	13.22	13.32
19	14.06	14.19	14.32	14.44	14.56
21	15.18	15.34	15.48	15.62	15.75
23	16.26	16.44	16.60	16.75	16.90
25	17.31	17.50	17.68	17.85	18.01
27	18.31	18.52	18.72	18.91	19.09
29	19.29	19.51	19.73	19.94	20.14
31	20.23	20.48	20.71	20.94	21.15
33	21.15	21.41	21.67	21.91	22.14
35	22.04	22.32	22.59	22.85	23.10
37	22.91	23.21	23.50	23.77	24.03
39	23.75	24.07	24.37	24.67	24.94
41	24.58	24.91	25.23	25.54	25.84
43	25.38	25.74	26.07	26.40	26.71
45	26.17	26.54	26.89	27.23	27.56
47	26.94	27.33	27.70	28.05	28.39
49	27.69	28.10	28.48	28.85	29.21
51	28.43	28.85	29.25	29.64	30.01
53	29.16	29.59	30.01	30.41	30.79
55	29.87	30.31	30.75	31.16	31.56
57	30.56	31.03	31.47	31.90	32.32
59	31.25	31.73	32.19	32.63	33.06
61	31.92	32.41	32.89	33.35	33.79
63	32.58	33.09	33.58	34.05	34.51
65	33.23	33.75	34.26	34.74	35.21

* BARROWS, W. B., *Jour. Forestry*, **16**, 747 (1918).

113. Errors of Diameter Instrumentation. a. Use of calipers. The case of the calipers will be taken up first. Common errors with calipers are as follows:

1. Movable arm too loose, leading easily to an underestimate of diameter by 0.1 to 0.5 in., with the error increasing with the size of the

tree. It is difficult to get an overestimate with the calipers, since if the movable arm is tightened too much, it will not slide on the graduated beam. The error is therefore systematic, negative, and significant. It is easily controlled by repeated checking of the calipers, particularly if they are old and worn.

2. Calipers placed too low on the tree. The error depends upon the taper of the butt and the amount to which the caliper is dropped, which may easily be 0.5 ft. The calipers are rarely placed too high on the tree. Diameters are overestimated by 0.1 to 0.4 in. The error is systematic, positive, and significant. It is corrected through the use of a stick exactly 4.5 ft. in length placed against the tree until the estimator gets into good habits.

3. Calipers placed at an angle not perpendicular to the axis of the tree. The error depends on the angle of departure from the horizontal, the thickness of the caliper arm, and the taper of the tree. With taper eliminated, the size of the error is shown in Table 36.

TABLE 36. ERROR IN DIAMETER MEASUREMENT DUE TO PLACING CALIPER NOT PERPENDICULAR TO TREE BOLE

Angle of departure from the horizontal, degrees	Errors due to slant alone, % of diameter	Errors due to thickness of caliper arm, in.
2	0.06	0.026
4	0.24	0.052
6	0.55	0.079
8	0.98	0.105
10	1.54	0.132
12	2.24	0.159
14	3.06	0.187

For example, with a 10-in. tree, an error of 4 degrees would mean an error in diameter of $10 \times 0.0024 + 0.052 = 0.076$ in. With an angle of 10 degrees, it would be $10 \times 0.0154 - 0.132 = 0.286$ in. For a 20-in. tree, the errors would be respectively 0.100 and 0.442 in. With ordinary care it should be possible to keep within a 2-degree error, in which case the error in diameter is held within 0.1 in. The error is systematic, positive, but not alarming, and easily controllable.

4. Elliptic cross sections. Two classes of error are here involved. The first occurs when only one diameter is taken on the tree. The error can run up to several inches or more, depending upon the degree of ellipticity. It tends to be compensating, but its nature is extremely

erratic. Almost full correction is obtained by measuring two diameters on the tree, usually the largest and smallest diameters, and taking the average. This practice leads to a minor error, which is systematic, always positive, and increases with the differences between the long and short radius, but within one such difference class the error decreases with increasing diameter. In most cases the error will be insignificant, since a plus 0.1-in. error is incurred only when the difference in the two diameter measurements exceeds 2 in. for trees 8 in. in long diameter, 4 in. for trees 24 in. in long diameter, and 6 in. for trees 48 in. in long diameter.

b. Use of diameter tape. Some of the errors incurred in instrumentation with the diameter tape are of the same class as those incurred with the calipers; others are somewhat different.

1. The error incurred by placing the tape too low on the tree is the same kind as with the calipers. It is systematic, positive, and significant.

2. The error of wrapping the tape around the tree at a slant is equal to the amount given in column 2 of Table 36. The error is systematic, positive, and significant only if the angle from departure from the horizontal is over 2 degrees.

3. In the case of elliptical cross sections, the error is only insignificantly different from that of the calipers when the long and short diameters are averaged.

4. Corresponding to the error of placing the calipers loosely on a tree is the error of not wrapping the tape tightly around the tree. This is easily controllable, but if not controlled is systematic, sizable, and positive.

5. Protuberances and bark irregularities tend to make the tape give readings which are consistently higher than those of the calipers. A comparative test in Douglas fir[1] showed that the diameter tape measurements resulted in an overestimate of about 1.5 per cent in basal area, as compared with calipers. The comparative error of tape from caliper measurement is therefore systematic and positive but not striking. The tape has a definite advantage in being more consistent between repeat measurements and between operators, a great advantage in research. It is also a much more compact instrument and can be used to measure large trees outside the scope of calipers.

c. Use of the Biltmore stick. The Biltmore stick is the most erratic of the diameter-measuring devices and is subject to serious abuse.

---

[1] McArdle, Richard E., Relative Accuracy of Calipers and Diameter Tape in Measuring Douglas Fir Trees, *Jour. Forestry*, **26**, 338–343 (1928).

Causes of errors in its use are completely different from those of errors with the tape and calipers.

1. **Stick held short.** This is a common error in practice. The nature of the error can be deduced from Table 35. If, for example, a stick is graduated for 25 in., but the operator has the bad habit of holding it only 24 in. from the eye, it will be seen that the error increases rapidly with increase in diameter. The error can be expressed in terms of the percentage number of trees in a tally which will be upscaled by 2 in., since the stick provides for only a 2-in. classification and does not permit reading to the nearest 0.1 in. of diameter. The upscaling is as follows:

D.b.h., in.	Number upscaled, % of total
9–11	3.5
19–21	11.7
29–31	24.2
39–41	36.5
49–51	51.3

The error is systematic, positive, and serious, especially in diameters 20 in. and over. It is easily controllable.

2. The effect of ellipticity of the trunk section is difficult to reduce to simple terms. If the stick is held against the flat side of a tree, a gross overestimate of diameter occurs; if held against the narrow side, a gross underestimate. If the two readings are averaged, the mean will be slightly larger than that obtained by tape or calipers. For a specific case the following occurs:

Assuming a tree with a 30-in. diameter in one direction and a 20-in. diameter at right angles to the first. Measurement on the flat side gives a Biltmore reading of 34.5 in., on the narrow side of 17 in. The average of the two is 25.8 in. Tape and calipers will give 25 in., while the true diameter of a circle of equal cross section is only 24.5 in.

Assuming a less extreme case, *viz.*, a tree 20 in. in diameter in one direction and 17 in. in the other, the Biltmore gives readings of 20.5 and 16.5 in., respectively, averaging 18.5 in. Tape and calipers give also 18.5 in., and the true diameter is 18.45 in.

The technique of averaging Biltmore readings is therefore sound for moderate degrees of ellipticity and medium-sized trees since the error is considerably less than 0.1 in., but with extreme ellipticity and large trees it can be substantial. The error is systematic and positive, but only very slightly so, except for extreme conditions.

3. **Eye not horizontal with the stick, but above it.** This gives read-

ings that are slightly large because of the taper of the tree and the introduction of an artificial ellipticity, whose effect, however, is negligible.

It is seen from the above discussion of the calipers, tape, and Biltmore stick that all the errors of instrumentation with the exception of one are positive in character, that they are all systematic in nature, and that all of them can be controlled by good practice.

References

BEHRE, C. E.: Comparison of Diameter Tape and Caliper Measurements In Second Growth Spruce, *Jour. Forestry*, **24**, 178–182 (1926).

BRUCE, DONALD: The Biltmore Stick, and the Point of Diameter Measurements, *Soc. Amer. Foresters Proc.*, **11**, 226–229 (1916).

———: A New Dendrometer, *Univ. Calif. Pubs.*, **3** (4), 55–61 (1917). Review, *Jour. Forestry*, **16**, 724 (1918).

———: Notes on the Biltmore Stick, *Soc. Amer. Foresters Proc.*, **9**, 46–58 (1914).

BURTON, R. G.: The Biltmore Pachymeter, *Forestry Quart.*, **4**, 8–9 (1906).

CLARK, J. F.: A New Dendrometer or Timber Scale, *Forestry Quart.*, **11**, 467–469 (1913).

FERREE, M. J.: The Pole Caliper, *Jour. Forestry*, **44**, 594–595 (1946).

JACKSON, A. G.: The Biltmore Stick and Its Use on National Forests, *Forestry Quart.*, **9**, 406–411 (1911).

KRAUCH, H.: Comparison of Tape and Caliper Measurements, *Jour. Forestry*, **22**, 537–539 (1924).

McARDLE, R. E.: Relative Accuracy of Calipers and Diameter Tape in Measuring Douglas Fir Trees, *Jour. Forestry*, **26**, 338–342 (1928).

McCARTHY, E. F., and H. KRAUCH: Comparison of Tape and Caliper Measurements, *Jour. Forestry*, **22**, 537–539 (1924).

MERRILL, F. G.: The B-10 Tree Stick, *Jour. Forestry*, **43**, 814–816 (1945).

ROBERTSON, W. M.: Review of the Case of Diameter Tape vs. Calipers, *Jour. Forestry*, **26**, 343–346 (1928).

SCHERER, N. W.: Relative Accuracy of Calipers and Steel Tape, *Soc. Amer. Foresters Proc.*, **9**, 102–106 (1914).

CHAPTER 15

THE MEASUREMENT OF HEIGHTS OF STANDING TREES

114. The Determination of Heights of Trees. Whenever standard tree volume tables are used in estimating the volume of standing timber, it is necessary to determine the heights of trees or their merchantable lengths as well as their diameter, since the tables give volume for each diameter and height class.

The height of a small tree can be measured directly with a graduated pole and that of a large tree by climbing the tree with a tape attached to a light pole to reach the tip. This direct measurement is reliable and may be required for scientific studies of height growth or on permanent sample plots. On the whole it is too time-consuming for practical application. Hence height is commonly estimated by indirect means, usually involving the functions of a triangle or two similar triangles. If the observer stands at a measured distance from the tree, lines of sight from the eye to any two points on its bole, such as the base and tip, form two sides of a triangle, which is completed by the axis of the tree.

Several principles of triangle relationship are used in the construction of *hypsometers*, which are instruments for measuring height, as follows:

1. The tangents of the angles above and below the horizontal multiplied by the horizontal distance from the tree, as with the Abney hand level (Sec. 115); or the ratio of legs of similar right triangles.

2. The ratio of the legs of two similar isosceles triangles, as with the staff hypsometer (Sec. 116).

3. The ratio of the legs of two similar triangles, not isosceles (Sec. 117).

In all procedures, it is assumed that the tree axis is perpendicular to the earth's surface, but since this is not always the case because trees commonly lean in one direction or another, a special technique general to all must be employed to get the best estimate of tree height. With a leaning tree, the line of plumb bob held to intersect the tip of the tree does not follow the axis of the tree but intersects the ground at some distance from the tree. If it were possible to suspend the

TABLE 37. CORRECTION OF MEASURED HEIGHTS OF LEANING TREES TO OBTAIN TRUE HEIGHT*

Slope of tree, %	Correction to obtain true height	Slope of tree, %	Correction to obtain true height
5	1.00125	22	1.02392
8	1.00319	24	1.02838
10	1.00499	26	1.03324
12	1.00719	28	1.03845
14	1.00974	30	1.04404
16	1.01272	32	1.04995
18	1.01607	34	1.05621
20	1.01979	36	1.06283

* Derived from FALCONER, J. G., A Method of Accurate Height Measurement for Forest Trees, *Jour. Forestry*, **29**, 746–749 (1931).

plumb bob directly from the tip and locate its point of intersection on the ground, a good but slightly short estimate of height would be obtained by measurement with a hypsometer from any point on a circle with the point of intersection as a center. Figure 38 shows the effect of ignoring this precaution when the instrument is located on the side either toward or away from the lean. If the lean is away from the observer, the height will be underestimated; if toward the observer, overestimated, with the degree of the lean and the distance from the tree base being the controlling factors (Sec. 121). If the observer is at right angles to the lean, only a small underestimate will be incurred, which will not need correction if the slope of the tree from the vertical is less than 10 per cent. Table 37 gives the factors which can be used for any amount of slope.

In scientific measurement of tree heights on permanent sample plots, errors in measurement may cause serious discrepancies, and it may be necessary to adopt a rigorous practice. This is done by locating first the point directly below the tip, then measuring the vertical height to the tip from the ground, and finally correcting by means of Table 37. The point on the ground below the tip can be located precisely if the observer holds a Biltmore stick or compass staff vertically to pass through the tip and sights in his assistant along the vertical plane. He then moves approximately 90 degrees, again raises his stick, and has his assistant shift along the first vertical plane until he intersects the second vertical plane. When this point is established, the tree height can be measured from any point on a circle with this point as the center. The finally corrected estimate of height is found as follows:

196   FOREST MENSURATION

The height of the tip above ground is measured as 90 ft. The distance of the point directly under the tip from the center of the bole is 9 ft. This gives a lean of 9 ft in 90, or a 10 per cent slope from the vertical. Opposite 10 per cent in Table 37 is the factor of 1.00499, which when multiplied by 90 gives 90.449, rounded off to 90.5 as the actual tree height. The correction factor, incidentally, is the secant of the angle of lean.

Wide or rounded crowns present a similar difficulty, which is corrected if the observer strikes the line of sight through the crown directly to the actual tip, and not over the apparent edge of the crown. Measuring such trees from a long distance, such as 100 ft., is helpful.

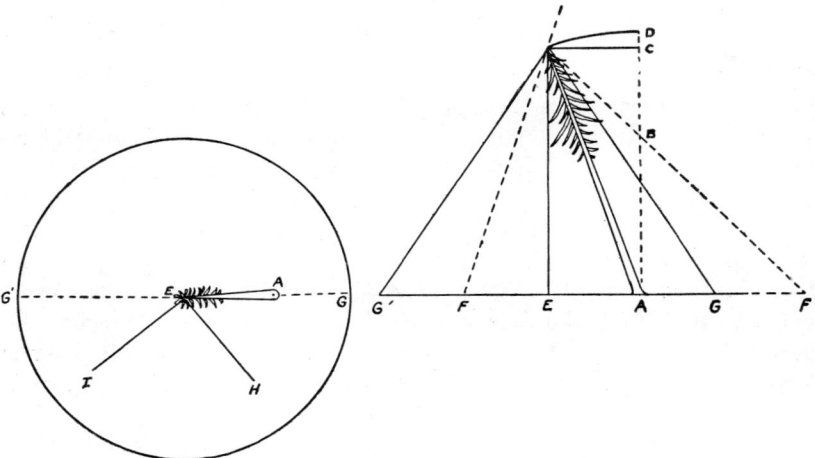

FIG. 38. Measurement of height of a leaning tree. If distance from tree is measured from its base, A, the height of tree when the instrument is sighted on the tip will be read as AB from position F'. From position F, in direction of G', the height read would be too high. From a position at right angles with AF the error approximates CD.

If, on the other hand, the point E is determined by intersection of two vertical planes passing through the tip, as EH and EI, the height AC can be measured from any point on the circle whose radius is EG or EG', incurring the small error CD which can be corrected in scientific measurements by use of Table 37.

115. Height Determination by Tangents of Angles—The Abney Hand Level—Forest Service and Faustmann Hypsometers. In this method, the observer places himself at a known horizontal distance from the tree and by means of an instrument (Abney hand level or other device) measures the angle to the tip of the tree and the angle to its base. A horizontal line to the tree forms a common leg of two adjacent triangles, whose apexes are at the observer's eye (Fig. 39). The tangent of the angle of the upper right triangle times the horizontal distance gives the length of the tree above a point on the bole located at the same level as the observer's eye, and the tangent of the

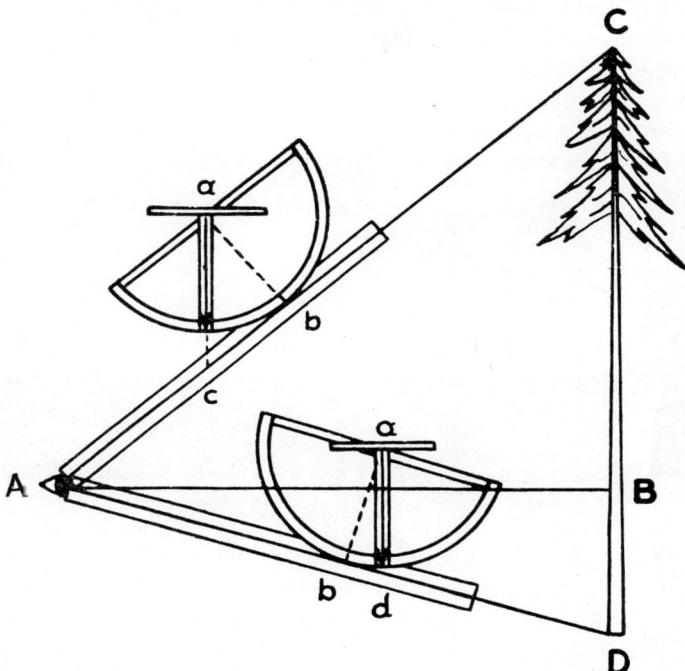

Fig. 39. Use of Abney level for height measurement. Pointer (*ac* and *ad*) is used with level bubble (Abney level) to obtain total height of tree *CD* by adding the measurements *BC* and *BD*. The scale of distance is laid off on the hypsometer as *ab*, and in the Abney level it is 100. The scale of height is read on the lines *bc*, corresponding to *BC*, and *bd*, corresponding to *BD*. In the Abney level these scales are placed on the corresponding graduations of the arc instead. The distance, *AB*, is measured horizontally to the tree.

angle of the lower triangle times the horizontal distance gives the length below this point. The sum of the two lengths gives the total height of the tree. It is essential that both angle readings be taken, because measuring the angle to the top alone can lead to a serious discrepancy. If the base of the tree is above the observer's eye, the second reading of length must be subtracted to get the net height of the tree.

The Abney hand level, when its arc is graduated in percentages, gives readings directly in tree heights at a distance of 100 ft. It can also be equipped with a "topographic" arc, which gives the rise in feet for each chain, or 66 ft., of horizontal distance. The distance can be shortened to some simple fraction of 100 ft., in which case the height is equal to the sum of the readings times the fraction. For example, if the distance is 50 ft. (½ of 100 ft.), and the upper angle is 55 per cent and the lower angle 9 per cent, the height of the tree is computed as ½ (55 + 9), or 32 ft. The Abney hand level is the instrument that is most commonly used in the United States for height determination, owing to its ruggedness, easy adjustment, simple operation, and high

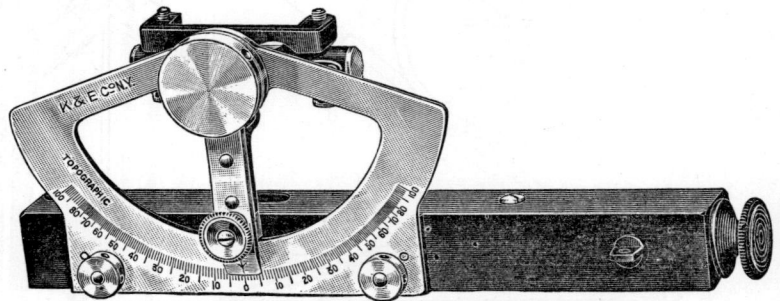

Fig. 40. "Topographic" Abney level.

accuracy. It is advisable not to use the instrument at angles greater than 100 per cent. Since this is the case, the instrument man should move back to a distance at which the reading will be less than 100 per cent. Thus, in the Douglas fir region, heights must often be measured from a distance of 200 ft. or more. In construction the Abney hand level consists of a square peep tube mounted with an arc and a level bubble (Fig. 40). The tube has a peep sight at one end, and at the other end, which is open, is a horizontal metal hairline. A graduated arc is attached firmly to one side of the tube, and from the center of its circle a movable arm is placed on which the level bubble is mounted. An index arrow on the lower end of the arm gives the reading of slope on the arc. The level bubble is reflected by a prism in the tube, on which a horizontal line is etched, so that the bubble can be brought in line with the hairline in the objective. When the instrument is held horizontally, the level bubble should coincide with the hairline, and the index arrow should point at 0. When the instrument is tilted up or down, the arm is swung until the bubble is brought in line with the hairline. The arrow then gives a reading of the slope.

If the instrument gets out of adjustment, there are three ways of correcting the defect. The ends of the level bubble can be raised or lowered by screws. The arm can be shifted slightly forward or backward. The hairline and prism, which are both mounted on a sleeve within the peep tube, can be shifted forward or backward. It is of course necessary that the instrument be kept in true adjustment constantly.

The *Forest Service hypsometer* is similar in principle to the Abney hand level, although quite different in appearance, being a round boxlike affair with a floating arc, whose 0 graduation is kept on the horizontal by weighting of the arc. It is graduated for direct readings at 100 ft.

Both the Abney hand level and the Forest Service hypsometer are simple compact instruments and give direct readings of tangents of angles. With both the theory conforms to that of similar right triangles, wherein the small triangle is formed within the instrument itself. These two instruments conform to the general rule of using similar triangles, which is applicable to all hypsometers. In this case, these are right triangles. Similar right triangles are the basis of an easily constructed homemade hypsometer, which consists of a rectangular piece of wood or heavy cardboard ruled off with scales representing various distances from the tree (Fig. 41). The upper edge of the board is used as the sighting line and near its farther end has a point of attachment for a light plumb bob. Separate scales are then graduated for different distances from the tree to show the height of the tree above and below the horizontal. In use the board is held to the eye and the top of the tree sighted along the top edge. The plumb bob, which should hang free, is clamped into position by the thumb, and the height is read directly on the scale corresponding to the distance used. The same is done to the base of the tree, and the two readings are added. The *Faustmann hypsometer* is a refinement of this idea, in which the fulcrum of the plumb bob can be raised or lowered to accommodate different distances, and the readings on the scale can be seen in a reflecting strip mirror; hence the instrument need not be taken from the eye in order to get a reading.

With all the above instruments the distance should be taped for accurate work, although, in timber estimating, with experienced men and with one-man crews, pacing may be used.

116. Height Determination by Similar Isosceles Triangles—Staff Hypsometer. In this method, the distance from the tree is not standardized nor is it determined before the tree is measured. After the observation is taken, measurement to the tree along the ground from the observer, regardless of slope, equals the total tree height. The simplicity of the principle permits it to be used with only the simplest equipment yet with good precision. One of the equal legs of the smaller isosceles triangles is formed by the observer's arm, the other by a staff held vertically. The corresponding legs of the large triangle are the distance from the observer to the tree and the height of the tree.

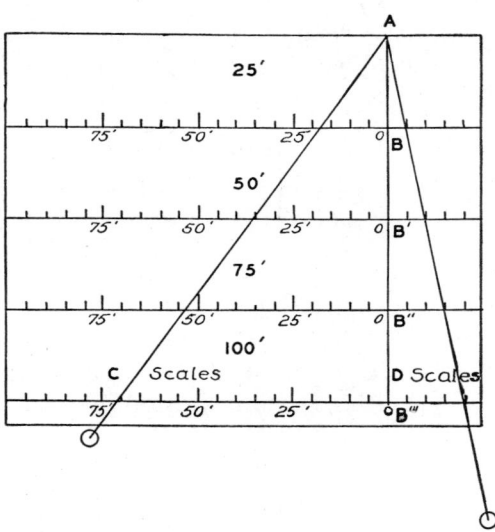

FIG. 41. Cardboard hypsometer. The principle is identical with that of the Abney level. The plumb bob is substituted for the level bubble and pointer. The distance scales AB, AB', etc., from tree are shown as 25, 50, 75, or 100 ft. The height above eye corresponding to the distance used is shown to left and that below eye to right of the axis AB'''.

Figure 42 demonstrates the principle. Here bc is made equal to ac; and, since bc is parallel to BC (the tree) and aC coincides in position with ac, aC must equal BC.

The *staff hypsometer* (Fig. 42) is the simplest of the instruments based on this principle. A straight staff, for example, a Jacob's staff, is chosen which is long enough to span the distance between the outstretched hand, which holds the staff, and the eye, and still have additional length below the hand to help in plumbing the staff. The lower end can be weighted to assist in plumbing. The thumb and finger hold the staff at a pivot point, the distance of which from the top of the staff is equal to the distance from the eye to the staff, so held. In measuring and permanently marking this distance on the staff, it must be held horizontally with tip to the eye and the shoulders square. In measuring tree heights, with the staff held at this distance, the line of sight between the eye and the mark on the staff must hit the base of the tree; the staff must be vertical; the line of sight over the top of the staff, with the arm held fixed and the head not tilted, is then observed. If it falls below the treetop, the observer steps backward slowly (and if high, forward) until the two lines of sight intersect the top and base of the tree from the same point. This is the point at which the similarity of two isosceles triangles is obtained. The height of the tree is then determined by measuring the distance from the observer to the tree as the ground side of the triangle. Proof is found in the following proportion:

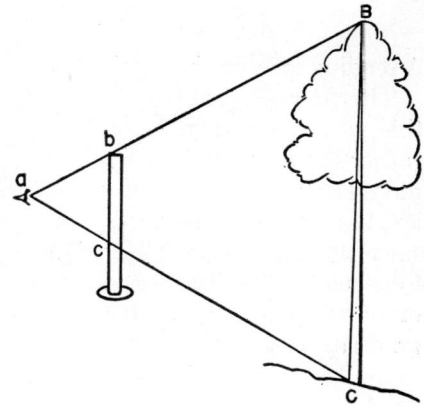

FIG. 42. The staff hypsometer. The distance bc on staff is first made equal to distance ac from eye to point grasped at lower end of bc. Then by finding point on ground where upper and lower lines of sight intersect tip and base of tree, the distance aC to the tree is made equal to BC or height of tree and when measured gives this tree height.

$$ac : aC = bc : BC$$

because of the similarity of the triangles. But ac is deliberately made equal to bc. Hence aC must equal BC.

Proper allowance must be made when the base of the tree is concealed by undergrowth or high weeds. In this case, it is often advisable to set a stick of known length at the base of the tree. Then sight at the top of the stick for the base point and later add the length of the stick to the observed height. This precaution is valid for any other instrument.

117. Height Determination by the Ratio of Legs of Two Similar Triangles, Not Isosceles—Merritt Hypsometer—Christen Hypsometer. The similar triangles need not be isosceles but may be of any other form, provided that one leg is kept vertical or parallel to the tree axis. Referring again to Fig. 42, assume that ac is no longer equal to bc but that the horizontal distance from a to bc is fixed and the stick is graduated to coincide with a fixed horizontal distance to the tree. The triangles abc and aBC are still similar; hence the proportion holds that

$$\frac{ac}{aC} = \frac{bc}{BC}$$

A reading of bc on the vertical staff will then indicate the length of BC.

The *Merritt hypsometer* (Fig. 43) is a simple instrument based on this principle. This instrument is used for estimating log lengths, rather than for total height, and is ordinarily combined with the Biltmore stick. Equal distances on the bole of a tree, such as 16-ft. log lengths, intersect equal distances on the vertical, or measuring, arm, since there are two similar triangles with corresponding sides. In this hypsometer the distance to the tree is usually standardized at 1 chain, or 66 ft., but any desired distance can be used, provided that the graduations on the vertical stick are marked appropriately. The length of reach is commonly taken as 25 in. The stick is graduated in intervals corresponding to reach and log lengths in the tree, as follows:

$$\frac{\text{Reach}}{\text{Distance from tree}} = \frac{\text{graduation interval}}{\text{log length}}$$

For example, if reach is 25 in., distance from tree 66 ft., and log length 16.3 ft.,

the graduating interval

$$= \frac{25 \times 16.3}{66} = 6.174 \text{ in.}$$

$$= 0.514 \text{ ft.}$$

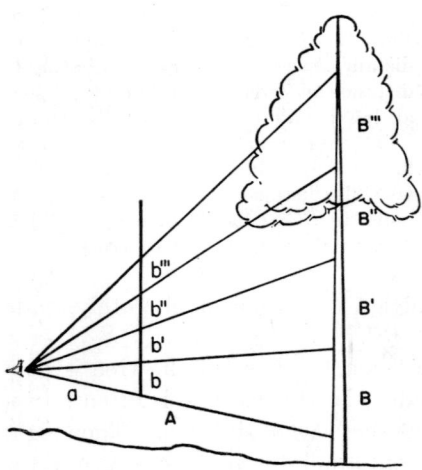

FIG. 43. The Merritt hypsometer. The staff graduations b, b', etc., are brought into proportion with the lengths B, B', etc., on the tree when A is the fixed distance from tree required by a, the length of reach.

In use the Merritt hypsometer is held out at the horizontal distance from the eye (25 in. in the above case), the lower end of the stick is dropped until the line of sight through this point intersects the stump height, the merchantable height of the tree is then sighted with head and arm in the same position, and a reading is taken from the hypsometer at the point where this line of sight crosses.

Each of the even graduations on the stick (6.174 in. in the above case) intersects a 16.3-ft. log on the tree. This is true since the major triangle formed by the eye, stump, and merchantable top is split into equal proportional parts in which the vertical subdivisions have a constant ratio.

The above fact has been used in the construction of a *modified Christen hypsometer* that eliminates the necessity of measuring the distance to the tree. This instrument is a short flat bar, 10 to 12 in. in length, with sighting notches at each end. The interval between the two sighting notches is divided into exactly 10 equal parts. The instrument is short enough so that it can be hung from a string and thus assume a vertical position automatically as a plumb bob would. When it is so held, the arm is moved backward or forward until the eye sights the base of the tree through the lower sighting notch and the tip through the upper sight. Then a sight is taken across the lowest tenth graduation and the point noted or marked on the tree. The distance of this point from the base, measured with a tape or ruler and multiplied by 10, gives the total height of the tree.

The *Christen hypsometer* in its usual form works somewhat differently. It is of the same length and has the same sighting notches at

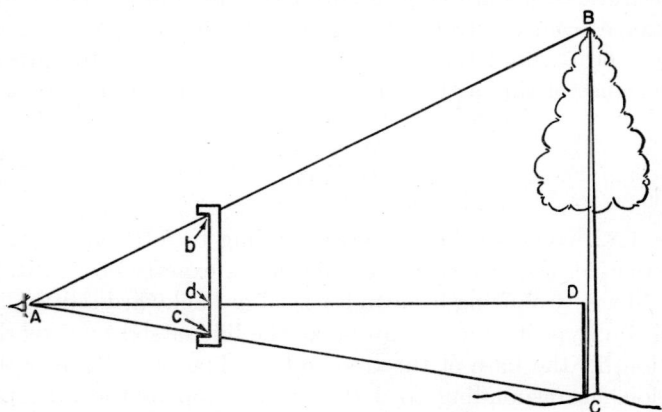

FIG. 44. Christen hypsometer.

the upper and lower ends, but the intervening graduations are based on that of a pole of fixed length set at the base of the tree. Figure 44 demonstrates the principle. Here

$$\frac{cd}{CD} = \frac{cb}{CB}$$

but CD is fixed at an arbitrary length, say 10 ft., and cb is a fixed length, *viz.*, the effective length of the hypsometer, say 12 in., or 1 ft. The above proportion with these known values is now

$$\frac{cd}{10} = \frac{1}{CB} \quad \text{or} \quad cd = \frac{10}{CB}$$

The interval cd for any height CR can be computed readily; for example, for a height of 90 ft., $cd = 10/90 = 0.111$ ft.; for a height of 10 ft., $cd = 10/10 = 1.00$ ft. The graduations are therefore marked off from the lower sighting notch of the hypsometer. By this method the height factor is read on the hypsometer instead of on the tree as in the first method.

118. The Chapman Hypsometer. The advantages of the control stick of fixed length used with the Christen hypsometer, and of the proportionate graduations of the Merritt hypsometer, are combined in the *Chapman hypsometer* (Fig. 45). This method consists in graduating a straight staff, such as a Biltmore stick, with equally spaced lines as is done with the Merritt hypsometer, which will represent 5-ft. intervals or any other chosen length on the tree. The intervals between graduations are computed on the basis of an assumed reach and distance from the tree. If the reach is 24 in., the distance from the tree is 66 ft., and the length of the pole is 5 ft., the graduation length because of similar triangles is found from the proportion

$$\frac{x}{24} = \frac{5}{66}$$

and $x = 1.82$ in. on the scale, corresponding to 5 ft. on the tree. In actual practice, the observer need not stand exactly 66 ft. away from the tree, since by shifting his arm forward or backward a little, as with the Christen hypsometer, he can make the line of sight across the zero graduation hit the base of the tree and the line of sight over the first graduation, corresponding to 5 ft., hit the top of the 5-ft. pole set against the tree. Then, if he sights to the top of the tree or to any lower point, his line of sight will cross the hypsometer at a graduation corresponding to this height.

In application, in a two-man crew, each man could be measured from the top of his head or hat to a point 5 ft. below, and the latter point

MEASUREMENT OF HEIGHTS OF STANDING TREES

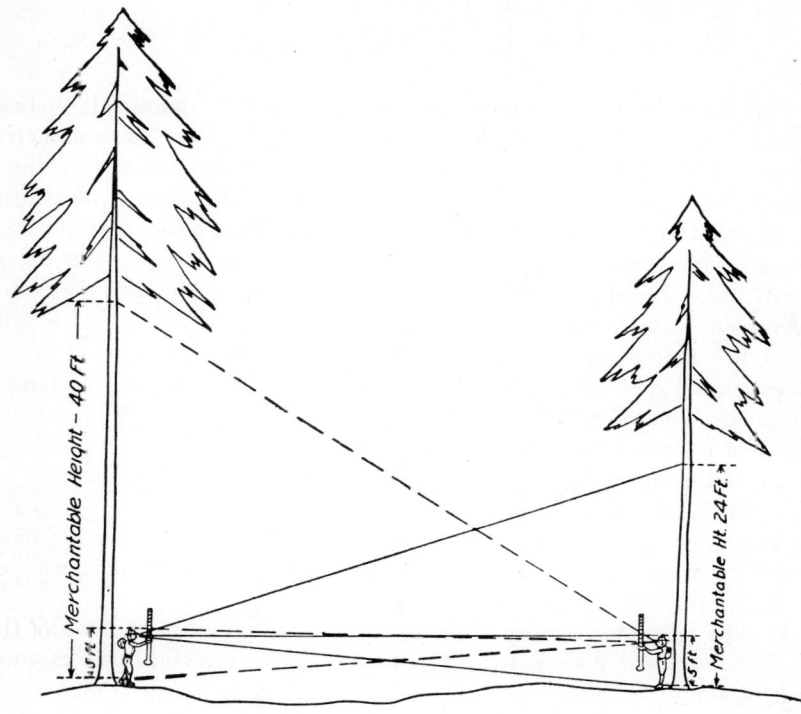

DISTANCE NOT MEASURED

FIG. 45. Chapman hypsometer. The staffs are graduated at 5-ft. intervals. The first 5-ft. graduation is made to coincide with the top of the observed person, and a point 5 ft. below. The height of tree is then read directly on the staff. To insure vertical position, a weighted portion should extend below the zero of the scale. Total height or any portion of the bole can be measured from same position.

clearly marked, thus eliminating the 5-ft. pole. By this means, heights can be taken by all men in the estimating crew with little loss of time.

119. The Preparation of a Curve of Height on Diameter. In determining the average height of trees of each d.b.h. class, a curve of height on diameter is drawn, the procedure for which is the same as that described for the curve of volume on d.b.h. (Sec. 73). Since height curves are of limited local application, diameters are recorded merely in the class in which they fall and not to $1/10$ in.; *viz.*, a 10.7-in. tree would be classed as 11 in. Heights are read to the nearest foot and recorded under the diameter class to which they belong; *viz.*, for 7 trees in the 14- and 15-in. classes, the record might be

D.b.h., in.	Heights, ft.
14	50,55,71
15	50,50,55,61

Curves of height on diameter do not represent a definite biological relationship, such as height on age or diameter on age. When such a curve is taken in stands containing several age classes and embracing a variety of sites, the individual trees will show a wide range of height for the same diameter. The height curve is drawn in order to arrive at a fair average height for each diameter class from reasonably few measurements. Height would be expected to increase with diameter, since increased diameter means tree growth, and this in turn infers height growth. Curves of height on diameter will therefore always be so drawn as to give an increase in height, however small, with increased diameter, except in the case of overgrown wolf trees, which tend to be short of normal height for their diameters.

The simplest method of plotting a height curve is as follows:

1. Plot the height of each tree directly over its diameter on cross-section paper. When two or more trees of the same diameter have equal heights, this should be indicated by the proper numeral, or weight, opposite the plotted point.

2. The curve may then be sketched in from ocular inspection of the plotted points. With a little practice in curve drawing, this method will be found satisfactory for timber estimating on small tracts.

3. Whenever greater accuracy is desired, the average height for each diameter class should be determined and these average values plotted. This is most conveniently done by counting the squares to each point on the cross-section paper from the nearest even point, totaling the squares, and dividing by the number of points involved; *viz.*, in Fig. 46 for the 12- and 13-in. diameter class, the procedure beginning at the 50-ft. ordinate value is

$$5 + 14 + 15 + 11 + 40 = 85 \qquad 85/5 = 17$$

The average height is $50 + 17 = 67$ ft. Average diameter for two diameters equals $(3 \times 12 + 2 \times 13)/5 = 62/5 = 12.4$ in. The point is therefore plotted at 67 ft. over 12.4 in. The average diameter can be found accurately by counting squares horizontally.

4. Connect the plotted averages by straight lines lightly drawn.

5. Draw a trial curve by eye, weighting its position by the number of trees in each average.

6. If necessary the curve can be checked:

a. By computing the total plus and minus deviations of the points from the curve, to see if it is too high or too low (Sec. 73).

b. By dividing the curve into two or three parts and checking the deviations for each part separately. It will be found, however, that irregularities in curves based on 50 trees or less are frequently so great that only the total check as in *a* above will prove useful. The proper trend can be secured in such cases only by reliance on personal judgment.

Figure 46 shows an example of this procedure. The heights are of hemlock trees at Devil's Hopyard State Park, Connecticut. The

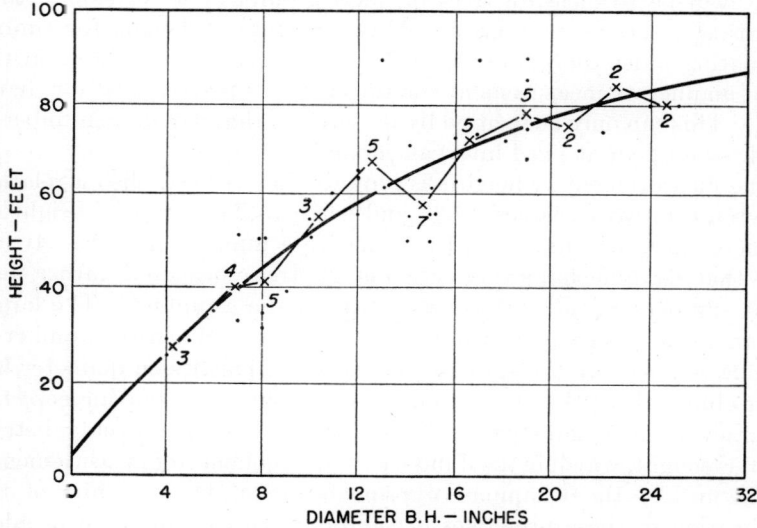

Fig. 46. Curve for heights of hemlock trees of all ages, based on d.b.h.

trees were of many ages; thus the resulting curve progresses rapidly from 25 ft. for 3-in. trees without noticeable flattening to about the 13-in. class. Beyond this point the curve is less steep, indicating less height growth. Had the stand been even-aged the smaller trees would have been proportionately taller and resulting curve much flatter (Sec. 182).

120. The Principle of Sampling, as Applied to Height Measurements. The purposes of measuring heights are twofold: first, as a scientific record on sample plots, in which case every height may be taken; second, as a means of getting a curve of average heights on diameters to permit the accurate derivation and use of local volume tables in timber estimating.

In selecting trees to measure for a height curve, it is important to observe the basic principle of sampling, which applies uniformly to all problems where a part is measured and applied as an average to the whole. This principle is the complete elimination of voluntary choice or bias in taking the sample. Malformed trees or trees with tops broken out should be avoided.

Where heights are taken by persons working independently of the survey party, it is practically impossible to avoid bias unless these men stay constantly with the other party as it moves over the area. Heights differ enormously with site, and the tendency of a detached party will be to measure trees on level ground or lower slopes, thus incurring a serious plus error. Measurements of height for timber estimating must therefore be evenly distributed over the area in the same manner as measurement for the strips or plots tallied for diameter. This can only be secured by prescribing that a definite number of heights be taken at fixed intervals along the strip.

The measurement of heights is required only to establish a relation between the two variables d.b.h. and height. The curve of height on diameter is clearly indicated by a moderate number of trees. It follows that rigid mechanical selection of the trees measured, rather than large numbers, is the important feature of the sampling. The larger the area, the less frequent can be this selection. When the number of heights required by the standard deviation of heights on diameter has been obtained, further measurement of heights does not increase the accuracy of the height curve. A total required number, evenly distributed, is sought, which in itself indicates the frequency of measurements.

The use of the Chapman hypsometer (Sec. 118) permits of the application of these principles of sampling with a minimum of additional cost. By this system, height measurements can be distributed uniformly over an area. With a two-man crew, the compassman can take his stand by the nearest tree and the estimator can read its height. With three or four men, the two estimators, or calipermen, can each read the height of the tree by which the other man stands. Heights can thus be taken rapidly at fixed intervals, such as at the end of every 10 chains, or as often as necessary to secure the proper number. With a stick properly weighted and swiveled and permitting a reading at a distance of from 66 to 100 ft., a sufficient degree of accuracy is attainable to serve the purpose in timber estimating.

On permanent sample plots, if the height of every tree is to be taken and the measurement repeated in 5 or 10 years, consistency in the repeat measurements is best obtained by measuring all heights from the same cardinal direction at each repetition. The establishment of fixed points for each measurement is thereby approximated.

121. Errors in Height Instrumentation. a. Due to lean of tree. When a tree leans toward the observer and no correction is made for lean, the height will be overestimated, and when the tree leans away, the height will be underestimated. Methods to eliminate this source of error have been described previously (Sec. 114). When leaning trees are measured for height by the observer standing at right angles to the lean, the error is small and of the following nature:

Angle of lean, %	Height error, %
Up to 13	Less than 1
Up to 20	Less than 2
Up to 25	Less than 3

For ordinary purposes and conditions, therefore, the effect of lean is negligible, if the tree is measured at right angles to the lean.

When leaning trees are measured with the lean toward or away from the observer, the error can be large, depending upon the angle to the tip and the angle of the lean, as shown in Table 38.

TABLE 38. ERROR IN OBSERVED HEIGHT DUE TO LEAN OF TREE*

Angle to tip, %	Direction and angle of lean, %							
	Toward observer				Away from observer			
	5	10	15	20	5	10	15	20
30	−1	−2	−3	−4	+2	+4	+6	+8
40	−2	−3	−5	−6	+2	+5	+6	+10
50	−2	−4	−6	−7	+3	+6	+9	+13
60	−3	−5	−7	−9	+3	+7	+11	+16
70	−3	−6	−8	−10	+4	+8	+13	+19
80	−4	−7	−10	−12	+4	+9	+15	+21
90	−4	−8	−11	−14	+5	+10	+17	+24
100	−5	−9	−12	−15	+5	+12	+19	+27

* Error rounded off to nearest per cent.

b. In use of the Abney hand level. Aside from the errors incurred in measuring leaning trees, other types of error creep in, largely through careless practice. A number of the errors listed in the following are not confined to the Abney but are general in nature.

1. Wrong horizontal distance to the tree. A short distance increases

the observed angles to the tip and to the base; a long distance decreases these angles. The relation holds that

$$\frac{\text{Observed height}}{\text{True height}} = \frac{\text{correct distance}}{\text{incorrect distance}}$$

Thus, if the distance is 5 per cent short, or 5 ft. in 100, as can easily happen in pacing, the true height is 5 per cent shorter than the observed height. The error is therefore significant; it may or may not be compensating, but it can be controlled easily.

2. Measuring slope distance instead of horizontal distance. This error has the effect of shortening the horizontal distance to the tree.

$$\text{Horizontal distance} = \text{slope distance} \times \cos \text{ of angle of slope}$$

An error of 1 per cent is obtained with slopes of 14 per cent, of 2 per cent with slopes of 20 per cent, and of 5 per cent with slopes of 33 per cent. The error is systematic and always positive but is not excessive with slight degrees of slope.

In some forested regions, *viz.*, hilly and mountainous regions, it is a definite advantage to use slope distance, and hence a special chart was prepared by McArdle and Chapman.[1] With the method of isosceles triangles, slope distance constitutes the correct measurement corresponding to height of tree; hence no error is incurred when the staff hypsometer is used.

3. Level bubble out of adjustment. If the level bubble is out of adjustment so that the line of sight hits the tree below the true horizontal, the height will be overestimated. It is a small but systematic error, reaching 1 per cent when the instrument is 4 per cent out of adjustment, which is inexcusable in good Abney practice.

c. In use of the Merritt hypsometer. This instrument is subject to careless use, possibly because it is simply made and employed for rapid work.

1. Holding stick short. The Merritt hypsometer is always graduated for a definite distance from the eye. When the stick is held short, the apparent log is longer than that of nominal length, such as 16.3 ft. When the stick is held long, the apparent log is shorter than it should be. If the stick is graduated for a 25-in. reach and 66-ft. distance from the tree, 1 in. change in reach means an error of about 0.7 ft. in a 16.3-ft. log, or about 4 per cent. It is apt to be a systematic

[1] McArdle, R. E., and R. A. Chapman, Measuring Tree Heights on Slopes, *Jour. Forestry*, **25**, 843–847 (1927).

error, often positive, since an observer frequently gets into the habit of holding the stick short.

2. Wrong distance from the tree. Distances are practically always paced or guessed when the Merritt hypsometer is used. If the distance is short or long, the error in determination of log length is equal to 1.52 per cent for each foot. Thus, if the distance is paced short by 4 ft., the log will appear slightly over 6 per cent shorter than it actually is or, if the distance is long by the same amount, the error is plus 6 per cent.

3. Tilting stick. If the hypsometer is held at the correct distance but allowed to tilt backward or forward, the log lengths will be measured off wrong, with the error increasing rapidly with increasing merchantable height of the tree. If the stick is tilted toward the observer, the apparent logs will be longer, but a maximum overestimate is reached when the stick makes a perpendicular to the line of sight to the merchantable top. When the stick is tilted away from the observer, the apparent log is too short. In the first case, too few logs will be counted in a tree, in the second too many. Table 39 shows the error in percentage of merchantable height due to this cause for 2 degrees of tilt, 5 and 10 per cent from the vertical.

TABLE 39. EFFECT OF TILTING MERRITT HYPSOMETER* ON HEIGHT MEASUREMENT

Merchantable height of tree (16.3-ft. logs)	Error in % of observed height from true height			
	Tilt toward observer		Tilt away from observer	
	5%	10%	5%	10%
1	+1.1	+ 2.0	−1.3	− 2.9
2	+2.4	+ 4.6	−2.5	− 5.2
3	+3.7	+ 7.4	−3.7	− 7.5
4	+5.1	+10.4	−4.8	− 9.4
5	+6.4	+13.5	−5.9	−11.4
6	+7.9	+16.7	−7.0	−13.3

* Graduated for 25-in. reach and 66-ft. distance from tree.

The above three sources of error, commonly incurred because of the opinion that the Merritt hypsometer is in any event only a crude instrument, have combined to discredit unjustifiably the use of the instrument.

These errors, of course, are the same for other hypsometers based on the same principles as the Merritt. They are serious errors and may be systematic in either a positive or a negative direction, depending upon the observer, but they are easily controllable through careful practice.

d. In use of the Christen hypsometer. Since distance to the tree is not involved, one source of error is eliminated immediately with this instrument. This is true also of the Chapman hypsometer. Its errors are in fact not subject to ready formulation, since they consist largely in inaccurate reading and manipulation of the instrument. Precision must be gained in correctly intercepting the base and tip of the tree and then in reading the interception of the stick held against the tree, without shift in the position of the head and arm. Large errors are possible. If the head is slightly low when the base is read and is raised slightly when the tip is sighted in, an underestimate of tree height will result. The three lines of sight should be focused not only once but at least twice or more in making a reading.

References

BARRETT, L. I.: Accuracy of Forest Service Standard Hypsometer, *Jour. Forestry*, **27,** 587–588 (1929).

BENNETT, C. H., and S. SEYMOUR: A Study of the Relative Accuracy of Various Hypsometers with the Transit as a Standard, *Idaho Forester*, **11,** 25–26, 46 (1929).

CALKINS, H. A., and J. B. YULE: Abney Level Handbook, U.S. Forest Service, 1935, 44 pp.

ERICKSON, M. L.: The Use of the Abney Hand Level, *Forestry Quart.*, **12,** 370–375 (1914).

HOLDSWORTH, R. P.: Adaptation of the Christen Hypsometer, *Jour. Forestry*, **39,** 721–722 (1941).

KRAUCH, H.: Some New Aspects Regarding the Use of the Forest Service Standard (Gradimeter) Hypsometer, *Jour. Forestry*, **16,** 772–776 (1918).

LIMING, F. G.: A Sectional Pole for Measuring Tree Heights, *Jour. Forestry*, **44,** 512–514 (1946).

NOYES, D. K.: Comparative Test of the Klaussner and Forest Service Standard Hypsometers, *Soc. Amer. Foresters Proc.*, **11,** 417–424 (1916).

SPURR, S. H.: The Navy Hypsometer, *Jour. Forestry*, **43,** 517–518 (1945).

STEVENS, R. C.: A Sliding Scale for Height Measurements, *Jour. Forestry*, **27,** 550–553 (1929).

CHAPTER 16

THE BOUNDARY SURVEY AND DETERMINATION OF TOTAL AREA

122. Retracement and Reestablishment of Surveys—Identification of Monuments. The first principle of retracement of boundaries established under either the United States public land survey or original surveys made by any other authority is the understanding of the fact that the physical objects established on the ground to mark the original position of the corner will always determine this position as long as any evidence remains to identify these objects. This principle will hold regardless of obvious errors of the original surveyor in the carrying out of the plan, resulting in glaring discrepancies between his notes of survey and the measured distances or actual bearings of the lines as so established. Only if the Federal government itself has carried out a resurvey, in which previous corners should be destroyed unless reestablished, can an original identified corner be set aside. This same principle holds good for any class of survey. The identification of the original monuments or witnesses is therefore of first importance in retracement and reestablishment of boundaries. Monuments consist of evidence placed at the exact spot which marks the corner and of bearing objects whose description, distance, and direction are recorded. In the United States land survey, corners are thus established only at $\frac{1}{2}$-mile intervals along the outer boundaries of sections, or at points of intersection of these section lines with "meanderable" streams or lakes or with private grants which form the boundaries of the public lands.

Monuments used by the United States public land surveys are described in Sec. 126. In prairie country, earth mounds with accompanying pits and buried glass or other foreign material were used.

Bearing objects consisted for the most part of trees growing nearby, on which were inscribed either on the outer bark if suitable or on a face freed of bark the designation of the township, range, and section on which the bearing tree stood.

In addition to the corners and bearing objects, the course of the line was originally marked, or blazed, where it went through timber, by placing fore and aft blazes on line trees which were intersected. On trees standing to right or left of this line if within a certain prescribed distance, two blazes were also placed on the side of the trees facing the line, being closer together as the distance from the line increased. In the absence of nearer trees, others might be blazed to a maximum distance of 33 ft.

123. The Field Notes. Any *general* system of public land surveys whether by the U.S. Land Office or earlier, by states or colonies, followed a prescribed plan, which serves as a guide to their retracement. In much of the old colonial territory, such plans were nonexistent, and reestablishment depends solely on monuments reinforced by clearings, rail fences, stone walls, or other evidence of boundaries, described in the notes of the original survey but difficult to interpret at present.

By contrast, United States land surveys are always accompanied by field notes, which follow a prescribed form and are the basis of the plats, or maps, of the survey. These field notes, in addition to recording the bearing and distance of the boundary courses, contain descriptions of all corners and bearing objects, intersection of the line with streams, changes of elevation or slope, character of forest cover, and quality of soil. The bearing and distance of each bearing tree from the corner, its species, and the size in inches are noted. These field notes are indispensable in retracing lines and identifying the monuments. It usually happens that wooden corners disappear long before the "witness" trees are gone. The identification of even one such tree, by reason of its species, size, and marks upon it, permits reestablishment of the corner. This is done by reversing the bearing and measuring out the distance in chains (links) called for from the original face on the tree to the corner. Where two trees can be found, the corner is set by intersection of the old bearings, checked by the distances. Whenever the original evidence is found, its position, if undisturbed, determines the exact location of the "corner."

The year in which the survey was made, as recorded in the field notes, gives a check on the authenticity of old tree blazes. The number of rings formed since the blaze was made should check with the elapsed time since survey.

When distances have been rerun carefully and the search is made in the known vicinity of an old corner, this corner may frequently be located by remains of the post or evidence of its former presence found below the surface. Sometimes it may be located by stump holes left from the destruction of the bearing trees by wind, fire, or rot. Still better evidence are stubs or stumps remaining and identified by species,

BOUNDARY SURVEY AND AREA 215

former size, and especially by position with relation to the corner and each other. Such identification is impossible in the absence of the field notes. For United States public land surveys these notes may be obtained in each state from the secretary of the state for a fee, or from the U.S. Land Office.

In certain localities and soils, the courses of stream beds may shift or a new channel may be formed, but ordinarily stream crossings or old channels give reliable checks on the probable position of the old lines.

124. The United States Public Land Survey—Essential Data. The plan of the United States public land survey, briefly, is the subdivision

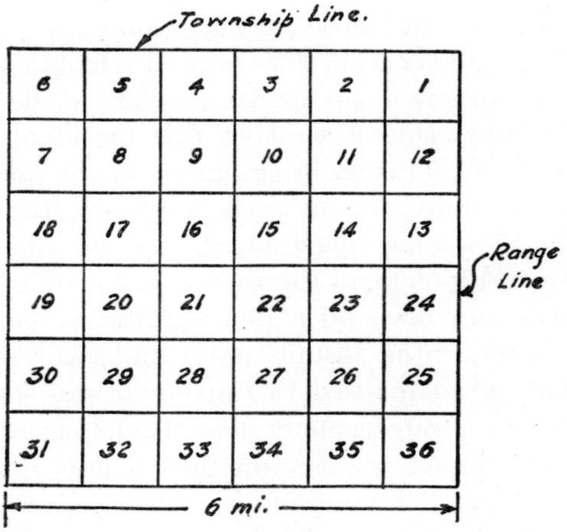

6	5	4	3	2	1
7	8	9	10	11	12
18	17	16	15	14	13
19	20	21	22	23	24
30	29	28	27	26	25
31	32	33	34	35	36

FIG. 47. Subdivision of a township into sections one mile square, in public land survey, and enumeration of sections.

of the surface into townships 6 miles square, divided into 36 sections each 1 mile square, and numbered as in Fig. 47.

The townships are originally laid out north, south, east, and west of an initial point by means of meridians or true north and south lines, and parallels, running east and west on parallels of latitude. The meridian intersecting the initial point is termed the *principal meridian*. The parallel intersecting this point is termed the *base line*. The parallels run at 6-mile intervals forming the north and south boundaries of the township and are the *township lines*. The meridian lines run at these intervals, forming the east and west boundaries, are the *range lines*. The enumeration of townships is by numerical series from the initial point, in tiers, north or south of this point, and in ranges, to east or west. For example, T 19 N, R 3 E designates the township lying in the nineteenth 6-mile tier north of the origin and in the third 6-mile range to the east. Many different surveys exist, each designated by its principal meridian; *viz.*, Louisiana meridian (La. Mer.).

This plan would work perfectly on a flat surface, but on the curved surface of the globe the straight meridian lines converge toward the pole and the parallels form curved lines, except at the equator, to offset the convergence and adhere to the 6-mile dimension for a township. A correction is required in east and west distances, which is secured by designating every fourth parallel, at 24-mile intervals, as a correction line, or *standard parallel*. In older surveys this interval was 36 miles. Along each of these correction lines the converging meridians and accompanying section lines close on one set of corners, and the new or corrected meridians and section lines originate from a second set, lying east of the closing corners if the township is east of the principal meridian, and on the west side if west of this meridian. These double corners often lead to confusion in ignorance of the plan of the survey.

In the original subdivision of each township, the section corners already established a mile apart along the meridians and parallels bounding the township control the section lines. The west boundary of each section is run first, on a line parallel to the range line. A quarter corner is set at the ½-mile point and the section corner at 1 mile. These lines are supposed to run north and south and are the most reliable lines for retracement (Sec. 128). The north boundary of each section is run next as a random line to determine the bearing of the corner 1 mile east. A true line was then supposed to have been run between these two corners and the mid-post, or quarter corner, set halfway between and on the line. Actually in old surveys this was seldom done. The line was surveyed either eastward or westward from a section corner, the monument and witnesses put in at exactly ½ mile and the remaining ½ mile not run at all, or run only in part, with no effort to correct the random. Yet such a corner must be accepted in determining boundaries and property lines.

On the north and west tiers of sections, the section lines are all run by random bearings to the corners on the township boundaries and established theoretically by rerunning the true lines between the respective corners. On both these tiers all the errors due to survey and losses from convergence are thrown into the extreme north and west rows of "forties." The first ½ mile to the quarter corner is full, the second short or long.

125. Meander Surveys and Township Plats. Whenever the survey line on the boundary of a township or section intersects a lake containing 25 acres or more, or a stream more than 3 chains wide, a *meander corner* is established and a monument set at this point. The boundary of the land is then considered as following the shore line of these bodies of water. For the sole purpose of determining the approximate area, and not to define this boundary, a meander survey is then run by straight courses along the shore, to tie into the next meander corner on the section boundary. Similar corners are established on the boundaries of grants such as exist in the former French and Spanish possessions. The boundaries of the latter are surveyed and the areas excluded, leaving irregular areas of the public lands adjoining them.

These surveys are then platted, and on this plat theoretical subdivisions are laid out, consisting of four quarter sections of 160 acres, each of which is divided into four square parcels of 40 acres. Whenever these rectangular lines encounter meandered boundaries, an arbitrary *lot* is platted, approximating as closely as possible 40 acres, but varying from 20 to 60 acres. Since no such lot may extend into two or more sections or include portions of two islands, lots in these situations may be smaller than 20 acres. The last tier of forties on the north and west of the township whose area is greater or less then 40 acres are likewise termed lots.

The quarter sections and forties are officially described by their cardinal location, beginning with the smaller subdivision; *viz.*, the forty in the extreme northeast corner of a section is the NE 1/4 of the NE 1/4 of, for instance, Section 35, Township 11 North, Range 12 East, 5th Principal Meridian, and its designation will be NE NE S 35, T 11 N, R 12 E, 5th P. Mer. The simplicity of these descriptions has greatly reduced the cost of land records and transactions. The lots, on the other hand, are given serial numbers on the plat, which become their official designation. These numbers run from the northeast corner of the section westward, thence south and east in the same order as for the enumeration of the sections on a township (Fig. 47), except that only the actual irregular lots are given numbers, all full forties being described as indicated above.

On these official plats, the theoretical or computed area of each lot is entered. The plats also show the bearing of all lines run, and their distances, including meander lines, the intersections with streams and swamps, and the date of surveys, declination used, and name of surveyor. They can be obtained from the U.S. Land Office. The government does not guarantee the accuracy of these land surveys, or plats, but bases its records and transactions on the official areas thus established. None of the lots or subdivisions of a section are run by the

218   FOREST MENSURATION

United States land survey. Although only on paper, the area of each subdivision is officially determined by the plat alone.

126. The Marking of United States Land Survey Corners and Accessories. In older surveys, in wooded country, the corner was usually a post of durable wood hewed square and set diagonally with edges facing the section lines. Notches were placed on the south and east edges, whose number equals the distance in miles to the township and range lines, respectively. No other marks were placed on the post. The newer corners made of metal with a brass cap have the township, range, and section stamped on the face of the cap.

The accessories in wooded country were ordinarily bearing trees, commonly called *witness trees*. At section corners, four were marked, one standing in each section and two at quarter corners. The inscription was placed on the face of the tree bearing toward the corner.

For smooth-barked species such as beech or hornbeam the characters were cut in the bark. For conifers or rough-barked hardwoods

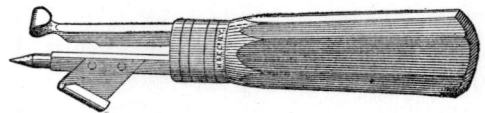

FIG. 48. Timber scribe used to make inscriptions on bearing trees, as seen in Fig. 49.

the bark was stripped from the face at a convenient height. The inscription was cut into the wood by means of a scribe (Fig. 48). The point of the instrument, with its appended cutting edge, makes characteristic figures which serve to identify the tree later on. At the base of each tree, a smaller blaze is made and inscribed "B.T." for bearing tree.

On the bearing tree the township and range numbers are placed, together with the number of the section on which it stands (Fig. 49).

The trees at the quarter corner bear merely the section number and "1/4 S"; *viz.*, 1/4 S 8. Meander corner witnesses, in addition to the numbers of the section on which they stand, bear the letters "M.C." Closing corners, on a standard parallel, are inscribed "C.C."

Rocks are more durable than wood, and rock corners, reinforced by rock piles, are inscribed as above with a chisel. Boulders or rock ledges may be selected as bearing objects and are marked "B.O."

On colonial surveys, established by France, England, or Spain, various practices were followed, many of which lacked system. Knowledge of the character of these surveys, the instruments used, the

Fig. 49. Beech marked in 1883 as a bearing tree for T 7 N, R 1 E, S 15, in La Salle Parish, La. (*Photograph by H. J. Lutz.*)

methods of establishing corners, and the field notes or plats is the basis of success in reestablishing boundaries.

127. Reestablishment of Lost or Obliterated Corners. A corner is obliterated when no trace of its position or existence is left, but it is not lost as long as any accessories or witnesses remain from which its position may be redetermined. When such accessories are identified as those originally marked or described, the bearing and distance of the corner from said "witness" as described in the field notes serve to reestablish it. It should then be marked by some durable monument such as a stone properly inscribed, a short section of iron rail, or a cement post.

A lost corner is one for which not even a trace of either corner or accessories remains, capable of identification. Such corners must be reestablished from known corners. For interior sections, four corners lying north, south, east, and west of the lost corner, respectively, must first be found. When on the township boundaries, two corners, lying east and west on the township line and north and south on the range line, serve as a basis for reestablishment.

The survey proceeds in each case from the nearest known corners. Each of the intersecting lines is retraced in the cardinal directions, following the courses and distances set forth in the field notes. At the point indicated for completion of each line, a peg is set. These four or, on township boundaries, two pegs will seldom if ever coincide, presumably because of the errors in the old survey. These errors may be in either direction or distance or both, and were caused by using worn lengthened chains or an inaccurate declination (Sec. 128). The correct position of the corner is now determined by proportioning the error in distance alone, disregarding indicated errors in bearing. The error in the north and south distances, or difference between the two pegs, is measured in a north-south direction only and is divided or proportioned according to the length of the lines run or distances to the known corners; *viz.*, if the corner to the north is 1 mile distant and that to the south 2 miles, the effect of the southern line is twice that of the northern, and the corner would fall at a point two-thirds the cardinal distance from the peg set for the corner from the south. This fixes the latitude, or north and south position, of the reestablished corner but not its departure, or east and west position. The same process is now applied to the east and west pegs. A north-south line is then drawn through the resulting position, to intersect with an east-west line of latitude passing through the first peg. This intersection reestablishes the lost corner (Fig. 50).

No one but a county surveyor or his deputy can legally reestablish a lost corner. If property lines of long standing are disturbed by discovery or reestablishment of lost corners, it does not follow that such owners lose title. The courts are the final arbiter and may confirm such lines on the basis of use and acceptance for a term of years.

128. Retracement of Old Lines. The bearings of lines given on old surveys either refer to a true meridian, or north and south line, or are given as the magnetic bearings existing at the time of survey. In either case, the present declination must be found which will correspond to that used by the original surveyor.

The magnetic pole is not stationary but has shifted from east to west, thus increasing the declination of the needle at points east of the iso-

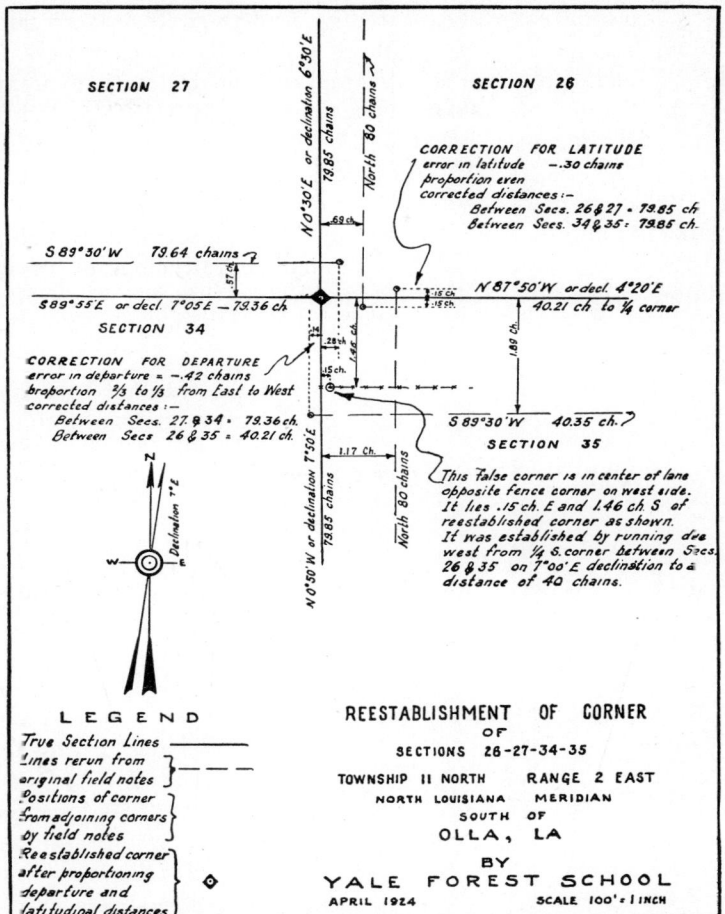

Fig. 50. Reestablishment of lost corner.

gonic line on which the true and magnetic pole correspond, and decreasing this declination on the west. A declination given in 100-year-old surveys as 5°W. may now be 8° to 8°30′W., while an east declination of 13° may now correspond to 10°E. The rate of change is not uniform in time or place. Actual redetermination of the meridian is necessary. In theory, this is best done by observation on Polaris. In practice, the local declination may already be known, and the old surveyor may have used a meridian slightly in error. On United States public land surveys, therefore, the best method is to retrace the line carefully between two known corners located on a north-south line, by running a random line on a given declination between these corners. The error, or falling of the line, is noted and its true declination computed; *viz.*, at the distance of 5,300 ft. the random line falls 25 ft. to left of the corner. The tangential offset is 25/5,300, or 0.0047 ft. per foot of distance. The offset for 1° is 0.0175; hence 0.0047/0.0175 indicates an error of 0.27° or 16′. A simple rule of thumb is to convert the offset into links per mile of line and multiply by three-sevenths, which gives the correction in minutes. It is customary in retracing old surveys to determine and set off not the true declination, but the one which will coincide with that used in the original survey, including such error as may exist.

When the magnetic declination is east of the true north, a random falling to the left of the true corner indicates too great a declination,

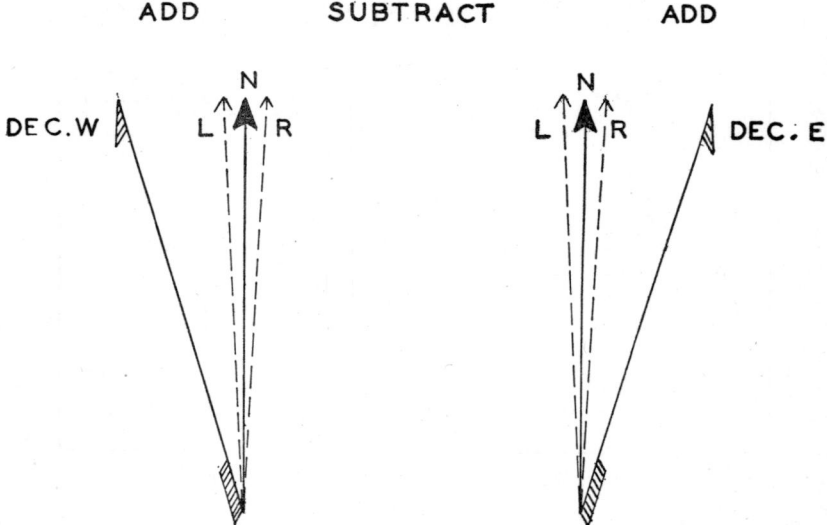

Fig. 51. Method of correcting the declination of a random line to conform to that of a true north line already established, when magnetic declination is east or west respectively.

and the error is subtracted, while a random falling to the right requires an addition to the declination used. These relations are reversed when the declination of the needle is west of north. In the above example, if the location were in Louisiana and the random line run on 7°E., the declination adopted for this retracement would be 6°44′E., while in Florida if run on 7°W. it would indicate a correction to 7°16′W (Fig. 51).

With the declination even approximately determined in this manner, the old lines are retraced as far as possible by running the identical courses and distances. Until and unless the true position of the line is known, no attempt is made to reblaze or otherwise reestablish the boundary. Random, or trial, lines should not be blazed. When necessary, true lines must be rerun between the established corners, and blazed or marked.

In irregular surveys, or where it is not necessary to reestablish the boundaries, random lines are run approximating the true courses and tied in by measured offsets to the true corners. From this survey, the position and direction of the true boundary lines are computed and plotted and the area determined.

Old lines are best retraced by using the same methods as were employed by the original surveyor. Preceding the use of the transit, the large staff compass with raised detachable sights, mounted on a tripod, with plumb bob, was employed. Declinations were presumed to have been originally determined for the initial survey of each township. Distance was measured with the half chain of 33-ft. length and 11 pins (the tally) horizontally regardless of slope, though in level country the full chain might be used, each 10 pins representing 2 tallies.

Errors in the original survey may be detected as arising from the following causes:

1. The chain became worn and its length increased. This led to overrun in distance, which varied from zero, for a new chain, to as much as 60 ft. per mile. Where chains of different actual lengths were used for surveying the meridians and parallels, and for subdivision of townships, the sectional survey was obliged to close on the established boundary corners. This led to various improvisations in the interior of the townships, sometimes difficult to detect, yet necessary to follow in retracement.

2. Common practice was to neglect running the randoms from west to east for the north boundaries of the sections. Instead, the ½ mile was run on a true east course on the declination being used, and the quarter corner established, the remaining line not even being blazed. Sometimes this ½ mile was run from east to west instead. When section corners were out of line, this resulted in hexagonal shapes of sections, since corners as established are permanent unless lost and reestablished.

3. The declination used for the boundaries and for the interiors of

the townships was not always the same and might vary 10 to 15 minutes. Accumulating errors might be arbitrarily adjusted somewhere in the survey, instead of in the prescribed manner at the closing on the west or north.

4. Occasionally, errors in chaining of from 1 to 2 chains occurred and were in some instances carried across the township. Subsequent county surveyors might fail to detect these errors and establish false corners on the basis that a mile was supposed to be a mile. Later identification of the "lost" corners led to confusion of property rights and fence lines.

5. The bearing of witness trees was sometimes recorded erroneously, owing to misreading the compass, resulting in errors up to 10 degrees, which, when only the single bearing tree remained, might misplace the corner by an error in proportion to the distance from the tree.

Errors occurring in running the meridian lines were the most serious. In one instance, in Missouri, the error due to a long chain accumulated to $2\frac{1}{2}$ miles, resulting in diamond-shaped sections.

6. The surveyor occasionally departed from the authorized procedure in order to avoid obstacles, such as deep streams not meanderable, and thus escaped observing accumulating errors until these reached considerable proportions.

In tracing such surveys, the consistency of the overrun in chained distances within the township should be worked out, compared with that of the meridians and parallels. Finally, the resurvey should endeavor to use all indications as to the actual procedure used by the original surveyor. The only possible final check on the survey is the identification of the corners. Lost corners can then be reestablished between all pairs of corners thus established, but it should be remembered that a quarter corner, regardless of "error," has the same validity as a section or meander corner.

129. Determination of Areas. The areas of forties and lots as given by the United States public land survey may be taken as correct unless resurvey or retracement shows the existence of large errors affecting area and consequently volume in timber estimating. In such cases the true area should be determined from the resurvey.

For irregular areas, the first step is to plot the survey by courses, determine the error of closure, and adjust this error by distributing it back according to the length and direction of each course, first eliminating any chance errors of plotting.

If no errors can be found in the map or field notes of survey, and the error of closure exceeds certain standards agreed upon, the survey cannot be accepted. Nothing remains but to search for this error in the field. Ordinarily it is sufficient to measure the distance from the final point of closure on the map to the initial or identical point and divide this distance by the total length of the lines run. For a paced map, an error of 1 in 50 is allowable. For chained distances, errors should come within 1 in 500 to 1 in 2,500, depending on the relative accuracy demanded.

The method of taking up or distributing this error over the entire boundary is to move each plotted intersection or corner in a direction parallel to a line drawn between the final plotted position of the initial corner or origin of the survey on the map and the initial, or true, position of this corner. The distance moved, beginning with the second corner, is proportional to the length of the boundary from the initial point to that corner. The final, or closing, corner is thus moved to take up the error completely. Instead of computing the lengths of these corrections, it is possible in rough maps, made merely for timber estimating, to make this adjustment by eye. The new or adjusted points are placed on tracing cloth, which is shifted in the indicated direction, between each two corners. In more accurate work, each correction should be computed.

The area may be obtained from the map by one of four methods:

1. By the use of the polar planimeter (Fig. 16). The scale of the map in feet per inch is converted into acres per square inch. The planimeter readings are then multiplied by this factor to get the area.

2. By using a sheet ruled in small squares. The equivalent area of the squares in acres must be known or determined. The squares may then be counted, adjusting for fractional squares by approximation. This is nearly as accurate as method 1 and often will suffice for timber estimating.

3. By dividing the tract into parallelograms and triangles. By scaling their dimensions from the map, their area may be quite accurately determined. The area of a triangle is one-half the product of the lengths of its base and altitude, or one-half that of a parallelogram of these same dimensions.

4. By computing the area from the field notes of survey by means of adjusted latitudes and departures. This is commonly called the *method of double meridian distances*. This method is used when exact areas are required as a basis of sale or transfer of the property. It may be obtained from any standard manual of surveying.

References

ABELL, C. A.: A Method of Estimating Area in Irregularly Shaped and Broken Figures, *Jour. Forestry*, **37**, 344–345 (1939).

CLARK, F. E.: A Treatise on the Law of Surveying and Boundaries, 2d ed., Bobbs-Merrill Company, Indianapolis, 1939, 896 pp.

LOCKE, S. B.: The Use of the Plane Table in Making Forest Maps, *Forestry Quart.*, **13**, 445–456 (1915).

NEWMAN, S. A.: The Grid-area Method for Determining Areas of Irregular Figures, An Application of the Mean Ordinate Principle, *Jour. Forestry*, **38**, 632–635 (1940).

SEWALL, J. W.: Woods Surveying, *Forestry Quart.*, **8**, 299–301 (1910).

U.S. Department of Interior: Manual of Instructions for the Survey of the Public Lands of the United States, Government Printing Office, Washington, D.C., 1947, 613 pp.

CHAPTER 17

TIMBER ESTIMATING FOR CUBIC AND CORD MEASURE

130. Timber Estimating in General—Ocular Estimating. Standing trees are measured in order to determine the merchantable contents of the entire stand of timber on a tract of forest land, either for sale on the stump or for purposes of forest management. This volume must be found in terms of salable products.

Cordwood stands may be estimated by eye, based on experience in cutting similar stands. An average number of cords per acre is estimated and is multiplied by the area in acres to obtain the total. Such an estimate is an approximation, its closeness depending on the local experience and judgment of the individual. All measurement of standing timber, no matter how painstaking, contains more or less error. These errors fall into three classes. Biased errors are controlled by personal skill, care, and judgment. Random errors of sampling are limited by statistical control as determined by the method used and the character and extent of the sample (Sec. 17). Errors of prediction are concerned with future changes in standards of utilization and in the forest itself, which render present estimates obsolete. Ocular estimating incurs the maximum of biased errors. The practice thus becomes an art rather than a science and is known as *timber estimating*, or *timber cruising*, and the estimators are called *timber cruisers;* this is probably due to the arduous trips made alone in the early days, guided only by a compass and with all equipment and provisions in a packsack.

While lump-sum ocular estimates may also be applied to board-foot contents of entire stands, the greater value and diversity of the products require a system in which individual trees are estimated or counted separately.

131. Measurement of Every Tree, or 100 Per Cent Estimate with Volume Tables. The highest degree of accuracy in timber estimating, in contrast to an ocular estimate, is obtained by measuring every tree of merchantable size and deriving its contents from a volume table, usually a local table. For cubic or cord measure, the average total heights of trees of each d.b.h. class are required.

In this method, it is not necessary to know the area of the tract in order to get the total estimate. The two essentials are, first, to locate the boundaries and, second, to obtain a tree tally of all the trees without duplication or omission. Boundaries are best located by the aid of some local resident acquainted with them. They may, of course, be identified from a map and description, if such exist. If for any reason the stand per acre is required, the total area must also be known.

If the area is small and is cut up by such topographic features as ravines, it is often possible to cover the ground by small blocks or irregular subdivisions without marking the timber. For larger tracts, the area may be completely covered by parallel adjoining strips on which all merchantable trees are calipered. These strips need not be of a regular width or be run to a true compass course. They should be run in a direction parallel to one of the boundaries. They may be from 60 to 200 ft. wide, depending on the size of the crew, character of the timber, and visibility or amount of underbrush present. The trees on the border of each successive strip should be given a light bark blaze on the outer side facing away from the strip so as to be seen on the return strip, or lime in a sock may be lightly tapped on the tree for a temporary mark.

One man can make such an estimate, but it is better to organize a crew of from two to three men, one of whom keeps the record, or tally, of the diameters and may guide the direction of the strip with a hand compass. The dot-and-dash tally is usually employed (Sec. 10). A more cumbersome form of tally consists of groups of five dashes. This form is less subject to error.

Trees may be separated in the tally by groups of species such as white oaks, black oaks, ash, hickory, and maple, or all may be thrown together. The basis of separation is stumpage value. If hickory is worth more than oak and there is sufficient hickory to pay to keep it separate in the cut, it should be tallied separately; the same for oaks as contrasted to birch or maple. The plan for timber estimating is always based on the market value of different species and products. It is thus an effort to get a quantitative basis for valuation of the standing timber.

From the completed tally, the volume on the area is found by multiplying the volume of a single tree for each d.b.h. class by the number of trees in that class and totaling the results. Volumes are totaled separately by species or groups of species. If the cubic foot is the standard used, reduction to cords is made by applying a factor, cubic feet per cord, adapted to the species and size of the material (Secs. 21

and 22). If area is known, the stand per acre may also be given. Even in a 100 per cent estimate with volume tables based on total height, it is seldom that the height of every tree is measured. Heights are selected for measurement according to the principles outlined in Sec. 120.

If a large area is covered, the samples of height are taken during the cruise, or, for a small wood lot, on its completion. For each height measured, the d.b.h. is recorded, for no attempt is ever made to get a single average height for all the trees in any stand. From 10 to 15 heights for each diameter class are usually sufficient for a curve of height on diameter even for large areas. For small lots, or plots, 15 to 20 trees may suffice for the entire range of diameters.

132. Factors Determining the Cost of Timber Estimating. A crew of three men can caliper and tally from 10 to 40 acres of timber per day of 8 hours. The amount of area covered depends on the number of trees, the density and kind of underbrush, and the flatness or roughness of the topography. With labor cost at $6 per day the cost of the field work would be from 45 cents to $1.80 per acre. Land bare of timber may frequently be bought outright for these prices. A stand of 20 cords per acre would cost, for estimating, from 3 to 9 cents per cord. The stumpage may be worth but 25 cents per cord. Too much money can easily be spent on obtaining an accurate estimate of standing timber of low value. Costs which exceed 3 per cent of the stumpage value of the timber may be considered high.

On the other hand, expenditure in estimating timber is certain to effect real economies. Buyers of small tracts of timber usually offer to uninformed owners a lump sum considerably below actual value, which is too often accepted. It is indispensable for owners to know the quantity owned, for otherwise they cannot expect to secure a fair price or their share of the margin representing true value of the stumpage. The buyer knows that he needs an estimate, since both his probable logging costs and his profit depend to a considerable extent on the quantity and price of the stumpage. But his expenditure for securing the estimate is wasted if he does not succeed in buying the timber, and it must therefore be kept to a low figure. For this reason, owners can afford a more accurate estimate than purchasers.

Timber estimating is thus required by the necessity for arriving at an agreement as to value and to permit sales of standing timber without awaiting the cutting and scaling of the logs or their manufacture (Sec. 4) where sales are made in a competitive market. Time within which to secure the estimate may be limited to the short period of an option. Trained men capable of doing reliable work may be scarce. The above factors indicate that the cost of timber estimating should be kept as low as will secure an estimate close enough to enable a buyer to meet competition. At the same time he must be assured that his bid permits him to operate with a margin of profit. Accuracy in estimating is relative and is dependent on a comparison of timber values with cost per acre for obtaining the data.

133. Estimating by Sample Areas—Partial Measurement. A reduction in the time and labor of timber estimating, in order to bring it within the limits of cost required, may be secured by sampling the area, or measuring only a part of the tract. The sample must be so distributed as to secure a result within the permitted limit of probable error. This limit is usually set at about 5 per cent for very close estimating and 10 per cent for average commercial cruising. Much greater errors were constantly incurred in past periods, when values were low and the operator or purchaser demanded larger margins for safety when buying timber. The margin of error should be below rather than above the actual volume, since the purchaser assumes all future risk of loss from maintenance and operation. For this reason, he should not be required to risk incurring an initial loss from overestimates. An increase in allowable margin of error permits a corresponding reduction in expense, with the result that partial estimates covering from 1 per cent up to 20 per cent of an area are standard practice in timber estimating.

The rules governing the sampling are three in number:

1. The percentage of area covered must vary inversely as the area of the tract for which the separate final total estimate is desired; as the size of the tract diminishes, the percentage covered must increase in order to secure an equal degree of accuracy (Sec. 17).

2. The sampling for timber types or stands of similar character must be controlled purely by mechanical location. This is accomplished by regular spacing of plots in order to eliminate unconscious choice of stands for measurement that are better than the average sought. These two rules are satisfied by the use of one of two systems: first, parallel strips of definite width, placed at fixed distances apart, and, second, plots of standard size, usually circular, placed at fixed intervals on lines run at definite distances apart. This second system is merely a modification of the first.

3. The probability of accuracy is greatly increased, in either case, by a third rule. The general direction of the strips must intersect belts, types, or stands of timber of similar character at right angles

rather than traverse them lengthwise. Very large errors have occurred through running strips along stream bottoms or in coves instead of intersecting them. On slopes or in mountainous regions, soil depth, moisture, fertility, and exposure change with elevation and slope, resulting in changes in timber types and tree heights. It follows that under such conditions the strips or lines should always run up- and downhill rather than along the contours at the same elevations.

In theory, the best method of sampling is the location of plots at equal distances in both directions, *i.e.*, at right angles to each other and equally spaced. This method is apt to reduce the total percentage of area covered or else increase the cost of the work beyond the desired limit. The strip method, by sampling in one direction only, over-emphasizes the effect of any influence which runs parallel to the direction of the strip, such as timber belts. The horseshoe method, by which a continuous strip is run in courses alternately in the two cardinal directions, approaches an ideal system of sampling for timber estimating (Fig. 52).

To apply the horseshoe method to strip estimating, the equivalent of two strips per forty is run as follows: Proceeding from the corner of a forty, or starting point, usually a section corner, an offset of 5 chains or 20 rods is made along its boundary to point of beginning. The strip is run successively to the center of each of the four 10-acre plots and finally to the opposite boundary; *viz.*, for the southeast corner of the forty, offset west 5 chains; run the strip north 5 chains, west 10, north 10, east 10, then north 5 to boundary, whence the northeast corner lies 5 chains east and is checked on. In this way, for a 4-rod strip, a 10 per cent estimate is obtained, and each forty is completed before the next is entered;

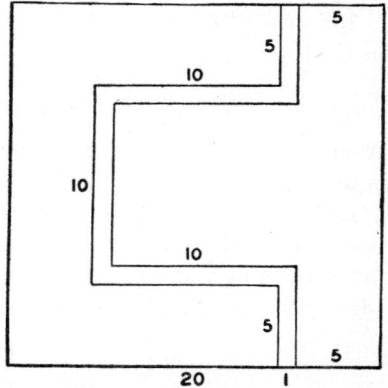

FIG. 52. The horseshoe method of cruising. Figures on the strip indicate chain distances in the center of strips to secure a 10 per cent cruise of a forty.

while for two parallel 4-rod strips it is customary to run the first strip across four forties before starting the second strip, a procedure apt to cause errors in tally.

From a statistical standpoint, all elements of sampling should be based on random selection, including location of each strip, plot, and tree selected for measurement. The additional work involved in adhering to this procedure compared with the slight gains in accuracy works against its employment in preference to regularly spaced plots, strips, or sample trees (Sec. 167).

134. Units of Land Measurement—Linear Measurement. The standard unit of linear measurement in most English-speaking countries is the foot of 12 inches. Tenths of feet are used frequently in engineering measurements. In continental Europe the meter (39.37 inches) is the basis for a decimal system of linear, areal, and cubic measure.

Based on the English foot, the units of linear measure in use are

1 mile = 5,280 feet = 320 rods = 80 chains = 16 tallies
1 tally = 330 feet = 20 rods = 5 chains
1 chain = 66 feet = 4 rods = 100 links
1 rod = 16.5 feet
1 link = 0.66 feet, or 7.92 inches

The terms *rod* and *perch* are identical.

135. Area Measurement. The standard unit of area measurement in land surveying is the acre. The equivalents are

1 acre = 43,560 square feet = 10 square chains

A square acre measures 208.7 feet on the side. In Quebec an "acre" is used to indicate a *distance* of 208.7 feet, or the side of a square acre.

Except in the thirteen original colonies, Kentucky, Tennessee, and portions of the former Spanish and French possessions covered by private land grants, the public land surveys of the Federal government (Secs. 124 and 125) are based on a rectangular system as follows:

1 section = 1 square mile of 640 acres
1 quarter section = 160 acres
1 quarter-quarter section, or "forty" = 40 acres
1 lot = 1 to 60 acres, irregular in form, but corresponding to a "forty"

The linear dimensions of the square subdivisions indicated above are

1 section = 80 chains = 320 rods = 5,280 feet = 16 tallies
1 quarter section = 40 chains = 160 rods = 2,640 feet = $\frac{1}{2}$ mile = 8 tallies

TIMBER ESTIMATING FOR CUBIC AND CORD MEASURE

1 forty = 20 chains = 80 rods = 1,320 feet = ¼ mile = 4 tallies

10 acres = 10 chains = 40 rods = 660 feet = ⅛ mile = 2 tallies

The convenience of the standard surveyor's chain of 4 rods, or 66 ft., is evident, since it coincides with the linear dimensions of square subdivisions of land down to 10 acres. The 10-chain unit thus bounds a 10-acre block on one side, and two such units bound one side of a forty. The chain of 4 rods is a standard unit for width of strips in timber estimating. Since 10 square chains equals 1 acre, a strip 1 chain, or 4 rods (66 ft.), wide and 10 chains, or 40 rods (660 ft.), long covers 1 acre of strip. On strips of this width the number of acres covered by a linear distance is equal to one-tenth of the number of chains. One mile of such strips covers 8 acres, or 2 acres for each forty. Since ½ chain was originally used in land surveying, one tally coincided with the use of 11 pins, the odd pin marking the initial point of the tally. Where a full-length chain is used, 11 pins cover 2 tallies.

In determining the area that should be covered in timber estimating, 4 acres gives 10 per cent of a forty. This area is secured by running two 4-rod or 1-chain strips across the forty. One such strip run straight across gives 5 per cent, four strips 20 per cent. The horseshoe method, with strips of a given width, is always equivalent to two such strips run straight; hence one horseshoe strip doubles the percentage run.

For straight strips of 1 chain width, the factors are given in the accompanying table.

Distance between strips		Number of strips per forty	Area covered, %
Feet	Chains		
1,320	20	1	5
660	10	2	10
330	5	4	20
165	2.5	8	40
132	2	10	50

The percentages of area covered are also in direct proportion to the width of the strips. For strips 32 ft., 2 rods, or ½ chain wide, the above percentages are halved, and for 2-chain widths, doubled.

The direction of the strips in a rectangular survey is either east and west or north and south. This is dependent on which direction will best serve to cross the timber belts and obtain the most representative sample of the stand. In unsurveyed or mountainous territory, directions may be chosen for each watershed or valley. The horseshoe method eliminates the choice.

136. Choice of Percentage of Area to Be Covered. In sampling a portion of a timbered area, the accuracy secured is relative and not absolute. The "random" error which results from sampling only a part of the area may be determined from the standard deviation of individual plots, as described in Sec. 17, for which formulas are available. In addition, the number of measurements required to attain the desired goal may be computed from the same formulas.

If a goal of accuracy is set which will admit of a given percentage of error, the question is, what percentage of the area must be covered to give reasonable assurance of this result? (It must be assumed that the work itself is done accurately and that the timber standing on the strip is correctly measured.) The answer depends on two factors. One is the character of the timber itself. The other is the size of the area unit.

1. Inaccuracy in sampling is caused directly by variation in the volumes of the stands to be sampled. Factors causing this variation in volume are species, diameter and height distribution, age, spacing or density of stand, and form of trees. If these factors are uniform in the timber on all parts of any given tract, regardless of its size, a single plot taken anywhere within its boundaries will give a true sample of the whole. Instead, the forest is composed of many different stands, no two of which may be alike in all particulars. Composition by species varies; the stands are of different ages; they reach different heights; and, more than all else, the number of trees per acre or stocking varies from open or sparse to dense and overcrowded, with resulting differences in volumes of trees as well as volume per acre. With increasing diversity, the percentage of area required for a given degree of accuracy correspondingly increases in direct relation to increase in standard deviation of plot volumes.

2. If a number of 1-acre sample plots are taken within an area as small as 40 acres, the volume of the total actual stand on the forty may then be measured as a check, by 100 per cent tally with a volume table. It will be found that the plots deviate from the volume of the average acre by percentages which depend, first, upon the amount of diversity in the timber on the forty and, second, upon the chance result of sampling. Deviations up to 50 per cent or more in volume might

TIMBER ESTIMATING FOR CUBIC AND CORD MEASURE 235

easily occur on any acre of plot. These variations shown on one forty may resemble those on many other areas of similar size.

A strip containing 5 per cent of a forty is located mechanically across the center line of the tract. Strips covering 5 per cent of a thousand forties are located in a thousand different places on the total area. If on one forty an error of plus 50 per cent is incurred and on a second the error is minus 50 per cent, and the actual stand is the same on each forty, the sum of the two strips measured when multiplied by 10%, or 20, is correct for the 80 acres.

Thus, as the sample covers an increasing area and number of plot units, the plus and minus deviations of these measured plots tend to balance each other, and the true average is approached by the method of greater dispersion of the sample over the whole. The standard deviation of the plot (acre) units from the average acre is determined directly as a function of the *number* of plots measured, irrespective of whether these units are taken as the sample of a small or a large area (Sec. 16c). Unless there is a greater diversity in the larger area and consequently a larger standard deviation, the same total number of plot units will serve equally well for either area. It is assumed that the same standard of accuracy, *viz.*, within 5 per cent, is required in either case. It follows that the larger the total unit of area on which the average and total estimate is required, the smaller the percentage of total area that need be covered to ensure any given standard of accuracy (Sec. 171). Hence an estimate by sections or even larger units requires far less area covered in the sample than if each 40 acres had to be estimated to the same limit of error as for the larger units.

Since one of the main factors governing the timber estimate is its cost, it becomes desirable to make the estimate with as small a sample or as low a percentage of area as possible, provided that the relative accuracy desired is attained. For any given area, therefore, if the goal of accuracy is set at ± 5 per cent in 20 out of 21 instances (Secs. 168 and 170), it is necessary to determine exactly how much sampling should be done to attain an average whose error will fall within ± 5 per cent of the true average acre. A smaller sample than necessary may not yield the desired results, and one that is too large will add unnecessarily to the cost of the estimate.

The process of determining this figure involves the use of statistical measures (Chap. 21), the theory of which is based upon the mathematical laws of error and of chances of variation in samples. These measures have to do with average, or mean, values and the distribution or dispersion of individual measurements around the mean of the sample.

Clearly this is the type of problem at hand in sampling or estimating a portion of a tract of timber and averaging the measurements as a basis for the volume on the entire tract.

137. Determination of Width of Strips. Young stands contain a smaller range of diameters and a much greater number of trees per acre than old timber. To adjust the sample to the timber, the strip should be wider in old than in young stands, and wider for large timber such as old-growth southern yellow pine than for the smaller species such as spruce or balsam in the North. This is an adjustment of dimensions of the strip to the difference in the "yardstick" of tree dimensions and number of trees per acre.

A more serious factor is the presence of underbrush. Dense thickets of laurel or of similar growth impair visibility and impede movement. In such cases it is not wise to attempt to cover wide strips, and if a larger percentage of area covered is necessary, the number of strips rather than their width should be increased.

Doubling the width of strips doubles the percentage of area covered. Doubling the number without changing the width accomplishes the same result with better distribution and improved accuracy. But this requires twice the mileage of line and hence increases the expense. The standard widths of strip are tabulated in Table 40.

TABLE 40. RELATION OF WIDTH AND NUMBER OF STRIPS TO AREA COVERED

Width of strips			Area covered by one strip per 40 acres or 4 per mile		Strips per ¼ mile to cover entire area, number
Ft.	Rods	Chains	Acres	%	
33	2	½	1	2½	40
66	4	1	2	5	20
110	6⅔	1⅔	3⅓	8⅓	12
132	8	2	4	10	10
165	10	2½	5	12½	8
330	20	5	10	25	4

138. Errors of Measurement on Strips. The percentages of deviation or error due solely to distribution and size of samples or strips (Sec. 17) are based on the assumption that the measurement of the strip itself is free from error. This is seldom the case. Errors are possible in connection with measurement of the following: width of strip, d.b.h., heights, and tree form or volume tables (Sec. 2).

1. Width of strip. The probability of error increases with the width of the strip and with the presence of underbrush and side slope and is controlled, first, by choosing a width suitable to the conditions and, second, by rigid precaution in checking this width whenever in doubt. Any tree whose bole centers outside the border should be excluded. In the Finnish national forest survey, a bamboo pole was used on narrow strips to test the distance to any doubtful tree. Ordinarily this check is by pacing from the center line or chain.

2. D.b.h. Where calipers are used, the most common errors are the lowering of the point of measurement (Sec. 113) and lack of adjustment of the calipers. The probability of error is increased by substitution of the Biltmore stick and becomes serious when ocular estimating of diameters is used. This last method makes a tallyman unnecessary, and the saving effected ensures its continued use. But unless the eye is rigidly trained and constantly checked by the use of calipers, large and inconsistent errors result. The loss of 1 in. in diameter is approximately equivalent to 8 ft. in height in its effect on volume.

3. Heights. Improper or insufficient sampling may give a height curve considerably in error. Use care in mechanical distribution and in avoidance of leaning trees. The number of heights required is indicated by the approach of the height curve to a regular form with few eccentric averages (Sec. 119). Where merchantable heights are tallied by eye, the same precautions are required in checking as for diameters.

4. Tree form or volume tables. Where local volume tables are constructed from trees grown on the tract, this error is controlled by the extent of the sampling done for the table. Where a standard volume table is the basis for the local table, it is quite possible that the average form of the local timber may vary above or below that possessed by the timber from which the table was constructed (Sec. 81). In extreme cases this error for the stand may amount to 10 per cent. It is avoided by felling and measuring a number of trees; the total, or sum, of actual volumes of the felled trees is compared with their total as found from the volume table, and the percentage of deviation is computed. This percentage may be applied to the estimate, or the volume table itself may be corrected as indicated by the local check (Sec. 98).

Every effort must be made to ensure complete accuracy in measurement of volume on the strip, since the inevitable error of sampling must not be increased by avoidable errors in the work itself. Yet such errors will occur in all cases, and the accuracy of the result depends on these two factors: sampling, and the precision of execution of the work.

139. Organization of a Strip Survey for Relatively Small Areas—Instruments Used.

The boundaries of the area to be covered must be known in advance, and their location on the ground must be familiar to the cruiser or conspicuously marked. The best general direction in which to arrange the gridiron of strips will be determined by previous knowledge of the topography and lay of the timber. A base line must then be laid off, either along the boundary or within the area. This line is preferably straight, but it may be a meandering traverse along a road or stream, composed of successive straight lines with different bearings. On a straight base line, stakes should be set at the intervals or points through which or from which the strips will be run. On a meandering base line, the positions of the stakes are computed from a map or by calculation and are placed so that the parallel strips will intersect them but be separated, as before, by the uniform intervals desired.

Ordinarily this base line may be run by a compass with raised sights, with socket to mount on a Jacob's staff, an ironshod staff reaching nearly to eye height. The compass, with needle clamped and sights folded, is dismounted while traversing and is mounted on the staff after the latter has been thrust into the ground. If local magnetic attraction is encountered, the compass must be oriented by backsighting. Magnetic declination of the compass must be known and can be set off on staff compasses of standard make, so that, with needle pointing to the magnetic bearing, the line of sight coincides with the true or corrected bearing. To set off the declination, set up the compass and turn the declination arc so that the zero, or northern, point on the graduated circle falls to the right of the line of sight for east declination, or to left for west declination. Then adjust the zero to the degrees and minutes required, by use of the vernier (Sec. 128).

For running the strips, the staff compass is the standard instrument. Except where local attraction is encountered, the errors of the magnetic needle are compensating. A pocket or box compass is useful in dense brush thickets, where it is better to take frequent sights than to attempt long shots. Practice is required in the use of box compasses before an accurate line can be run.

The distance is measured with a steel tape 100, 132, or 200 ft. long, the forward end of which is fastened to the compassman's belt. These tapes have replaced the old surveyor's chain. All distances must be reduced to horizontal measurement; but, instead of attempting horizontal chaining, corrections are made in slope distance for different percentages of grade. A second man signals the distance or chain length, measures the slope percentage with an Abney level, and sends the compassman ahead for the additional distance required. This is measured on an extension of the tape when specially constructed "trailer" tape patterns are used. Slope corrections are negligible under 15 per cent but increase rapidly for steeper grades (Table 41).

In a crew of four men, the compassman runs the line, mapping topography when required (Sec. 160). The tallyman records diameters, reads slope, secures the proper chained slope distance, and keeps the record of height readings. The two remaining men caliper the diam-

TABLE 41. TABLE FOR SLOPE CORRECTION

Slope, %	Slope distance for 100 ft. horizontal, ft.	Slope, %	Slope distance for 100 ft. horizontal, ft.
5	100.1	55	114.0
10	100.5	60	117.0
15	101.0	65	119.5
20	102.0	70	122.0
25	103.0	75	125.0
30	104.5	80	128.0
35	106.0	85	131.0
40	108.0	90	134.5
45	110.0	95	138.0
50	112.0	100	142.0

eters and take height samples. In a three-man crew, the duties of compass and tallyman are consolidated, one of the two calipermen acting as rear chainman. When ocular estimating is substituted for measurement of diameters on strip, a single man measures and tallies the trees, thus reducing the crew to two men. Frequently in this case pacing is substituted for the chain or tape. When circular plots are used, the crew can consist of one man.

A single cruise may, in addition to running line and measuring trees, include the collection of other classes of data, such as growth, topography, and wildlife. Some of the extra jobs are simple and consume little time; others are more demanding. The different members of the crew may easily vary in their efficiency and speed of following several routines simultaneously. It is therefore important to distribute the various jobs among the different members so that each one is working constantly and the work flows methodically without interruption. A great saving in time and cost can thus be achieved. To gain this efficiency, it is necessary for each party to study the flow of work and redistribute the small jobs accordingly. If one man has too much to do, all the other members of the party must stand idle, waiting for him to finish. No one constant distribution can be advised for all parties, but each must be fitted to time, place, and personnel.

140. The Measuring of Distance by Pacing. Pacing, or measuring distances by counting paces, can be reduced to an average error of less than 1 per cent, but only by practice and the observance of certain rules. Its advantage lies in the elimination of the need for pausing to measure each chain. It permits of one-man cruising. It is better adapted to level or gently rolling country but can be used on steeper slopes, with practice.

The first essential for acquiring accuracy in pacing is to determine, on measured courses preferably divided into units of $\frac{1}{4}$ mile, the natural pace, or swing, of one leg. Counting steps doubles the count with no advantage. The length of the natural pace is a function of the length of the leg, which is approximately correlated with the height of the person. For a height of 5 ft. $8\frac{1}{2}$ in. the natural pace is 1,000 paces for 1 mile, or 250 paces per $\frac{1}{4}$ mile. When speed is increased, the tendency is to lengthen the pace instead of increasing the number of paces per minute.

A natural pace is one in which no extra effort is used to extend the pace beyond the natural pendulum swing of the leg. The factor of distance of one pace in proportion to height is about 0.925, or for a 5 ft. $8\frac{1}{2}$ in. height it is 5.28 ft.

Table 42 gives the natural paces per $\frac{1}{4}$ mile, based on this factor. In acquiring a standard natural pace, the first practice should be on smooth unobstructed level ground, such as a road or path. A pace should be chosen corresponding to the nearest 5-pace division for $\frac{1}{4}$ mile, or 20-pace difference for 1 mile, and errors noted. The effect of freshness is to reduce the number, and of fatigue to increase it, unless the pace is spoiled by reaching out or by keeping step with someone with a different pace. While the pace is being standardized, practice should be through natural ground cover, including brush. Here it will be found that any unnatural lengthening of the pace (a 6-ft. pace is often mistakenly attempted) will be lost, especially when one is tired, but the natural pace can be maintained by stepping firmly and purposely to clear obstructions. When this adjustment has been made on measured courses, the third stage can be undertaken, that of pacing on slopes of different degrees of steepness, roughness, and cover. On slopes the pace falls short not only by reason of the slope error but still more through the greater exertion required to take a full pace when the

TIMBER ESTIMATING FOR CUBIC AND CORD MEASURE 241

natural swing is restricted in climbing. It is therefore a mistake to add a correction to the total distance covered. Instead, the amount of correction should be added to the count of paces continuously so that the count is correct for horizontal distance at all points. On slopes under 15 per cent this can be done by slightly increasing the effort for each step. Slope pacing should be tested for distance to determine the

TABLE 42. NATURAL PACES PER ¼ MILE, ACCORDING TO HEIGHT OF PERSON

Height	Paces	Height	Paces
5′	286	5′9″	248
5′1″	281	5′10″	244
5′2″	276	5′11″	241
5′3″	272	6′	238
5′4″	267	6′1″	235
5′5″	263	6′2″	232
5′6″	259	6′3″	229
5′7″	255	6′4″	226
5′8″	252	6′5″	223
5′8½″	250		

degree of adjustment needed. As steeper slopes present themselves, the number of paces for one count should be increased—the first graduation is 5 steps to 2 paces; next, 3 steps for 1 pace; and, on slopes approaching 100 per cent, 2 paces or more are taken for each count. The steps and paces can thus be shortened to make easy walking, and at the same time the distance is approximately accurate. If pacing is to be used in steep topography, practice on horizontally measured slopes, for short distances, will acquaint the cruiser with the proper modifications necessary to get the desired results. On precipitous slopes, a distance equal to 1 pace can be marked on the Jacob's staff and measured horizontally for each pace in the climb.

When paces are counted and memory relied on for the hundreds, error in dropping 100 paces may occur. Of course, the tally whacker can be used for the count, but it is a simple matter to make an accurate and almost automatic count by using a device which is easily perfected by practice, *i.e.*, counting only every tenth pace. For 1 mile and 1,000 paces, a count of 100 completes the distance and one-fourth of this completes ¼ mile. No count is made of the intervening 9 paces, which are ticked off in a rhythm centering on the fifth pace, as tick, tick, tick, tick, *tick;* tick tick, tick, tick, one; the next count of 10 paces being 2, and continuing. It is possible, when in practice, to count by this method and conduct a (desultory) conversation at the same time, but concentration is required to ensure accuracy. It can be seen that pacing is an acquired art capable of astonishing accuracy if sufficient interest and patience are exerted to master it. It is an ability which will be of lifelong service.

Horses can be trusted to pace reliably on level ground but need a "slope" correction.

The secret of accurate pacing on brushy or swampy ground is to lift the feet and pace deliberately. Obstacles can be passed either by right-angled offsets or by estimating the distance across them in paces.

Pacing reduces the cost of timber cruising but like ocular measurements can be used only by men willing to train themselves to the point of reliability.

141. Determination of Total Area and Estimate. The sample or percentage estimate thus secured must be applied to the total area of the tract, if it is assumed that it represents a true average stand per acre for the entire area, and if whatever random error this sample may conceal (Sec. 17) is accepted as unavoidable. To accomplish this, both the total area of the strips run and the actual acreage of the tract must be known or determined. When the area is rectangular and a fixed percentage is obtained from the strips a multiple can be applied to the strip totals, which is determined by the percentage run; *viz.*, for a 5 per cent estimate, $100/5 = 20$; and for $2\frac{1}{2}$ per cent estimate $100/2.5 = 40$.

For lots with irregular outlines, the cruiser either must possess a map, or he may undertake a boundary survey giving the shape and lay of the area. In either case he must find or be shown the true boundaries as a preliminary to the forest survey. The boundary survey is a distinct undertaking in itself and should never be included as a part of the ordinary cost of timber estimating.

When the boundary survey and map are available, or have been made previous to the estimate, they will control the strips run as to true length and direction. The base line may coincide with one boundary, or on larger tracts it may preferably bisect the area. But in either case the intersection of each strip with the boundary should be noted, and the length of the strip recorded and checked against the distance scaled from the map. If not in agreement, the mapped and not the strip distance is taken; otherwise the error in distance, if in excess, would dilute or reduce the average stand and, if short, would increase it.

In offsetting from one strip to another from the intersection of a boundary which does not run at right angles to the course of the strips, the simplest procedure is to make the regular right-angled offset

TIMBER ESTIMATING FOR CUBIC AND CORD MEASURE

to the proper distance, and then offset to the new intersection of the boundary before beginning the tally.

On returning to the base line, the strip should intersect the stake next in order. Any error is recorded and may be used in plotting the true position of the strips. Such errors of direction have no effect on the acreage sampled, provided that the true distance of the line run is obtained. This fact permits of the use of the hand compass when preferable for other reasons.

When no map of the boundary exists and no boundary survey is to be made, the party must plan on securing a crude sketch map of these boundaries as an incident of running the strip lines. It is assumed that they can locate the boundaries at each point of intersection with the strip lines. In such cases, the strips are run as accurately as possible, and their length and direction are accepted as correct. The base line forms the backbone of the map. At each intersection, the compassman should sketch or record the bearings of the boundary in both directions. From these notes the boundary map is completed as an appendage of the skeleton map formed by the strips (Fig. 53).

The total area of the tract as determined from the map must now be divided by the area covered by the strips to obtain the multiple to be applied to the strip volume for total estimate; *viz.*, a wood lot is found to contain 117 acres. The area of the strips was planned to cover 10 per cent but, owing to irregular outline, measured 14 acres instead of 11.7 acres. The multiple is $117/14$, or 8.36, instead of 10; or the strip estimate can be divided by 14 for the average acre and multiplied by 117 for the total area.

The method thus outlined assumes that the area is not separated into two or more divisions, types, or stands. If this is done, the further procedure is described in Secs. 155 to 157.

142. Substitution of Plots for Strip Estimate. If circular plots are substituted for strips, it is possible to reduce the crew to one man, though even with this method two men are often used. It is not a safe procedure for one man to work alone in wild country because of the possibility of falls or other accidents, when the whereabouts of the party would be only approximately known. Nevertheless this chance is frequently taken, especially by experienced men. This method consists in running a compass line in a manner similar to that for strips. At fixed intervals, with centers on this line, circular plots are measured by the same procedure as that used on the strips.

The size of the plot is determined by its radius. The percentage of total area covered by plots of different sizes will be fixed by the two

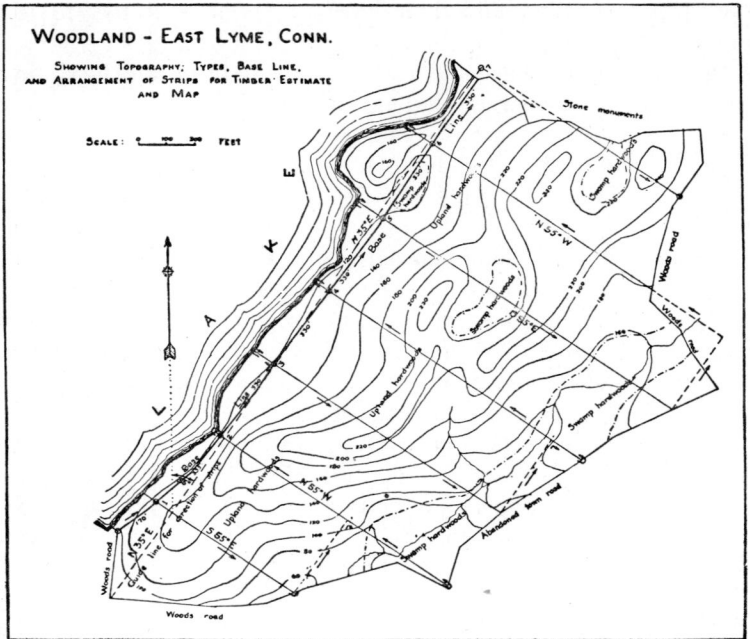

Fig. 53. Arrangement of strips for timber estimating, on an irregular lot. Strips are here spaced 330 ft. apart, at right angles to an imaginary line (dotted) whose course approximates that of the base line but is on a convenient bearing, *i.e.*, N 35° E, from which to run the strips at right angles. The base line followed an old railroad survey. The positions and intervals for stakes, on this base line, were indicated by intersections and staked out. The direction of the compass courses on the strips is indicated by arrows, which show the method of intersection used for the boundaries of the tract.

factors: distance apart of the plots on the strip, and interval between strips. Plots are used of different sizes, uniform for any one tract, and depending on the character of the timber and the visibility, or brush cover.

The sizes of circular plots are shown in the table on page 245.

Plots may not be placed closer together between centers than a distance equal to their diameter. Ordinarily they will be farther apart than this. The interval customarily used is a factor of the length of a forty in chains, or rods. For plots of $\frac{1}{4}$ acre, or 4 plots per acre, a location at intervals of $2\frac{1}{2}$ chains, 10 rods, or 165 ft. is about as close as is usually attempted. This gives the same area as a single 4-rod strip, or 5 per cent for 8 plots constituting one strip. The first plot would be placed at 1¼ chains, 82½ ft., or 5 rods from the boundary and the remainder at the full interval. One-quarter or one-fifth acre plots are suitable for dense brush or for small timber such as sprout hardwoods, spruce, or old field pine.

Size of plot, acres	Radius, ft.	Diameter	
		Ft.	Rods
1/10	37.2	74.5	4.05
1/5	52.6	105.2	6.4
1/4	58.9	117.8	7.15
1/2	83.3	166.6	10.0
1	117.8	235.5	14.3

Half-acre or commonly 1-acre plots are required for larger timber, but strips are preferable. One-acre plots placed at intervals of 330 ft. give a 10 per cent estimate on one strip, or double the area covered on a 1-chain strip.

Plots so arranged are merely strips with only a portion of the area measured at regular intervals. All directions for running strips with regard to base lines and tying in the survey therefore apply to plots.

143. Plots Arbitrarily Chosen. When plots are laid out arbitrarily by choice, independent of control by strip lines, it should not be for the purpose of obtaining an average stand per acre from the sum of their volumes. Any such effort would be subject to large errors by introducing the factor of human judgment or bias superimposed on that inherent in mechanical sampling. In order to select by eye plots that will give an average of the variable conditions of the timber even on 40 acres, almost as much time must be spent in examining the area as would suffice to run out regular plots evenly spaced. If the timber is uneven, no single plot may contain the average volume of the area, and several may be required for this average.

The usual procedure, adopted frequently by old timber cruisers, is deliberately to select an area of maximum or full stocking and measure a sample acre in this stand. With this as a guide, the remainder of the area is estimated by the process of discounting from this standard. Plots thus used are an adjunct to ocular estimating (Sec. 130).

References

BARTON, W. W., and C. B. STOTT: Simplified Guide to Intensity of Cruise, *Jour. Forestry,* **44,** 750–754 (1946).

CANDY, R. H.: Accuracy of Methods in Estimating Timber, *Jour. Forestry*, **25**, 164–169 (1927).
CHAPMAN, H. H.: Timber Estimating, *Soc. Amer. Foresters Proc.*, **4**, 114–128 (1909).
COOPER, W. E.: A Common Denominator Method of Estimating Timber, *Jour. Forestry*, **40**, 397–400 (1942).
DUNSTON, C. E., and C. R. GARVEY: An Efficient System for Computing Timber Estimates, *Forestry Quart.*, **14**, 1–2 (1916).
GIRARD, J. W., and S. R. GEVORKIANTZ: Timber Cruising, U.S. Forest Service, 1939, 160 pp.
GOODSPEED, A.: A Modified Plot Method of Timber Cruising Applicable in Southern New England, *Jour. Forestry*, **32**, 43–46 (1934).
GRASOVSKY, A.: The Use of the Median in Estimating Standing Timber, *Jour. Forestry*, **23**, 71–77 (1925).
HASEL, A. A.: Analysis of Sampling Methods for Volume Determination in a Ponderosa Pine Forest, California Forest and Range Experiment Station, 1937, 43 pp.
———: Arrangement of Cruise Plots to Permit a Valid Estimate of Sampling Error, California Forest and Range Experiment Station, 1937, 13 pp.
LYFORD, C. A.: Factors Involved in Timber Cruising, *Timberman*, **31** (1), 44, 48, 52 (1929).
MACON, J. W., and S. R. GEVORKIANTZ: Estimating Volume on the Spot, *Jour. Forestry*, **40**, 652–655 (1942).
MUDGETT, B. D., and S. R. GEVORKIANTZ: Reliability of Forest Surveys, *Jour. Amer. Statis. Assoc.*, **29**, 257–281 (1934).
ROBERTSON, W. M.: The Line Plot System, Its Use and Application, *Jour. Forestry*, **25**, 157–163 (1927).
SCHUMACHER, F. X., and H. BULL: Determining the Error of Estimate for a Forest Survey, with Special Reference to the Bottomland Hardwood Region, *Jour. Agric. Res.*, **45**, 741–756 (1932).
SEWALL, J. W.: Cruising for Quick Values in the Northwest, *Jour. Forestry*, **22**, 65–68 (1924).
U.S. Forest Service: Standard R-7 Computation Procedure, Management Plan Preparation, Region 7, 1938, 82 pp.
———: Timber Management Handbook, California Region, Part V, Timber Surveys; Parts A–I, 1947.
———: Timber Management Handbook, Northern Rocky Mountain Region, Part V, Timber Surveys, 1947.
———: Manual, VI, Acquisition of Lands; I, Valuation, Inventory and Presentation; LA 1, 11–18, 110–115, amended to 1948.

CHAPTER 18

TIMBER ESTIMATING FOR BOARD FEET BY USE OF LOG RULES

144. The Log as a Unit in Timber Estimating. The total merchantable volume in board feet of a stand of timber is the sum of the scaled volumes of the logs which can be cut from the trees that are merchantable in the stand.

When the contents of the logs contained in individual trees of given d.b.h. and height classes can be totaled, the tree unit may be substituted for the log, as being more serviceable, accurate, and timesaving (Chap. 19). But this is not always possible or convenient. First, there may be no tree volume table available for the log rule desired and no time to make one. Second, the merchantable portion of the tree may be restricted to the bole below the crown, and, as in the case of old hardwood stands, this may vary greatly in length and top diameter for trees of the same d.b.h. class. In such cases, the log rule (log volume table) is still available, and, by dealing directly with the log unit, the estimate is obtained.

145. Rough Methods of Log Estimate. Two primitive but effective methods of timber estimating utilize the log as the basis. In the Lake states and elsewhere, the cruiser depended on three items to obtain his total: first, a count of trees; second, the average number of 16-ft. logs per tree; and, third, the number of logs required to scale 1,000 bd. ft.

The count, in open timber, might be made to cover the entire area, or 100 per cent, especially for large old trees. Otherwise the count was applied to 1-acre plots used as samples to arrive at the average stand per acre for the forty (Sec. 143), or strips were run and the trees counted only on the strips (Sec. 133).

If the timber was of two different classes, large old-growth and smaller or younger stands, separate counts would be made of each class.

For the trees in the separate classes thus counted, the number of 16-ft. logs per tree was guessed at; *viz.*, "3-log" timber would cut out an average of 48 ft. of merchantable length. The total tree count was then multiplied by the factor 3 to get the total log tally.

The third step, that of selecting the log divisor, *log run*, or number of logs whose scale would total 1,000 bd. ft., was based on familiarity with log scaling. Scalers always noted the log run on their scale sheets, and timber of given sizes and character was known to "run" so many logs per thousand. The log run is merely the contents of the average log divided into 1,000 bd. ft., and is found directly by dividing the total count of logs by their total scale in thousands; the smaller the average log, the greater the number per thousand. The "run" is inverse to the size of the logs. Old-growth southern pine or white pine may run 5 logs per thousand, while pole timber may go 30 to 40. In estimating standing timber, the division of the total log count by the estimated log run gives the estimate in thousands of board feet for the trees counted. This figure is then applied to the area on which the tree count was made, depending on whether 100 per cent, sample acres, or strips were run. From this basis the estimate of the entire area, usually 40 acres, was calculated.

In southern pine, the use of the Doyle rule for 16-ft. logs permitted the cruiser to compute the contents of logs by rule of thumb. For 16-ft. logs this is "subtract 4 inches from the diameter of the log inside the bark, and square the remainder." (Sec. 52.) This led to the computing of the total contents of a tree assumed to be of about the average dimensions and volume, and the substituting of this tree of average volume for average logs. Accuracy thus depended largely on the minimum diameter to which the stand was estimated, since large numbers of small trees pulled down the actual average volume for the tree count. The system worked fairly well when the estimate was confined to large old-growth timbers, but it depended on judgment in selecting the average tree volume. In open stands of old pine timber a total count was usually made. Neither the Doyle log rule nor the method is suitable for estimating second-growth timber.

146. Tally of Log Diameters. With old large hardwood timber in which utilization is not close and most of the limby top is left in the woods, the number of merchantable logs for trees of the same diameter will vary considerably. Where tree volume tables were not available based on number of merchantable logs per tree, and where several species must be separated because of differences in quality and value per thousand board feet, cruisers resorted to a tally of each merchantable log in the sample. In such stands a "1-log" tree might mean a single butt log of large dimensions and small taper, with the upper bole unmerchantable, or a small slender tree, merchantable only to 1-log height. Under such circumstances, neither the number of logs per tree nor the size and contents of the logs could be averaged. Each log was tallied under its actual diameter at the small or top end inside the bark, separately by species or classes as desired.

The width of bark, doubled, must be deducted from ocular observation of diameter for this record. The log lengths chosen must coincide with the average length into which the timber is sawed in the region, but it is seldom necessary to tally logs of varying lengths. Instead, should a tree contain an odd length such as 8 ft., when 12-ft. logs are being tallied, it is better to tally by approximation a second 12-ft. log with a smaller diameter, to equal the extra 8 ft. of merchantable length.

Such a log tally is much slower than the method of counting and log run, and it is not usual to apply it to 100 per cent of an area unless the trees are very large and scattered. Instead, it is confined to strips or plots, whose combined area will not run the cost of the estimate beyond the desired limit. On separate forties, with large scattered trees, a total tally is required. Being entirely ocular, such a survey is conducted by a single estimator, who may have a compassman, making a two-man crew for running strips. If circular plots are used, a single cruiser can do the work.

In old-growth hardwoods reasonably free from brush, a crew can cover by tally from 16 to 24 acres of strip a day by this method. The area estimated is determined by the percentage covered by the sample; viz., if 20 acres of strip is tallied, on a 5 per cent estimate, the area estimated will be $20/0.05 = 400$ acres. For all young stands of pole timber or second growth and for close utilization in the tops, the advantages of a volume table based on d.b.h. and total height are so great as to require its use in the place of tallying individual logs.

147. Training in Ocular Estimates of Log Diameters. Familiarity with bark thickness can be attained by measurements of bark in scaling logs, especially on felled trees not yet skidded. On standing trees the length of the standard log may be measured. The pole with crossarm, described in Sec. 102, can be used in preliminary ocular training. After a short period of practice the pole is discarded and the upper diameter judged solely by eye. The first log contains a large percentage of the entire tree contents, and if this log is correctly tallied, the remaining logs may be recorded without serious error. The d.i.b. at the top of the last log will depend on standards of utilization and will be determined by the quality or form of the bole up to a certain point. Limbs might give rise to large knots and excessive or sudden taper. The top diameter is variable and tends to increase with larger d.b.h. With the position of upper point, its approximate diameter, and its height in terms of standard logs judged by eye with the aid of the measurement on the butt log, the distribution of diameters for the intervening logs is a matter of proportion. Given a tree with a butt log measuring 20 in. i.b. at 16 ft. above stump height, and a merchantable length of 64 ft. to a diameter of 10 in. inside bark, there is a 10-in. taper to distribute. Based on the information obtained by scaling logs and measuring tree dimensions, the taper of the second log might be set at 2 in., giving an 18-in. top, and the remaining 8 in. divided as 3 and 5, giving the third log 15 in. and the fourth the 10-in. top as originally estimated. The tally for such an estimate is by the dot-and-dash system under log diameters and tree species, as given below.

D.i.b. at small end, in.	Species		
	White oak	Black oak	Hickory, etc.
6	⌐.	⊠ : ˙	: :
7	⊠	☐	⊠
8	: :	⎮ :	˙ ˙

As an aid in judging taper, the d.b.h. may be calipered but need not be recorded. This d.b.h. measurement is a useful check on tapers but has no other purpose when logs alone are tallied.

148. Deduction for Defect and Cull. Timber estimating is intended to give only the sound contents of standing trees, such as would be measured in log scaling (Sec. 39). In tallying logs, defect must be noted and allowance or deduction made so that it will be excluded from the computed volume of the logs tallied.

In trees which contain no merchantable logs, no logs are tallied or counted. Those containing one or more cull logs are tallied only for the sound logs they contain. This leaves the question as to how logs which are partly unsound but still merchantable should be recorded. The ability of the cruiser to make deductions from such logs depends on his familiarity with the appearance, character, and consequent amount of defect, as acquired in scaling (Chap. 5). The tally is not by volume but exclusively by diameter. Where only a standard fixed log length is tallied, the defective logs would have to be omitted altogether or recorded as logs of smaller diameter. Tallying odd lengths if practiced would permit of shortening the recorded length of a defective log.

A second method is to tally merchantable logs, omitting only culls, and then, from the final total volume of the tally, deduct an average *per cent* for defect. This deduction should agree with the total later deducted from the scale of the logs when brought into the mill. Hidden defects in some species such as cypress or incense cedar make such a percentage deduction almost necessary (Sec. 43).

CHAPTER 19

TIMBER ESTIMATING FOR BOARD FEET BY USE OF VOLUME TABLES

149. Factors in the Choice of a Volume Table. If it is the intention to use a volume table in estimating the board-foot contents of standing timber, the following factors must be considered.

1. The species. Are tables available which have been made for the species? If not, are there tables that have been shown to apply although constructed for other species? (Sec. 71.)

2. The unit of estimating. Are tables available in the log rule by which the logs are to be scaled and by which they must therefore be estimated? (Sec. 76.)

3. The standard of utilization. Do the tables coincide with the top diameters and stump heights to be utilized? If not, can they be corrected to conform to these standards? (Secs. 77 and 105.)

4. Is the estimate wanted in terms of standard 16-ft. logs as scaled by a standard log rule, or is it desired on some other basis, such as 12-ft. logs, or lumber 2 in. thick? If so, are volume tables available in these units? (Sec. 78.)

5. Are the existing tables reliable? Are their authorship, basis in trees, the method of construction, stump heights, top diameters, and average form quotient known? (Sec. 104.)

6. Is height in the table expressed in the form most suitable for estimating the timber, whether as total height or merchantable log lengths? (Sec. 80.)

If tables are not available in suitable form, there are four possible methods of procedure: first, the volume table may be dispensed with and replaced by a log tally (Sec. 144); second, a local volume table may be constructed from felled trees (Sec. 72); third, the existing table may be modified to the form required (Sec. 105); or, fourth, a new standard table may be constructed (Chap. 12). The choice of these methods depends largely on the acreage and value of the timber to be estimated and the availability of funds for such work as the construction of volume tables. Ordinarily, standard tables are made by governmental and research agencies and seldom by private parties, while local tables are either derived from the former or made on the ground if no other source is available.

150. The Tree Tally. The use of a tree volume table calls for a tally of the d.b.h. of each tree whose volume is to be measured. This will either be every tree, for a 100 per cent estimate, or every tree on strips or plots whose areas cover a percentage less than 100. A third possible method is to count the trees on a strip and tally the d.b.h. of a fixed proportion of the trees counted.

The last method works well in timber composed largely of one species or of similar species which are not to be separately estimated, such as pines or red oaks. It is especially useful for fairly large open timber where a larger percentage or possibly all the trees should be actually counted. It is a better method than the total count and log run or total count and average tree estimated (Sec. 145), since it employs a tree volume table. The proportion to be tallied will vary from 1 in 3 to 1 in 5 in large timber or 1 in 10 in small timber. Where timber is small and uniform, however, it is better to use a narrow strip and tally the d.b.h. of all the trees upon it. If there are a number of different species or products to be accounted for, a 100 per cent tally will usually be required, at least of the trees on the strip or sample. The system of count and partial tally is adapted only to stands where there is a possibility of using wide strips such as 2 chains or $2\frac{1}{2}$ chains. Its advantage lies in covering a larger proportion of the area, in old valuable timber, with very little extra cost. For 1-chain strips, all the trees on the strip are tallied.

Where conditions permit the use of this method, it is absolutely necessary to choose the trees to be tallied by an unbiased mechanical method of selection (Sec. 12). The purpose of the tally is to obtain, from the total volume of the trees tallied, an average tree volume which will be identical with that of the total number counted. This can never be done successfully if the cruiser insists upon exercising his judgment in selecting, out of 3, 4, or 5 trees, the one that seems to represent the average of this small group. The law of chance, if allowed to operate, will fall upon the proper percentage of large and of small trees, provided that there are enough trees on the unit to permit of its operation. The method works well with 500 trees or over per forty. The tree to be tallied for d.b.h. may be taken as the nearest tree out of each group count, or as the one passed last in the count. Several groups may be counted before tallying if more convenient, as, for instance, counting 15 trees and tallying the last 5. The volume of the trees tallied must finally be multiplied by the count factor of 3, 4, or 5, for the estimate of the trees counted. The total estimate will then depend on the percentage of area covered.

This method compared with the 1-chain strip has been found by experience to permit the inclusion of from two to three times the area, by covering strips of proportionally greater width. It substitutes a rhythm which relieves the physical strain of tallying every diameter and height and secures more accurate work for both of these reasons.

151. Heights. When total heights are the basis to be used in preference to merchantable heights for securing volumes from a standard table, seldom if ever is an attempt made to tally the height of every tree measured for d.b.h., since heights would then have to be estimated by eye to save time. Even the simplest height check such as the use of the Merritt hypsometer requires a paced distance from each tree. This would mean from five to ten times the work required for the tally of diameters, making its cost prohibitive. Rather than guess at total heights, when the tips are frequently concealed by dense foliage, it is better to measure a few heights carefully by paced or taped distances and depend upon the curve of height over diameter (Sec. 119). From this height curve the volumes from the standard table can be taken for a local volume table based on diameter alone, but whose average heights agree with those in the stand (Sec. 86). When volume tables based on total heights are to be used, tree heights are always measured by the method of sampling and use of a height curve based on d.b.h., and the standard volume table is always converted into a local table applicable to the area to be estimated.

In old-growth timber, cruisers when they do employ tree volume tables prefer to use merchantable heights, tallying the trees according to the number of logs of standard lengths into which they will be cut (Sec. 152). This tally of merchantable height is practically always made by eye, occasional trees being checked especially in starting the day or at intervals, as on beginning a new forty. Usually the bole can be seen to the point at which it will be cut. In old growth, there is apt to be considerable variation in the merchantable length of trees of the same d.b.h. The cruiser then has the choice of separating the trees in each d.b.h. class into their respective "number of logs" as merchantable height classes, such as 1, 2, 3, or more logs of the standard length, per tree; or he can use the same technique as is employed for total heights *viz.*, take the merchantable "heights" of enough sample trees to permit a curve to be drawn to apply only to the timber which he is cruising. But on such a curve, the merchantable height must inevitably fall for most trees at points not coinciding with the heights used in the table; *viz.*, for 24-in. trees, perhaps 61 ft. is found, the nearest height in the table being 64. It would be necessary to interpolate volumes for these heights. The curve would show the number of standard logs and half logs in each d.b.h. class, plus an odd number of feet in the top log. Interpolation for the actual average curved merchantable height would be by proportioning the difference in volume to that in height, as usual; *viz.*, a tree with 67 ft. of merchantable length would be given a volume three-eighths greater than the board-foot difference between the volumes of a 64-ft. and a 72-ft. tree. By this method a local volume table is constructed for the timber on each area.

When close estimating of small areas such as separate forties or wood lots is desired as a basis of sale, and the trees vary widely in their merchantable lengths within the same d.b.h. class, separate tallies of the merchantable height of each tree are preferred. In this case the length is usually recorded to the nearest half log, above or below the merchantable limit, and only rarely to closer approximations.

Merchantable heights are seldom used for the record of permanent sample plots, most of which are in young stands. Even on such plots and for total heights, limitations of time and cost indicate the use of height curves from sample heights and local volume tables, rather than tally of heights of all trees, though the latter method gives a complete scientific record of the relation of individual trees to the development of the stand as a whole.

The 10 dot-and-dash system of tallying is used (Sec. 10), the tally being recorded separately under each d.b.h. and height, as 1 log, 1½ log, 2 log, etc. This tally calls for use of the standard volume table, which can then be applied directly in computing the total volume from the tally. The computation of the tree volumes is more laborious than that for the local volume table and height curve, because it involves several calculations for each d.b.h. class, corresponding to the spread of heights.

When large areas are to be covered which lie within recognized types and site classes, local volume tables based on heights of about 1,000 trees can be derived, which will permit the estimate to be based on diameter only, as was done by the U.S. Forest Survey. When a local volume table has been made and accepted, it makes no difference whether it was derived from total heights or from merchantable heights. The volumes, in the desired unit, such as board feet by the Scribner Decimal C log rule, cubic feet, or crossties, are given for average trees of each d.b.h. class, and all that is then required is the d.b.h. tally on the sample areas.

Local volume tables thus represent the use of inclusive averages by which the separation of heights within d.b.h. classes is rendered unnecessary. Tally by separate heights breaks down this average into the series of averages in which height as well as diameter is the basis, and it is thus more accurate as well as more expensive. Tally of the actual merchantable length of each tree to the nearest foot still further restricts averages and is of course the most accurate method. It is used only for hardwoods of high individual value per tree. In the choice of these methods the determining factor is always the balancing of cost against the value of the timber, and the permissible standard deviation from accuracy (Sec. 16).

152. Ocular Tally of Dimensions. When a total tally of merchantable heights or of separate logs is made, the substitution of ocular tally for actual measurement of d.b.h. and of height becomes a practical necessity. Ability to do this work satisfactorily comes as the result of progressive education of the eye to read these dimensions with accuracy. Scalers in the old days became so skillful that they seldom measured the diameter of any but the largest logs. The practice of ocular scaling is not encouraged by the U.S. Forest Service owing to the presumption of avoidable error which it entails. But in timber estimating, if it is found possible to judge diameters by eye, it may result in eliminating two men out of a four-man crew and cutting the cost in half. Since accuracy and cost are so closely related in estimating, this is an advantage that cannot be ignored. By tallying his own diameters, one cruiser can progress almost as fast as two calipermen with a tallyman. A compassman, who can also make the type and topographic map, completes the two-man crew (Sec. 160). An alternative practice is to use the compassman also as tallyman and require the cruiser to make the topographic sketch.

Training in ocular judgment of d.b.h. should consist in tallying by eye a large number of trees. Each tree should then be immediately checked for diameter with calipers, in the direction in which it was observed, to avoid the discussion arising from trees with eccentric shape. A total check on the estimate of diameters may also be made by tallying by eye all the diameters on check plots whose trees have all previously been measured. The comparison of the count with the record for each separate d.b.h. class will reveal the tendency of the cruiser to over- or underestimate diameters. In the case of overestimates the larger classes will show too many tallies, and the smaller classes too few, while the reverse is true for the underestimates.

The same principles of checking apply to heights. Cruisers are especially apt to be erratic in the measurement of heights on different days since the ability to estimate log lengths is more difficult to acquire than the ability to estimate diameters. It is necessary even for experienced estimators to check heights every day before starting work if they would avoid astonishing and unexpected errors in height tally. A source of considerable error in height estimates is the changing from relatively tall stands of timber to those of shorter dimensions, or vice versa. Careful checks of ocular estimates are indicated in each case.

153. Checking the Accuracy of Estimates. The theory of all strip or plot methods of timber estimating is that the stand measured (the sample, Sec. 12) will have a volume and composition as nearly identical as possible with the true average of the total area for which the estimate is sought. The precise error is never known unless, as a check on the methods used, the 100 per cent volume is compared with the scale of the logs cut from the area. Such checks by cutting are invaluable but can be made for large areas only if estimates are separated by logging units, the scale for which is also kept separate. Natural logging units do not follow rectangular lines; hence the logs from different forties are mingled together and are usually scaled on the landing, not in the woods. A logging unit with natural boundaries permits of separate scale records, but these will be kept only when insisted on. In sales where a portion of the timber is left standing because of inaccessibility or forestry regulations, the check is not complete until this residual timber has been measured and added to the scale. In clear cutting, the check is complete when logging is finished. Checks may thus be by detached forties, by logging units, or by entire sales. Years may elapse before the check is obtained, and frequently no checks are attempted. It may happen also that the timber is utilized either more or less closely than the estimate called for.

For these reasons, estimating should always be checked instead by the method of sample plots, forties, or strips, in which every standing tree is carefully measure as to diameter and a large percentage for height. Measurements are made preferably with the calipers and hypsometer. Thus not only may the gross, or total, error of the cruiser be checked but the sources of inaccuracy may be identified. They may consist of errors in diameter, in heights, or in the count, not to mention differences of opinion as to cull.

Although in log scaling the preferred method of checking is by individual logs, without comments on the total scale (Sec. 46), the method giving best results in checking the ocular ability of timber cruisers involves not merely testing individual trees but over-all results. In his training period the cruiser should use the total tally method on measured areas of 20 to 40 acres. His tally by d.b.h. and height (number of logs) classes is then compared with that of the record, which has been measured with calipers and hypsometer. At least one-fifth and preferably more of the heights in the record should have been measured, and the remainder allocated, *within each d.b.h. class,* by distributing them in proportion to those tallied. Two height crews for each caliperman can measure about this proportion of trees.

From this record, the number of trees in each d.b.h. class gives the basis for the check of ocular tally of d.b.h. by the following method. The numbers in each class are compared with his tally, and he records his surplus or deficiency, in number, opposite his d.b.h. By the distribution of the plus and minus signs he can tell, first, whether his errors are erratic or consistent; second, whether he tends to average diameters in groups or classes larger than those designated (single inch classes should be used in eastern timber and 2-in. classes in large western timber for training the eye); and, third, whether, if consistent, he is over- or underestimating the diameters. So far the check itself is "ocular" by the mere arrangement and grouping of his "differences."

A similar check of heights is next made. No attempt is made to check separately the tally of each height class *within* each d.b.h. class. This is not necessary. Instead, heights for each height class (preferably in ½-log classes both for training and estimating) are totaled at the bottom of the page, and differences are recorded as plus or minus, as for diameters. Here will usually be found a very clear indication of tendency either to over- or underestimate heights. Surplus trees in the higher classes indicate overestimates; minus signs in the lower heights mean the same thing; and vice versa.

It is even more important, in ocular estimating, to succeed in counting and tallying all the merchantable trees on the area sampled. This involves not only the ability to keep the count straight and know what area has been covered, and to be able to judge accurately by eye the width of the tallied strip or plot, but it also means the distinction between merchantable and unmerchantable sizes. The inclusion or exclusion of trees respectively below or above the minimum diameter, by the estimator, affects both the count and the average volume of the trees. The total count in the sample areas has been accurately recorded, and from the tally the actual "standard" volume on the check area is *accepted as correct*—a most important psychological necessity if any progress is to be made by the cruiser in correcting his errors. This same psychology applies to the record of diameters and heights;

the recorded heights are most apt to be suspected by the trainee, since the tally was by sample and not complete.

Either preceding or following the check of the cruiser on his d.b.h.'s and heights, three major checks are made against the totals: The first is the estimate itself, in which the cruiser's error is expressed in percentage of the measured volume. The second is the total count, again expressed in percentage of the tree count. The third is the volume of the average tree compared, in percentage, with that on the check area. The sum of the second and third percentages must equal the first percentage.

Where an error due to miscounting is shown, that due to combined errors in d.b.h. and height constitutes the remaining weighted, or percentage, source of error, the significance of which is complete if the count is correct, and is obscured in proportion to inaccurate counting. The distribution of this residual error between diameters and heights is again a matter not for computation but for appraisal from the tally. A 20 per cent error may be plainly indicated as occurring solely in overestimating of heights, or a minus error of the same magnitude may be due obviously to underestimating diameters.

By this system of isolating and identifying the three classes of error, *viz.*, area and total count, diameters, and heights, and insisting on the complete check at the close of each day, the cruiser learns his tendencies and sources of error and can then strive to correct each one separately. By contrast, when the only check employed is that on total volume, the inevitable result of checking is a worse error in the opposite direction, just as it is in scaling. Similar vacillations occur in trying to "correct" tendencies to over- or underestimate heights or diameters. The *eye* should be corrected, instead, by checking individual trees, as is done in check-scaling logs (Sec. 46).

Rudimentary training in ocular estimating can be given in about 10 days by employing this method. The steps, as applied to check plots of 20 and 40 acres, on old-growth southern pine, were as follows:

1. Measurement of check plots, accompanied by ocular estimate of each tree previous to measuring it.

2. Complete ocular tally of trees on 20 acres, by running unmarked strips of $1\frac{1}{4}$ chain widths, counting only in one direction, with checkers who are not estimating and who indicate the outer boundary of the strip (checking its width, by pacing), and check-measure numerous test trees for diameter and height *after* the estimators have recorded them (measuring diameter in the same direction as viewed).

3. Similar to 2 but on 40 acres, counting in both directions, with

checkers eliminated, thus doubling the width of strip to 2½ chains, and depending on the eye for distance, with occasional pacing.

4. One tree in three tallied on 40 acres by running ocular 2½ chain strips, without checkers, selecting each tree by a fixed procedure which eliminates bias, such as taking them in the order of the count (Sec. 150).

5. Alternate 2½ chain strips run on 40 acres, with estimate of the volume correction factors for the adjoining strips. Then the alternate strips run to check the correction factor (Sec. 158).

By the end of the 10-day period, the relative ability of the trainees to acquire the art of ocular estimating by further practice is clearly shown, and the methods of rapidly improving their technique are established. The chief accomplishment is to dispel the "delusion of infallibility" which was the bane of timber cruisers in the older days and which, with the absence of accurate records of the cut by forties, prevented most of them from identifying and correcting their errors.

Mere mechanical measurements of diameters and sample heights, in systems of timber estimating dependent on strips or plots, will never take the place of ocular skill and judgment (especially of quality and defects) for close appraisals of valuable timber, and ability in this line is fully as important for the forester as for the old-line cruiser who specializes solely in this work. Since the forester also has other things to do, his training must be aided by a systematic approach such as is indicated above.

154. Deduction for Defect. In timber estimating, deductions for the defects described under scaling and log estimates are made in d.b.h. tally by the omission of cull trees. For defective trees containing some merchantable logs, deduction may be accomplished by one of two methods. Either these trees may be tallied as of smaller diameter or shorter heights, both deductions being subject to inaccuracies; or else special tests can be made on selected plots of about 10 acres in size. On the plots, by log tally, the actual amount of the deductions can be made for each log (Sec. 148), the total can be compared with the total volume of the stand when all trees are tallied as sound, and the deduction can be expressed as a percentage of sound volume. These percentages of deduction can then be used as guides to apply to other stands of timber to indicate proper reductions of the initial estimate.

155. Forest Types as Units for Timber Estimating. Aside from endeavoring to secure the greatest possible accuracy in strip measurement and to place the strips in such a manner that an average stand is most apt to be secured (Sec. 133), the possibility exists of further improving the statistical probability of accuracy of the average volume of the samples taken. This may be accomplished by confining these averages to subdivisions of the area which will have greater uniformity than the whole. Of these possible subdivisions, the forest type is of first importance (Stratified Sampling, Sec. 12).

The problem of the timber cruiser is purely one of expediency as far as estimating is concerned. His purpose in separating different areas on the basis of type is to obtain a more accurate total estimate on the area by attaining greater accuracy in his sampling. If by separating two areas the average stand on each is more closely obtained than the average of the whole would be if these areas are combined into one, he will make the separation.

More far-reaching reasons exist for this separation of type when it comes to the treatment of the forest as a crop, requiring the measurement of growth. Here it is indispensable that types be separated, and with this end in view these distinctions may be made when on the basis of the estimate alone they might be dispensed with.

Decision on the separation of two or more types will depend, first, on the need for distinguishing species in the estimate, because of differences in utilization, markets, and stumpage values. Conifers are usually separated from hardwood types on this basis. In the second place, the different types may have different quantities of timber per acre. Neither of these factors may require type separation solely for the estimate. If the cruiser could be assured in advance that the strips or plots would take the same percentage of each type, *viz.*, 20 per cent, then a correct sample for the whole area would also be secured. It is for this reason that types are ignored in many cruises, when there is good reason to assume that this distribution will be attained, and there exists no special need for growth and management data.

A third and more fundamental reason for separating two or more types in timber estimating is the probability that average heights based on d.b.h. will differ within each type. By separation of the tally, it is possible to apply separate height curves to the timber within each type. Volume computation from these separate curves will give a much more accurate average volume per tree within the type and permit of the use of a local volume table based solely on diameter within each type. This advantage still holds even when the percentage of each type sampled by the strips is identical with that existing on the total area cruised.

A fourth gain in accuracy is attained when it is possible actually to map the areas occupied by the forest types. When this is done, the percentage within the respective types which is included in the sample taken in each type may be accurately determined. Knowing this sampled area, the proper factor may be computed by which to convert the sample estimate or percentage measured within each separate type into the total estimate for that type. The sum of the type estimates will give the total estimate for the tract.

For very large areas, the airplane map serves this purpose (Sec. 277). This constitutes its principal function in timber estimating. By means of aerial photographs, it is possible to recognize and eliminate from the strip survey all unproductive areas, such as peat swamps or muskegs, burns, young reproduction, barren rock, or grasslands. The strips may thus be planned to sample only the remaining timbered areas, and the strip location, direction, and spacing may be laid out prior to actual field work on the ground.

Lacking the aerial photographs, attempts to determine the actual type boundaries may be made while the strip survey itself is being run. The boundaries of the types are mapped from the points of intersection of strips or compass lines. This method is subject to considerable error through inability to see beyond a certain distance, which is limited by the density of the trees or brush or by change of slope. Even lake boundaries may be inaccurately mapped by this method. Unless the stand is sufficiently open and the type boundaries plainly visible, their location is difficult. Some type boundaries are not definitely marked, as when one type gradually merges into another over a distance of several chains. Attempts to complete type boundaries by sketching or connecting the type lines between strips will not add to the accuracy of the type areas unless the strips are run at sufficiently frequent intervals to permit of correct observations. Type boundaries between timbered and nontimbered areas, especially if the latter be open and not brush-covered, are more easily mapped.

Where strips are run twice across a forty, at 10-chain intervals, it is usually possible to complete the sketch map of the type boundaries. Completion of such a map makes type areas simple of determination, thus giving the same advantage as that of the aerial photograph. This result is sometimes possible when strips are run at 20-chain intervals. In extensive surveys of cutover areas undertaken by Michigan and Wisconsin, strips were run from 40 to 80 chains apart and the type boundaries sketched as well as possible. On the other hand, when there is much diversity of type and strips are run at wide intervals, no attempt is made to complete the sketch map, and the percentages of type area to total area, found on the strips, must be accepted for the entire area. This premise will hold good for any large areas by this method, because of the tendency of the areas to balance and approach the true average (Sec. 12). Examples of this kind of estimating are the nation-wide forest surveys of the Scandinavian countries and Finland and the National Forest Survey in parts of the United States, in which surveys the strips were run from 3 to 10 miles apart.

Type maps are therefore possible when small areas such as 40 acres are being estimated closely, when the timber cover is badly broken up by former cuttings, cultivation, or topography and at least two strips per forty are run, or when aerial photographs are used.

The type area may be determined from the map by planimeter or by a tracing ruled off in small squares whose area equivalent is known (Sec. 129).

For the purpose of determining the volume on a forest type, the application of a type correction is as follows: On an area, whether rectangular or irregular, it is planned to run a 10 per cent estimate in the form of strips 1 chain wide and 10 chains apart. On the completion of the strips and type map, the area covered within each type is found by dividing the length in chains of the strips run within the type by 10, since 10 square chains equal 1 acre. If 5 acres are run, there should be 50 acres in the type to give the 10 per cent. The map may show only 35 acres. Therefore $5/35 \times 100 = 14.3$ per cent is covered, while for the other type or types the percentage run will be correspondingly smaller. Some such result is inevitable since type areas are irregular in shape.

To get the estimate of the type, reduce the strip estimate to the stand per acre on strip. This average is then multiplied by the area within the type. The remaining types are treated in a similar manner. The estimates of all the types within the area unit are totaled. Only this total need appear in the report on the tract, with the stand per acre as an additional figure if desired. The subtotals for the types are tabulated when needed for growth or management.

156. Site Qualities. Qualities of site vary from optimum to minimum for a given species. Their differences may be such as to produce entirely different types, such as are found, for instance, in swamps or on dry sandy thin or rocky soils as contrasted with better loams. These site qualities are of importance in timber cruising because they influence the volume per acre and especially the average heights of the trees in the stand. So marked is this effect on rate of height growth and consequent total height attained at any given age that height at specific ages has been selected as the best index of site qualities (Sec. 219).

In timber cruising, relative height is compared only with diameter. Even so, differences exist on different sites, for the average heights of entire stands are greater for trees of all diameter classes on good than on poor sites.

When curves for average height on diameter are depended on for securing local volume tables based on d.b.h., it is dangerous to combine extreme variations in site quality. The result would be an average height to be applied in turn to tall and short timber alike. Thus the stands on the poor sites will be overestimated and those on the good sites underestimated. An average sample might give a proper balance for the whole tract, but, as in the case of forest types, the chance for accurate sampling is improved by stratifying sites which show marked differences in average heights on diameter, even though this adds to the expense of computing the estimate. Not over three such site qualities are usually distinguished for any given forest type, although for the entire range of a species from seven to eight site qualities may be found. The increased accuracy of sampling, when the areas sampled are reduced to a relatively narrow dispersion of tree heights and volumes, is clearly indicated by the consideration of the standard error of estimate (Sec. 17).

Site quality is as important in the study of growth as are forest types. For all practical purposes, site quality constitutes an area division to be used in the same manner as forest types, whenever a type is so broad as to cover two or more such qualities which should not be averaged together. Unless the sites are separated in the timber estimate, there is no accurate basis on which growth may be computed.

157. The Stand or Forest Cover. A stand of timber is an area occupied by trees which have similar characteristics. These resemblances include, first, *species or composition*. Types may be composed of stands having marked differences in composition within the type. When such differences are consistent and widespread, they may form the basis of additional separate types as shown above.

The second factor is *age*. Volumes of stands differ more with age than with any other factor, and the separation of stands of distinctly different age is frequently necessary in timber estimating and is fundamental in growth and yield studies. For even-aged stands it will be found that, with advancing age, heights become consistently greater for trees of the same diameter, since height growth of surviving trees is more nearly equal than diameter growth (Secs. 199 and 200), in which sharp differentiation occurs except in stands repeatedly thinned (Sec. 202). Separate height curves and local volume tables are therefore required, within the same species and site, for different age classes. Form may also differ as between ages, tending to a higher average form quotient for older fully stocked stands (Sec. 61). The Girard butt log form factor for second-growth southern pine (Sec. 101) is 0.78 and for old-growth pine it is 0.84, equivalent to 18 per cent difference in tree volumes for equal diameters and heights.

The third characteristic is *density*. Stands may be open-grown, with spreading branches and rapidly tapering bole of low form quotient, or dense and clear of branches with little taper.

Fourth comes *forest form*, or distribution of ages in individual stands. Stands may be even-aged, with characteristics peculiar to that form; or many-aged, with old and young trees intermingled.

Stands may differ in any one or in any combination of these four characteristics. As the result of origin after fire, a forest type may contain large areas of uniform species and age but vary greatly in density of stocking. Whenever possible, differences, especially of stocking, are taken care of by the sampling or strip without separation of areas of different density. Differences in age are more serious, and timber cruisers nearly always estimate separately old and young trees whether separate or intermingled in one all-aged stand or in a two-storied stand.

Area mapping, following the plans outlined for the separation of forest types, is resorted to in order to separate stands of young immature timber containing little or no merchantable volume from older stands in the same forest type which are merchantable. Burned, cutover, or denuded areas within a forest type are also separately mapped. The area of merchantable timber is thus isolated on the map, and the tally or strip estimate may be confined within its boundaries.

To sum up, the functions of the area or map corrections in timber estimating are (1) to eliminate areas containing no appreciable quantity of merchantable timber and (2) to stratify the remaining area into portions within which is found a relatively high degree of uniformity of stand as to composition, average heights, and volume per acre, for which separate areas the percentage covered by the strips is accurately determined, and the standard error of estimate (Sec. 173) which applies to the average acre is minimized.

158. Volume or Density Correction. The map of forest types, site classes, and age classes takes care of all major variations among different stands except that of varying *density of stocking*. This variation is usually pronounced because of the operation of natural factors over long periods of time. Portions of a stand may be overstocked, while

ESTIMATING BY USE OF VOLUME TABLES 265

nearby areas are open or parklike. Fire and insects may have made holes by destruction of individual or groups of trees. Wind has taken toll. If any forty were to be divided into square acres and each acre measured, the variation even in the most regular even-aged stand would be surprising. It may be overcome in one of two ways, the first of which is by increasing the percentage of area covered (Sec. 136). This is the chief reason for counting all trees on a forty, in large timber, since a 100 per cent estimate eliminates error in the count at least. The second method, applied to sampling, is to attempt to correct or modify the estimate obtained on the strips, where it appears to the cruiser that the sample is in error and that a 100 per cent estimate if made would give a different result. It may be questioned whether such an attempt to "correct" a sample should be made, but it is not so difficult to gain accuracy by this method as by the one frequently employed, of guessing at the total stand per acre without measuring a single tree or with the aid of a plot or two taken in the heaviest timber (Sec. 130).

The basis of any correction of strip estimates, short of running more strips, must be an ocular guess at the stand per acre. With the strip estimate as a reference, this guess need be only a *comparison* between the timber on and off the strip. If by inspections on brief offsets to the edge of the tallied strip it appears that the timber off strip averages twice as heavy as that tallied, the estimate for the area not tallied should be double that of the strip. It is much better to express this in terms of the strip estimate than to guess at it independently, for this would make no use of the strip *tally* except for the strip area. The tallied strip is presumably more accurate than the ocular estimate, and this advantage cannot be disregarded without introducing the full error of ocular estimating.

Several such observations are needed if any approach to an accurate comparison is to be secured. If done at all, the comparison must be systematic. This requires two conditions. The observations must be made at regular intervals and must be expressed on a strictly comparative basis for the area on strip and off strip at each point, and for each successive pair of observations. The comparison must be made before the strip volume is known. Expressing each observation off strip in terms of percentage of the estimate on strip at that point will not accomplish the latter objective, because it violates the principle of a weighted average, *viz.*, for two successive observations:

1. Off strip plus 50 per cent of on strip.
2. Off strip minus 50 per cent of on strip.

The mean of the two is apparently zero. But suppose the ocular estimate on the strip at the first point of observation is 10,000 bd. ft. (disregarding the tallied volume, which is not then known), while at the second point it is but 5,000 bd. ft. per acre.

The true comparison is
1. Off strip 15,000 bd. ft., on strip 10,000.
2. Off strip 2,500 bd. ft., on strip 5,000.

Weighted average:

Off strip 8,750, on strip 7,500.

The true correction ($+16\frac{2}{3}$ per cent for the area off strip) is then applied from the computed strip estimate.

The required method of weighing all the observations on the same basis is therefore to set down at each point an *ocular* estimate both *off* and *on* the strip. Then total both columns and divide the off strip by the on strip total.

Since the purpose of this weighted comparison is merely to get the percentage by which the actual tally on the strip must be modified in order to obtain the stand per acre on the area off strip which is not tallied, the ocular estimates both on and off strip may be low or high, yet the correct *percentage* will still be obtained, provided only that the two ocular estimates are consistent or contain the same percentage of error.

Obviously this correction percentage cannot be applied to the entire area of the unit, for this would raise or lower the estimate already determined for the strip or sample, as well. Separate calculations are required for the strip and for the remaining area. In the case of rectangular area units such as a forty, the correction percentage can first be reduced by the percentage of the unit covered by the strip, and then correctly applied to the entire area. With irregular areas or varying percentages covered, the method should be that of first obtaining the average stand per acre on the strip, applying the percentage correction, computing the volume off strip by applying this correction, and then totaling the on and off strip estimates. If 25 per cent of the area is covered in the strip, the volume correction would add not over 10 per cent to the field cost, instead of 300 per cent for a complete tally.

159. Varying the Percentage Covered by Strips in Different Cover Types. The need of area or map corrections, and of supplementary volume corrections for density of stand, becomes greater the smaller the area unit for which accurate separate estimates are required (Sec. 171). As the area unit increases, the effects of the laws of sampling are such that volume corrections for density of stand (Sec. 158) become unnecessary. Type area or map corrections of percentages cruised become impossible unless an unnecessarily large percentage of the total of such large areas is covered or airplane mapping is resorted to. This principle of reducing the percentage of area cruised can be applied in cases where sections of 640 acres or even quarter sections are evidently barren because of fires or topographic and soil factors. On such areas the strip interval may be doubled or quadrupled with proportionate reduction in miles of line run, or it may be omitted altogether if controlled by air photographs. Likewise in types of heavy timber such as are characteristic of coves in the Appalachian hardwood types, a system by which the percentage of this type covered is increased over that run on the slopes above, which are more sparsely timbered, will increase accuracy where most needed.

160. Maps Showing Stocking or Stand per Acre, and Topographic Maps. In addition to the mapping of the above classes of area based on types and ages, it is frequently even more useful to an operator or owner to possess a map showing the distribution of the stands by volume. When this is done, it is customary to select the forty, or in extreme cases 10 acres, as the unit for mapping this distribution. A scale of volume differences is adopted, based on utility with reference to logging operations, and each forty or subdivision containing an estimate which falls within the same class is given the same color or symbol. For example, a Southern lumber company mapped their cutover lands as containing 0 to 1 M bd. ft., 1 to 3, 3 to 5, 5 to 10, and over 10 M bd. ft. per acre and used the map in planning operations for a second cut.

Location and volume of the stands, however, do not finally determine the proper location of roads or other improvements necessary for timber removal. For such purposes, a topographic map is most useful. Rough topographic maps may be made from notes and sketches obtained in conjunction with the timber estimate at little additional cost, or from air photographs.

In mountainous regions, topographic maps with a large contour interval such as 50 ft. may be made by the use of an aneroid barometer. When the slopes are gradual, however, this method becomes less and less accurate because of the inaccuracy of the instrument itself, which reflects differences in pressure caused by both altitude and barometric changes, without distinction. In areas of low relief, the difficulty is obvious.

The best method of obtaining data for a topographic map by groundwork alone is by use of the Abney level graduated in percentages, and a 200-ft. trailer tape with extensions marked for varying percentage of slope. In making maps of considerable accuracy, such data must be acquired on strips run fairly close together because of the limitation of vision as in forest typing. Either the relative or absolute elevations of the strip origins must be known prior to cruising. This means the running of altitude control around the area or along each of the base lines, either with a spirit level or by vertical angles with transit for base control. Abney control can be kept within an error of 20 ft. per mile, if the instrument is kept in adjustment.

Where strips are run and tallied, sketch maps of topography of sufficient accuracy for purposes of logging location may be made in conjunction with the timber estimate and type map at a total cost for all classes of data which does not add materially to the cost of a timber survey.

References

BARNES, G. H.: Relative Accuracy of Various Methods of Compiling Timber Estimates by Volume Tables, *Forestry Chron.*, **4**, 18–22 (1928).

BERRY, S.: Determining the Quality of Standing Timber, *Jour. Forestry*, **15**, 438–441 (1917).

GEVORKIANTZ, S. R.: Guide for Estimating Defect in Northern Hardwoods, *Jour. Forestry*, **45**, 128–130 (1947).

Lake States Forest Experiment Station: Scaling Cull in Standing Trees in Board Feet, Scribner Rule, *Tech. Note* 107, 1936, 1 p.

LEXEN, B.: The Determination of Net Volume by Sample Tree Measuring, *Jour. Forestry*, **45**, 21–32 (1947).

ZILLGIT, W. M., and S. R. GEVORKIANTZ: Estimating Cull in Northern Hardwoods with Special Reference to Sugar Maple, *Jour. Forestry*, **46**, 588–594 (1948).

CHAPTER 20

MEASUREMENT OF PIECE PRODUCTS IN THE TREE

161. Estimation of Poles and Piling. Both poles and piling are usually taken to lengths sufficient to utilize all of that portion of the tree bole which is free from large limbs, excessive taper, or butt swell. Every additional 5 ft. of length adds a greater proportional increment to value than is expressed by the additional length. Hence each tree contains one pole or pile, plus a small amount of cordwood in tops and limbs. Volume tables for poles and piling may be obtained best by using middle area, though the Smalian formula (Sec. 28) is also used. In the latter case, the averaging of the top and butt cross sections, in flaring butts, results in an overestimate of volume. On the other hand, the averaging of the top and butt diameters, as they stand, especially in the lack of butt flare, results in an underestimate of volume. The merchantable cubic-foot volume table to a 5- to 6-in. top would coincide closely with the total volume in piling and the solid contents of cordwood.

Poles and piling are estimated by tallying each suitable tree for a piece under its proper length. The points to be noted are top diameter limit inside bark, length to that point, taper or circumference at butt, sweep or crook, and defects. For poles or piling, trees must be free from sudden change in taper due to limbs and must have reasonably small butt measurements, requiring that they be cut above the butt swell. The specifications for sweep are rather exacting (Sec. 56). Visible defects such as rot will usually disqualify the piece. Trees whose merchantable top diameters as limited by branching exceed by 3 or 4 in. the minimum specified for the pole are better manufactured into lumber or sawed ties.

162. Dendrometer for Top Diameters of Piling or Poles. In tallying the number of poles or piling in standing timber, it becomes necessary to determine the point on the tree having a diameter equal to or slightly greater than that required for the product. This point on different trees may be from 18 to 90 ft. from the ground, depending upon the individual trees in the stand. The ocular estimation of diameters at such distances from the eye is subject to serious error. Con-

270   FOREST MENSURATION

tinuous training of the eye will tend to improve the ability to estimate closely. An instrument intended for use in the measurement of diameters at a distance from the ground is termed a *dendrometer*.

The Biltmore stick or a similar straight stick may be given a width which will serve as a guide in estimating given upper diameters. When this stick is held vertically at a fixed horizontal distance from the eye, and at a distance from the tree for which the stick is graduated, the principle of similar triangles will work for upper diameters as it worked for diameters at the base (Sec. 116). The top of the stick will

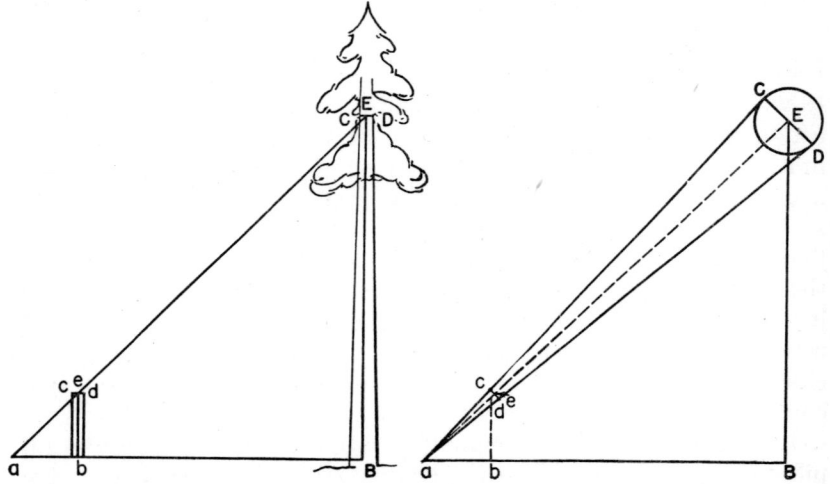

FIG. 54. Method of graduating the thickness of a stick as a dendrometer to measure a fixed top diameter for poles or piling. In the triangles shown on the right side of the figure, the scale of inches, cd and CD, has been exaggerated.

intercept the same horizontal or diameter measurement on the tree regardless of the height at which the measurement is taken. In Fig. 54, it is evident that the lines ae and aE are proportional to the horizontal distances ab and aB, since the triangles are similar triangles, or

$$\frac{ae}{aE} = \frac{ab}{aB}$$

Since the distances ab and aB are fixed, the ratio must be constant. Now ae and aE are the altitudes of a second set of similar triangles acd and aCD, in which the following proportion must hold:

$$\frac{ae}{aE} = \frac{cd}{CD}$$

MEASUREMENT OF PIECE PRODUCTS IN THE TREE 271

and, by joining the two proportions, one gets

$$\frac{ab}{aB} = \frac{cd}{CD}$$

which proves the above statement.

To find the width of the stick, the last proportion will be converted and accepted values introduced for the solution. Assuming that the stick is made for a 25-in. reach, or 2.083 ft.; that the observer will use it at a constant horizontal distance of 66 ft., or 1 chain from the tree; and that the top diameter will be 8 in., the proportion is changed to

$$cd = \frac{CD \times ab}{aB} = \frac{8 \times 2.083}{66} = 0.252 \text{ in.}$$

It will be noted that the horizontal distances are given in terms of feet and the diameters in inches, but this does invalidate the calculation. A stick of 0.252 in. in thickness held vertically 25 in. from the eye of the observer who stands 66 ft. away from the tree will therefore show the location of the 8-in. top. The Biltmore stick can be notched at one end to give the required diameter. This method must be used with care, since with careless manipulation it is subject to great bias.

For measuring the lengths of poles or piling, the Merritt hypsometer graduations may be made at 5-ft. intervals and read preferably from the top down. From a single position the length of the pole from its minimum top diameter is shown on the hypsometer. The 5-ft. graduation for the stick for a distance of 66 ft. is

Reach, in	5-ft. graduation on stick, in.
23	1.74
25	1.89
27	2.04

For reading at a distance of 99 ft., or $1\frac{1}{2}$ chains, these values must be reduced one-third.

163. Measurement of Sweep in Standing Trees. The Biltmore stick or any straight stick may also be used to measure the amount of sweep and to determine whether it exceeds the specifications. Sighting the stick along the concave side of the sweep so that its edge intersects at the same time the middle of the top and that of the butt, the observer can readily tell whether the axis of the pole or pile falls within the piece or, if not, the approximate amount of divergence beyond this.

For poles with a one-way sweep, the stick will be held to intersect the profile of the pole at the top and the point at the butt which is called for in the specifications for the particular species and size (Sec. 56).

Two measurements at right angles should be made, in order to find the maximum sweep. As these observations can be taken at any distance from the tree, the cruiser can for this purpose stand close to the base and thus take the positions with least distance to cover. If the tree qualifies as to straightness and soundness, it may then be measured for its length, at the required distance.

Where piling is scarce and commands a high stumpage price and the trees which qualify are few, the estimate on small areas should cover 100 per cent of all pile trees, which may be done in rough strips and by marking each tree as it is tallied. On large areas, or for smaller material such as short poles, the strip system may be used.

The tally for piles or poles does not require d.b.h. but will read

Piling	Lengths, ft.				
	20	25	30	35	etc.
Number of trees..........................	—	—	—	—	—

A similar tally will be kept for poles, and care should be taken to reject those whose minimum circumference at butt exceeds the specifications.

164. Tree Volume Tables for Ties, by Grades. To substitute tree volume tables for tie log rules would result in a saving of effort similar to that of using volume tables instead of logs in estimating for board-foot volumes. Just as in the case of board feet, the volume contained in trees of standard d.b.h. and total height classes will depend upon the number of logs or bolts in the tree up to the limit of merchantable diameter, the d.i.b. at the small end of each of these bolts, and the sum of the tie contents read from the tie volume table.

For trees of any given d.b.h. and total height, these bolt diameters will depend on the taper, form quotient, and consequent form class of the tree. Those with full form, falling into classes 0.70 to 0.85, will yield a much larger number of ties than poorly formed trees from 0.60 down. To fall below the minimum size means rejection of the tie, or degrading to a smaller size class. In the former case the entire volume is excluded.

Two possibilities are open for the construction of such tree volume tables for ties. The first is to average the total volumes in ties of different grades for all trees falling in the same diameter and height class, regardless of form. The results would be expressed in fractions of ties per tree, of each grade. This method would give an accurate estimate, provided that the average form quotient or class of the trees used in the table coincided with that of the stand to be estimated.

The second method is first to determine the average form class for the timber and then to apply it to a tie table for trees of this same form class.

165. Estimation of Mine Props and Mine Timbers. Were it possible to sell mine props on the basis of their cubic contents, it would greatly simplify the estimation of the standing timber. The top diameter to which props are merchantable would give the basis for the cubic contents of the tree, and a standard "merchantable" cubic volume table would give the contents of the tally. Where props are purchased by the piece, the yield of trees of different size depends on the specifications as to length of prop and diameter at the small end. Props may be split where there is no demand for the large size in round form; but, with a broad market, nearly the entire tree of the smaller sizes will find use in the form of round props.

The number of props is often coordinated with weight, carloads being purchased on this basis. Green weights are in this case usually the basis of sale.[1]

For any extensive work in estimating the number of mine props of different sizes in standing timber, tables of taper are useful in the same manner as for crossties. From the tapers of trees of different diameters and heights, it is possible to read, for the desired lengths and top diameters, the number of props of each class which can be cut from a tree of each size class. In this way a tree volume table can be made for mine props or, in a similar manner, for mine ties or combinations of the two.

166. Estimating Post Timber. Posts are similar to mine props in character and may be estimated in the same manner. Posts are cut more uniformly to standard lengths and are commonly estimated by counting the trees and estimating the number of round posts which may be cut from the average tree. Split posts, as in the case of split props, can be computed from the contents of bolts of proper length. Tree volume tables, where considered useful, could be made in the manner outlined for crossties or mine props. In the case of any

[1] McIntyre, A. C., and G. L. Schnur, The Measurement of Mine Props, Linear Feet, Top Diameter, Weight, and Volume Tables, *Pa. Agric. Expt Sta. Bul.* 269, 1931.

round or shaped products, the principles of sampling the area by strips or plots hold good.

References

ANDERSON, I. V.: Western Larch and Douglas Fir Volume Tables, *West Coast Lumberman*, **58** (11), 40 (1931).

CUMMINGS, L. J., R. M. VARNEY, and R. E. SWANSON: Taking the Guess cut of Cedar Pole Inventories, *Rocky Mountain Forest and Range Expt. Sta. Applied Forestry Note* 88, 1935, 8 pp.

D'ABOVILLE, P.: Determination of the Middle Diameter of a Standing Tree, *Jour. Forestry*, **17**, 802–806 (1919).

GIRARD, J. W., and U. S. SWARTZ: A Volume Table for Hewed Railroad Ties, *Jour. Forestry*, **17**, 839–842 (1919).

GRAETER, G. C.: How to Cruise Timber for Cross Ties, *Cross Tie Bul.*, **8** (4), 8, 10 (1927).

TIMM, J. L.: Conversion of Standard Pole Classes to Tree Diameters in Lodgepole Pine, *Northern Rocky Mountain Forest and Range Expt. Sta. Res. Note* 39, 1946, 2 pp.

CHAPTER 21

FUNDAMENTAL STATISTICAL TECHNIQUES

167. Estimating by Sampling. In Chap. 2 the computation of the measures of central tendency and of dispersion was briefly outlined, together with a listing of the more pertinent definitions of terms involved. In Sec. 12 it was shown that, whenever an estimate of the total volume of timber on a given area was sought, the measurement of the timber on only a sample of the area cannot be mathematically accurate, except by chance. Here is a specific case for the application of the principles of standard deviation and its associated measures.

When a portion of the tract is covered, the average stand per acre for the entire tract must be calculated from the area and volume of the portion cruised, and if the sample is truly representative of the total area, the total volume on the tract will be obtained by multiplying this average volume by the area. Owing to the variability of stands, such as the range of volumes on individual areas and the variation in forest type, age, stocking, composition, and site class, precise accuracy in the estimate of the total volume cannot be expected as the result of any partial or sample cruise. If a single stand is sample-cruised by 10 different estimators, it is highly probable that 10 different estimates of total volume will be obtained, even if each estimator employs accurate techniques of measurement. This variation or error of sampling is controllable to a certain extent and can be brought within predetermined limits of accuracy, since definite mathematical laws have been developed covering such cases. These laws apply whenever the units subject to measurement belong to a single population or universe (Sec. 12), or, in other words, are definitely related to each other and spring from a common origin.

Three conditions are necessary in any job of sampling. If they are not met, the results may be unreliable and even meaningless.

1. The universe or population must actually constitute an organic or related group. Since forest stands are commonly a composite of several populations of trees, owing to differences in type, age, stocking, and site, each with a different average volume per acre, the first effort in estimating the total volume must lie in dividing the forest into its

unit populations in so far as feasible. Then each population can be sampled and the results of the sample be applied to the area occupied by the population in question. Such subdividing is usually called *stratifying*, and the sample is a *stratified sample* (Sec. 12). Stratifying is the first and most essential step to be taken in improving the accuracy of estimate of volume of a nonhomogeneous forest.

2. The selection of the units to be measured, whose sum constitutes the sample, must be absolutely free from bias or deliberate choice, or, in statistical terminology, must be *random*. A random selection means that the observer has no choice in selection but that one unit has as good a chance of selection as any other. In timber estimating, it has proved most feasible to allow place or location alone to dictate selection of the area. The estimator predetermines his design of placement before he sees the timber stand and then abides by this placement. Such selection is commonly called *mechanical* sampling, which the pure statistician is apt to abhor, since it is not totally random. Foresters as a whole deem this form of sampling sufficiently random in nature and besides are more or less forced to adopt it because of the technical difficulties in applying true random selection to the field and because other information, for example, topography and improvements, is collected on the cruise. Mathematical statisticians are now developing mechanical sampling theory, but it has already been demonstrated by foresters (Sec. 133) that at least some forms of space location in forestry work qualify well under the usual random sampling theory.

3. The measurement of the units of the sample must conform to a required standard of accuracy. The statistical theory takes into account only the variations inherent in the population and does not make provision for errors of inaccurate technique. Such errors in timber estimating would consist of measuring diameters too high or low on the bole of trees, using instruments that are out of adjustment, and inaccurately determining the width of the strip or the borders of the plot. In addition to these would be the fundamental error of using a volume table unsuited to the conditions. Such errors are usually cumulative rather than compensating (Secs. 113, 121, and 149).

168. The Normal Frequency Distribution. Distributions have been previously defined (Sec. 16) and an example was shown, but no effort was then made to discuss the basic laws governing distributions. It was seen that the individuals of a population varied from one another but that in their aggregate they formed a distinctive pattern of spread. This pattern is related to a definite mathematical *law of probability*.

Simple evidence of this universal law of probability lies in a test of tossing a given number of coins and tallying the resulting numbers of heads or tails for each toss. With 10 coins the possibilities are as follows:

Heads: 10 9 8 7 6 5 4 3 2 1 0
Tails: 0 1 2 3 4 5 6 7 8 9 10

It does not follow that, with 1,100 tosses of the 10 coins, there will be 100 occurrences in each of the above 11 combinations. An actual test of 1,268 tosses has resulted in the following order:

2 12 50 151 253 330 252 149 54 14 1

Thus combinations of 10-0 or 0-10 rarely occur, but occurrences of 6-4, 5-5, and 4-6 are commonly obtained. These numbers of occurrences or frequencies can be plotted over the combinations, as in Fig. 55, from

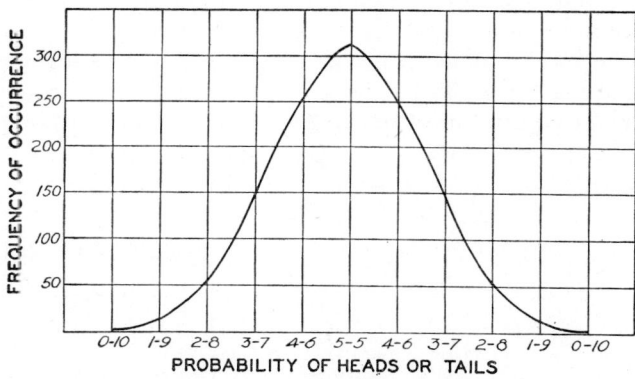

FIG. 55. Normal curve of error or probability obtained from tossing 10 coins 1,268 times and tallying heads and tails.

which it is readily seen that the *frequency curve* is symmetrical and bell-shaped.

If only 100 tosses had been made, the symmetry would not have been so pronounced, and there would have been irregularities, yet with the same tendency to group near the center. With each additional 100 tosses, the irregularities would have been gradually removed until the comparatively smooth curve of the 1,268 tosses was obtained. It is to be expected that, if the numbers of tosses are increased indefinitely, complete regularity will be obtained.

A second feature of the above series is that it is *discrete* or *discontinuous*, in other words, that frequencies exist only for specific points and not for all intervening graduations. Figure 55 errs in that the continuous curve may convey the idea that the series is *continuous*, but here the error has been deliberately incurred to bring out the symmetry.

If now the total number is made to approach infinity with frequencies put on a percentage basis, rather than in terms of absolute totals, and the series is made continuous, the normal frequency distribution is derived, which in one of its mathematical forms is expressed as

$$Y = \frac{1}{\sqrt{2\pi}} e^{-\frac{x^2}{2}}$$

where Y is the ordinate or percentage of occurrence for any *normal deviate* x from mean and e is the natural constant or the basis of the Naperian logarithms and is equal to 2.71828. The normal deviate is a variation from the center of the series expressed in terms of standard deviation. The curve is commonly called the *normal curve of error* and has an area under the curve equal to 1.0000.

When the total number of occurrences is set at N and the standard deviation of a given series has been computed as s (Sec. 16d), the normal frequency equivalent of the given series is obtained by the formula

$$Y = \frac{N}{s\sqrt{2\pi}} e^{-\frac{x^2}{2s^2}}$$

where x is now not the normal deviate but a deviate in terms of the variable X, and Y is the ordinate, or the number of individuals expected at the value of X. Since the calculation of normal frequencies from known series is a common practice, tables have been set up which eliminate the necessity of going through the detailed and complex manipulation of the formula. The table shown (Table 43) is considered adequate for most elementary purposes and is based upon the first of the above two formulas. It shows the relative height of the ordinate at regular intervals of standard deviation from the mean, as well as the cumulative area under the curve, working from the left extreme.

One may consider the normal frequency curve a curve of perfection. In many instances, unknown factors, operating against pure chance, enter into the problem and introduce distortions in the curve. For example, if in the example of tossing coins, one or more coins were not exactly balanced about their central plane, a tendency would be introduced for those coins to fall on their heavier side, and a bias in frequencies in that direction would be obtained. The frequency curve loses symmetry and becomes *asymmetrical* or *skewed*. In biological phe-

TABLE 43. AREAS AND ORDINATES OF THE RIGHT HALF OF THE NORMAL CURVE OF ERROR*

x/σ	A	Y	x/σ	A	Y	x/σ	A	Y
0.00	0.5000	0.3989	0.78	0.7823	0.2943	1.56	0.9406	0.1182
0.02	0.5080	0.3989	0.80	0.7881	0.2897	1.58	0.9429	0.1145
0.04	0.5160	0.3986	0.82	0.7939	0.2850	1.60	0.9452	0.1109
0.06	0.5239	0.3982	0.84	0.7995	0.2803	1.65	0.9505	0.1023
0.08	0.5319	0.3977	0.86	0.8051	0.2756	1.70	0.9554	0.0940
0.10	0.5398	0.3970	0.88	0.8106	0.2709	1.75	0.9599	0.0863
0.12	0.5478	0.3961	0.90	0.8159	0.2661	1.80	0.9641	0.0790
0.14	0.5557	0.3951	0.92	0.8212	0.2613	1.85	0.9678	0.0721
0.16	0.5636	0.3945	0.94	0.8264	0.2565	1.90	0.9713	0.0656
0.18	0.5714	0.3925	0.96	0.8315	0.2516	1.95	0.9744	0.0596
0.20	0.5793	0.3910	0.98	0.8365	0.2468	2.00	0.9772	0.0540
0.22	0.5871	0.3894	1.00	0.8413	0.2420	2.05	0.9798	0.0488
0.24	0.5948	0.3876	1.02	0.8485	0.2347	2.10	0.9821	0.0440
0.26	0.6026	0.3857	1.04	0.8508	0.2323	2.15	0.9842	0.0395
0.28	0.6103	0.3836	1.06	0.8554	0.2275	2.20	0.9861	0.0355
0.30	0.6179	0.3814	1.08	0.8599	0.2227	2.25	0.9878	0.0317
0.32	0.6255	0.3790	1.10	0.8643	0.2179	2.30	0.9893	0.0283
0.34	0.6331	0.3765	1.12	0.8686	0.2131	2.35	0.9906	0.0252
0.36	0.6406	0.3739	1.14	0.8729	0.2083	2.40	0.9918	0.0224
0.38	0.6480	0.3712	1.16	0.8770	0.2036	2.45	0.9929	0.0198
0.40	0.6554	0.3683	1.18	0.8810	0.1989	2.50	0.9938	0.0175
0.42	0.6628	0.3653	1.20	0.8849	0.1942	2.55	0.9946	0.0154
0.44	0.6700	0.3621	1.22	0.8888	0.1895	2.60	0.9953	0.0136
0.46	0.6772	0.3589	1.24	0.8925	0.1849	2.65	0.9960	0.0119
0.48	0.6844	0.3555	1.26	0.8962	0.1804	2.70	0.9965	0.0104
0.50	0.6915	0.3521	1.28	0.8997	0.1758	2.75	0.9970	0.0091
0.52	0.6985	0.3485	1.30	0.9032	0.1714	2.80	0.9974	0.0079
0.54	0.7054	0.3448	1.32	0.9066	0.1669	2.85	0.9978	0.0069
0.56	0.7123	0.3410	1.34	0.9099	0.1626	2.90	0.9981	0.0060
0.58	0.7190	0.3372	1.36	0.9131	0.1582	2.95	0.9984	0.0051
0.60	0.7257	0.3332	1.38	0.9162	0.1539	3.00	0.9987	0.0044
0.62	0.7324	0.3292	1.40	0.9192	0.1497	3.10	0.9990	0.0033
0.64	0.7389	0.3251	1.42	0.9222	0.1456	3.20	0.9993	0.0024
0.66	0.7454	0.3209	1.44	0.9251	0.1415	3.30	0.9995	0.0017
0.68	0.7517	0.3166	1.46	0.9279	0.1374	3.40	0.9997	0.0012
0.70	0.7580	0.3123	1.48	0.9306	0.1334	3.50	0.9998	0.0009
0.72	0.7642	0.3079	1.50	0.9332	0.1295	3.75	0.9999	0.0004
0.74	0.7704	0.3034	1.52	0.9357	0.1257	4.00	1.0000	0.0001
0.76	0.7764	0.2989	1.54	0.9382	0.1219	4.30	1.0000	0.0000

* x/σ is the *normal deviate*, or the difference of any specific value of variable X from its mean value divided by the standard deviation. A is the area under the curve from the left extreme. Since the curve is symmetrical about the mean, or normal deviate 0, the areas of a minus normal deviate equals that of the corresponding plus deviate subtracted from 1.000. Y is the ordinate of the curve. Since the curve is symmetrical, the ordinate of a minus normal deviate equals that of the corresponding plus deviate without change.

nomena, asymmetry is common, and special statistical techniques have been developed to take care of it or to reduce the asymmetrical form to the normal form. An extensive treatment of this problem is out of question here, since it becomes quite involved.

To use Table 43, it is necessary to compute the arithmetic mean, or average, of the sample according to the methods of Sec. 15a and to get the best estimate of the standard deviation according to Sec. 16d. As an example which repeats the techniques of the named sections and which gives a concrete case for discussion, one can turn to a typical problem, such as that faced by a paper company using several thousand cords of pulpwood bolts per year, which desires to ascertain the average size of these bolts and also the variation in size it may expect.

In solving the problem, several questions arise: It will obviously cost too much to measure all sticks in the several thousand cords of wood. Then, how many sticks must be measured to get a good sample of the whole? What is a fair sample of the whole, or in other words

TABLE 44. COMPUTATION OF ARITHMETIC MEAN AND STANDARD DEVIATION IN A PULPWOOD SAMPLE

Size or diameter, in.	Frequency, number of sticks f	Deviation from assumed mean, in. x_a	fx_a		fx_a^2
3	15	−3		− 45	135
4	238	−2		−476	952
5	441	−1		−441	441
6	416	0	0		0
7	295	+1	+295		295
8	151	+2	+302		604
9	56	+3	+168		504
10	13	+4	+ 52		208
11	2	+5	+ 10		50
Total = S	1,627	...	+827	−962	3,189

Assumed mean = nominally 6, actually 6.45 in.

$$\text{Arithmetic mean} = x_a + \frac{S(fx_a)}{S(f)} = 6.45 + \frac{827 - 962}{1,627} = 6.45 + \frac{-135}{1,627}$$

$$= 6.45 - 0.083 = 6.367 \text{ in.}$$

$$\text{Standard deviation} = \sqrt{\frac{S(fx_a^2) - \{[S(fx_a)]^2/S(f)\}}{N - 1}} = \sqrt{\frac{3,189 - [(-135)^2/1,627]}{1,626}}$$

$$= \sqrt{\frac{3,189 - 11.20}{1,626}} = \sqrt{\frac{3,177.80}{1,626}}$$

$$= \sqrt{1.9544} = \pm 1.398 \text{ in.}$$

what accuracy is desired in the sample? How should the sticks be selected for measurement?

In this case, the simplest choice will be a mechanical selection, wherein every hundredth stick entering the pulp mill on the endless chain for a certain period of time each week will be measured, provided that all the sticks come from the same general forest locality. The sticks are classified by diameter into 1-in. classes, the classification being 3-in. class = 3.0 to 3.9 in., 4-in. class = 4.0 to 4.9 in., etc. The average diameter for the class is therefore not quite at the mid-point of the inch, but slightly below; for example, for the 3-in. class, it is 3.45, and for the 4-in. class 4.45, etc. The computation of the arithmetic mean and the standard deviation proceeds according to the principles of Secs. 15 and 16 in Table 44.

169. Fitting of the Normal Curve of Error to Observational Data. In many problems it may be found necessary to fit the normal curve of error to observational data. In the above pulpwood problem, this question was not raised in the original statement of the case, but a fitting will be demonstrated here for purposes of illustrating the technique. The values of the arithmetic mean, standard deviation, and number of trees are sufficient to define the curve completely. The fitting can be made in two different ways: (1) by use of the normal ordinates of Table 43 and (2) by use of areas from the left extreme of the same table. In the first case, the diameter class of the bolts must first be expressed in terms of the *normal deviate*, or the quotient of the differences between the class centers from the true mean divided by the standard deviation, or $(X - M)/s = x/s$. This is done in the first four columns of Table 45. For example, the 3-in. class with its class center of 3.45 is converted to the normal deviate as follows: $3.45 - 6.367/1.389 = -2.087$. This means that the center of the 3-in. class lies at a distance of 2.087 times the standard deviation (1.398) from the mean or center of the distribution (6.367 in.). Now, since the table of ordinates applies to the equation $Y = (1/\sqrt{2\pi})e^{-\frac{x^2}{2}}$ but we want a fitting in the equivalent equation $Y = (N/s \sqrt{2\pi})e^{-\frac{x^2}{2s^2}}$, the ordinate as read from the table must be multiplied by N/s. In this case, $N/s = 1{,}627/1.398 = 1{,}163.80$. Table 43 shows that the normal deviate -2.087, corresponding to the 3-in. class in the calculation, has a normal ordinate of 0.0452. N/s times the normal ordinate is therefore $1{,}163.80 \times 0.0452 = 5.26$, which is the fitted, or theoretical, number of bolts for the 3-in. class. Table 45 is a consolidated calculation of the entire series.

TABLE 45. FITTING OF THE NORMAL CURVE OF ERROR BY THE ORDINATE METHOD

Size class, in.	Class center, in.	Deviation from mean, $X - M$	Normal deviate $(X - M)/s$	Normal ordinate*	Normal ordinate × $N/S = 1{,}163.80 =$ theoretical frequency
2	2.45	−3.917	−2.802	0.0079	9.2
3	3.45	−2.917	−2.087	0.0452	52.6
4	4.45	−1.917	−1.371	0.1559	181.4
5	5.45	−0.917	−0.656	0.3217	374.4
6	6.45	+0.083	+0.059	0.3982	463.4
7	7.45	+1.083	+0.775	0.2955	343.9
8	8.45	+2.083	+1.490	0.1315	153.0
9	9.45	+3.083	+2.205	0.0351	40.8
10	10.45	+4.083	+2.921	0.0056	6.5
11	11.45	+5.083	+3.636	0.0005	0.6

* From Table 43.

A satisfactory visual comparison of the actual distribution and the fitted distribution is obtained by plotting the two sets of values over the respective class centers, as is done in Fig. 56. The actual distribution has obviously excesses of numbers in the 4- and 5-in. classes and deficiencies in the 6- and 7-in. classes in comparison with the normal distribution. In other words, the actual distribution is somewhat skewed to the smaller size classes.

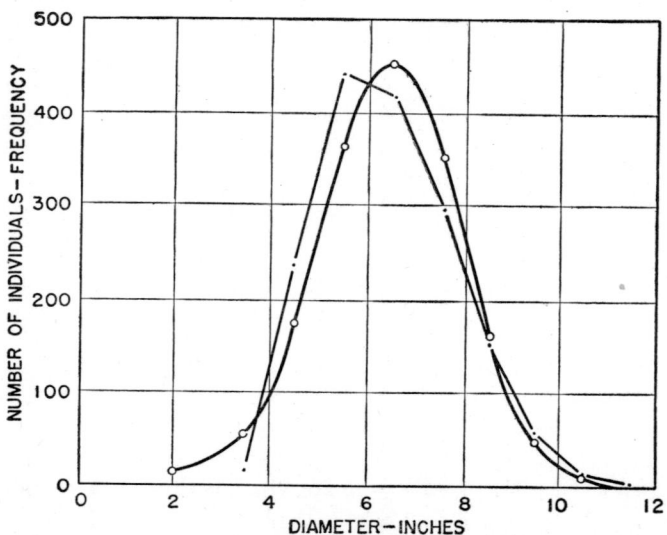

FIG. 56. Curves of frequency of a pulpwood sample.

The second method of fitting, *viz.*, that by area, proceeds in somewhat the same fashion with the exception that upper class limits rather than class centers are dealt with. The upper class limits are expressed in terms of the normal deviate, and then the areas from the left end corresponding to these deviates are read from the table. The next step is to subtract consecutive area readings to get the net area in the class. Finally the net area in the class is multiplied by the total number to get the number in the class. Take, for example, the 3-in. class. The upper limit of the class below is 2.9 and of the class in question 3.9 in. In terms of standard deviates, these are

$$2.9 - 6.367/1.398 = -2.48,$$

and $(3.9 - 6.367)/1.398 = -1.76$, respectively. In reading areas for negative normal deviates from Table 43 it is necessary to read the area for the equivalent positive deviate and subtract it from 1.000. The corresponding areas from the left end as interpolated from the table are 0.0066 and 0.0392, respectively. The difference between the two is 0.0326, which, when multiplied by the total 1,627, gives 53. The consolidated calculation appears in Table 46.

TABLE 46. FITTING OF THE NORMAL CURVE OF ERROR BY THE AREA METHOD

Size class, in.	Upper class limit, in.	Deviation from mean $X - M$	Normal deviate $(X - M)/s$	Area from left end*	Area in class	Net area times N = number in class
2	2.9	−3.467	−2.48	0.0066	0.0066	11
3	3.9	−2.467	−1.76	0.0392	0.0326	53
4	4.9	−1.467	−1.05	0.1469	0.1077	175
5	5.9	−0.467	−0.33	0.3707	0.2238	364
6	6.9	+0.533	+0.38	0.6480	0.2773	451
7	7.9	+1.533	+1.10	0.8643	0.2163	352
8	8.9	+2.533	+1.81	0.9649	0.1006	164
9	9.9	+3.533	+2.53	0.9943	0.0294	48
10	10.9	+4.533	+3.24	0.9994	0.0051	8
11	11.9	+5.533	+3.96	1.0000	0.0006	1

* From Table 43.

The discrepancies between the theoretical frequencies of the calculations by the two methods are accounted for by the lack of sufficient decimal places throughout the calculation.

170. Probability of Occurrence. It has been shown in the above that an estimate can be made of the numbers which should fall in each class of the data. A more general case is that the numbers can be predicted between any two limits or above and below any limit. If these numbers are expressed in percentages of the total, one can readily derive the probability of occurrence. For example, the question may be raised as to how many sticks 10.0 in. and over in size may be expected in a lot of 100, or in other words what is the probability of a stick being 10.0 in. or larger? The answer to the question is found again by means of the average diameter, the standard deviation, and the table of areas of the normal curve. The limit of 10.0 in. in terms of standard deviate is $(10.0 - 6.367)/1.398 = 2.60$, for which the table of areas show 0.9953 lying to the left, or 0.0047 in sizes larger than 10.0. Hence 0.47 sticks out of 100 are expected to be in these large sizes, or the chance of any one stick chosen at random of being in this size class is 0.47 out of 100. This form of interpretation has interesting applications. A more general statement of the problem is to ask what the chance is that a stick chosen at random will lie with the range of twice (or three times) the standard deviation, or conversely will lie outside these limits. Table 43 shows that, for a normal deviate of $+2$, the area to the right is $1.0000 - 0.9772 = 0.0228$; since the curve is symmetrical, the area to the left of a normal deviate of -2 must also be 0.0228. The combined areas are therefore 0.0456. This means in turn that there are about 4.6 chances out of 100 that a stick chosen at random will lie outside the limits of the mean plus or minus twice the standard deviation, or $6.367 \pm 1.398 \times 2$. Obviously there are 95.4 chances out of 100 that the stick lies within this range. If the limit of three times the standard deviation is chosen, the chances of exclusion are only 2.7 out of 1,000. These criteria are often used for the purpose of rejecting from a lot of supposedly homogeneous material excessively high or low values which may be considered to belong to another population.

171. Standard Error of the Mean. In Sec. 17, the formula for the standard error of the mean was given. This concept now can be applied to the case in question. The average size of the pulpwood sticks has been computed as being 6.367 in., but since only a sample of the entire lot was measured, one can expect that the precise average of the total lot will be somewhat different. The question immediately arises as to the extent of the difference. One laborious way to get a solution would be to take additional samples of 1,627 sticks, compute averages from each sample, and compare the results. As mentioned previously, this laborious process need not be followed, since the standard error of the mean is adequately estimated from the standard deviation of a single sample through the formula $s_M = s/\sqrt{N}$. The greater the number of observations, therefore, the smaller is the standard error of the mean, and consequently the smaller the range in which the true mean may lie. The accuracy of the mean is consequently directly controllable through the number in the sample, and if the standard deviation of the population is once known, any limit of accuracy may be set for the mean and attained through a computed number of observations. For example, in the case of the pulpwood sticks, suppose that it had been considered sufficiently accurate to estimate the average size with ± 2 per cent standard error. This error is s_M and corresponds to 0.02×6.367, or 0.127 in. Introducing the known values into the above formula and solving for N, we get

$$0.127 = \frac{1.398}{\sqrt{N}}$$

or

$$\sqrt{N} = \frac{1.398}{0.127} = 11.01$$

Therefore

$$N = 11.01^2 = 121$$

Only 121 sticks, chosen without subjectiveness or bias, need have been measured out of the total lot to get an estimate of average diameter which was within the range of ± 0.127 in. of the true mean. It must be remembered, however, that a range of plus and minus one standard deviation (derived from Table 43) includes an area of only 0.6826 or in other words covers only 68 cases out of 100, and conversely excludes 32 cases out of 100. It has therefore become accepted practice to use the limit of twice the standard error (in some cases three times) and to term this multiple the *maximum error*. Thus the question would commonly be stated: How many sticks must be measured to ensure a maximum error of 4 per cent in average diameter? In this case the first step would be to reduce the maximum error standard of 4 per cent to the single standard error basis by dividing by 2 and then proceeding as above.

Accuracy of the mean is now seen to vary with the square root of numbers of observations. To get double the accuracy of a preliminary test, four times as many observations must be taken in the second test. It is also seen that no accuracy standards can be set until there is knowledge of the standard deviation. Since accuracy gains only relatively slowly with increasing numbers, it follows that the undue intensifying of time and effort becomes uneconomic. If the standard deviation is large, it is often useless to set up high standards of accuracy, and conversely if the standard deviation is small, it is relatively simple to attain good accuracy. These considerations must always be borne in mind when one sets up specifications for accuracy in any mensurational job. If the maximum error of average diameter of pulpwood sticks had been set at 1 per cent and if the more rigorous limit of three times the standard error, instead of two times, had been specified, the number of sticks to be measured would have been calculated as follows:

$$\text{Maximum error} = 3s_M = 1 \text{ per cent}$$
$$s_M = 0.33 \text{ per cent} = 0.33 \times 6.367 = 0.021 \text{ in.}$$
$$0.021 = \frac{1.389}{\sqrt{N}}$$

or

$$N = \left(\frac{1.389}{0.021}\right)^2 = 66.6^2 = 4{,}436 \text{ sticks}$$

Has it been worth while in this case to reduce the standard of accuracy from a maximum error of 4 per cent (on $2s_M$ basis) to a maximum error of 1 per cent (on a $3s_M$ basis), since it has increased the work from measuring only 121 sticks to measuring 4,436 sticks? This is a typical question and can be answered only on the basis of practical considerations of time and cost.

An aspect of great importance in respect to this formula is that it shows that accuracy of estimate is dependent not upon the percentage of coverage but simply upon the basis of total units of sample taken. Thus in estimating volume in a forest stand, a common procedure is to accept a definite percentage of coverage, such as 10 per cent, and to apply this portion of coverage to all stands regardless of the size of the tract. The accuracy of the estimated volume per acre and hence of the volume on the tract is then variable with the size of the tract. Small tracts may be poorly estimated, while large tracts of exactly the same character may be closely estimated. In order to get the same degree of accuracy in all sizes of tracts, it is theoretically necessary to take only the same number of plots in each tract regardless of size of tract, provided that the sample units are so scattered through the area as to give the best representation of the conditions. Under practical circumstances, however, this theory is often disregarded, since it may be desired to have fairly reliable volume information by small subdivisions, such as each forty, or quarter section, in the entire tract. The subdivision then becomes the unit for calculation, and all sizes of tracts are reduced to a common basis. In this case the volume per forty or quarter section maintains a constant level of accuracy.

172. Regression. There has been little attempt in previous chapters to define the relationship between variables other than in a graphical sense. Sometimes it is convenient and even essential to give the relationship a mathematical expression. Although the fitting of curves to complex data involves complicated techniques, the fitting of a straight line to the relationship of two variables is fairly simple and not beyond scope of this text. The derived line of fit is commonly called a *regression line*. It may be a straight (*rectilinear regression*) as is often the case of bark thickness plotted on diameter, or it may be curved (*curvilinear regression*) as is the case of heights of trees plotted on diameter.

A rectilinear regression between two variables takes the equation form of $Y = a + bX$, where Y is the dependent variable; X the independent variable; a the value of Y when $X = 0$, or the *intercept;* and b the *slope* of the line, or the *regression coefficient*. In order to develop the method of fit, the case of 20 measurements of diameter growth for the last 20 years in an even-aged stand of ponderosa pine 65 years old growing on a poor site will be taken. The data are as follows:

Diameter of tree, in.	20-year growth, in.
5	1.0
6	1.2, 1.4
7	1.3, 1.7
8	1.4, 1.5, 1.9
9	1.6, 1.9
10	1.8, 2.0, 2.3
11	2.0, 2.1, 2.2
12	2.6
13	2.3, 2.5
14	2.7

From inspection of the listing, it is obvious that increase in diameter is associated with increase in growth rate. The first question to answer is: Which is the dependent variable and which is the independent? Since this is an even-aged stand, it may be argued that the large trees are large simply because they had fast growth and that tree diameter should therefore be the dependent variable, because it depends upon growth rate. In the field, however, the forester is not going to bore the trees to find out growth rate and from this deduce the diameter of the tree; he is going to measure the tree and try to estimate the growth rate from relatively few samples. Hence diameter growth becomes the dependent variable and tree diameter the independent. For this reason, diameter growth is plotted on diameter. The plotting appears to reveal a rectilinear relationship between the two. The problem now is to pass the best straight line through the scattered points. Graphical fitting has been described before (Sec. 14) and could be used for many purposes. Mathematical fitting will be more precise and in addition will permit deductions beyond those conveyed by the graphical fit.

Since the graphical fit has many useful applications and since an equation can be deduced from it, it will be described in further detail than has been done previously. A satisfactory method is as follows:

1. Compute the average diameter and the average growth rate:

Average diameter = 9.4;   average diameter growth = 1.87

2. Split the data into two groups, one group containing all trees smaller, and the second all trees larger than the average diameter. Compute the average diameter and the average growth for each group:

Small trees: Average diameter = 7.3;   average growth = 1.49
Large trees: Average diameter = 11.5;   average growth = 2.25

3. Plot these three points. A straight line will run through all three points (see Fig. 57). This straight line will balance the points perfectly. By this method, different operators will get the same result; by pure freehand methods, they will probably get slightly different results.

4. Derive the values of a and b to insert in the regression equation $Y = a + bX$, or if the variables of the example are used, diameter growth $= a + bD$.

(1) b is found first by taking the difference between the average growth rates of the two groups of trees and dividing it by the difference between the two average diameters:

$$b = \frac{Y_a - Y_b}{X_a - X_b} = \frac{2.25 - 1.49}{11.5 - 7.3}$$
$$= \frac{+0.76}{+4.2} = +0.181$$

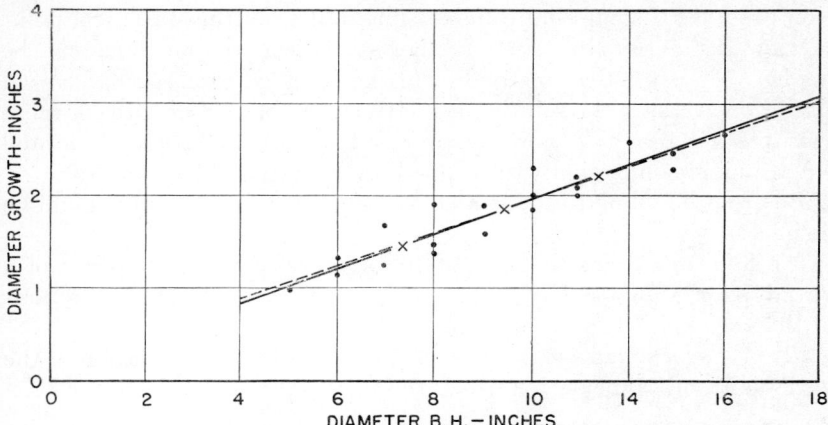

Fig. 57. Fitting a straight line to plottings of diameter growth on d.b.h. for a 65-year-old stand of ponderosa pine on a poor site. Full line is obtained by graphical methods. Dashed lines are obtained by method of least squares.

This is the slope of the line and indicates that, for each increase of 1 in. in diameter, the diameter growth increases 0.181 in.

(2) a is found by subtracting b times one of the average diameters from the average diameter growth associated with it.

$$a = 2.25 - 0.181 \times 11.5$$

or

$$= 1.49 - 0.181 \times 7.3$$

or

$$= 1.87 - 0.181 \times 9.4$$
$$= 0.169$$

Inserting the values in the base equation, one gets

$$\text{Diameter growth} = 0.169 + 0.181D$$

The statistical method of fitting a rectilinear regression line is somewhat more complex and rests on a different basis. In the above described method, it was found that the plotted line balanced the points perfectly, or in other words the sum of the deviates from the line equals zero. With the statistical method, there is still a perfect balance, but of a different sort, viz., that the sum of the squares of the deviates of each point from the line is a minimum, in addition to the sum of the unsquared deviates being equal to zero. The added restriction produces a somewhat different line, but from the statistical point of view a superior line, since further statistical computations can be made. The statistical method of fitting is called the *least squares fit*. For two variables, the solution is relatively simple. For three or more variables, it becomes more complicated (see the solution of normal equations in connection with logarithmic volume tables, Sec. 89).

Before proceeding to the statistical solution, several terms should be defined:

(1) *Sum squares* is the sum of the squares of the full values of a variable, or $SS_x = S(X^2)$, where SS_x is the symbol for sum squares of the variable X, and S is a sum.

(2) *Net sum squares* is the sum of the squares of the deviation of the variable from its mean, for which the formula is

$$\text{Net } SS_x = S(x^2) = S(X^2) - \frac{[S(X)]^2}{n}$$

where x indicates the deviate from the mean. It should be noted that the deviates from the mean need not be found but that the net sum square can be computed directly from the full values, or gross scores.

(3) *Sum products* is the sum of the products of the paired full values of two associated variables, expressed by

$$SP = S(XY)$$

(4) *Net sum products* is the sum of the products of the deviates of paired values from their respective means.

$$\text{Net } SP = S(xy) = S(XY) - \frac{S(X)S(Y)}{n}$$

Again it is not necessary to work with the deviates, since the net sum products can be computed from the gross scores. The expressions using the gross scores appear to be more complicated, but in fact they are the opposite and make a solution easier.

In addition to the above expression, those for $S(X^2)$, $S(x^2)$ and similarly for Y as given previously will be needed.

For the problem of ponderosa pine diameter growth using X as diameter and Y as diameter growth, the values for each are as follows:

$$S(X) = 188 \qquad S(X^2) = 1{,}890$$
$$S(Y) = 37.4 \qquad S(Y^2) = 74.34$$
$$S(XY) = 373.2$$

Then

$$\text{Net } SS_X = 1{,}890 - \frac{188^2}{20} = 122.2 = S(x^2)$$

$$\text{Net } SS_Y = 74.34 - \frac{37.4^2}{20} = 4.402 = S(y^2)$$

$$\text{Net } SP = 373.2 - \frac{188 \times 37.4}{20} = 21.64 = S(xy)$$

If desired, the standard deviations of X and Y can be computed from the first two by dividing by $n - 1$ and taking the square root (Sec. 16d).

$$s_x = \pm 2.52 \text{ in.}$$
$$s_y = \pm 0.482 \text{ in.}$$

Now referring to the equation $Y = a + bX$, b is computed as follows:

$$b = \frac{S(xy)}{S(x^2)} = \frac{21.64}{122.2} = +0.177$$

This means that for each inch increase in diameter there is an increase of 0.177 in. of diameter growth. a is computed as before, *viz.*,

$$a = M_Y - bM_X = 1.87 - 0.177 \times 9.4 = 1.87 - 1.66 = +0.21$$

The final equation, after the values for a and b are inserted, will read

$$Y = 0.21 + 0.177X$$

or in terms of the example

$$\text{Diameter growth} = 0.21 + 0.177D$$

If this line is placed on Fig. 57 (see dashed lines), it is seen to be slightly less steep than the line obtained by the first method. For many purposes, the difference can be considered negligible.

173. Standard Error of Estimate from a Regression Equation. The term *standard error of estimate* must not be confused with standard error of the mean (Sec. 171). It is more analogous to the standard deviation since it indicates a form of dispersion of the dependent variable, *viz.*, dispersion about the regression line. In the case of the standard deviation, there is a question of only one variable. In that of the standard error of estimate, it is a question involving two or more associated variables. The standard deviation is computed from the deviates of the observations from the mean, while the standard error of estimate is computed from the deviates of the dependent variable from the regression line. It is a measure of dispersion to show the variability of the dependent variable after the effect of regression is removed. The formula is given as

$$s_{y \cdot x} = \sqrt{\frac{S(y^2) - bS(xy)}{n-2}} = \sqrt{\frac{S(y^2)}{n-2} - \frac{[S(xy)]^2}{(n-2)S(x^2)}}$$

where $s_{x \cdot y}$ is the symbol for the standard error of estimate of variable Y on variable X, and the remaining terms have been previously defined (Sec. 16). No new computations need be made, other than the insertion of values already in hand into the formula. In the case of ponderosa pine growth, this is

$$s_{y \cdot x} = \sqrt{\frac{4.402 - 0.177 \times 21.64}{18}} = \sqrt{\frac{0.572}{18}} = \pm 0.178$$

or

$$s_{y \cdot x} = \sqrt{\frac{4.402}{18} - \frac{21.64^2}{18 \times 122.2}} = \sqrt{0.0317} = \pm 0.178$$

By mere chance in this case the regression coefficient and the standard error of estimate happen to have the same numerical value. Originally the dispersion of diameter growth about its mean was expressed by a standard deviation of 0.482 in. The new dispersion is only 0.178 in. This means that, if the estimator allows his knowledge of the relationship between diameter and diameter growth to affect his judgment, he can estimate the growth rate of single trees much more closely.

It has been fairly common practice in forest mensuration to compute a standard error of estimate from a freehand regression line by finding the individual residuals, squaring them, and handling the squares as in the case of standard deviation. If the line is well located, the final result will come close to that obtained above.

174. Standard Error of a Difference and of a Sum. If samples are taken from two populations which cannot be proved to be the same, the differences between pairs of means will plot in a distinctive frequency pattern, which when the number of observations increases approaches the normal curve of error. This fact has given rise to a method of distinguishing whether two samples that have been drawn separately represent the same population or two different populations.

To illustrate, the average height of 100 Caucasians is 68 in. and that of 100 Asiatics is 62 in. Is this difference of 6 in. important, or is it so small that for practical purposes Caucasians and Asiatics can be considered as of the same general height average? Analysis of the measurements showed that the standard deviation (s_C) for the Caucasian heights was 5 in. and that (s_A) for the Asiatic heights was 4 in.

The standard error of the difference (s_D) between these two samples is found by the formula

$$s_D = \sqrt{\frac{s_A{}^2}{n_A} + \frac{s_C{}^2}{n_C}}$$

Since each fraction under the radical is equal to the square of the standard error of the mean, it can be stated that the standard error of a difference is equal to the square root of the sum of the individual standard errors squared. In this case,

$$s_D = \sqrt{\frac{5^2}{100} + \frac{4^2}{100}} = \sqrt{\frac{25}{100} + \frac{16}{100}} = \sqrt{0.41}$$
$$= \pm 0.64 \text{ in.}$$

The actual difference is 6 in., and now it has been found to have a standard error of only 0.64 in. The quotient of the two $6/0.64 = 9.37$ indicates that it is well outside the realm of reason (see Sec. 170) that this difference will occur by chance. The two populations of height must therefore be considered as being different. The factor of 9.37 is much larger than the 2 or 3 which has previously been described as being used for purposes of rejection. Even if the factor had been only 3, Table 43 would have shown that there were only 3 chances out of 1,000 that the two samples could have been drawn from a homogeneous population of heights. The standard error of $+0.64$ multiplied by 3 gives 1.92. Therefore only 1.92 in. of differences between the two collections of height measurements would have been sufficient to discard the hypothesis that the two could be considered as forming a common group.

The standard error of a sum is computed in exactly the same way and by the same formula. To illustrate, we may have found that, in measuring 100 trees, the standard deviation in volume incurred solely by variations in diameter (such as that caused by elliptic cross sections in comparison with circular cross sections) is 5 per cent; that the standard deviation in volume due to inaccurate height measurements is 7 per cent; and that the volume table itself has a standard error of estimate of 10 per cent. When these errors are combined, or added in the measurement of the 100 trees, the combined standard error of the final volume determination will be

$$s_S = \sqrt{\frac{5^2}{100} + \frac{7^2}{100} + \frac{10^2}{100}} = \sqrt{\frac{174}{100}}$$
$$= 1.32 \text{ per cent}$$

175. Correlation. In many cases the question arises as to the closeness of fit or the reliability of the regression equation. Without the computation of the standard error of estimate or other measures to be described later, the only decision that can be made must be stated in general terms, such as a "strong" fit when all the points lie close to the regression line, or a "weak" fit when there is a wide scatter about the line. The standard error of estimate serves the purpose in part by showing to what degree the variation in a dependent variable is reduced by removing the effect of the independent variable. A still more effective way is to compute the *correlation coefficient*. When there is no association between two variables, the correlation coefficient is 0; when there is perfect positive association, the coefficient is $+1$; and when there is perfect negative association, it is -1. Thus the coefficient varies from a maximum value of $+1$ through 0 to a value of -1. Perfect positive correlation would be the case when each plotted point lies exactly on a rising regression line. In the case of the ponderosa pine growth, the points fall in general on a rising line, but not exactly on it; hence the correlation is something less than perfect. A perfect negative correlation would be one where there is a constant and steady decrease in the value of the dependent variable with each unit increase in the independent variable.

The computation of the correlation coefficient again makes use of the values already computed in Sec. 173.

$$r^2 = \frac{S(xy)^2}{S(x^2)S(y^2)}$$

where r is the correlation coefficient and the remaining terms have been previously defined. In the case of the pine

$$r^2 = \frac{21.64^2}{122.2 \times 4.402} = 0.8705$$

and

$$r = 0.934$$

The correlation between diameter growth and diameter has therefore been shown to be a close positive correlation, coming nearly to the value of $+1$ for a perfect correlation.

176. Coefficient of Alienation. The *coefficient of alienation* is a measure with a meaning opposite to that of the correlation coefficient since it shows the lack of full relationship between the two associated variables. The larger the alienation coefficient, the more widely do the plotted points scatter about the fitted line. The two coefficients are related through the expression

$$\text{Correlation coefficient}^2 + \text{alienation coefficient}^2 = 1$$

Consequently if either one is known, the second can be computed easily. In the above example where $r^2 = 0.8705$

$$\text{Alienation coefficient}^2 = 1 - 0.8705 = 0.1295$$

or

$$\text{Alienation coefficient} = 0.360$$

This relationship has been used frequently, since in some cases it is more convenient to compute the alienation coefficient first. This is due to the fact that it can be derived independently from the expression

$$\text{Alienation coefficient}^2 = \frac{\text{net sum squares from the regression line}}{\text{net sum squares of dependent variable}}$$
$$= \frac{S(y^2) - bS(xy)}{S(y^2)}$$

In the example this is

$$\frac{0.572}{4.402} = 0.1299$$

and the alienation coefficient is then 0.360. The two results are identical. A practical application of the formula permits the computation of the coefficient without knowledge of the equation of the fitted line, since the numerator can be found by reading off the deviates of the plotted points from the curve, squaring each value, and adding them. The denominator is computed from the original values of the dependent variable according to Sec. 172.

The practical application just noted is even extended to the case where the fitted line is not linear but is curved. In the case of the curvilinear relationship, the final calculated value is called the *alienation index*, and the correlation measure derived from it through the first formula above is called the *correlation index*. It is not considered sound practice statistically to make these computations from a curve which has been fitted freehand, yet it is satisfying in that they give an approximation of the strength of the relationships.

By referring back to the equation

$$\text{Correlation coefficient}^2 + \text{alienation coefficient}^2 = 1$$

it can be concluded that correlation coefficient^2 represents the percentage of the net sum squares of the dependent variable that has been accounted for or removed by the regression lines and that alienation coefficient[2] is the percentage still remaining unaccounted for.

177. Multiple and Curvilinear Relationships. Many of the problems in forest mensuration involve multiple or curvilinear relationships, separately or in combination. The relation between volume and diameter and height is one example. Many specialized graphical and computational techniques have already been demonstrated for various classes of problems. These are considered to be effective substitutes for rigorous statistical computations. The development of multiple and curvilinear relationships in a statistical sense requires a training in methodology which is beyond the scope of this elementary text. It is a field requiring the services of the expert mensurationist. Even then there are still many problems which cannot be handled satisfactorily, owing to the limited knowledge as to the basic curve forms and equations involved. The mensurationist must continue to rely heavily on sound graphical practices.

References

BEAN, S. H.: A Simplified Method of Graphic Curvilinear Correlation, *Jour. Amer. Statis. Assoc.*, **24**, 386–397 (1929).

BICKFORD, C. A.: A Simple Accurate Method of Computing the Basal Area of Forest Stands, *Jour. Agric. Res.*, **51**, 425–433 (1935).

BLYTHE, R. H., JR.: The Chi-square Test in Frequency Curves, *Jour. Forestry*, **33**, 759–760 (1935).

BRUCE, D., and L. H. REINEKE: Correlation Alinement Charts in Forest Research, A Method of Solving Problems in Curvilinear Multiple Correlation, *U.S. Dept. Agric. Tech. Bul.* 210, 1931, 87 pp.

DAVIS, D. S.: Empirical Equations and Nomography, McGraw-Hill Book Company, Inc., New York, 1943, 200 pp.

DEMING, W. E.: Statistical Adjustment of Data, John Wiley & Sons, Inc., New York, 1943, 261 pp.

ELDERTON, W. P.: Frequency-curves and Correlation, Charles and Edwin Layton, London, rev. 1917, 192 + 22 pp.

EZEKIEL, M.: Methods of Correlation Analysis, 2d ed., John Wiley & Sons, Inc., New York, 1941, 531 pp.

FISHER, R. A.: The Design of Experiments, Oliver & Boyd, Ltd., Edinburgh and London, 1935, 252 pp.

———: Statistical Methods for Research Workers, 7th ed., Oliver & Boyd, Ltd., Edinburgh and London, 1938, 356 pp.

GOULDEN, C. H.: Methods of Statistical Analysis, John Wiley & Sons, Inc., New York, 1939, 277 pp.

MATHER, K.: Statistical Analysis in Biology, Interscience Publishers, Inc., New York, 1943, 247 pp.

ROBERTSON, W. P.: Statistical Analysis of Line Plots, *Jour. Forestry*, **25**, 157–163 (1927).

RUNNING, T. R.: Empirical Formulas, John Wiley & Sons, Inc., New York, 1917, 144 pp.
SCHUMACHER, F. X., and R. A. CHAPMAN: Sampling Methods in Forestry and Range Management, *Duke Univ. Forestry Bul.* 7, 1942, 213 pp.
SNEDECOR, GEORGE W.: Statistical Methods Applied to Experiments in Agriculture and Research, 4th ed., Iowa State College Press, Ames, Iowa, 1946, 485 pp.
TIPPETT, L. H. C.: Random Sampling Numbers, Cambridge University Press, London, 1927, unnumbered pages.
WRIGHT, W. G.: Application of Statistical Methods to Cruising, *Forestry Chron.*, **3** (4), 27–29 (1927).
———: Statistical Methods in Forest Investigation Work, *Canada Dept. Int. Forestry Branch Bul.* 77, 1925, 36 pp.
———: Suggested Applications of Statistical Methods in Forestry Practice, *Jour. Forestry*, **22**, 372–385 (1924).

CHAPTER 22

INFLUENCES AFFECTING THE GROWTH OF TREES

178. Purpose of Studying the Growth of Trees. During the period of exploitation of the original forests, the timber on the land was either destroyed to make room for farm crops and pasture or was regarded as a mine of raw material awaiting the saw. Though it was evident that trees were alive and growing, the idea that the processes of growth and decadence were significant factors affecting investments in standing timber or forest land found no place in the minds of operators until most of the virgin timber had fallen. The carry-over of this attitude caused much of the second growth, which had sprung up in spite of fires and adverse conditions, to be slaughtered in turn. These processes still prevail over a majority of the remaining forest areas in private, especially in small, ownerships.

Many old loggers neither knew the significance of tree rings nor even identified young seedlings as of the same species as the old timber. Aside from indifference as to the future, this attitude was due to the comparatively slow change which the annual growth produced in the outward appearance of trees.

Yet growth and decadence of trees and forests is the one characteristic which sets them apart from minerals and makes possible their management as a renewable crop, capable under intelligent procedures of producing continuous yields of wood and its products. This can be done on most areas submarginal for agriculture and on which grazing furnishes a smaller income or profit than timber.

Like all living forms, trees germinate and find a foothold; the survivors develop to mature size and finally must die, thus making room for other younger trees. They are not immortal, though their life span may so far exceed that of man that in parks and reservations they are often regarded as everlasting until some tragedy occurs and the "Washington elm" or the "Charter oak" is gone.

The growth of trees takes place, first, at the tips, or growing points, of the leaders and branches, which increases the height and spread of the live crown; and, secondly, in the tissue of living cells covering the entire surface of the tree between bark and wood, called the cambium, which annually deposits a new layer of inner bark, the phloem, and of sapwood, the xylem. On the thickness of the ring of wood formed annually, and the rapidity of growth of leader and branches in height and length, depends the rate of increase in total and in merchantable volume of the tree. In temperate climates, or in the tropics where the year is marked by a wet and a dry season, growth occurs seasonally each year, and the early wood, or springwood, usually has a more porous and lighter quality than that formed later, termed summerwood. When the new growth starts again, it may form a sharp contrast in color and density with the late wood of the year before. Annual rings on such species are easily distinguished, as in pines. The more even the texture of growth, as in aspen, maple, and basswood, the more difficult it is to count the rings by eye. But except in cases where accurate historical records and maps are available, the identification and counting of annual rings is the sole dependence of the forester and landowner for determining how long it takes to grow a crop of trees.

In Chaps. 1 to 20, inclusive, the methods of measurement were concerned wholly with what in business is termed the inventory. It is the purpose of an inventory to determine the amounts on hand of all classes of goods at a given date. In this respect it is static and remains accurate only so long as no more goods are either acquired or sold. It is indispensable, but, compared with rate of production and turnover, it is but a single item in the data needed to conduct the business successfully. In the forest, the inventory consists of the stock of raw materials on which future operations will be based, rather than the finished goods, and its importance is greatly enhanced by this fact. An inventory could be kept reasonably up to date were it possible to keep accurate records of all goods acquired and sold. So, in the forest, an accurate estimate can be modified annually to account for changes in the total volume of merchantable standing timber caused by logging and by the natural processes of growth and loss, or mortality. But, as will be seen, this process, though fairly accurate as to the measured volumes of the trees cut and scaled since the estimate was made, is far more difficult when applied to changes in the forest caused by growth and natural losses.

Nor is it sufficient to determine what has happened since the last inventory, or timber estimate. This can be accomplished by a second inventory in the present year, a process as indispensable for forest management as it is for business but seldom attempted at intervals shorter than 10 years because of the expense involved. If all the information needed in the business could be supplied by these successive inventories, or estimates, the actual rates of growth of trees and stands of timber would never need to be studied at all. The growth of a body of timber, over a decade, is most accurately ascertained by contrasting its total merchantable volume at the initial date with the existing volume now, and adding to the present volume the amount cut during the period. The difference between the two totals is the net growth in volume. All changes due to growth, decay, and mortality, and to logging, are automatically accounted for by the successive inventories. In fact, the so-called "recurring inventory" method is an accepted means which European foresters have used to measure *past growth* of their forests (Sec. 266).

There is just one compelling reason for going further than the recurring inventory in an effort to determine how fast trees and stands of timber grow, and that is the need for predicting the *future*. Inventories alone will not do this for a business. Future output is not gauged by amount of goods sold or on hand. Rate of production depends on size of plant, future expected markets, prices, labor, skill, and anticipated profits. In business, past results are only of value as they enable management to forecast the future. So in forestry, although past performance of trees and stands is the only sure foundation for prediction of future growth, yet it is just this prediction of the future volume of stands of timber which is the real objective of all growth studies. Only when the methods of studying past growth are capable of producing fairly accurate forecasts of the growth of entire stands and forests in the future can they be of any real value to the owner and investor in forest property. By the test of this rule, the validity of all growth studies must be gauged.

This orientation of growth studies is emphasized by the fact, often forgotten, that the present worth of productive property depends entirely on the future net income it is capable of yielding. An inventory of present volume does not alone determine value. This is dependent on what the forests, stands, or trees will yield when they are cut, hence on the physical changes which are certain to occur before that time and must be predicted as accurately as possible, and not merely upon future prices, costs, and margins of profit per unit of product. Reduction of these future values to their present worth is the subject of forest valuation.[1]

[1] CHAPMAN, H. H., and W. H. MEYER, Forest Valuation with Special Emphasis on Basic Economic Principles, McGraw-Hill Book Company, Inc., New York, 1947, 521 pp.

Trees and their products do not necessarily constitute the entire value of forest property. The land which produces the crop has a value, just as does farm land. This value cannot be found unless future yields of wood products can be predicted. In the past, when the *growth* of trees was not valued, neither did the land itself have a value when cut over, except for other purposes. It was either sold or abandoned.

The values of forests and forest land for recreation and watershed protection are not measured by the inventory of tree values or growth. These values are appraised by other means.

The importance of growth studies in forest mensuration, and their relation to timber estimates, or inventories, can be summed up as follows:

1. Inventories are indispensable in forestry, and without them no accurate predictions of growth can even be attempted. This will be brought out more fully in later chapters.

2. Prediction of growth and of mortality are equally indispensable for determining the present value of forest property, land and timber together.

3. Predictions of future growth give the only possible basis for sound management of forest property on any other than an exploitation basis.

4. Prediction of future growth must be based on facts determined by the present inventory and past growth, but, unless properly interpreted and applied, data on past growth will not give accurate future results and may even be widely misleading.

Mistakes, involving cost and wasted effort, have been made in the past in interpreting growth data, because of failure to grasp the true objectives of growth studies. The controlling principles in growth studies may be stated as

1. Future growth per acre, not past growth of individual trees, is the end sought.

2. Growth of individual trees in diameter, height, and volume is useful chiefly as a means of determining net growth per acre for the future period.

3. Net growth for a future period is the resultant of the changing rates of increase in height, diameter, and volume of surviving trees in a stand, offset by the death of a portion of the trees during the period.

4. Net growth per acre is easily determined for a past period by recurring inventories, but for a future period it requires careful interpretation of natural laws of growth, not merely of trees as individuals, but of trees as they behave in stands, under competition with other trees.

Determination of volume growth of individual trees by cross-sectioning them and working out the successive volumes which the tree had for each decade of its life, from the rings in each section—the so-called "method of stem analysis" (Chap. 26)—was pursued with enthusiasm in the early days of forest mensuration. But these "stem analyses" have as yet found only a limited use in answering the three vital questions of *volume, area,* and *time* required to grow a *crop* of trees. Furthermore they are too costly. Their utility has been confined to research to determine the effect of changes in form of the tree bole caused by altering the density and therefore the competition of trees with each other. For this purpose, stem analysis has a place. Only when these limitations were recognized did forest mensuration progress toward the solution of the immediate practical problems of predicting the future yields per acre of tree crops.

A tree which grows in an isolated spot, free from competition, takes a form characteristic of the species. The crown attains a full symmetrical development. Usually the live branches persist to a point near the ground. For shade and scenic effects such trees have a high value; for forest products they are relatively worthless. By contrast, the form, height, quality, rate of growth, and chance for survival of forest-grown trees are modified by the struggle for existence, in competing with other trees of the same or of different species in even-aged stands or with older and larger trees in stands of mixed ages. Were it not for this competition, not only would the yield of wood, from scattered trees, be much lower per acre, but valuable products could not be grown, and forests as a crop would not be profitable.

The behavior of different tree species and individuals in their struggle for existence and the means by which natural forces can be controlled so as to produce tree crops of greatest value and profit to man are treated in the subjects of silvics and silviculture. The objective of silviculture is to *grow tree crops*. The measurement of the tangible results of silvicultural treatment lies in the field of mensuration. Hence an understanding of certain of the fundamental laws of tree growth constitutes a common ground of these two branches of applied forestry. The laws of tree growth may briefly be considered under the headings

1. The life pattern of tree species, as such, which in general determines their behavior and ability to survive under competition (Sec. 179).

2. The effect of competition on the individual tree and stand, which falls into two well-defined divisions:

a. Behavior of trees in even-aged stands (Chap. 27 and 28).

b. Behavior of trees in uneven-aged stands (Chap. 29).

179. The Life Pattern of Tree Species. The life pattern of certain tree species is so constituted that they survive best or at all only in even-aged stands. For other species, survival under shade in uneven-aged stands is favorable or even habitual. Each tree species, whether conifer or hardwood, has a characteristic growth pattern from seedling to maturity. This growth pattern for closely related species in the same type, such as red oaks, or for species occupying similar sites in entirely different climatic regions are often very much alike. For instance, individual trees of certain species habitually growing in even-aged stands, such as jack pine in the Lake states, lodgepole pine in the Far West, Virginia scrub pine on old fields in the Middle Atlantic states, and pond pine in the swamps of the South, show a tendency for rapid initial diameter growth, followed by a falling off of the growth rate as maturity approaches, on a comparatively short life cycle of 80 to 120 years. This habit is also characteristic of slash pine, which can grow in dense stands, and to a lesser extent of shortleaf pine and Norway pine in the Lake states. Somewhat more tolerant "even-aged" pines, notably all the species of white pine, tend to maintain a consistent growth rate in the trees which remain dominant, throughout their entire life. Individual hardwood trees, such as oaks, which survive as dominants, develop large spreading crowns which serve to maintain the rate of diameter growth to large sizes.

In addition to the life curve, or trend of growth, species differ inherently in their growth rates. Longleaf pines grow consistently slower than other southern pines. Individual loblolly pine trees grow 50 per cent faster that longleaf pines. Slash pine in turn in its younger stage grows faster than loblolly pine under similar circumstances. The red oak group grows about 50 per cent faster than white oak on the same site. Tulip poplar on suitable sites grows faster than all its competitors. Dogwood has a rapid initial growth for 8 to 10 years and then subsides. Because of these differences in the life curves and inherent rates of growth, accurate growth predictions usually require a separate study of the more important species and forest types.

The life patterns described above are characteristic of tree species which demand full sunlight for their survival and maturity and die when shaded for any length of time. Tolerance, or ability to survive under shade, varies with the different species, through a wide range. For instance, at one extreme, we find aspen and other light-seeded species; at the other, beech, maple, hemlock, and balsam fir. Shade always retards the growth of any species, even if it does not kill the trees. The slowness of growth under shade and the advanced age of small suppressed tolerant trees are astonishing. Yet these same trees, when freed from shade, may grow at rates that compare favorably with less tolerant species. Hemlock illustrates this behavior. Though difficult to establish as a seedling except in sheltered sites, it is capable of growing as fast as white pine when in the open, on the same site.

Life patterns of trees of tolerant species therefore differ fundamentally from those requiring constant light. To the ability to survive in spite of almost no annual growth is added that of recovery of a normal growth in height and diameter following release. This gives rise to the questions, how "old" is a tree and what determines its maturity, decadence, and ultimate death? Within the normal life cycle of the species, what trees reach the greatest age?

In every case it will be found that the oldest trees for a given species, whether it be sequoia, white pine, hemlock, or oak, are those which have undergone a long period of suppression or shading but have finally survived to reach full normal size. The pattern for sequoia may thus produce trees 1,000 years older than the average 2,000-year life span. In one rare case, a 200-year-old tamarack was found with a suppressed core of 75 years. Hemlock trees 400 years old may show two to three periods of prolonged suppression followed by recovery, and they nearly always have a core of slow growth which may count up to 75 years. When small hemlocks of 100 to 150 years of age are freed, growth rates increase rapidly. But the "release" of trees of average mature size produces little change in growth. These patterns can easily be discovered by anyone who will study and count the rings on stumps. The conclusion is important to the grower of timber. Maturity, in trees, is primarily *dependent on size* or volume of the trunk. Actual age is not the determining factor. Age is important only to the extent that it is accompanied by normal growth. Size, expressed as d.b.h., must be substituted for age in all such stands, as the basis of growth studies (Chap. 29 and 30).

180. Normal Mortality—Its Control. In the struggle for survival, undergone by trees as individuals and species, mortality, or losses by death, must occur at all stages of tree development, or else the species itself would be destroyed. To go on living, trees must annually increase their growth and volume by a new ring of wood and hence require a constant progressive increase in space for crown and roots in order to mature and bear seed.

Just as with animals, mortality in trees may occur in the young, the mature, and the overmature stages of the life cycle. Juvenile mortality in trees gives the space essential for growth of crown to the survivors and is the direct result of close competition between too numerous seedlings, saplings, and poles. Mortality of trees of vigorous maturity is low. It is in this class that logging produces the same kind of mortality, for utilization, as does the stockman who slaughters his beef animals in the prime of life. The mortality in decadent stands may be long drawn out but, as indicated above, is closely correlated with the sizes reached. The ultimate death of individual redwood trees has been found to result from a slow shift in the center of gravity of the enormous trunk, due to weakened roots, ending in the ultimate fall of the tree.

The importance of mortality in its effect on survival, maturity, reproduction, and value of the tree crop is very great. The processes of nature are primarily adapted to one ultimate end, which is the survival of the species. Man, on the other hand, is interested in producing values for his own benefit. Natural and human objectives do not harmonize when it comes to the consideration of overmature timber. In nature, it takes too long for the rugged veterans to die, and their spreading crowns for decades hold back or prevent the growth of new stands. Nature finally disposes of these survivors, usually by violent means. Mortality records on permanent sample plots indicate that the larger the tree the greater is its chance of death through wind throw or insect attack, the two primary causes of "normal" losses. Heart rot increases, and lowers the vitality and resistance to wind and beetles. Roots lose their vigor. Fire scars may finally burn through and lay the tree low. Future losses from mortality of overmature trees come, in time, to overbalance growth. Salvage and prompt removal of trees above certain diameter limits become a controlling factor in management. Mensuration, as usual, guides the practice and indicates the range of vulnerable sizes. Then the new crops can be successfully started and the growth rate on the area restored.

In the initial stage, that of reproduction and establishment, mortality is also important. The objective here is not to limit but to encourage mortality at a rapid rate. But this presupposes that abundant reproduction of seedlings is first obtained. For, with scant reproduction, parklike stands are formed with no mortality; the only economic benefit to be hoped for is the early production of seed and the correction of the initial failure by later bountiful reproduction. With this kind of start, rapid differentiation in height growth will not only thin out the overcrowded stand but will ensure good form and clear boles later on.

Here again, man must assist nature in the process of suppressing surplus trees. Again natural processes take too long and are wasteful, no matter how ultimately effective they are. If the strongest, or dominant, trees do not grow considerably faster than the others, the entire stand may become overstocked, with small crowns, arrested height growth, and insignificant growth in diameter. But even in normally differentiated stands the dominant trees seldom grow fast enough to escape a loss of side branches greater than their best growth requires. In no stand is it safe to let the proportion of green crown be shortened to less than about 40 per cent of the total height of the tree. Yet in dense or full stocking this shortening, even of the most vigorous crowns, usually occurs before the competing trees, which are doomed to die, finally give up. Hence man-made thinnings are desirable to forestall and control the timing of this mortality from suppression, to further human ends. When thinnings are properly made, natural mortality from suppression among merchantable size classes is practically eliminated.

181. Abnormal Mortality—Calamities. To the three classes of "normal" mortality so far named, *viz.*, decadence, suppression, and that controlled by human operations, must be added a fourth, termed calamities. Partial or complete destruction of forest stands may occur through tornadoes, fire, insects, or disease, or may be brought about by forest devastation through unregulated cutting followed by one or more of these other destructive agencies. In some instances, following wind or fire, natural reseeding is provided for, but it is often seriously handicapped after clear cutting. In any case, complete removal of the standing timber means making a new start, possibly by the only means left, that of planting. Although nature has taken care of these emergencies, or even turned them to its own advantage, as in the case of jack pine opening its cones after a fire, there is no natural protection against the invasion of imported insects and diseases. Of what use was it to make an accurate estimate of the growth of chestnut, when it was known that the species would be wiped out in 10 years?

Akin to these destructive agencies, young trees face the hazards of climate, to which they are quite vulnerable in the first years. Extremes of cold or heat, and of moisture or drought, may decimate or even wipe out an entire seed catch. Within the life span of the natural forest, reproduction of species acclimated and adapted to the soils and environment is certain to succeed. With man's interference it may permanently fail, and the land will require artificial restocking. Mensuration is concerned with predicting the growth of stands on which full stocking has been secured (Chap. 27), with determining the effect of incomplete stocking on the resultant yields (Chap. 28), and with estimating the volume of possible salvage which may be recovered after destruction by fire, wind, insects, or disease.

182. Characteristics of Height Growth. Survival in the struggle for existence between established seedlings, of the same species and of different species, depends primarily upon their relative rates of height growth, beginning in the first year and continuing until the tree is assured of complete dominance over its competitors. Height growth comes first in importance in predicting the future composition of an immature stand and in selecting the trees which should survive to maturity in a pure stand of a single species. Only if the individual tree succeeds in overtopping its immediate competitors can it obtain the free space required for the lateral expansion of the crown, necessary to give it enough leaf surface to support the demands of the increasing area of live tissue in the cambium of the bole and the spread of the roots in the soil. If the rate of height growth is insufficient to capture and retain a dominant position, the only other recourse for the tree is to get along with less light until its overshading competitors reach maturity and are destroyed. The most tolerant can survive with very slow growth rates.

The typical course of height growth for any species forms a curve reflecting these conditions. For dominant intolerant species, height growth after a year or two of slow development as seedlings forms a rapidly ascending curve, which gradually tapers off as the tree obtains mature volume and size, though it never completely stops as long as life remains. For tolerant species, capable of recovering from suppression, the curve of height growth based on total age will still be similar, provided that these trees originate (by accident) in open ground and are not suppressed. Under shade the height growth of such species as maple, beech, balsam fir, or spruce may show as little as 2 to 6 in. per year, which has no relation to their normal possible growth in height. Some of the more tolerant pines, such as the loblolly and the spruce pine of the South, may survive with a height growth one-fifth of normal and recover vigor if released before the inherent vitality of the tree is lost. The process of recovery and the length of the period of survival under shade are directly dependent on the relative tolerance of the species. Rates of height growth and the normal curve of height on age are therefore significant only for even-aged stands and for trees growing free from overhead shade throughout their entire life cycle. Otherwise the relation of height to age is of no importance except as an indication of tolerance.

In an even-aged stand, the individual trees which secure a slight advantage at the start maintain and increase their lead through a more rapid height growth than their competitors. They then develop a large spread of crown and dominate trees of lesser vigor. Such trees, with most of the crown in full light and shaded only along the lower sides of the crown, are classed as *dominants*. As the stand grows, and each crown requires more room, competition sets in between the young dominant trees, and those of lesser vigor begin to lose ground. As long as the leaders and part of the upper crown still receive full light, these trees are called *codominants* and, together with the dominants, form the main upper canopy of the stand (Chap. 27). Where these codominants in turn become overtopped, they are termed *intermediates* (receiving a little overhead light but with most of the crown shaded) and then *suppressed* when completely overtopped. Death follows in about 5 years for intolerant species.

In all-aged stands dominant trees are usually older trees that have survived earlier periods of suppression because of their tolerance, or trees growing in small openings with plenty of light. Tolerant trees preserve a greater proportion of crown under shade than do intolerant trees, which is the reason for their ability to recover their growth when freed by removal of larger trees. Tolerant trees are killed by too much shade, but the process takes longer. Crown classes in uneven-aged stands are therefore dependent more upon the condition of the individual tree than upon the crown canopy characteristics of even-aged stands.

As long as they remain in the sun, the rates of height growth of the dominant and codominant individuals do not differ greatly. Hence the average height growth of these two dominant tree classes, omitting intermediate and overtopped trees, gives a reliable index of growth rate, in height, of the entire even-aged stand.

The average height growth of an even-aged stand, so determined, varies in direct relation to the fertility of the soil and the presence of an adequate supply of moisture. On sterile or dry soils on the one hand, or with too much water, and peat soils lacking mineral elements on the other, height growth of the stand falls off and the maximum heights attained are less. This fact is used to distinguish differences in site quality for the same specie, or for forest types of similar composition, and is made the basis for classification of yields by separate site classes. Height growth is intimately and directly related (though not in a straight line) to growth in volume attainable on the average acre from fully stocked stands of timber as expressed in "normal" yield tables (Chap. 27).

The most rapid growth in height, in even-aged stands, is attained by dominant trees in stands which are neither too dense nor too open. The excess in height growth in optimum stocking over open-grown trees may be as much as 10 per cent but seldom exceeds this figure. In overstocked stands characteristic of certain species on poor sites, such as jack, lodgepole, shortleaf, and ponderosa pine, initial overstocking may cause almost complete stagnation of height growth. In one instance, a stand of lodgepole pine 70 years old was but 7 ft. high, with over 500,000 stems per acre. On the other hand, certain species such as longleaf pine possess such marked ability to differentiate between individuals and to express dominance and early selective mortality that it is practically impossible to secure too dense an initial stocking.

In any case, the rate of height growth of all trees which form the upper crown canopy in an even-aged stand is relatively the same and begins to fall off noticeably for individual trees only when the crown becomes overtopped. The dominants maintain their lead, but by only a few feet. When, as in plantations, each tree has an even chance, height growth does not become differentiated soon enough for the best vigor of the dominant trees, and early thinnings become important.

183. Characteristics of Diameter Growth. Growth in d.b.h. is merely an indication of diameter growth on the entire bole and of growth in total cubic volume of the tree. Growth in volume is in direct proportion to the spread and surface of the functioning crown of the tree. The loss of crown due to shading and later death of the lower branches occurs before the tree is overtopped. Hence height growth, at the tip, continues unchecked long after the volume of the green crown is reduced by shade. But diameter growth of the tree falls off in response to this loss of functioning foliage. Hence it is retarded much sooner than is height growth. What applies to the crown is also true of the root spread, which must remain in balance with the crown. Increased diameter growth of released trees is due as much, or more, to the added space available for roots as it is to increased light. This is especially true in semiarid regions, for such species as ponderosa pine.

For these reasons, in fully stocked stands only the dominant trees can maintain an even normal rate of diameter and volume growth. In even-aged stands, the task of management is to make thinnings of a character and frequency which will preserve sufficient length and spread of crowns on dominant trees to maintain an even rate of growth, while leaving enough trees to utilize to fullest extent the entire capacity of the site. The more rapid growth attained by trees which have too much space is never enough to give the maximum yield per acre of which the site is capable. Repeated measurements on permanent plots (Chap. 32) are necessary to determine the effects of thinnings.

Mensuration is also necessary to determine the effect of pruning on diameter, height, and volume growth. Excessive pruning of a portion of the trees may cause these trees to fall behind and drop out of the stand. Measurements have shown that, when 50 per cent of the green crown is removed, diameter growth is retarded but that 25 per cent can be removed without loss of growth. Height growth is not retarded until about 75 per cent of the crown length is removed.[1] It has also been shown that, except in overdense stagnant stands, the diameter growth of young dominant trees is not retarded below a normal rate as long as a satisfactory ratio of green crown to total height is retained. This indicates that thinnings can be postponed as long as the dominant trees retain this proportion of crown to height of tree, by which time thinnings may be merchantable for fuel or pulpwood. In unnaturally overstocked stands and in even-aged plantations spaced too closely, or wherever dominance is slow to assert itself, the resulting stagnation of diameter growth of the entire stand is due to the loss of the required crown ratio of the *dominant* as well as the remaining trees. The trend of diameter growth is therefore a sensitive response to leaf surface as determined by crown spread in both length and breadth.

But the direct relation of growth to crown is expressed as growth in volume, not in diameter. Volume is the product of cross-sectional areas and lengths of the measured sections of the bole. The relation of the cross-sectional area in square inches to its diameter in inches is $A = \frac{1}{4}\pi D^2$. Area varies as D^2 and volume as $D^2 L$. A tree which is growing annually in cross section at the same increase in area must show a steady diminution in the width of the annual rings, as shown in the table on page 311.

184. Slowing Down, or Deceleration, of Diameter Growth. Deceleration in diameter growth which does not reduce the area of the cross-

[1] CRAIB, I. J., The Silviculture of Exotic Conifers in South Africa, British Empire Forestry Conference, Great Britain, 1947, 35 pp.

sectional annual increase is characteristic of very young trees and of the first few annual rings, which, except in shaded tolerant seedlings, are usually much wider than later growth. If the rate of diameter growth is maintained, without deceleration, the area of each added ring is successively greater. This is the objective of good silviculture and is often attained in European practice. Wood of even texture is produced. When fully stocked even-aged stands are not thinned, a

Area, sq. ft.	Diameter, in.	Growth in diameter equal to 0.1 sq. ft. in area, in.	Decrease, in.
0.100			
0.200	6.05		
0.300	7.40	1.35	
0.400	8.55	1.15	0.20
0.500	9.57	1.02	0.13
0.600	10.49	0.92	0.10
0.700	11.33	0.84	0.08
0.800	12.11	0.78	0.06
0.900	12.84	0.73	0.05
1.000	13.53	0.69	0.04

rate of deceleration greatly in excess of that indicated in the above table sets in as soon as the crown ratio of dominant trees falls below about 40 per cent, and this condition occurs much sooner with overtopped and even with codominant trees. Old field stands of loblolly pine uniformly show abrupt deceleration of diameter growth from this cause between the twentieth and twenty-fifth year, falling from a normal rate of 1 in. in 3 years to between one-half and one-fifth of this rate.

Pruning, whether natural or artificial, is ineffective in producing adequate percentages of clear lumber from the growth subsequently laid on unless the growth is fairly rapid. Craib's results indicate that the ultimate zone of clear wood must have a radius of over 4 in. in order to make pruning profitable. Although natural pruning occurs more quickly and to greater heights on crowded trees than on trees with adequate growing space, this advantage is entirely offset by the loss in the growth rate. In southern pines natural pruning in fully stocked stands will continue after the crown canopy is opened up by 50 per cent and will still prevent the retention of more than 40 per cent as the green crown ratio. It is delayed or absent only on "wolf" trees, which have started as isolated trees and overtop the main crop.

312   *FOREST MENSURATION*

From these relationships it follows, first, that in the untreated forest and for even-aged stands of intolerant species retardation of diameter growth normally occurs in all except the very largest classes and sizes of trees and is more rapid, leading to death, in the smaller sizes. In the second place, normal deceleration must occur unless the rate of growth in cross-sectional area increases. In managed forests with full stocking, the rate of growth in diameter can be regulated by means of thinning. The effect of thinning seldom goes beyond the point of maintenance of an even growth unless the number of trees per acre is reduced below that necessary to maintain a maximum yield per acre of high-quality timber.

185. Effect of Logging and of Mortality on Diameter Growth. This picture changes in dealings with forests and stands composed of tolerant species, in which attempts are made to maintain a distribution of several age classes of trees on the same acre. Mensuration is the handmaid of silviculture. Only by measuring the resultant growth and yields of stands managed on the principle of commingled age classes (and species) can the relative merits of different management practices and silvicultural systems be determined. The rate of growth of the individual trees comprising the stands is the first point of attack, but the ultimate answer must be annual growth in volume per acre.

Species possessing the capacity to survive under shade, some of which actually need shaded conditions for establishment (hemlock), depend for their ultimate maturity on the release of growing space by the death of the dominant species or individuals which shade them. This is at best a slow and tedious process, prolonging or even doubling the age required to reach mature sizes. Under these conditions, the growth rate of suppressed trees is comparatively meaningless until release occurs. The resultant growth rate may then bear no relation to the previous growth, which it may exceed by 500 to 1,000 per cent.

Semitolerant pines such as loblolly pine, though not possessing the persistent capacity for recovery shown by spruce, balsam, or hemlock, will attain full vigor when given free space and sun if released before the age of about 30 years. For older suppressed trees, the possibility and rate of recovery are progressively diminished.

Intolerant species, such as longleaf pine, as well as other southern pines, will recover immediately in the year of release provided that the surviving crown is fairly well distributed in length, regardless of its width. The shorter the live crown and the less the crown-height ratio, the smaller will be the response to release.

In shortleaf pine and in northern species, response may be delayed for a short period of years, varying with the species, the condition of the tree, and the site. Young trees capable of renewed height growth may increase the length of live crown and with it the rate of diameter growth. Where height growth has ceased, response depends solely on the existing ratio of crown to total height.

What is true of trees growing under natural conditions applies with even greater force to those released by cutting operations. Natural reproduction following clear cutting may depend upon the leaving of seed trees. If these trees will show increased growth and are clean boled, they may put on clear lumber at a rate which will return all the expense of raising a new crop of timber, as has been demonstrated for longleaf pine. A more extensive problem lies in adopting a practice of selective cutting under which, by contrast to thinnings, the larger trees are removed and the smaller ones left. This practice should result in an increase in the rate of growth of all released trees in the residual stand, provided that these trees possess the capacity for recovery. In general, the cutting of the larger trees in the stand, under any system of management, should speed up the growth of the remaining trees and tends to an acceleration of the average rate of diameter growth. For this reason it is sometimes advocated that the larger dominant trees in an even-aged stand of an intolerant species should be cut first, leaving the better pruned codominants to produce a higher quality of lumber. This is a doubtful practice except for rough wolf trees, for the increase in growth rate can seldom equal the rate of the dominant trees if left. When of good form, the more rapid growth of dominants may produce more high-grade lumber than will be grown on the smaller codominants in an equal period. Here again mensuration supplies the answer.

Removal of trees by logging thus has the same effect upon growth predictions as the death of similar trees. The cutting of intermediate and suppressed trees, or thinning from below, has practically no influence on the growth of the remaining stand, since these trees are on the way out. This fact has been demonstrated on permanent sample plots. Only when codominant competitors in the upper crown classes are removed will the future growth of the residual dominant trees be affected. Release of suppressed trees is a different story, but it must be actual release and not a thinning which leaves a stand with enough dominant trees to utilize the growing space fully. Over a large area both logging and mortality are usually continuous in time. If mortality is salvaged when it occurs, the volume of cut is affected in the same way as if the trees had remained alive but were cut in a regular operation. To the volume which these trees had at the beginning of the logging operation is added a quantity equal to one-half the growth they would have put on had they remained alive to the end of the period required to log the area.

In a selection system of cutting, the trees marked for removal are chosen with the objectives, first, of removing all trees which are overmature or decadent or cannot be released because other trees are preferred for release; and, second, to remove a certain prescribed volume per acre from the forest. These preferences apply also to clear cutting in even-aged stands. In many-aged stands the selection system, except for heavy thinnings in young dense stands, tends to remove an increasing percentage of trees of successively larger diameter and all trees over a maximum size beyond which growth ceases to give a satisfactory return on the tree volume. Although the prediction of growth in stands marked for cutting, or those just cut, is concerned only with the residual stand, the problem of prediction for the entire area may require an arbitrary assumption of a diameter limit, above which there exists or will occur by growth a sufficient volume of cut to satisfy the requirements of management. Instead of a *minimum* diameter limit including all *merchantable* timber, there will be indicated a *higher* diameter limit, which may correspond with the zero margin of *profitable* logging. The latter has in the past often been confused with the minimum diameter of the merchantable tree and has caused much clear cutting. The limit may be set still higher and more of the timber reserved for future growth and management. In the absence of stand tables and marking tallies, the assumption must be made that the approximate volume which can be cut during a given period of years will be equivalent to the volume of trees now above the diameter limit chosen as a guide for cutting and marking, plus one-half of the growth that these trees would put on during the full period, plus in addition one-half of the volume of the trees than will grow into and above the chosen limit during the period (Sec. 249). The fact that many trees above this guiding diameter should be left because of their rapid growth and good quality, while an approximately equal volume of trees below the limit should be cut because of defect or for the sake of thinning, does not prohibit the averaging of the data. Stands which have been properly marked and cut under a selection system will show better average diameter growth than if a rigid diameter limit were used. This improvement in growth can be taken into account when the marking practices are known in advance.

In the prediction of future growth, it is evident that the influence of mortality in natural stands is superseded, under management, by the use of the ax and saw. No natural forest is ever "static," by which is meant that mortality would balance growth, year by year, and the total volume remain the same. A forest composed of all-aged stands of tolerant species may approximate this balance but only in the relative absence of wind, fire, insects, or imported diseases. This never happens over large areas. Hence a natural forest is either increasing or decreasing in volume at any given time. The saving of waste and destruction through removal of mature and defective timber and by thinning of the younger age classes is enormous and constitutes man's improvement on natural processes for his own benefit. Silvics and silviculture indicate the best methods of management, and mensuration, through prediction of results, gives the final answer.

186. Effect of Climatic Cycles on Periodic Growth. Whatever influence the sunspot cycle has upon climate and rainfall will affect the current growth of trees to the extent that conditions for growth in the given region are improved or impaired. Where rainfall is the limiting factor of tree growth, recurring cycles of light and heavy precipitation must have an effect on the cumulative growth in diameter and volume as well as on the establishment of reproduction and on mortality. It may well happen that current growth predictions for a future decade following a decade of dry years will be far too low, while a prediction dependent on a past wet decade will considerably overshoot the mark. Hence in regions where studies of current growth, extended back for several decades, reveal a correlation between growth and precipitation sufficiently cyclic to be dependable, this factor cannot be neglected in short-term growth predictions but calls at least for a factor of safety when the growth trend is downward.

References

BEHRE, C. E.: Notes on the Cause of Eccentric Growth in Trees, *Jour. Forestry*, **23**, 504–507 (1925).
BOHANNAN, R. D.: Laws of Tall-tree Growth Investigated Mathematically, *Jour. Forestry*, **15**, 532–551 (1917).
BRIEGLEB, P. A.: Growth of Ponderosa Pine by Keen Tree Class, *Pacific Northwest Forest Expt. Sta. Forest Res. Note* 32, 1943, 15 pp.
BULL, H.: Increased Growth of Loblolly Pine as a Result of Cutting and Girdling Large Hardwoods, *Jour. Forestry*, **37**, 642–645 (1939).
CHAPMAN, H. H.: The Recovery and Growth of Loblolly Pine after Suppression, *Jour. Forestry*, **21**, 709–711 (1923).
——— and R. BULCHIS: Increased Growth of Longleaf Pine Seed Trees at Urania, Louisiana, after Release Cutting, *Jour. Forestry*, **38**, 722–726 (1940).

COPE, J. A.: Loblolly Pine in Maryland, Maryland State Department of Forestry, 1923, 95 pp.
DOUGLASS, A. E.: Climatic Cycles and Tree Growth, *Carnegie Inst. Wash., Pub.* 289, Vol. 1, 1919, 127 pp.; Vol. 2, 1928, 166 pp.; Vol. 3, 1936, 171 pp.
———: Crossdating in Dendrochronology, *Jour. Forestry*, **39**, 825–831 (1941).
DUNNING, D.: Relation of Crown Size and Character to Rate of Growth and Response to Cutting in Western Yellow Pine, *Jour. Forestry*, **20**, 379–389 (1922).
———: Tree Classification for the Selection Forests of the Sierra Nevada, *Jour. Agric. Res.*, **36**, 755–771 (1928).
GLOCK, W. S.: Growth Rings and Climate, *Bot. Rev.*, **7**, 649–713 (1941).
———, A. E. DOUGLASS, and G. A. PEARSON: Principles and Methods of Tree-ring Analysis, *Carnegie Inst. Wash., Pub.* 486, 1937, 100 pp.
HAWLEY, F. M.: Relationship of Southern Red Cedar Growth to Precipitation and Run Off, *Ecology*, **18**, 398–405 (1937).
HOUGH, A. F., and R. F. TAYLOR: Response of Allegheny Northern Hardwoods to Partial Cutting, *Jour. Forestry*, **44**, 30–38 (1946).
HUNTINGTON, E.: The Climatic Factor as Illustrated in Arid America, Carnegie Institution of Washington, 1914, 341 pp. Review, A. B. Recknagel, *Soc. Amer. Foresters Proc.*, **9**, 546–551 (1914).
JESSUP, L. T.: Precipitation and Tree Growth in the Harney Basin, *U.S. Geol. Survey Water-Supply Paper* 41, 1939, pp. 10–13.
KEEN, F. P.: Climatic Cycles in Eastern Oregon as Indicated by Tree Rings, *Monthly Weather Rev.*, **65**, 175–188 (1937).
KOEHLER, A.: Heredity vs. Environment in Improving Wood in Forest Trees, *Jour. Forestry*, **37**, 683–687 (1939).
———: Rapid Growth Hazards Usefulness of Southern Pine, *Jour. Forestry*, **36**, 153–159 (1938).
———: Wood Quality, A Reflection of Growth Environment, *Jour. Forestry*, **36**, 867–869 (1938).
MACKINNEY, A. L.: Increase in Growth of Loblolly Pines Left after Partial Cutting, *Jour. Agric. Res.*, **47**, 807–821 (1933).
MARSHALL, ROBERT: The Growth of Hemlock before and after Release from Suppression, *Harvard Forest Bul.* 11, 1927, 43 pp.
MEYER, H. A.: Growth Fluctuations of Virgin Hemlock from Northern Pennsylvania, *Tree-Ring Bul.*, **7**, 20–33 (1941).
MUNGER, T. T.: Ecological Aspects of the Transition from Old Forests to New, *Science*, **122**, 327–332 (1930).
MURPHY, L. S.: The Red Spruce. Its Growth and Management, *U.S. Dept. Agric. Bul.* 544, 1917, 100 pp.
PAUL, B. H.: The Application of Silviculture in Controlling the Specific Gravity of Wood, *U.S. Dept. Agric. Tech. Bul.* 168, 1930, 19 pp.
PERSON, H. L.: Increment of Residual Redwoods, *Jour. Forestry*, **40**, 926–929 (1942).
PINCHOT, C., and H. S. GRAVES: The White Pine, Century Company, New York, 1896, 111 pp.
ROE, A. L.: A Preliminary Classification of Tree Vigor for Western Larch and Douglas-fir Trees in Western Montana, *Northern Rocky Mountain Forest and Range Expt. Sta. Res. Note* 66, 1948, 6 pp.

SAMPSON, A. W., and W. S. GLOCK: Tree Growth and the Environmental Complex; A Critique of "Ring" Growth Studies with Suggestions for Future Research, *Jour. Forestry*, **40**, 614–620 (1942).

SCHREINER, E. J.: Improvement of Forest Trees, *U.S. Dept. Agric. Yearbook*, Separate 1599, 1937, 56 pp.

SCHUMACHER, F. X., and H. A. MEYER: Effect of Climate on Timber-growth Fluctuations, *Jour. Agric. Res.*, **54**, 79–107 (1937).

——— and B. B. DAY: The Influence of Precipitation upon the Width of Annual Rings of Certain Timber Trees, *Ecol. Monog.*, **9**, 387–429 (1939).

SIMMONS, E. M., and G. L. SCHNUR: Effect of Stand Density on Mortality and Growth of Loblolly Pine, *Jour. Agric. Res.*, **54**, 47–58 (1937).

STICKEL, P. W., and R. C. HAWLEY: Comparative Basal Areas, *Jour. Forestry*, **22**, 302–305 (1924).

TARBOX, E. E., and P. M. REED: Quality and Growth of White Pine as Influenced by Density, Site, and Associated Species, *Harvard Forest Bul.* 7, 1924, 30 pp.

TAYLOR, R. F.: The Application of a Tree Classification in Marking Lodgepole Pine for Selective Cutting, *Jour. Forestry*, **37**, 777–782 (1939).

———: A Tree Classification for Lodgepole Pine in Colorado and Wyoming, *Jour. Forestry*, **35**, 868–875 (1937).

WADSWORTH, F. H.: Value of Small-crowned Ponderosa Pines in Reserve Stands in the Southwest, *Jour. Forestry*, **40**, 767–771 (1942).

WESTVELD, R. H.: The Relation of Certain Soil Characteristics to Forest Growth and Composition in the Northern Hardwood Forest of Northern Michigan, *Mich. State Col. Agric. Sta. Tech. Bul.* 135, 1933, 52 pp.

WOODWARD, K. W.: Tree Growth and Climate in the United States, *Jour. Forestry*, **15**, 520–531 (1917).

ZON, R., and J. L. AVERELL: Drainage of Swamps and Forest Growth, *Wis. Agric. Expt. Sta. Res. Bul.* 89, 1929, 27 pp.

CHAPTER 23

THE AGE OF TREES AND STANDS

187. The Age of Trees. For species growing in temperate zones marked by annual seasons of growth or for some tropical species for which the same condition holds good, the age of a tree may be found by counting the number of annual rings (Sec. 179).

Difficulties in counting rings arise from three sources. For certain species, such as aspen, basswood, birch, and maple, this differentiation is not easily discernible, requiring great care and the use of a lens. For suppressed or slow-growing trees, even those having well-marked rings, it is easy to overlook some of the less conspicuous annual layers. Again a lens is required, together with the smoothing of the surface of the cross section. Slanting cuts increase the area of each ring exposed. The third difficulty is the formation of extra, or *false*, rings, and its opposite, the failure to form rings at the stump section, which occurs occasionally on badly suppressed trees. False rings are more apt to be found on vigorous trees and are caused by periods of drought followed by later rains. They are most commonly found in species inhabiting a semiarid region, such as junipers in the Rocky Mountain region. False rings may usually be detected by tracing them around their circumference. At certain points the summerwood of the false and true rings may join. In plantations whose age is definitely known, the presence of false rings may be detected by the excess in apparent age over the historical record. False rings will usually occur in the same year on all trees, counting inward from the outer ring, or cambium layer. This also aids in their detection.

Suppressed rings occur in years of unusual drought in which practically no summerwood is formed. Again, a knowledge of the existence of this phenomenon correlated with climatic data serves to avoid the error of dropping this ring in the count and reducing the true age of the tree. The drought of 1924 produced a notably suppressed ring in longleaf pine in the South, in both that and the following year.

False rings are common in the first 2 or 3 years of rapidly growing southern pines. Glock[1] gives a rule for distinguishing false rings as

---

[1] GLOCK, WALDO, The Principles and Methods of Tree-ring Analysis, *Carnegie Inst. Wash.*, *Pub.* 486, 1937.

follows: If a thin band of summerwood lies close inside a thick band of summerwood, it is part of a double (false) ring, but if the thin band lies immediately outside the thick band, the thin one constitutes a separate annual ring. This rule will aid in identification on increment cores (Sec. 202).

For tropical trees without annual growth rings the best method of determining growth is but an approximation. It consists in the repeated measurements of sample plots at regular intervals to note the increase in diameter of the different species of trees. The increase is correlated with crown class, exposure to light, and growing space. In this way the period required to produce trees of merchantable size is determined.

188. Seedling Age and Total Age of Trees. The total age of a tree is the period elapsing from its origin as a seedling or sprout to the present year. To determine this age requires two steps. The number of rings may be counted on the stump section; the lower the stump, the more nearly will this number approximate the age of the tree. To this count must be added the years required for a normal dominant seedling or sprout to reach stump height. This varies from 1 year for hardwood sprouts to 20 years or more for seedlings of coniferous species growing on a dry site. None but the most vigorous dominant seedlings should be measured for height growth, since, out of the large number which may germinate, running to over a million per acre in some cases, only a few hundred will reach maturity and those will always be the most rapid growers. To include any others would therefore give too long a seedling period. For the same reason, shaded or suppressed seedlings of a species which normally grows in even-aged stands and is intolerant should never be selected because in all probability these will not survive. Only when the species normally reproduces itself under shade should such seedlings be measured.

For the construction of a table of seedling growth, the selected seedlings should be cut off within 1 in. of the ground and the annual rings counted and height measured. Seedlings whose heights do not greatly exceed average stump height should be chosen, though data may be obtained from older and taller specimens if no others are available. In observing the annual rings, care must be taken to distinguish the pith, which in species with large buds is of considerable diameter. This analysis of annual height growth is best made when the growth of each year can be separated by the presence of annual whorls of branches, or bud scars on the stem.

In plotting seedling growth, height becomes the independent variable, since the heights of stumps are the known factors. The age to be determined is the dependent variable. This reversal of the natural order, which for height growth would be height on age, is necessary since age and not height is here sought. Since both height and age will vary by small gradations, $\frac{1}{10}$ ft. and 1 year, respectively, will be used. Each seedling may give one point representing its total height and age, or a succession of points each representing 1 year's growth. The data, if not too numerous, can be plotted for each tree and a curve drawn and weighted from the resultant graphic pattern. Where data are very numerous, it is best to average the ages for different height gradations of 6 in. and plot the resultant age on the height scale laid out in these 6-in. intervals.

The addition of the average number of years required to reach a given stump height to the number of rings on the stump gives the total age of the tree.

189. The Age of Even-aged Stands. The age of a stand of timber is recorded in the ages of the trees of which it is composed. These ages will vary according to the history and composition of the stand. If it has originated in a single year, as sprouts after a clear cutting, lodgepole pine or aspen after fire, or southern pines in a seed year, all trees whether large or small in diameter will be found to have the same age. If the process of seeding has been somewhat delayed and a full stocking of all the remaining blanks has not been completed for from 10 to 20 years, the trees with largest diameters will be found to be several years older than those of medium size. The smaller trees may in turn be a few years younger than the average. As long as these trees develop together, *forming a single crown canopy*, the stand can still be regarded as even-aged. In this case a differentiation into dominant and suppressed trees will be very marked, a desirable condition favoring rapid growth of the stand. When the difference in age becomes great enough so that the younger trees develop as an understory and come up in openings made in the older crown canopy, the stand as a whole cannot be called even-aged though the dominant portion may be so. Two distinct intermingled age classes, each of even age, are frequently encountered with intolerant species, where the older stand is not fully stocked.

The average age of an even-aged stand which has a spread in the actual ages of its trees should be based upon the trees which make up the larger part of the volume or crop. Small or suppressed trees should be disregarded as well as any understory or later crop. There may be occasional large trees which are older than the main crop.

These should not be considered in determining age of stand. The average age must be obtained from the age of a number of trees taken from the dominant crown class of the principal stand. Where difference in age appears in these main crop trees, there is a choice between averaging these ages or accepting the oldest age as that of the stand. The latter practice is preferred as the more conservative. The greater the apparent range of age the more care should be taken in getting test trees for age. The European definition of average age for such stands is "the age at which an even-aged stand would produce an equal *volume* of wood."

190. The Increment Borer. The age of trees may be found if they are felled, by counting the rings on the stump and adding the average

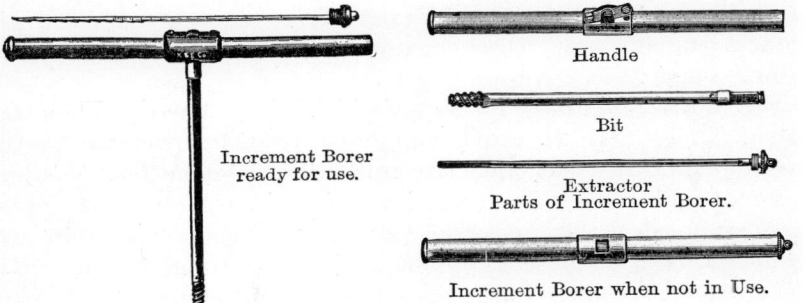

Fig. 58. Swedish increment borer.

age of dominant seedlings of height equal to the stump. But when plots are taken in standing timber these ages must usually be determined from the standing tree. For this purpose the increment borer may be used. This instrument consists of a hollow auger fitted into a handle. When it is bored into the tree, a plug or core of wood is obtained, which is removed by an extractor inserted after the boring is completed (Fig. 58).

The standard form of increment borer, manufactured in Sweden and obtainable in this country, is suitable for northern conifers or for hardwoods of soft texture but has not given complete satisfaction with woods of hard texture.

In the use of the borer, the initial penetration into the wood requires strong pressure accompanied by a very slow turning of the crossbar handle, after which the threads of the screw give the purchase and the auger penetrates rapidly. The extractor is crescent-shaped in cross section and slides alongside the core without injury to the rings. After this is pressed in by hand, the crossbar is given one or two turns to the left, and the extractor is drawn out, removing the core.

In the use of this instrument to determine the age of a tree, the borings should be made at 1 ft. from the ground on all trees whose diameter permits of reaching the pith with the boring. For larger trees, it must be made at b.h., necessitating a further correction for age of seedlings to this height. The boring must reach the pith; and, since this may not be in the exact center of the tree, two or more borings may be necessary, because the first reveals the approximate position of the pith, through curvature of the rings. With a 12-in. instrument, trees up to 20 in. in diameter at b.h. may be bored successfully.

The rings are counted on the extracted cores, sometimes with the aid of a reading glass. These cores are also used in measuring diameter growth (Chap. 26).

191. Correction of Age for Suppressed Trees. A tree can have but one actual age. In the case of suppressed trees, however, this age far exceeds that of a tree of equal size growing on the same site, but which has had access to light from the start (Sec. 179). Once the suppressed trees are freed, if they are of species capable of recovering, they may assume a normal rate or growth and subsequently produce a crop in the usual time required for free-growing trees of the size corresponding to those released.

When relative growth rates can be found between trees growing in full light and trees of the same species whose early period was one of suppression, it is thus possible to assume that, instead of the long period with slow growth, the *normal* age may be found by substituting the growth rate of the dominant tree of an equal diameter; viz., if the core of a suppressed spruce is 2 in., and spruce on the same site free from shade reaches a diameter of 2 in. in 20 years, the normal or *possible* age of the suppressed spruce is but 20 years. It may actually be from 60 to 80 years old. What has happened is that another crop such as aspen or birch is being produced on this land during the period of spruce suppression. The actual age of the spruce has no significance, while the corrected or normal age may measure the period required for spruce, unhindered, to produce a crop. In this way the age of an even-aged stand grown initially under shade may be roughly correlated with yield in spite of the extension of age due to this suppression. "Normal" ages for stands of tolerant species do not have much significance, and age of stand for these species is seldom a factor in management, giving way to diameters as the best indicator of maturity of the crop (Chap. 29).

However, where tolerant conifers such as spruce and balsam normally reproduce themselves in even-aged stands (Sec. 189) under aspen or birch, the yield, at different ages, for such stands, may be based on that of its origin irrespective of suppression of the conifers. Thus age is truly representative of the period required to grow crops of spruce and balsam, under these conditions.

References

ANDREWS, S. R., and L. S. GILL: Determining the Time Branches on Living Timber Trees Have Been Dead, *Jour. Forestry*, **37**, 930–935 (1939).

CUNO, J. B.: Increment Borer and Core Technique, *Jour. Forestry*, **36**, 1234–1236 (1938).

DUNNING, DUNCAN: An Instrument for Measuring Increment Cores, *Jour. Forestry*, **23**, 183–184 (1925).

FENSKA, R. R.: Device for Measuring Increment Borings, *Jour. Forestry*, **23**, 540–542 (1925).

FRITZ, E.: Increment Core Handling, *Jour. Forestry*, **37**, 491–492 (1939).

HALL, R. C.: Handling and Filing Increment Cores, *Jour. Forestry*, **33**, 821–822 (1935).

HORNIBROOK, E. M.: Further Notes on Measurement and Staining of Increment Cores, *Jour. Forestry*, **34**, 815–816 (1936).

LORENZ, R. C.: Discolorations and Decay Resulting from Increment Borings in Hardwoods, *Jour. Forestry*, **42**, 37–43 (1944).

MEYER, W. H., and S. B. HAYWARD: Effect of Increment Boring on Douglas Fir, *Jour. Forestry*, **34**, 867–869 (1936).

PATTERSON, J. E.: Micrometer Slide Adapted to Core Measurement, *Jour. Forestry*, **24**, 691–693 (1926).

REINEKE, L. H.: A New Increment Core Instrument and Coring Wrinkles, *Jour. Forestry*, **39**, 304–309 (1941).

CHAPTER 24

THE GROWTH OF TREES AND OF STANDS

192. Nature of Forest Crops. Trees are a crop produced by the soil but requiring a relatively long period of years to mature. Unlike annual yields of agricultural or horticultural crops, the tree crop has no fixed time of maturity, nor need it be harvested all at once. Each tree may be treated as a separate crop unit that may be utilized whenever it best suits the owner as long as the tree remains alive and after it has reached a size permitting its use for some purpose.

Although the unit of crop production is the individual tree, just as, in the case of field crops, this unit is the single head of grain, fruit, or stalk, yet tree crops must be measured by the two standards applied to agriculture, a measure of quantity, and a measure of area. These constitute the subjects so far dealt with in this text. To these must now be added the measure of time. In field-crop production this may be 1 year or but part of a year as when two or more crops may be grown of the same or of different kinds in a single season on the same land. Or it may take 2 years for a crop. In horticulture, a period of several years may elapse before fruit is borne, and the land may or may not be made to yield other crops during this period. After fruit trees come into bearing, this crop is an annual one, with conspicuous fluctuations from year to year. It must be harvested each year or go to waste. The initial period of waiting must be compensated for by yields of sufficient value to return a profit on all costs of establishment over this period unless defrayed by incidental income. The young orchard is valued on the basis of the net annual income expected after it comes into bearing.

In the forest or stand of trees, important differences appear. Each tree can be harvested but once. Forestry is therefore a process of constantly renewing the crop by natural or by artificial methods (planting), awaiting the maturity of the trees, and then cutting them. Instead of an annual harvest, a tree yields but one crop of wood in its lifetime. The wood is laid on in annual rings, but this wood remains and is added to the volume of the tree in the same manner that interest is added to a sum which cannot be touched by the owner but must be allowed to accumulate until a given year.

THE GROWTH OF TREES AND OF STANDS 325

On the other hand, the demand for wood is continuous. This demand must be supplied annually by cutting different stands of timber each year or by selecting a portion of the trees in a stand for felling and leaving the remainder for other years. With a sufficient area to draw from, it is possible by these measures to harvest a sustained yield of wood products annually from the same forest.

The adjustment is greatly aided by the fact, previously mentioned, that a tree may be held indefinitely before harvesting, or until it dies or becomes defective past any profitable use. Wood can literally be cut at any time after it reaches a profitable size. Thus harvested, it constitutes the crop from forest land.

193. Even-aged versus Many- or All-aged Stands. The mere estimating or cutting of existent timber gives a measure only of area and volume of the crop. The time element in production must still be determined. Volume or yield must be measured by the years required to produce it. In this way comparison is made between forest crops and other crops.

In seeking the rate of growth per acre for trees by contrast to annual crops, the difficulty is that the tree and not the acre is the real unit of production, although the yield must be expressed on an area basis. Area, or space available for each tree, profoundly influences the growth per acre.

Tree crops may be handled in two forms. First, a single main crop of trees, all of which are of approximately the same age—the so-called "even-aged" form of stand—may be produced on an acre. Second, trees may be cut or may die out singly or in small groups over a long period of time, permitting other trees to spring up in the openings thus made. When cutting can follow the natural pattern of such a stand, it produces a more or less continuous yield from the area. The form of this type of forest is many-aged, and the crown canopy is broken into many corresponding heights. Forest stands vary from the pure even-aged form, in which all trees originate in the same year, through the variations of even-aged form, in which the crop, though differing in age by several years, still forms a single crown canopy and behaves like an even-aged stand (Sec. 189), to stands in which the distribution of ages is more and more complex, and the total range of ages is found in smaller and smaller groups until, in a pure all-aged form of stand, trees of each year's origin might be found within an acre. A forest composed of even-aged stands is one in which stands having the character of even age exist on separate areas large enough to be treated individually and mapped separately. A forest composed of all-aged stands is one in which the age classes are so intermingled that they cannot be separated by area or mapped. The difficulty of distinguishing or mapping separate age classes increases as the size of the groups diminishes.

There is an important distinction between these two forms of stands. The even-aged stand permits of measurement of the average annual production of wood on an area basis for the entire period required to grow the crop, while the many-aged or all-aged form does not. No accurate method has been devised by which the various ages may be sorted out in all-aged stands and their area determined and yield per year found. The reason for this is that the *area* occupied by individual trees maturing at different times on the same acre does not remain constant but is continually increasing as trees drop out and adjoining trees extend their roots and crowns into the openings thus created. Thus for the trees composing any one age class, the area is small to start with and constantly expands until decadence sets in, while with even-aged stands, the area of a single age class remains fairly constant and the competition for space is confined to trees within this class.

194. Current and Periodic Annual Increment for Even-aged Stands. The growth in volume for a year is the *current annual increment* of a tree or stand. The sum of the current annual increments of all years since its origin gives the total volume of an even-aged stand.

The *average*, or *mean, annual increment* is found by dividing this total volume by the number of years required to produce it.

Trees are constantly dying out of all stands through suppression or shading in the earlier stages, and through natural causes such as insects, fungus diseases, wind, or lightning in the mature stands (Sec. 180). Thus the growth of a stand of timber occupying an acre of land cannot be found by assuming that all trees now living will survive. It is rather the *net* increase in the volume of the stand as a whole, after losses are deducted. During a given 10-year period, a portion of the existing stand either must be harvested or it will die and be wasted before another 10 years go by; *viz.*, a stand at 40 years contains 25,000 bd. ft. of timber and thus has averaged 625 bd. ft. per year for its life. But in computing the yield at 50 years, 2,500 bd. ft. of trees now alive will have disappeared either by death and decay or by cutting. Whatever volume the stand then has is represented by the surviving living trees alone. Suppose this volume amounts to 30,000 bd. ft., or an average of 500 bd. ft. per year for the 10 years. This was put on by the trees representing but 22,500 bd. ft. at 40 years. If the entire stand is cut at any given year, the average annual growth

THE GROWTH OF TREES AND OF STANDS 327

always includes the trees now alive which would die in the next period, and if no thinnings or advance cuttings have been made to date, this measures the total yield. On the other hand, if the crop is thinned and the losses averted, the total yield of any period is increased by the amount of these previous thinnings. Thinnings may thus constitute a distinct gain in crop production due to management or crop cultivation. In the above example, for instance, the yield of 50 years is 30,000 bd. ft. plus the thinning of the previous decade or a total yield of 32,500 bd. ft. When thinnings remove trees which will not die and which have room to continue a normal growth, these thinnings, if too drastic, *may reduce* the yield for the next decade.

The current growth of a stand is usually measured not for single years but for a period of 10 years. This *periodic growth* divided by 10 gives an average annual growth for the period which is the figure commonly substituted for current annual increment. For a stand, the *periodic annual increment* is measured by the net increase in the total volume of the living trees, divided by the years in the period. This would then be less than the growth put on by the surviving trees by an amount equal to the total volume of the trees which have died or been cut during the period. In the above example the net periodic annual increment of the unthinned stands is $5,000/10 = 500$ bd. ft., while for the thinned stand it would be $7,500/10 = 750$ bd. ft. including thinnings.

It is this element of loss in number and volume of trees from period to period which characterizes the progress or curve of current annual increment of stands. In the *initial* period of the life history of a stand, the trees are seedlings or saplings, putting on diameter and height growth (Fig. 59). They are still too small to possess any merchantable volume. Although in a full stand of reproduction the loss of numbers is large (Fig. 60), this is a benefit to the survivors. Even on a basis of total cubic-foot volume, the stand increases but slowly during the initial period.

The second period of current growth is entered when the larger trees of the stand reach a merchantable size. Each year during this period is marked by a rapid increase in current growth over the preceding year. This is due in part to the entry of numerous additional trees into the merchantable size classes and the increasing growth of those already merchantable. Although loss in numbers, by suppression, continues at a reduced but still high rate, most of the trees that die during this period are still too small to contain an appreciable volume of merchantable material or to pull down the net current increment of the merchantable portion of the stand.

In the third stage of current annual growth, the loss of numbers still continues at a fairly rapid rate. By this time, however, more and more of the trees that die have a merchantable volume. The deduc-

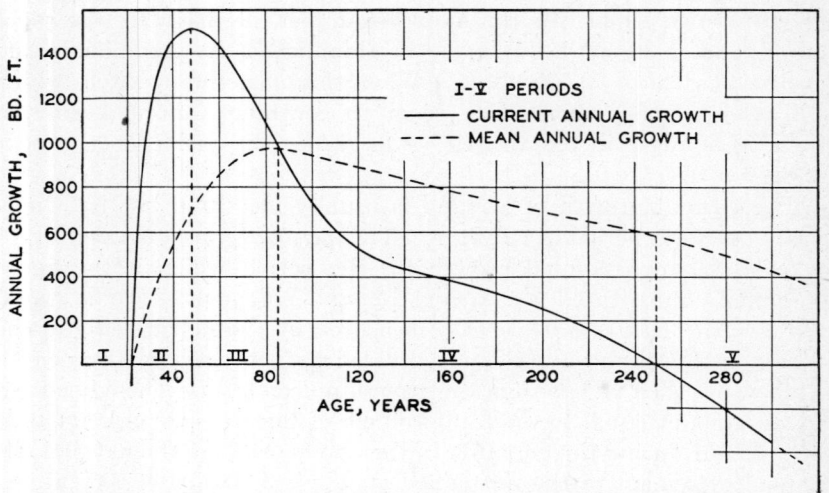

Fig. 59. Current and mean or average annual growth in board feet, International rule, for Douglas fir, Pacific Northwest, based on yield table to 160 years and extended to 300 years to indicate probable trend. Periods I to V show the successive stages in the life history of the stand.

tion of this loss as an offset against the growth of the survivors lowers the annual net gain in volume almost as rapidly as it increased during the second stage. But the total volume of the stand is still increasing at a satisfactory rate, which exceeds the *average* annual growth which the stand has put on since its origin (Sec. 195).

In the fourth stage, the trees have reached mature stature with large crowns and lessened height growth. The loss in numbers is now much slower, though each tree dropping out removes a relatively larger volume than in the previous stages. The deduction of this loss from the total volume of the stand still leaves a small net gain from growth. Thus the stand volume continues to increase slowly throughout this fourth stage.

In the fifth and final stage, this is no longer true. When this period is reached, the volume of the trees that are lost, plus increasing decay in living trees, exceeds the sound growth of the stand.

Thus the current growth for the period becomes *negative* and the stand begins to retrograde, a process continued in nature until the last survivor is gone. The completion of the fifth stage, that of dissolution, may take from 60 years for short-lived species, up to 2,000 years for the sequoias.

These five successive stages of net current annual increase or decrease in the volume of an even-aged stand may be plotted as ordi-

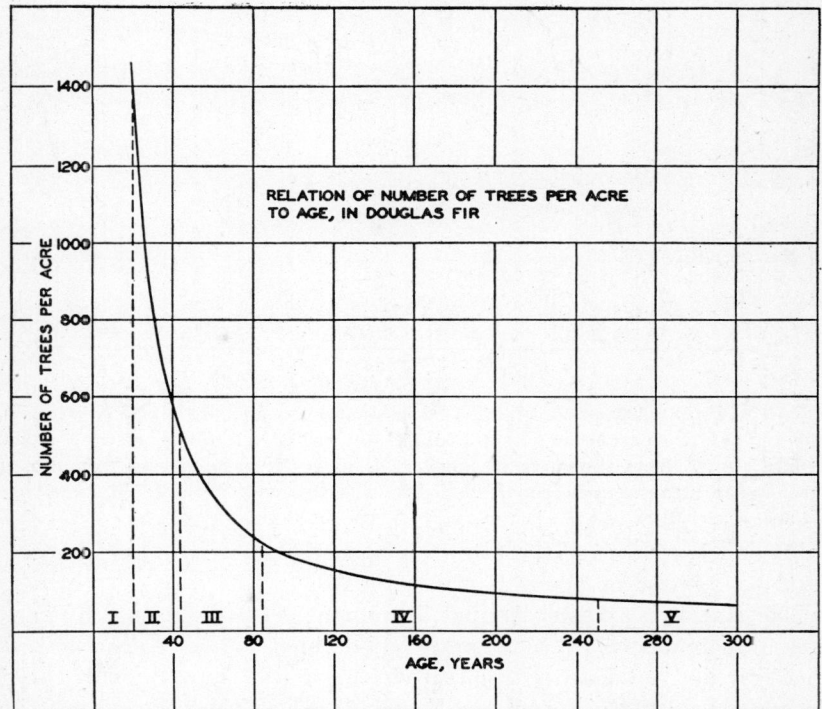

Fig. 60. For the five periods in the life history of Douglas fir, the number of trees per acre plotted on age. From yield table up to 160 years and extended to 300 years.

nates over age of stand as abcissas and will form a curve as shown in Fig. 59. The stages of current growth are clearly seen. The values are neither determined nor plotted for each year, but by plotting the average periodic growth over the mid-year for the period, the approximate current growth for all other years may be read from the resultant curve.

In the same manner, the numbers of trees plotted as ordinates over ages as abcissas show the stages in the reduction of numbers during the life history of the stand (Fig. 60).

Total yield is the sum of the successive current increments, or at the end of each period it is the sum of the total net increments for all preceding periods. In the first four stages of current growth, the curve of total yield based on age continues to increase, though in the

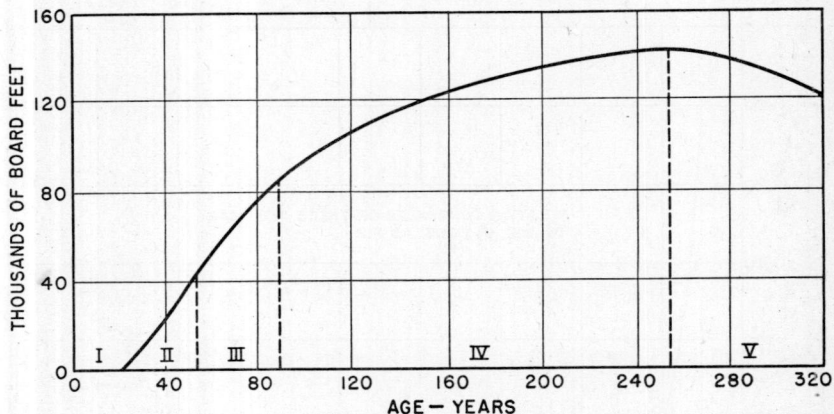

FIG. 61. Total yield in board feet based on age for Douglas fir, Pacific Northwest, showing the five periods in life of a stand, derived from the sum of the current annual growth from the curve in Fig. 59. This is the basis in turn from which the curve of mean annual growth in Fig. 59 was derived.

third and fourth stage the rate of increase is lessened. The curve of total yield based on age is shown in Fig. 61.

195. Average (Mean) Annual Increment. By definition, average annual increment is equal to total yield divided by age of stand. It is therefore derived directly from the curve or table of total yields based on age, as shown in Fig. 61, and thus indirectly from the sum of the successive current increments.

Each successive average differs from its predecessor by the amount of changing growth of the current year over that of previous year divided by total present age which is 1 year greater than the last divisor. As long as the current year's growth is greater than the average at the beginning of the year, the succeeding average will be increased by $1/n$ of this difference when n equals number of years in the divisor. This average thus continues to *increase* as long as the current growth for the year exceeds the previous average annual growth, even though this current net increase is *diminishing* rapidly as in the third stage above. In the year in which the current growth sinks to a quantity equaling the mean, the second curve, representing these successive and independent averages, reaches its highest point. This period marks the completion of stage three in the development of the stand.

From then on, the additions to volume by net current growth are less than the preceding averages, which must diminish each year by $1/n$ of this difference. This diminution steadily continues through stages four and five of current growth. The curve of *average* growth does not reach zero, however, until the sum of *minus* current growths equals that of the previous *plus* growths and the stand ceases to exist. Either cutting or heavy loss from natural agencies such as fire, wind, insects, or disease will cause both curves to take a sudden downward slump. The curve of mean annual growth is therefore not a true growth curve but merely a trend of arithmetic averages of volume growth for successive ages of the stand.

Based on volume production alone, and comparing this annual yield with that of other crops, the highest annual production per acre is not secured at the point on these curves where the net current increment diminishes to zero and the curve of total volume reaches its highest point. At this period the mean annual increment has already fallen considerably. Instead, maturity occurs in the year when the mean annual increment is at its peak, which is the year when current annual increment sinks to an equality with this average. This is comparable with judging the expediency of holding a security as long as its net yield exceeds the rate of interest on average investments. One would hardly wish to dispose of it "at cost" when the rate earned is at a peak similar to the culmination of current increment.

The curve of mean or average annual increment is shown in Fig. 59 in its relation to current increment.

Culmination of average annual increment in cubic volume or even of that for board-foot volume does not by itself indicate the proper age for harvesting a stand of timber. Price of the product; resulting value of the crop; the culmination of the average annual net yield in money per acre; and, finally, the highest rate of compound interest earned on the crop for the entire growth period are the considerations which determine the age planned for felling known as the *rotation*. These factors are treated in forest management.

196. Periodic and Current Increment of Many- or All-aged Stands. When it is impossible to determine the average age of a stand of timber by reason of the mixture of small groups or single trees of different ages that cannot be separated by mapping the areas of each age group, there is no basis on which to determine the average annual rate of increment of a forest crop from seed to maturity. It is often assumed that an acre of land in the long run will produce an average quantity of wood of about the same amount whether the trees are grown and harvested as a single even-aged stand once in the lifetime of the crop or are realized by the cutting of trees successively as they mature over an equal period of time. It is difficult to prove the equality of yields for these two extreme forms of crop production for the reason that intolerant species thrive best in even-aged stands or may not be able to grow at all in the shade of their own or of other species of trees. On the other hand, tolerant species such as hemlock seldom originate in the open but are naturally found springing up under the shade of young hardwoods or, if germinating at the same time, are outstripped and held back by these more rapidly growing species. When finally freed by the death of their competitors, these suppressed or tolerant species make rapid growth. But their total age contains this period of slow suppressed growth, and the yield, based on this age, is no indication of the possible productiveness of the site. The productive capacity would be better gauged by taking the total yield of all species over the entire period required to develop and harvest both the overwood and the succeeding understory (Sec. 192). Spruce under birch and poplar is a good illustration. Mixed hardwood forests are apt to be of all ages. In the old-growth, or original, forests, the development of young trees is held back for long periods by the survival of old, rotten, or badly formed trees long past their prime but capable of absorbing the light and moisture needed by the young stand.

When a stand is culled of these overmature trees, a normal rate of growth is restored. In such a forest, with a species capable of regeneration and survival under partial shade, the record of yields over long periods will finally establish the average current annual rate of wood production. The probable current growth of any stand may be predicted for one and possibly two decades in advance by methods outlined in Chaps. 29 and 30.

197. The Purpose of Growth Studies. The study of growth of trees and stands can have but one purpose, the prediction of the growth that will occur on existing stands of timber or on bare land for a definite *future* period. If the area and volume of the present stand are all that is required, and if it is of no interest what changes will take place in this stand or what its possible increase may be in the future, then the study of growth will be of correspondingly little importance. This is the case with lumber companies who expect to cut out their holdings in a few years.

Young stands will be held if the available data on growth show that it is profitable to permit them to reach larger sizes. Land will be planted if yield tables show that the possibility of valuable crops at given ages is sufficient to justify the expense and provide an adequate return for waiting. Past growth of stands and of trees can be measured and thus forms the basis of predicting what the future growth will be. The basis of such predictions is similarity. One cannot be sure what the history of a 30-year-old pine stand will be in the next 10 years, but there are two ways of approximating this future stand. One is to measure other pine stands of similar character which are 40 years old and to assume that this stand will behave in a similar manner. This method of *comparison* of younger stands with older measured stands gives rise to the use of tables of yield for even-aged stands. The basis of such comparison is yields per acre based on total age (Chap. 27).

For stands of all-aged form, prediction of future growth must also be based on past growth, but in this case *individual trees* must be used as the basis of study. Past growth of trees may be measured. Can it be used to predict future growth? One procedure is similar to that adopted in the case of yield tables, *viz.*, comparison of one tree with another older tree; for instance, a tree now 12 in. in diameter has in the past 10 years grown 2 in. It was a 10-in. tree 10 years ago. Its growth rate would then be applied to trees now 10 in. in diameter.

It is also possible by a second method to attempt the prediction of future growth by projection or extending the past growth of a tree or stand into the future by means of curves of diameter growth for all-aged stands. The method of prediction by comparison of younger with older trees or stands mentioned above is on the whole more reliable than this second method (Sec. 259). For stands of even age, in the absence of yield tables, the projection of growth of separate diameter classes by curves gives more accurate results (Sec. 257).

For even-aged stands measured on an area basis, the existing volume at any age gives directly the result of conflicting forces and subordinates the growth of the single tree to that of the stand, except as modified by cultural operations. In many-aged stands, where prediction of growth and mortality perforce depends on that of individual trees, neglect of the two factors of death and deceleration (Secs. 180 and 202) will invalidate any growth predictions based on past diameter growth of tree classes even when separately averaged for each d.b.h. class.

Although the use of frequent and adequate thinnings, as practiced in European forests, has shown that an even and constant rate of diameter growth of single trees can be maintained until harvest, only by this assumption can we use past diameter growth of existing trees as the basis for predicting future growth of American forests. To do so would result in an error, which, if accompanied by neglect of mortality, could easily reach 100 per cent or more in excess of actual performance.

CHAPTER 25

THE GROWTH OF TREES IN TOTAL HEIGHT BASED ON AGE

198. Objectives of the Study of Height Growth. The average rates of height growth of trees of different species are found by measuring the total height reached at successive ages of the tree; just as for diameter growth, a distinction is made between increment per acre on the one hand and behavior of individual trees on the other. The first of these objectives deals with the area and the total production of the crop, the second with the means of production, *viz.*, the trees which make up the stand. Trees can be grouped into classes based on similar characteristics and treatment, such as dominant trees, seed trees left after cutting, or trees in thinned stands. The effects of treatment, spacing, and selection of species, as well as of site or general factors affecting the entire stand, can thus be determined.

The two standard methods for determining increment per acre are yield tables for even-aged stands, and current growth based on diameter classes. In neither of these methods is it necessary to know how fast individual trees grow in height. In yield tables (Chap. 27), the site index of the stand (Sec. 219) is determined for each age by obtaining the average height of the dominant portion of the stand. Each successive age class will thus include a different and smaller number of trees in the average sought than any earlier age since trees are constantly dropping back into subordinate classes. The curves of a site index chart (Sec. 227) are therefore not true curves of height growth based on age and can be made without such a study. When current increment is studied, a stand table of diameters and current growth in diameter and height is used (Chaps. 28 and 29). It is concerned not with the total ages of the trees measured but only with their increase in height for a short period.

The forester seeks knowledge of the behavior of different species in mixed stands, however, in order to foresee what will happen and to be guided accordingly in the expenditure of funds for planting, weeding young stands, and releasing suppressed trees. A knowledge of height growth based on age, as affected by site, competition, and treatment, is at this stage even more important than a knowledge of diameter growth. In the seedling and sapling stages height growth rather than diameter growth determines the survival or suppression of the species. Differentiation in height must occur in young stands, or else diameter growth suffers and the stand may stagnate. Only after the completion of the period of rapid height growth and the approach of maturity does diameter growth assume a greater silvicultural importance than height growth. The primary purpose of studies of height growth is to learn how to tend and control tree crops in their initial stages, in order to obtain the best species and quality in the final crop.

199. Height Growth of a Single Tree Based on Ages of Sections. When height is based on age of trees, age is the independent variable, and the table shows the heights reached at given ages such as 10, 20, 30 years. The history of the growth in height of a tree is obtained by counting the annual rings at each cross section. At any given height above the ground, the first ring is not formed until the year after the tree has attained this height. If the tree is 30 years old and is sectioned at 8 ft. from the ground, the count of rings at this section indicates the number of years elapsed *after* the tree reached 8 ft. in height. The period of years required to grow up to this height equals the difference between the age of the section and the total age of the tree. If, for instance, the above tree showed 24 annual rings at 8 ft., it required 6 years out of the 30 to reach this height. Except in the study of seedlings, the first section is at the stump. The height-age curve is first computed from the age of the stump. To this curve is then added the period of years taken by the seedling to grow to stump height (Sec. 188).

With age as the independent variable, the theoretically correct method of analyzing the bole for height-age relations would be to cut the tree into many small sections so that height could be secured at the exact age sought, *viz.*, a section falling at 10, 20, 30, and 40 years in age above stump. Aside from destroying the merchantable value of the tree, this sectioning for specific ages is too laborious for the results obtained.

A method by which height growth for successive years can sometimes be directly based on age is applicable to young trees of species which form one distinct whorl of branches annually. Conifers such as white and Norway pine have this habit, which is shared by certain other conifers but by very few hardwoods. Other conifers, such as loblolly and slash pine, produce two and three whorls per year. Eventually these whorls, and even the knots, disappear, but on the young tree, especially if the total age of the stand is known as a check, the height growth for species producing one whorl per year can be measured for each single year or for groups of 5 to 10 years, as desired. The curve of height growth by years for five white pines and five Norway pines is shown in Fig. 62.

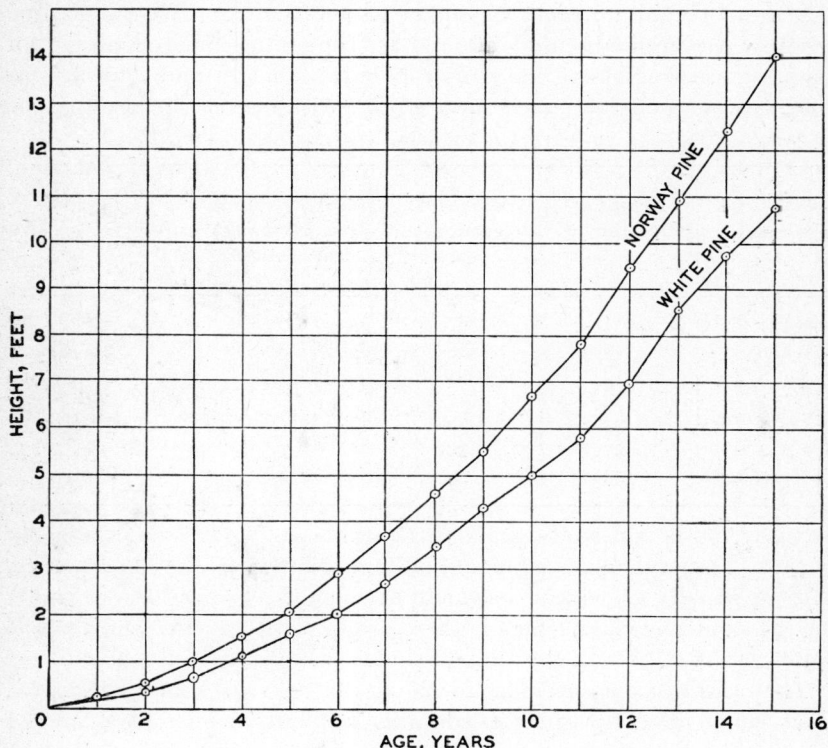

FIG. 62. Height growth of red or Norway pine and of white pine in plantations, by years, Milford, Pike County, Pa.

The process of determining the height growth of a single tree by sectioning it at intervals indicated and counting the rings on each section is shown in Table 47 for a chestnut oak, based on age at stump of 67 years. Since no averages are involved, either basis, height or age, gives accurate data for one tree.

200. Curves of Average Height on Age. Since the standard procedure in securing height-age data is to take readings of age at even height intervals along the bole of the tree, the plotting of smooth curves of average height on age by combining data from a number of

338   FOREST MENSURATION

trees is made difficult, for there is no direct way of getting average height for even age intervals. If the base is reversed, a curve of age on height would be obtained. This is not the desired curve, since the substitution assumes that a perfect correlation exists between height and age, which is not the case (Sec. 175). Hence an adjustment from the usual technique must be made to get the approximate desired result. A commonly used method is to plot the height and age for each 8-ft. section above the stump, plus total height and total age for each tree as separate points on a graph, with age as the scale of the

TABLE 47. HEIGHT GROWTH OF A CHESTNUT OAK

Length of section, ft.	Height of section, ft.	Rings on section, years	Years required to grow from last section	Age from stump at height of section, years
1	1	67	0	0
8	9	58	9	9
8	17	54	4	13
8	25	47	7	20
8	33	33	14	34
4	37	26	7	41
16	53	0	26	67

abscissa and height as the scale of the ordinate. It will be practically impossible to average the heights falling within successive age classes, because rapidly growing trees would often show two successive heights for the same age class. One is therefore forced to average the ages at 8, 16, 24, 32 ft., etc., even though this is technically not sound.

The results of measurement for 20 scarlet oaks are shown in Fig. 63. Each height level up to 40 ft. contains 20 points, which are averaged in a horizontal direction. At 48 ft., there are only 17 points, the missing 3 points being replaced by plottings of total height clustering near age 70, with a total height of slightly over 48 ft. When these three values are included, the average is shifted slightly to the right and upward. At 56 ft., only 8 section measurements remain, and a weighted point is obtained by considering the 9 trees for which total heights have been plotted. From this point on, the curve is guided by the averages for total heights within the successive 10-year age classes.

A sounder method of adjustment involves interpolation for each tree of the height at each decade of age, then plotting these interpolated values on the appropriate decade, and averaging heights by decades.

The curve is then drawn through the average values. The interpolation can be made in two ways, first, by straight-line interpolation between two successive height ages; and, second, by plotting directly height of each section measured over age at that point and making a curve of height on age for each single tree. The heights, from the curves, of all trees are thus averaged directly on age and the final curve plotted. By the first of these methods in Table 47, the age at 25 ft.

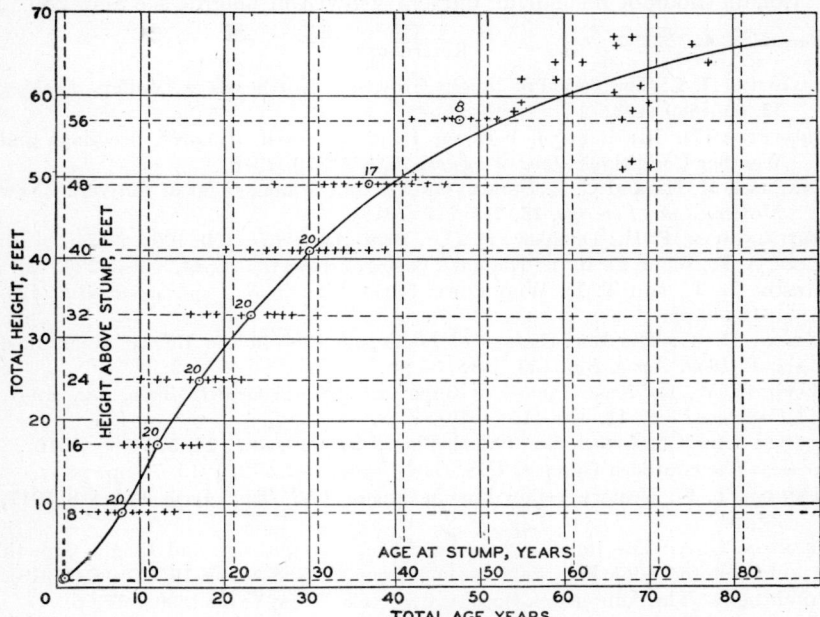

FIG. 63. Height growth for 20 red and scarlet oaks, ridge type, East Lyme, Conn. Based on 1-ft. stump and corrected for average age of seedlings 1 ft. high.

is 20 and at 33 ft. is 34. It is desired to get a height for 30 years. Straight-line interpolation gives

$$[(33 - 25)/(34 - 20)] \times 10 + 25 = 30.7 \text{ ft.}$$

The second method would tend to give a slightly different result, depending upon the nature of the height curve between the two points. Either of these adjustments involves a more time-consuming calculation but leads to a result which is defensible, which the first was not.

Typical curves of height on age show a slow initial growth for seedlings and a rapid growth for sprouts. The seedling growth soon becomes rapid, and the first half, approximately, of the curve is steep.

As maturity is reached, the height curve begins to flatten out until in old timber heights increase by very small additions per year, even by an inch or two. Intolerant and "weed" species, such as jack pine or aspen, often show very rapid initial height growth while longer lived trees, such as white pine, retain a slower rate of height growth for a much longer period, eventually reaching greater heights and remaining dominant for perhaps a century or more.

For methods of measuring current growth in height, see Sec. 255.

References

BALDWIN, H. I.: The Period of Height Growth in Northeastern Conifers, *Ecology*, **12,** 665–689 (1931).

BREWSTER, D. R.: Relation between Height Growth of Larch Seedlings and Weather Conditions, *Jour. Forestry*, **16,** 861–870 (1918).

CHAPMAN, H. H., and C. E. BEHRE: Growth and Management of Pinyon in New Mexico, *Jour. Forestry*, **16,** 215–217 (1918).

FROTHINGHAM, F. H.: Douglas Fir, *U.S. Dept. Agric. Cir.* 150, 1909, 38 pp.

KERR, A. F.: White Fir in the Klamath Basin, *Forestry Quart.*, **11,** 349–362 (1913).

LARSON, L. T., and T. D. WOODBURY: Sugar Pine, *U.S. Dept. Agric. Bul.* 426, 1916, 40 pp.

MASON, D. T.: The Life History of Lodgepole Pine in the Rocky Mountains, *U.S. Dept. Agric. Bul.* 154, 1915, 35 pp.

MATTOON, W. R.: Slash Pine—An Important Second Growth Tree, *Soc. Amer. Foresters Proc.*, **11,** 405–415 (1916).

———: Some Characteristics of Slash Pine, *Forestry Quart.*, **14,** 578–588 (1916).

———: The Southern Cypress, *U.S. Dept. Agric. Bul.* 272, 1915, 74 pp.

MUNGER, T. T.: Western Yellow Pine in Oregon, *U.S. Dept. Agric. Bul.* 418, 1917, 48 pp.

PEARSON, G. A.: The Relation between Spring Precipitation and Height Growth of Western Yellow Pine, Saplings in Arizona, *Jour. Forestry*, **16,** 677–689 (1918).

PINCHOT, G.: The Adirondack Spruce, Critic Co., New York, 1898, 157 pp.

WEBER, R.: Über das Hohenwachstum der Bäume (Laws of Height Growth), *Ztschr. f. Forst u. Jagdw.*, **40,** 152–159 (1908). Review, *Forestry Quart.*, **6,** 306–307 (1908).

WOOLSEY, T. S., and H. H. CHAPMAN: Norway Pine in the Lake States, *U.S. Dept. Agric. Bul.* 139, 1914, 42 pp.

ZON, R.: Balsam Fir, *U.S. Dept. Agric. Bul.* 55, 1914, 68 pp.

CHAPTER 26

THE GROWTH OF TREES IN DIMENSION AND FORM

201. Growth of Trees in Diameter. The two broad methods of study of increment are, as shown, first, yield tables for even-aged stands and, second, current growth for the past decade or two for stands of mixed or all-aged form (Sec. 193). In the application of yield tables for predicting growth on even-aged stands, the only field data required touching growth are the ages of representative trees. The *rate* of diameter growth is not measured. This method is therefore entirely independent of diameter growth (Chap. 27). In the investigation of current growth of all-aged stands, the growth in diameter is measured, but only for the outer decade or two (Chaps. 29 and 30). Hence in this method the ages of the trees measured and their growth from seedlings are not needed in determining current growth per acre. What is needed in both the above methods is the present volume of the stands whose growth is to be predicted and, for current growth of all-aged stands, the stand tables of diameter distribution, present and future (Secs. 245 to 247).

It is necessary to distinguish sharply the process of predicting increment of stands, which is a business problem connected with the inventory, on the one hand, from the equally vital and interesting problem of silvicultural treatment of stands or cultivation of timber crops, on the other.

It is in the latter, not the former, field that diameter growth based on *age of tree* as well as current diameter growth has its significant uses. In silviculture, while the entire stand finally constitutes the crop, it is the individual tree in relation to other trees which forms the unit of attention and cultivation. In this respect, forestry is midway between orcharding, in which each tree is isolated, and grain or hay crops, in which the individual plant is seldom distinguished. The forest tree has sufficient individual importance to justify the recognition of each tree in a stand and its treatment as a final crop tree or merely a temporary filler to be removed by thinning or allowed to die from suppression. The spacing of these individual trees and their resultant crown development profoundly influence their rate of diameter growth and the quality of the lumber and other products which they will yield (Sec. 183).

In the business problems of predicting increment, the average and total growth for the entire stand is desired regardless of the individual tree. In the studies of diameter growth required in connection with silvicultural problems, the rates of growth investigated are not the averages of all trees in stands but rather the average rates shown by special classes of trees all of which have the same characteristics and treatment. The effect of these characteristics and of this treatment can then be shown in the form of growth in volume of the tree, for the determination of which diameter growth is the first and most important step.

The problems in silviculture for which the study of diameter, height, and volume growth based on age of tree is required are:

1. Survival and dominance of species in mixed stands.

2. Crown differentiation into classes and reduction in numbers with age.

3. Relative rate of growth (in diameter, height, and volume) of dominant compared with intermediate trees.

4. Effect of spacing at different ages, including that of plantations, on the diameter and height growth of trees and on their form and quality.

5. Effect of thinnings or removal of trees on the diameter growth of the remaining trees, and indications, in unthinned stands, of the ages and conditions requiring thinnings.

6. Effect of early dominance or spacing on the ultimate sizes reached by the favored trees in an even-aged stand and indications as to methods of producing trees of these sizes.

7. Effect of crowding versus opening the stand on the number of rings and width of summerwood per inch and indications as to management for production of timber of high quality.

8. Ability of trees to recover after suppression and of seed trees or timber released by thinning to put on increased growth, and the relation of the character of crown of the trees so freed to their behavior.

9. Distribution of diameter growth on different portions of the boles of trees under different conditions of spacing, crowded and open growth, or freed from competing trees by cutting. Resultant changes of form and net increase in volume under different circumstances.

10. Comparative development of different species in diameter, height, and volume growth per tree.

There is hardly an important problem in silviculture, outside of mere initial establishment of the seedling and prevention of destruction of the trees by outside agencies, which does not find the measurement of its effectiveness and the regulation of its permissible cost in the effects produced upon the diameter growth of individual trees and consequent *influence* on yields per acre and value of the crop.

202. Influences Affecting the Diameter Growth of Trees. The two factors which exercise the greatest influence on diameter growth for a given species in a pure stand are quality of site and number of trees per acre at given ages. Diameter growth directly reflects the size of the live functioning crown. Number of surviving trees is in inverse ratio to the size and spread of the crowns. On young saplings the crowns extend to the ground. As long as this condition continues and there is differentiation in the crown classes giving dominance to part of the stand, the diameter growth will continue unchecked on these dominant trees. Open-grown trees develop wide-spreading or bushy crowns for most species, with maximum diameter growth below the crown but having a minimum of clear bole, and a poor form.

With advancing age, the competition in height growth and the normal increase in spread of crown brings the crowns into competition, progressively shading the lower limbs and killing them. The live crown is thus shortened and the greater the number of surviving trees the greater will be the effect of shading and the shorter the relative length of the crown.

Diameter growth will maintain itself at a fairly normal rate until the crown becomes less than one-half to less than one-third of the total height (the effect varying with the species and site) when it falls off quite suddenly, especially in even-aged stands of intolerant species. The more tolerant the species, the longer will be the crown, in spite of slow growth caused by shading from the sides and above.

Diameter growth varies with quality of site and on favorable locations may easily be two to three times as great as on unfavorable locations. On poor sites the trees develop slowly in height and diameter and do not differentiate rapidly in dominance. As a result, in unthinned stands greater numbers of trees survive to a given age on poor sites than on good ones, but a satisfactory rate of diameter growth requires *fewer* trees per acre (Sec. 183).

Since height growth for intolerant species must continue at nearly the average rate of the stand, or the slower growing trees will be suppressed and die, the codominant and intermediate trees, after the shading out of their lower crowns, must have the capacity to continue height growth, even at the expense of diameter growth, and thus taper off in diameter growth sooner than they do in height growth.

In even-aged stands the dominant trees always show the fastest diameter growth as well as slightly superior height growth, while the suppressed trees are those whose diameter growth has been slowest. It follows that the stand table of an even-aged stand shows an increasing spread between smallest and largest diameters as the stand increases in age. Where the diameter growth of individual trees of different diameter classes is plotted over age through the course of their life, different rates of increase are shown. The curve of growth for the smallest trees surviving at any age will be found to have maintained itself well in the earlier decades until this class becomes intermediate or suppressed, when the curve flattens out and the trees soon die.

If the growth in diameter of the average tree of a stand at successive decades is found, each decade will show that the average tree is a different tree, and one more nearly approaching the upper dominant group, until at maturity only the few trees survive which have secured full crown space. These dominant trees grew at first much faster than the average, but with the progress or increase of the average tree the curve of average diameter growth rate finally catches up with that of the dominants. These relations between the growth of diameter classes are shown in Fig. 64.

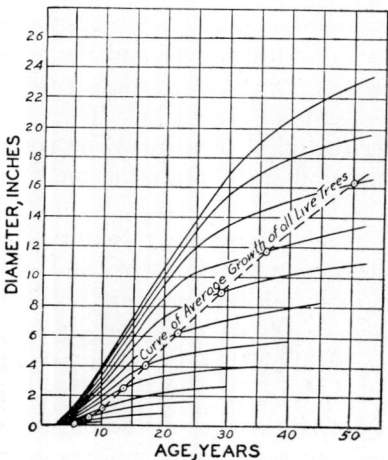

FIG. 64. Relation of diameter growth of different groups of trees from dominant to suppressed, in an even-aged stand, to the average diameter for the entire stand at different ages.

Shading or starvation of tolerant trees, and absorption of soil moisture by dominants, may reduce diameter growth to negligible proportions during the period of suppression. On release and the admission of light and adequate moisture, the growth in diameter may resume its normal course and even exceed the average dominant, because the length of crown retained by such tolerant trees can now function as a result of light and more root space. Plotting this suppressed growth on the basis of age reveals the effect of such suppression, but the age is of no significance in indicating yields or growth possibilities for the species (Sec. 193).

203. Diameter Growth on the Stump. The growth in diameter at b.h. outside the bark is the objective of diameter growth studies. The determination of this growth when not taken directly at b.h. in successive measurements of sample plots (Sec. 272) requires three steps: first, diameter growth on the stump; second, correlation of stump diameters inside bark with d.b.h. outside bark; and third, addition of age of seedling whose height is equal to that of the stump.

The direct study of growth at b.h. based on total age of tree presents difficulties, for on many trees the radius is too large for the increment borer, and it is never possible to obtain an *average* radius in one boring. To this growth, if taken, would have to be added the age of a seedling of $4\frac{1}{2}$ ft. in height. For these reasons, diameter growth *based on age* is usually measured on stumps and correlated with d.b.h. outside bark.

The butt swell increases downward; and, when stumps are cut very low, as in preparing the ground for logging spurs, the diameters at the surface of the ground are shown. The growth in diameter if measured on such low sections would be abnormally large. The higher the stump, the closer will growth on the stump section approximate growth at b.h.

Diameter-growth studies are usually conducted for a given problem in a specific locality. If for this purpose trees are felled by the crew taking the growth data, the heights of stumps can be made uniform if desired. When such studies are taken on a logging operation, the regulations and practice of sawing govern the heights of stumps. For any study, a fairly consistent relation will usually be found to exist between heights of stumps and d.b.h.'s of trees cut. The *average* diameter growth on the stump is then accepted, small variations in stump heights being disregarded.

On very few trees is the pith, or center of growth, in the exact center of the stump section even if the stump is circular. Yet the radius on which growth should be measured must be exactly one-half of the average diameter on this stump and must be measured to the nearest 0.05 in., corresponding to 0.1 in. for the diameter. On any section this length of radius may be found in at least two places. The securing of this average radius is the chief advantage of measurement on stump sections.

In Fig. 65 the method of locating the average radius is shown.

The annual rings, as in the case of current growth studies, are measured in decades. The rings are first counted from the cambium, or

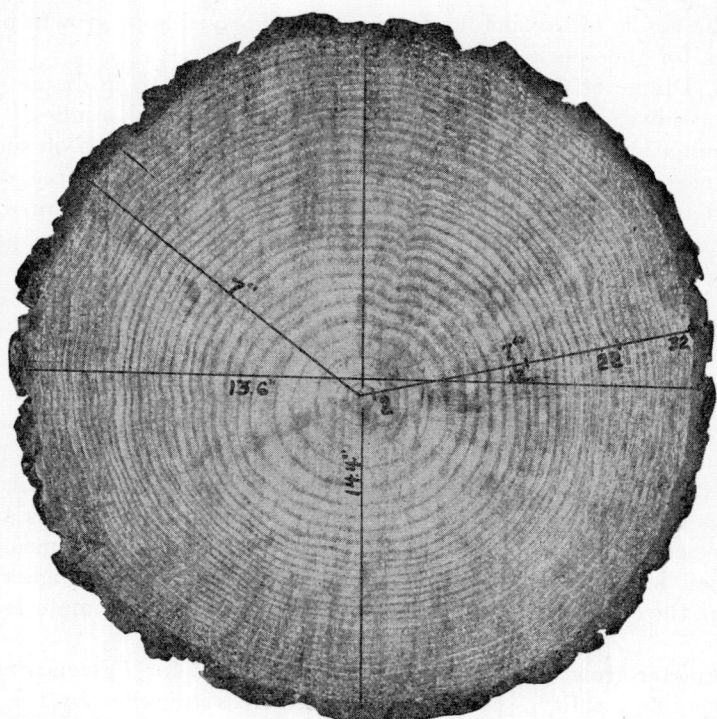

Fig. 65. Method of measuring growth on average radius on a tree section. Two diameters are measured at right angles, and the length of average radius is determined. This radius is then laid off from the pith. On full cross sections the desired radius can be found at two points, as shown. Decades are counted inward from the cambium, marked, and measured outward from the pith. This tree is 32 years old. The 2-year fractional decade is thus placed in the center and is the first measurement.

contact with bark, inward, along one of the average radii. The line of the radius may require smoothing with a sharp knife or chisel to reveal close or narrow annual rings in slow-growing trees or for suppressed periods. For very close rings a reading glass is used. A mark is placed on the inner edge of each tenth ring in succession.

The measurement of the radial growth is now made in the reverse direction, or from the pith outward. Unless the age on the stump is a multiple of 10, the count leaves a fractional decade at the center. This fraction is first measured, and opposite the record is placed this odd number of rings. The second and succeeding decades are full 10 years. For each, the total radius *from the pith* is recorded, *cumulatively*, and not the actual radial growth for the separate decade. This practice permits the scale to remain with its zero on the pith and avoids errors caused by shifting it and resultant possible loss of fractional quantities. The smallest unit recorded for total radial growth to any tenth ring is 0.05 or $\frac{1}{20}$ in. The most convenient graduation for the scale or ruler is in this unit. The final radial measurement must exactly agree with the average radius, or one-half the average diameter.

Accuracy in determining both total age at stump and growth per decade depends on the ability to distinguish each of the successive annual rings laid on during the entire life of the tree.

Where growth based on age is studied directly at b.h. from borings, practice will enable the operator to hit the pith of the tree about once in two tries or to come so close to it that its position relative to the axis of the boring is shown by the curvature of the rings. In such studies, or even in those for current growth if their importance warrants it, borings may be collected in proper receptacles, identified as to d.b.h. and species, and then treated in the laboratory, even to the employment of stains to bring out the wood texture and identify the rings.

Although failure to distinguish very faint annual rings in badly suppressed trees or difficult species always leads to overestimating the growth rate, the counting of two rings for a single season would have the reverse effect.

A check on the accuracy of total ages from ring counts is especially needed in otherwise easily handled species, when the trees have been slowed down in growth for one or more decades. In many regions and species, the method made familiar by anthropologists in the Southwest may be found helpful. The tree ring calendar, by whose aid the Pueblo ruins have been dated over a period of 1,000 or more years, to the actual year of construction, was established by matching the patterns formed by varying width of rings on beams cut at different times in the past. Peak growths in these arid regions occurred in wet seasons, which formed the points of reference. (In humid regions the reverse is sometimes true.) In the South, because of a 6-month drought in 1924, the most severe in 100 years, the annual ring on practically every pine was abnormally small and lacking in summerwood. Dearth of soil moisture produced a second stunted ring in 1925. These two separate annual rings form a point of reference for the ring count from there to the circumference, which is frequently more reliable than the counting of the remaining rings if suppressed. The same rings, incidentally, form a living proof, if any were needed, of the validity of ring counts as indicating age, a fact disputed at one time by certain scientists in the Southwest. Errors in ring determination are reflected as a percentage of the age and growth involved, one ring missed or added in a decade being equivalent to an error of 10 per cent in diameter growth with its accompanying distortion of growth in volume. Unless ring counts can be verified beyond the possibility of cumulative error, especially with difficult species, growth predictions should not even be attempted on this basis.

204. Compilation and Averaging of Diameter Growth on Stump. For all trees falling in the class whose growth should be averaged, the first step is to total the radial growth, first for the fractional decade at the center, and then for each of the remaining decades. These totals are divided by the number of trees to obtain average growth, which is cumulative. The averages are then plotted as growth in radius on the basis of age and are connected by a curve. The average for the first, or fractional, decade is plotted over the average number of years in the first measurement, and each succeeding average is placed 10 years to the right. The original averages for radial growth of 65 red and scarlet oaks are shown in Table 48.

TABLE 48. RADIAL GROWTH AT STUMP FOR RED AND SCARLET OAKS

Red oak, age	Growth at stump, i.b.	
	Radius	Trees
6	0.71	65
16	1.78	65
26	2.81	65
36	3.62	65
46	4.33	65
56	5.31	63
66	6.41	44

As long as the total age of all the trees used for the average is the same or is within the same decade, this curve stands without correction as representing the growth in radius. If the total age of the trees differs by a decade or more, the portions of the curve representing fewer trees will not represent the growth of all and will usually depart from this growth curve, sometimes quite abruptly. This deviation will be upward if different stands are measured in each of which the trees are even-aged, since only the more rapidly growing trees survive to greater age. If suppressed trees of tolerant species are included, the deviation will be downward, but curves based on age should not be made for such trees without special precaution (Sec. 191).

205. Correlation of Stump Diameter Growth with D.B.H. outside Bark. Diameter growth on the stump must be correlated with d.b.h. outside bark, since volumes of trees are standardized on the latter basis, and the sizes which trees reach in a given period are based on d.b.h. measurement. What is wanted is the d.b.h. outside bark for trees at given *ages*. The age of the tree at the stump has been found and its d.i.b. at stump. If the relation between stump diameter inside bark and d.b.h. outside bark is now determined, the corresponding d.b.h. for a tree *of this age at stump* is indicated.

Height of stump is a very important factor in influencing these relationships, and ratios which apply to stumps 1 ft. high will not hold true for 2-ft. stumps. But this condition is equally true of the growth on the stump. It is evident that in measuring either growth or stump taper the height of stumps should be fairly uniform. It is possible to prepare separate tables of growth for stump heights differing by ½ ft. and to take stump tapers at ½-ft. intervals to correspond. This gives the highest degree of accuracy. For ordinary purposes, however, a study in a given region for a single species will be based on the average stump height, and stumps will be thrown together in the same average.

A stump d.b.h. taper table is prepared as follows:

1. Paired measurements are taken, preferably on at least 300 trees, of d.i.b. at average stump height and d.b.h. These stump tapers may be taken from either felled or standing trees, since they are independent of growth and constitute merely a relationship of form or dimensions. On the felled trees the d.i.b. may be measured directly on the stump. The d.b.h. in all cases is taken with calipers *outside* the bark at the proper height. On standing trees, the d.i.b. at stump requires the caliper measurement outside the bark at average stump height, and the thickness of bark at that point. This bark thickness is doubled and subtracted to get d.i.b. at stump. It can be found by notching with an ax or by use of the Swedish bark gauge.

2. Classify the stump tapers on the basis of 1-in. d.i.b. stump classes. The d.i.b. stump is the independent variable from which the d.b.h. is deduced, hence must be taken as the basis of classification of the stump tapers.

3. Plot the resulting original averages of d.b.h. on the ordinates of a graph, whose scale of abscissas is d.i.b. on stump. In dealing with a small number of trees, it is necessary, as usual, to average the d.i.b. values within each d.i.b. class as shown in Table 49 and to plot these

350 FOREST MENSURATION

values on the proper ordinate rather than on the median class value, viz., at 8.2 in., not at 8.0 in.

In Table 49 the classification of 66 red and scarlet oaks for stump taper is shown to the nearest 0.1 in. for each 1-in. stump class, with its corresponding d.b.h.

TABLE 49. TAPERS OF RED AND SCARLET OAK FROM STUMP D.I.B. TO D.B.H. ORIGINAL DATA

Basis, trees	Stump d.i.b., in.	D.b.h., in.
1	7.5	6.9
5	8.1	7.2
8	9.1	8.2
13	10.1	9.0
13	10.9	9.6
6	12.1	10.05
5	12.8	10.05
1	13.8	11.7
5	14.7	11.7
7	16.2	12.3
1	17.1	12.3
1	19.2	13.8

4. Tabulate the stump tapers reading from the median class values, viz., at 8.0 in., etc., from the curve (see Table 50).

Figure 66 shows the stump tapers for 64 red and scarlet oaks in Connecticut. The increase in taper shown with increasing diameters is characteristic of practically all species. The degree of taper varies with the height of stump and with the species.

TABLE 50. TAPERS, D.I.B. STUMP—D.B.H. RED AND SCARLET OAK, CURVED DATA

D.i.b. stump, in.	D.b.h., in.
6.0	5.4
7.0	6.3
8.0	7.2
9.0	8.1
10.0	8.9
11.0	9.6
12.0	10.2
13.0	10.7
14.0	11.3
15.0	11.8
16.0	12.2

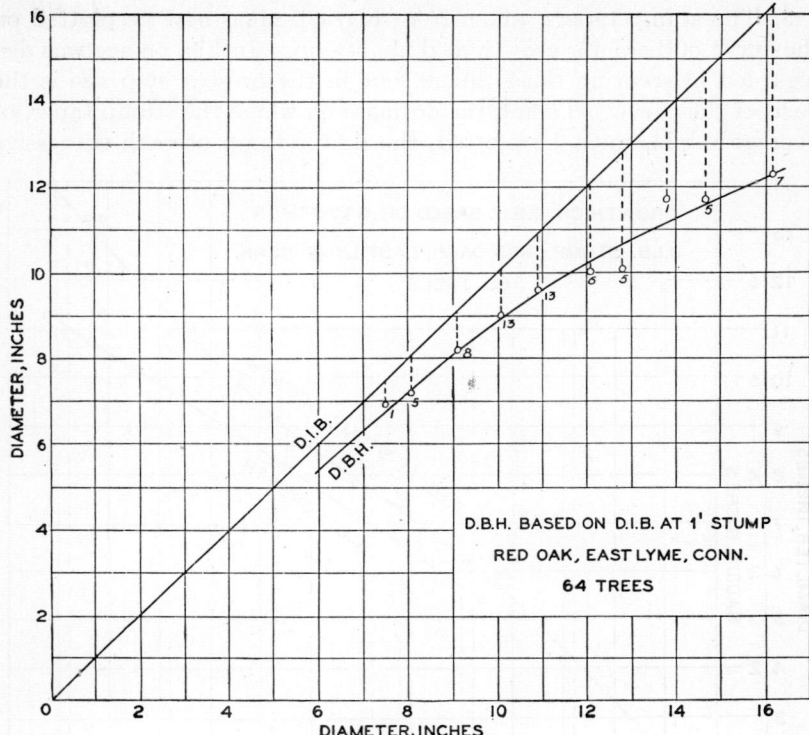

FIG. 66. D.b.h. based on d.i.b. at 1-ft. stump. Red oak, East Lyme, Conn.

206. Plotting the Curve of Growth in D.B.H.

1. On graph paper, the growth data computed in Secs. 204 and 205 are now combined in a curve of radial growth on stump based on age. The growth of the first fractional decade is plotted above the average number of years in that period, which in the example was 6 years. Each succeeding growth value is plotted 10 years to the right, *viz.*, at 16, 26, 36, etc., years (see Fig. 67).

It is not customary to smooth or harmonize a curve of growth based on age, except at the extremity whose irregularities are caused by the dropping out of numbers of trees. The plotted points are connected, and the result is accepted as the growth history of the trees studied.

2. The scale of ordinates, representing growth in inches of radius, is now doubled, thus reading growth in diameter. This must be done at this point in order to permit the next step, in which diameters are plotted.

3. The stump tapers, taken from Fig. 66, must now be plotted on the curve of diameter growth at d.i.b. stump. In Fig. 66 age was disregarded in securing these tapers, but in the present step age is the basis of the curve. To find the ordinate on which the stump taper, or average d.b.h., must be plotted, the d.i.b. stump at each successive

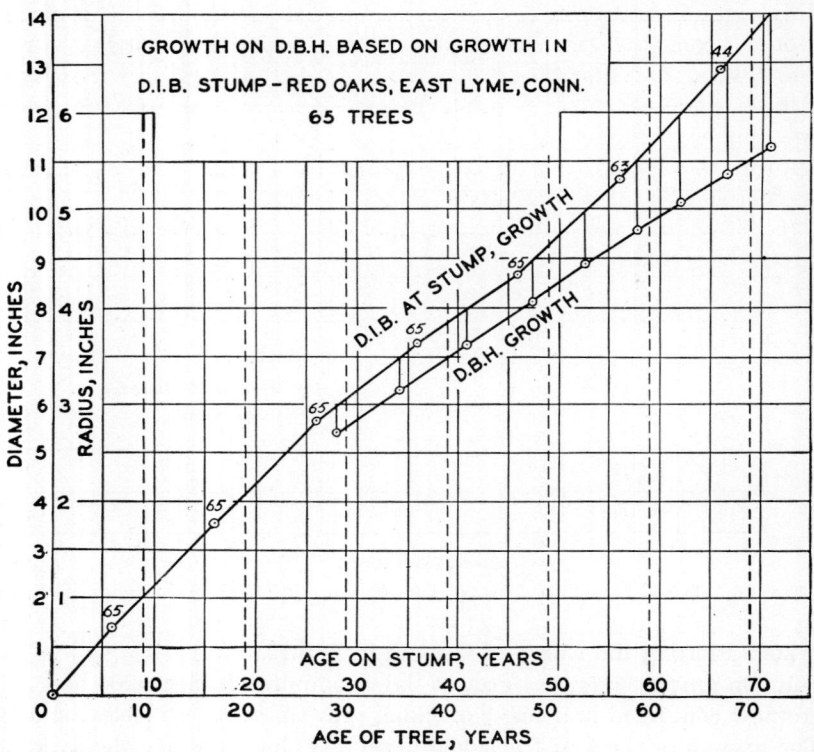

Fig. 67. Growth at b.h. based on growth on stump i.b. Red oak, East Lyme, Conn.

age, on the curve of d.i.b. on age, is found. From Fig. 66, the d.b.h. corresponding to this d.i.b. stump is read, to the nearest $\frac{1}{10}$ in. This d.b.h. is then plotted on the ordinate for the age in question. The process is shown in Fig. 67. In order to complete the curve of growth in d.b.h., stump tapers are required on trees of all the smaller sizes. This curve is usually completed by approximation, since it must terminate at a point equal to the age of a 4½-ft. seedling.

4. The final step in completing the curve is the correction of the age scale by the addition, to the age of stump, of the age of an average

seedling of stump height. This is accomplished by moving the zero of the scale to the left by the indicated number of years. In Fig. 67, which was for sprouts, 1 year is allowed to reach stump height. In many instances, for seedlings of species growing in dry climates, a much longer period, even up to 20 years, may be found as a correction for total age of tree. The lower age scale in Fig. 67 indicates this correction. From this scale the true age of the tree is read, and above it, at the desired intervals of 5 or 10 years, is taken the d.b.h. for these ages. Stump diameter growth, until corrected for d.b.h. and for age of seedling of stump height, is of no scientific or practical value, since it is subject to too great a variation from d.b.h. sizes and is therefore incapable of use for determining volume growth.

5. From the completed curve, the table is read showing d.b.h. at given ages (Table 51).

TABLE 51. D.B.H. AT GIVEN AGES, RED AND SCARLET OAK, EAST LYME, CONN.
(Basis 65 trees. Ridge type)

Age, years	D.b.h., in.	Basis, trees
10	....	65
20	....	65
30	5.4	65
40	7.0	65
50	8.4	65
60	9.8	63
70	11.0	44

The increase in the curve between 60 and 70 years is probably due to the dropping out of slowest growing or suppressed trees.

Unless diameter growth is confined to the averaging of trees of similar crown development and growth rates, it is subject to the same error that affects the determination of average diameters for a stand when based on the arithmetical average of the diameters instead of being derived from a tree of average basal area (Sec. 15). This error is always minus. The growth in volume of a stand, not the average growth in diameter, is the objective, and this is best accomplished by yield tables for stands of even age rather than by growth of individual trees. Diameter growth based *on the age* of the tree is therefore limited in its utility to averages for trees of similar growth rates.

207. The Derivation of Growth at B.H. from Current Growth in Diameter on the Stump. The method of measuring growth on the stump and relating it to growth at b.h. is also applicable when only the last 10 years growth is analyzed as is the case in current growth studies. This is sometimes necessary for species that are difficult to bore, such as oaks.

On the stump, it is possible to select the average radius, thus increasing the accuracy of each measurement (Sec. 203). The growth for the required periods is measured on this radius from the inner ring outward as at b.h. This growth must now be referred to d.b.h., which requires the use of the stump d.b.h. taper curve; *viz.*, it is found that for the past 10 years the average growth at the stump for a 10-in. d.i.b. stump class is 1.2 in. In Fig. 66 the d.b.h. of a 10-in. stump is 8.8 in. For a stump 1.2 in. larger, or 11.2 in., the d.b.h. is 9.7 in. The difference at b.h. for a growth of 1.2 in. i.b. on stump is 0.9 in. This figure indicates the current growth at b.h. for the 10-year period. Similar figures for b.h. are obtained for each stump d.i.b. class. These growth figures at b.h. are finally plotted as ordinates, above d.b.h. as abscissas, and harmonized by a curve, from which the table of current growth at b.h. is read by substituting the proper age of tree for each d.b.h. on the scale of abscissas.

208. Purpose of Determining Growth in Upper Diameters or Form. Cubic volume is not the sole measure of volume or value for the contents of individual trees. A second consideration is size and form, a third, quality, and a final element, price. The dimensions of the bole of a tree, its height, the diameters of the merchantable portion of the bole, and total merchantable length determine the contents in terms of board feet and are of importance in the production of crossties, poles, piling, and dimension lumber. Quality is in turn gauged by grading rules and by specifications as to minimum sizes and allowable defects (Chap. 7).

In general, trees of cylindrical form are more valuable per cubic foot for piece products than are those resembling a cone. The less the taper, the greater the unit value. Species with rapid height growth, produced in crowded stands, and with straight habits of growth, yield appreciably more poles and piling than do others which tend to develop bends or crooks. It becomes important to control the form as well as the quality of crops of second-growth timber and hence to know and measure the effect of density and of thinnings on the relative growth in diameter at different heights above the stump.

It has been shown by Girard (Sec. 101) that the greatest variation in form between trees as individuals, and also between species, occurs in the taper of the merchantable bole between the top of the first 16-ft. log and the d.b.h., and, secondly, above the merchantable limit usually used for saw logs. This means the base of the green crown for hardwoods and above a diameter corresponding to that of the last merchantable log for conifers. Growth in diameter does not remain uniform at different points on the bole during the life of the tree. It is fairly rapid in the crown portion, and with increasing size the characteristic butt swell is formed at the base. From the green crown down to the butt swell, the bole has a more constant growth in basal area than in diameter (Sec. 183). This produces proportionately wider rings on the upper narrower portion of the bole, which tend to give it a more cylindrical form. Open-growth trees, with long crowns, remain more conical, resembling the shape of the upper portions of closely grown trees.

209. Measurement of Diameter Growth at Upper Sections of the Bole. The technique of measuring diameter growth at upper sections is identical with that described for the stump section in Sec. 203. The method of counting in from the cambium toward the pith secures the coordination of groups of 10 rings corresponding to identical decades in the life of the tree for each of its cross sections. Analyses of single trees give insufficient evidence of the growth of *average* trees; hence a number of stem analyses must be made in such a manner that, first, the data taken *can* be averaged for the trees measured and, second, the results can be applied in a practical and accurate manner to standing trees and stands of timber.

The first of these objectives is secured by sectioning the sample trees at equal heights above the ground. This usually requires the cutting of the selected trees by the research crew, since trees cut into odd lengths by loggers will not fulfill this requirement.

Studies of the effect of release cuttings on the form of residual trees indicate a tendency for growth to redistribute itself in a manner, not to make the form more abnormal, but to approach more closely the average normal form instead.[1] Growth at the butt, however, tends to increase with the stronger wind stresses.

210. Averaging the Growth Data for Trees of a Given Group. For complete stem analyses, covering the trees from origin to present age, it is best to include in one average only trees belonging to the same age class or taken from an even-aged stand. These trees should be separated into diameter classes of not over 3 in. in spread. The diameter

---

[1] Meyer, Walter H., Effect of Release upon the Form and Volume of Western Yellow Pine, *Jour. Forestry*, **29**, 1127–1133 (1931).

Behre, C. E., Change in Form of Red Spruce after Logging and of Northern White Pine after Thinning, *Jour. Forestry*, **30**, 805–810 (1932).

growth of the successively earlier decades from cambium to pith is then averaged for the trees comprising each group, and the average diameter at the successive decades may then be plotted over an age scale, corresponding to the age of the stand. Age may be taken as the number of years that have elapsed since the stand originated, *viz.*, following a fire or planting, or it may be based on the average age of the trees in all the groups or preferably the average age of the dominant trees only, comprising those in one or two of the largest groups (Sec. 193). The average diameter of each group is plotted as the ordinate on the age of the stand as the abscissa, and the average diameters at the earlier decades are similarly plotted as ordinates on the age of the stand at these decades. The plotted points are curved out, with the curve originating at a point corresponding to the age which the stand had when the average tree of the group reached stump height. This age is found by subtracting the average age at stump for the group from the total average age of the stand.

211. The Graph of Tree Form Based on Age. The average growth in diameter for sections above the stump is constructed in the same manner, beginning with the total diameter plotted as the ordinate above total age. Since the average age of each section equals the total average age of tree minus the period required for it to reach the height

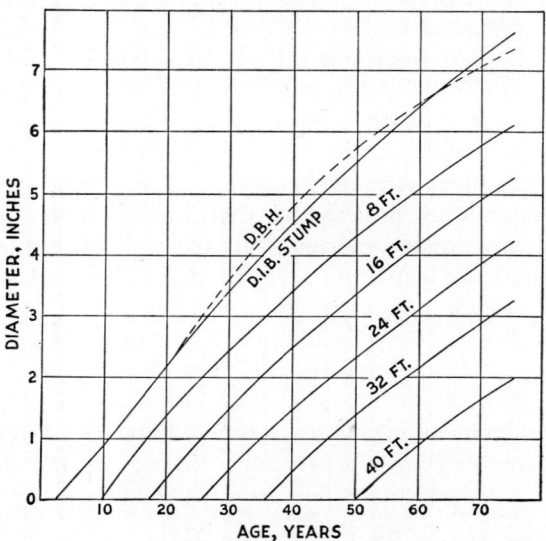

Fig. 68. Average diameters based on total age for 35 chestnut oaks, Milford, Pike County, Pa.

at which the section was measured, the zero origin of each successive curve will coincide with an average curve of height growth based on age for the group (Sec. 200). The resultant average curves of diameter growth at specific heights are shown in Fig. 68. The curve for growth at stump can be corrected as usual for d.b.h. outside the bark (Sec. 206, Fig. 67).

The curves shown in Fig. 68 give the diameters of the average tree only at the points sectioned, viz., 8, 16, 24 ft., etc. In order to reconstruct the entire form of this tree, these data can now be replotted on a graph in which height is introduced on the scale of ordinates and diameter on the scale of abscissas, with a separate curve of form or taper for the tree, inside bark, at each decade counting from the outside or total age inward. The data in Fig. 68 are shown in this form in Fig. 69. The base point of origin of each curve is the stump d.i.b., and its terminus is the approximate height

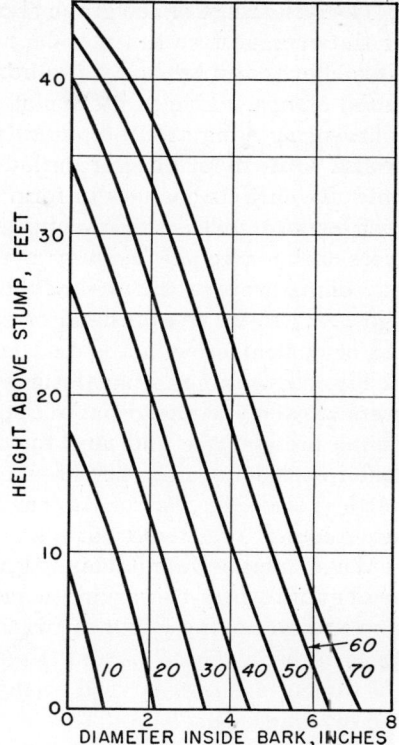

FIG. 69. Average tree from Fig. 68, plotted to show diameters or tapers at all points on the stem for each decade.

of the tree at the age indicated, which can be read off a curve of height growth or obtained by proportioning the height of the tip above the last section, on the basis

$$\text{Length of tip} = \frac{\text{net age of enclosed tip above lower cross section}}{\text{difference in ages at the two ends of section}} \times \text{length of section}$$

For an 8-ft. section, in which the cross section below the tip contains 13 rings and that above the tip 8 rings, the age of the enclosed tip is 3 years, or 13 − 10, and the difference in ages of the cross section above and below is 5 years. Then the length of the tip is $3/5 \times 8 = 4.8$ ft.

The d.b.h.'s corresponding to each stump d.i.b. are shown on the abscissas, or zero scale of ordinates.

The advantage of the graph shown in Fig. 69 is that the dimensions of the average tree can now be read for any height or length of log. This chart constitutes a taper table which shows the changes in form based on age. If accurately plotted, the form quotient for the tree at different ages may be computed from the chart.

212. Growth for Shorter Periods. A similar procedure is used when only the current changes in form are studied, which occur after thinning or after release from suppression or competition. In this case, trees of the same *past diameter* at time of release are averaged together, and diameter is substituted as the basis of a single set of curved averages, *viz.*, 16-in. trees instead of trees in the 100-year age class.

For current growth, the data can be averaged directly on the basis of Fig. 69, without going through the step shown in Fig. 68. Diameters at each fixed height above stump may be averaged, first for d.i.b. of the present tree and next for d.i.b. at the beginning of the period under consideration. The base, or scale of abscissas, is d.i.b., and the vertical scale, or that of the ordinates, is height above stump. The inner core of the tree is not measured, and its curves do not appear.

When trees are available for growth analysis at time of logging but are cut into logs of varying lengths, the dimensions for the present tree and for any period in the past may be plotted, for each tree singly, on form 558a (Sec. 73); and, from the curve formed by these points, the dimensions may be read at any fixed points and lengths desired for the investigation.

213. Volume Growth of the Average Tree. To determine the growth in volume for the average tree whose form is shown in the chart (Fig. 69) for any unit of volume, such as cubic feet, board feet by any given log rule, mine props, crossties, poles, piling, or fence posts, the diameters are read for each age at the required heights above stump, and the volumes are computed for the successive logs thus measured. Or a volume table can be applied to each successive tree form by decades, from Fig. 69, by reading its d.b.h., total or merchantable height, and average form quotient, and interpolating in the volume table to get volumes corresponding to the nearest $\frac{1}{10}$ in. in diameter and 1 ft. in height, and for the nearest form class. The first method is the most accurate.

214. Differentiation by Site Class. Where height growth indicates differences in site (Sec. 219) the stem analyses of trees should be separated by site classes. The same is true when and if site classes are separated on the basis of site indicator plants. Without such site classification, the points of origin of growth for upper sections for different trees in the group would be spread over a wide range with consequent loss of accuracy in growth data obtained from averaging these sections

Where considerable difference in height of the sample trees remains within a given site class, it is possible to resort to the use of percentile tapers (Sec. 109). In that case the trees can be sectioned at each tenth of their total height, the diameter growth of the corresponding sections averaged, and the results plotted on the assumption that the sections fall at each tenth of the height of the average tree of the site class and group analyzed.

References

BEHRE, C. E.: Change in Form of Red Spruce after Logging and of Northern White Pine after Thinning, *Jour. Forestry*, **30**, 805–810 (1932).

BENTLEY, J., JR.: Stem Analysis, *Forestry Quart.*, **12**, 158–166 (1914).

BROWN, C. T., R. C. ROSE, and S. H. SPURR: The Dial Gauge Dendrometer as a Tool in Silvicultural Research, *Jour. Forestry*, **45**, 102–104 (1947).

BRUCE, DONALD: A New Technique for Growth Studies by Stem Analysis, *Jour. Forestry*, **22**, 58–61 (October, 1924).

DAUBENMIRE, R. F.: An Improved Precision Dendrometer, *Ecology*, **26**, 97–98 (1945).

DWIGHT, T. W.: A Simplified Method of Stem Analysis, *Jour. Forestry*, **15**, 864–870 (1917).

FERREE, M. J.: Growth, Cull, and Mortality as Factors in Managing Timber in the Anthracite Region, *Northeastern Forest Expt. Sta. Paper* 15, processed, 1948, 21 pp.

FINCH, T. L.: Effect of Bark Growth in Measurement of Periodic Growth of Individual Trees, *Northern Rocky Mountain Forest and Range Expt. Sta.*, *Res. Note* 60, processed, 3 pp.

GATES, F. C., and G. E. NICHOLS: Relation between Age and Diameter in Trees of the Primeval Northern Hardwood Forest, *Jour. Forestry*, **28**, 395–398 (1930).

HALL, R. C.: A Vernier Tree-growth Band, *Jour. Forestry*, **42**, 742–743 (1944).

KEMP, P. D., and M. E. METCALF: Tables for Approximating Volume Growth of Individual Trees, *Northern Rocky Mountain Forest and Range Expt. Sta.*, *Sta. Paper* 11, processed, 1948, 14 pp.

KRAUCH, H.: Diameter Growth of Ponderosa Pine as Related to Age and Crown Development, *Jour. Forestry*, **32**, 68–71 (1934).

LODEWICK, J. E.: Effect of Certain Climatic Factors on the Diameter Growth of Longleaf Pine in Western Florida, *Jour. Agric. Res.*, **41**, 349–363 (1930).

MEYER, W. H.: Effect of Release upon the Form and Volume of Western Yellow Pine, *Jour. Forestry*, **29**, 1127–1133 (1931).

MLODZIANSKY, A. K.: Measuring the Forest Crop, *U.S. Dept. Agric. Div. Forestry Bul.* 20, 1898, 71 pp.

PEARSON, G. A., and A. D. FOLWEILER: Acceleration of Growth in Western Yellow Pine after Cutting, *Jour. Forestry*, **25**, 981–988 (1927).

PEGG, E. C.: Mechanical Aids in Stem Analysis, *Jour. Forestry*, **17**, 682–685 (1919).

REINEKE, L. H.: A Precision Dendrometer, *Jour. Forestry,* **30,** 692–699 (1932).

ROBERTSON, W. M.: Changes in Relationship between Diameter and Height Growth, *Forestry Chron.,* **17,** 162–169 (1941).

SCHUMACHER, F. X.: Effect of Partial Cutting in the Virgin Stand upon the Growth and Taper of Western Yellow Pine, *Calif. Agric. Expt. Sta. Bul.* 540, 1932, 32 pp.

WILLIAMS, A. S.: Difficulties and Errors in Stem Analysis, *Forestry Quart.,* **1,** 12–17 (1902).

ZON, R., and H. F. SCHOLZ: How Fast Do Northern Hardwoods Grow? *Univ. Wis. Agric. Expt. Sta. Res. Bul.* 88, 1929, 34 pp.

CHAPTER 27

NORMAL YIELD TABLES FOR EVEN-AGED STANDS AND THEIR CONSTRUCTION

215. Yield Tables, Definition. A yield table is usually a table showing the volume or yield per acre and other stand data at different ages for even-aged stands of trees growing on forest land of differing productive capacities. Yield tables for many-aged stands are not based on age and will be treated separately. Separate yield tables are constructed for different species and for each separate unit of merchantable product, such as cubic feet; standard cords; board feet for one or more given log rules, usually including the International $\frac{1}{4}$-in. rule as standard; and, rarely, for piece products.

From the data taken to determine yields, other accompanying tables are constructed that reveal the character and development of the stands with progressive ages. These show, on an acre basis for different sites and ages, averages for basal area, number of trees, heights, and diameters. Tables of mean and current annual growth, as well as stand and stock tables, may also be given. Table 52 shows yield in board feet.

Most yield tables are for pure stands of one species (Sec. 221).

Yield tables are also made for typical groups of species having similar characteristics, such as mixed oaks, mixed pines, or even mixed hardwoods. Studies tend to show that, with as low as 30 per cent of oak in a mixed hardwood stand, for example, the total yields are not seriously affected.[1] This fact extends the possibility of constructing yield tables for mixed stands.

216. The Normal Standard of Stocking for Yield Tables. Yield tables are of two classes, normal and empirical. A *normal* yield table shows the yields capable of being produced on forest sites when the normal capacity of the site is fully utilized by even-aged stands of forest-grown trees. These tables serve first as a standard or goal of management and second as a basis from which to determine the probable yields of understocked stands. *Empirical* yield tables are any

[1] SCHNUR, G. L., Yield, Stand, and Volume Tables for Even-aged Upland Oak Forests, *U.S. Dept. Agric. Tech. Bul.* 560, 1937, 87 pp.

TABLE 52. BOARD FEET (INTERNATIONAL RULE, ¼-IN. KERF) PER ACRE OF TREES 6.6 IN. AND MORE IN DIAMETER TO A TOP DIAMETER OF 5 IN. I.B. LOBLOLLY PINE, LOUISIANA*

Age, years	Site index					
	70	80	90	100	110	120
	Yield per acre, bd. ft.					
20	900	2,200	3,400	4,700	6,100	7,400
30	6,200	9,500	12,800	16,000	19,100	22,400
40	13,400	17,900	22,300	26,700	30,800	35,100
50	17,200	22,100	27,000	31,900	36,700	41,500
60	19,100	24,300	29,600	34,900	40,100	45,300
70	20,600	26,100	31,700	37,300	42,900	48,400
80	21,800	27,600	33,500	39,400	45,200	51,100
90	22,900	29,000	35,100	41,400	47,400	53,600
100	24,000	30,300	36,600	43,200	49,500	56,000

* MEYER, W. H., Yield of Even-aged Stands of Loblolly Pine in Northern Louisiana, *Yale Forest School Bul.* 51, 1942, 39 pp.

tables of yield which apply to stands that are partly stocked. They may be derived from normal yield tables or constructed from direct measurements of existing stands (Chap. 28). The natural tendency for all stands of timber is to reach as soon as possible a condition of full stocking by utilizing all the light and soil moisture available and needed for their crown and root requirements. When the trees are evenly but not too densely spaced, each tree has more room in which to expand its crown. If the spacing, in turn, is not too open to be ultimately absorbed by expanding crowns and roots, the ultimate yield in cubic volume of stands originally "understocked" in their number tends to equal that of more densely stocked stands.

Understocked stands thus tend in later life to become fully stocked, meanwhile losing but few trees by suppression. Although stands fully stocked at an early age lose trees rapidly, they tend to stagnate and hence may often be considered overstocked.

Ideal, or optimum, density would be found in a stand in which the entire crown area was occupied by trees so placed that each tree had sufficient growing space to develop and retain a thrifty crown extending downward from 30 to 40 per cent of its total height and composed of vigorous but not excessively large branches. With this degree of stocking, the stand would be dense enough to shade the lower limbs below this normal crown. In such a stand, diameter growth would continue at an even, fairly rapid rate on all trees, and large trees would be produced in the shortest time without sacrifice of yield in cubic volume. If the stand is less dense, the branches will attain large diameters before they are killed by shading. This means large knots and bad stubs to heel over and perhaps form focuses of infection for heart rot fungi.

In nature, ideal spacing is seldom found over any large area, because the space best adapted to the individual dominant "crop" tree constantly increases with age. Under natural conditions this additional space is obtained by the dominant trees crowding out and killing the weaker ones. In this struggle, especially with certain species of conifers, such as jack, lodgepole, or Norway pine, the trees may be too evenly matched to permit early differentiation in height growth. Thus the height growth but especially the diameter growth of all trees is greatly retarded, and this in turn stagnates wood production (Secs. 184 and 185). Early and marked differentiation of the trees into crown classes, dominant, intermediate, and suppressed trees, is a distinct benefit to a stand. Lacking this differentiation, the normal yield may not be attained even with full stocking.

On the other hand, stands with too few trees to utilize the site fully may never reach full stocking. Open-grown or pasture trees of certain species, like white or live oak, may produce enormous crowns with large short boles, entirely worthless for structural purposes. If wood is to be grown reasonably free from knots and with a maximum of bole wood rather than branches, trees must at an early age be spaced at least as closely as they should stand when nearing maturity. For best form, there should be at least three to four times as many trees as this in the juvenile, or sapling, stage. In selecting stands by the normal standard of density, it has not been the practice to attempt to distinguish those that are overstocked from those that were at one time understocked but have later closed their crown cover. All stands with a practically complete crown canopy have usually been accepted as being fully stocked or normal.

217. The Effect of Site Quality on Yields. Aside from density of stocking, the yield of tree crops is affected enormously by the quality of the site, *i.e.*, by all the factors of environment that influence the growth of trees. Certain of these factors, such as latitude, total amount of heat and rainfall during the year and its distribution, and the effect of altitude, have wide regional influence. Others, such as aspect, whether southern or northern exposure, and degree of slope, are more local. Soil and soil moisture, however, are the factors which exercise the greatest local influence. Soil, in turn, is influenced by relative elevation, slope, exposure, height of water table, and climate. The kind of soil and its texture and fertility are important, but its relation to available soil moisture is more so. Saturated, or waterlogged, soils cause complete changes in forest types as well as in site classes. The best sites are those supplied both with optimum moisture held in the soil or available from the water table and with drainage for soil aeration. With all other factors constant, sites become poorer with either too much or too little moisture. The poorest sites are found on thin soil overlying rock on hilltops or ledges, or in swamps which are constantly wet. There are all gradations between these best and poorest sites, depending on the combination of site factors present.

To use only average yields of all such widely different sites in order to predict the growth of fully stocked stands would be to incur the risk of errors of 50 to over 100 per cent in the use of such tables for individual sites which depart from the average, and with no means of knowing how wide the divergence may be. The better sites give a more rapid height growth, greater total height at maturity, and more rapid differentiation of trees, hence a faster growth in diameter and a greater range of diameters for surviving trees. They also produce a larger total yield, earlier maturity, and greater current and mean annual growth per acre. Fewer trees will form the final crop on the good sites, because each tree grows so much more rapidly than trees on a poor site. This condition holds throughout the life of the stand.

If average yields based on age were found only for the average site, no one could tell whether the difference between the yield of an actual stand and the yield table volumes was caused by variation in site or in number of trees per acre. If different sites are separated in the construction of yield tables, however, the other stand factors for yield for each age, such as average volume and basal area, number of trees per acre, average diameter, and average height of stand, may be found separately for each site class. A yield table in this form will offer a standard applicable to these different sites and may therefore be used as a basis for comparison with actual stands of different ages and growing on various sites. Of these factors, average volume or yield per acre is self-explanatory. A basic table of yields in cubic feet is compiled, and the yields in cords and board feet are obtained by conversion.

Total basal area is the area in square feet of the cross sections at b.h. of all the trees 1 in. and larger in the stand. This is a reliable measure of the relative cubic volume of different stands of a single species on identical sites and reaches its maximum in the United States of over 450 sq. ft. in fully stocked stands of Sitka spruce and western hemlock (site class 200 at 170 years).[1] Average height is taken as the height of a tree of average basal area for the stand.

The average diameter of the stand is likewise taken as that of a tree of average basal area and is consistently larger than a diameter obtained from averaging the actual diameters of all the trees in the stand. It could also be considered that the average height of the stand is the height of the tree of average basal area. But both these averages are unduly influenced by the large numbers of small trees usually present, which contribute little to the basal area or volume of the stand and will soon drop out through mortality. Hence for purposes of the classification of forest stands into site classes it has become accepted practice to exclude all intermediate and suppressed trees. This would leave the trees in the upper crown classes, the dominants and codominants which receive full overhead light, as the ones to include in totals for basal area and number of trees used for calculating the average height of the stand. Selection of correct crown class requires judgment, and the consistency of results depends largely on placing the work for a given yield table in charge of a single experienced field man.

218. Classification of Forest Soils and Environment into Site Classes. The purpose of distinguishing separate site classes is to permit the application of the yield table to specific stratified stands of different ages and sites for which the growth is to be predicted. The site classes given in the yield table must be based on some numerical site index which may easily be determined in the field. For this purpose the yields per acre will not serve because of the enormous difference in density of actual stands. It is difficult or impossible to identify sites by comparing actual yields or volumes with those in yield tables.

Depth, moisture content, soil type, and also plant indicators, *i.e.*, characteristic herbaceous or shrubby plants in the ground cover, are advocated and have been used as the basis of site determination in Finland and Canada; but they have not yet been shown to be feasible in this country for universal practice, though forming an interesting field for research.

The height attained at given ages, on the other hand, has been adopted in America as giving a workable basis for site classification.

[1] MEYER, W. H., Yield of Even-aged Stands of Sitka Spruce and Western Hemlock, *U.S. Dept. Agric. Tech. Bul.* 544, 1937, 86 pp.

Since the height of the largest trees, the dominants (including codominants), has been least affected by competition, the height of this crown class is a better measure of the ability of the area to produce wood. The height that average dominant trees attain on different sites at an arbitrary standard age, for example, at 50 years, has therefore been used as an index of site. Dominant heights on the various sites at any other age will be related (Fig. 70) to their heights at the stand-

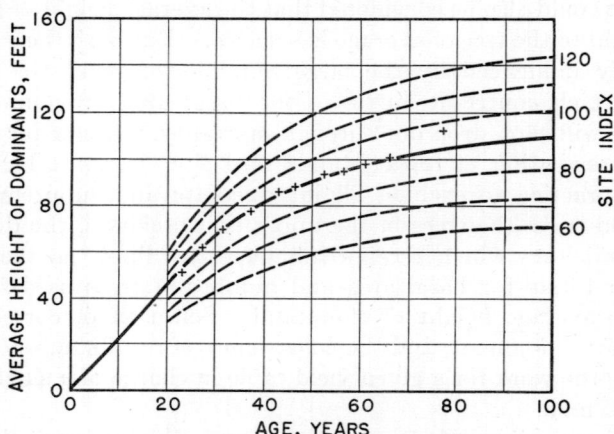

FIG. 70. Site index chart for loblolly pine, showing the relation of average dominant heights to age by 10-ft. site index classes.

ard age. This arbitrary age should be great enough to cover the period of most rapid height growth of the species and is usually taken about the middle of the age range. Fifty years has been used for most eastern species. For the larger trees of the West that may take a longer time to reach maturity, 100 years has been most frequently used. The standard age for Douglas fir in the Pacific Northwest was chosen as 100 years. In Virginia scrub pine, 30 years was chosen; for southern pine and Lake states species, 50 years was adopted.

219. The Site Index Graph. The site index of a stand is the average height that the trees in the upper crown classes are expected to attain, or did in the past attain, at the standard age used as the basis for site classification for the species. This extension of site index to stands younger and older than the standard age is accomplished by constructing a graph composed of bands starting from zero, each of which encloses the plots of all ages whose average heights at these different ages indicate that at the standard age they would fall within a given arbitrary height division.

NORMAL YIELD TABLES

With the standard age for site classification decided on for the species, the next step is to divide the range of site into arbitrary site classes based on height attained at this age. These fixed classes may differ by either 10, 20, or 30 ft. The usual practice is to use 10-ft. divisions, afterward grouping them into 20- or 30-ft. classes if preferred. The number of site classes depends on the total range of average dominant heights found to exist at the standard age. If 20 ft. is the final division, a stand falling at this age into the height class 90 to 110 ft. is placed in the 100-ft. site class, while one with a height between 70 and 90 ft. is in the 80-ft. class.

The *site index* for a given stand can be expressed to the nearest foot of height at the standard age; *viz.*, if a stand had a height of 93 ft. at a standard age of 50 years, its site index is 93. For any other age, the site index may be found from the chart showing the height for each site quality at each age (see Fig. 70) by a method described for a single plot (Sec. 228). For field classification of site and prediction of growth in the yield table it is sufficient to know in which standard site class the stand belongs, without determining the site index to the nearest foot.

220. Alternate Methods of Using Plots in Constructing Yield Tables. A yield table is supposed to show, theoretically at least, for each site class, the evaluation of the average fully stocked stand from the seedling stage to maturity. The best method of finding out what a stand of timber will yield at different ages is to measure permanent plots located in the stand, at intervals of 5 or 10 years over the entire period of growth until it is cut. The actual increase in the diameters and heights of the trees, the number of trees that die from year to year, and the net volume of the surviving stand at any age are then a matter of record. This entire process would take a long time to complete, and the stand might be destroyed at any time by fire, blowdowns, diseases, or other agencies.

The next best plan is to measure a number of permanent sample plots located in stands of different ages (Chap. 32) and to remeasure these plots at 5- or 10-year intervals. By the overlapping of the ages chosen, the trend of development for one or two decades for plots of all ages is obtained after the second or third measurement. Changes in site conditions, underbrush, ground cover, and even in soil may be noted on such permanent sample plots. The main essential for such plots is that the land should be in stable ownership, preferably in a public forest, so that the owner's whim may not interfere with the experiment.

These methods are too time-consuming. Some other method is needed if yield tables are to be prepared for immediate use. To fill this need a third standard plan has been extensively used in this country based on the principle of *comparison* of plots of different ages. The averages of stands for the same site but differing in age are combined into a curve assumed to show the trend of growth. If many plots of different ages on different sites are measured, each plot will show the yield obtained in a fully stocked stand at the given age and for the site on which it is located. For example, 20-year-old stands will show what may be expected at 20 years, and 30-year-old stands may be assumed to yield what the present 20-year-old plots will do in 10 years more; the same for plots 40 and 50 years old. If an entire age series of plots can be found and measured in the year of the study, proper compilation of these data should show with a reasonable degree of accuracy the probable behavior of average stands from the first decade to maturity. To do this, the plots are arranged first by site classes and within sites by 10-year age classes.

221. Normal or Fully Stocked versus Under- and Overstocked Stands. "Pure" stands of even age are seldom composed entirely of a single species and almost never persist in this condition but, if undisturbed by fire, wind, or logging, are later invaded by younger trees of more tolerant species. Hence a pure stand is defined as one in which 75 per cent or more of the *merchantable volume* of the crop is of the same species. Where a stand is pure in this sense, the volume of any minor species composing the remaining percentage of the main stand will be measured and included in the normal yield per acre. The percentage by volume for the minor species can be indicated, as when shortleaf pine, red gum, or oak are found in mixture with loblolly pine of the same age. When mixed stands are composed of similar species, such as the red, black, and scarlet oak group, normal yields may be shown for the mixed type.

The objective of the normal yield table is to determine the *average* yield or crop which each forest site is *capable* of producing when used to its full normal capacity by a stand and species adapted to it. Average stands over large areas are seldom fully stocked. An average production for an entire forest type and species or for an area under management is considered good if it attains 85 to 90 per cent of normality. The variations of local microenvironment within the site create a spread of normal production. As in the case of any average, about one-half of the area within the site class exceeds and one-half falls short of the average normal yield given in the table for that class.

The influences tending to reduce this average yield below normal for an even-aged stand of even the most favored species are numerous and unavoidable.

Influences which must be excluded from a standard for normal yield tables are the effects of destructive agencies that reduce the yields of the actual stands. These include insects and diseases, high winds, abnormal droughts, fire, flooding, or any other agency which damages the stand before it is ready for harvesting. The first function of a normal yield table is to set an average standard of possible yield for each age and site class for a species or mixture of species, which becomes the theoretical goal of management and silviculture. The second is as a vehicle for use in predicting the future growth of understocked stands (Chap. 28).

222. Progress of Subnormal Stands toward Normality. Regardless of the actual number of trees present at different ages, normality of stocking is attained at any age when the stand of living trees makes full use of light, moisture, and soil. But this concept fails to satisfy the forester. He seeks the best average development of each surviving tree in the stand, so that the crop on each acre will produce the greatest value in the shortest possible time. Each tree, from seedlings up, shows remarkable powers of adaptation to permit it to survive as long as possible. Only the strongest and most favorably situated trees reach maturity. The basic law of survival is that trees must grow in height and expand their crowns. Dominant trees of intolerant species succeed by suppressing or killing out their competitors. When trees are tolerant of shade, survival becomes an endurance test, and dominance awaits in turn the death of the oppressors (Sec. 185).

It follows that the individual trees in understocked stands tend to grow rapidly and if above a certain minimum number eventually will fully utilize the site. The stand has then become "normally," or fully, stocked. Where parts of an even-aged stand are destroyed by wind or other agencies, the stocking of this *separate age class* may be permanently reduced, the more so as the stand approaches maturity. If new reproduction is established in these openings, the capacity of the site as a whole is fully utilized for tree growth. The "normality" concept used in yield tables for *even-aged stands* is therefore restricted to the condition within stands of *single age classes* and not to the distribution of age classes among stands in a normal forest.

The degree or percentage of normality of stocking for any given age class is never static but varies with the passage of time. All stands that did not start as fully stocked reproduction thus tend to progress from a condition of understocking to one of normal, and then of overstocking. Fully stocked stands, which originate free from the competition of older trees, as is seen in old field pine and in species coming up after fire (aspen, jack pine, western larch, Douglas fir, white pine) may have many thousands of surplus seedlings which survive the first year and compete for space. Normal development is conditioned for each species by the ability it possesses to differentiate into dominant and suppressed trees. Species such as longleaf, loblolly, and white pines, which possess marked capacity to express early dominance, can do this successfully in most instances and thus develop normal stocking and normal natural yields. Those less capable of expressing dominance when in dense initial stands (ponderosa, shortleaf, lodgepole, jack, red, and Virginia scrub pines) undergo stagnation and become overstocked in numbers but subnormal in volume, with the yield dropping even to one-tenth or less (ponderosa and lodgepole pines) of the normal. Where stagnation is evident in the stand by such criteria as too close spacing, too many suppressed trees, and short crowns, plots should not be taken for the construction of normal yield tables.

At the other extreme, the problem is one of determining the initial density of stocking of actual stands that will produce a normal yield *at some future time.* Whenever a stand is initially stocked with established seedlings spaced so that when they are mature trees they will utilize the entire area, that stand will later become "normal." In selecting plots in young stands for normal yield tables, this criterion has not often been used. Plots are usually selected at each successive age which appear to be fully stocked for the age, *i.e.*, the trees now have practically full crown cover or fully utilize the site. In existing normal yield tables the size of the average trees in plots of the same age is not dependent on age alone but is inversely related to the number of trees per acre and the growing space per tree—the fewer the trees in a normal stocking, the larger the average tree.

As has been stated, in normal yield tables for natural unthinned stands no effort has been made to differentiate between stands that were initially fully stocked and those which later attained this condition. All plots fully stocked in a given age and site class are averaged for total volume, total basal area, number of trees per acre, average basal area, and average height, and the table is constructed accordingly.

223. The Need for Normal Yield Tables for Thinned Stands. The total future *cubic* volume of a fully stocked stand cannot be increased materially by thinnings, although cultural operations to secure full stocking, such as the early weeding out of inferior species, are of the utmost importance. When a stand that has been frequently thinned is compared, at say 50 years of age, with a natural fully stocked stand of the same age, the managed stand usually has a smaller total cubic volume. This will happen if the thinnings have reduced the number of trees to such a point that the full capacity of the site has not been *continuously* utilized. For overstocked stagnating stands, however, lack of well-timed thinnings will result in a loss in increment and total volume. Because of the ability of individual trees to expand and utilize all released site factors, the total cubic yield of managed stands may be increased by at least a portion of the amount removed in well-spaced properly timed thinnings. Early thinnings can be much heavier than later ones to the great benefit of the stand since the young trees with comparatively narrow crowns possess maximum power of expansion to form full coverage. Hence *light* early thinnings removing only intermediate or suppressed trees are poor practice. In loblolly pine at least three, sometimes four, thinnings can be made in young fully stocked stands removing 50 per cent of the upper crown cover without reducing the final yields; but similar practice with longleaf pine diminishes total cubic yield.

Aside from the financial gain in obtaining early returns from thinnings and the salvage of trees that would die naturally and be lost, the advantage of proper thinnings lies in the maintenance of normal even growth in diameter on all surviving crop trees. After 50 to 60 years under management, the yield in *board feet* should considerably exceed that of natural stands of the same age, owing to the larger sizes obtained and therefore the increased ratio of board feet to cubic feet. At the same time, quality need not suffer, since either natural pruning (southern pines) will continue at a satisfactory rate, or artificial pruning, for species such as white pine, could be practiced.

Normal yield tables for both cubic and board feet for thinned stands are therefore very necessary. The preparation of satisfactory yield tables for thinned even-aged stands must await the growing and maturing of managed stands. For this reason the method of permanent sample plots is obviously the best one to use. Meanwhile, with insufficient permanent plots available at present, those that are developed may be compared with natural stands of the same age on similar sites, previously measured, and percentages worked out to show the gain to be obtained from management. Normal yield tables for natural stands will continue to indicate the average productiveness of different sites and species, in cubic feet at different ages, for surviving trees at that age (and as a means to discourage excessive thinnings). The remaining discussion deals with the methods used in preparing normal yield tables for natural unthinned and unmanaged stands.

224. Selection of Plots for a Yield Table. From 100 to 300 plots are recommended for a normal yield table, covering the entire range of site and age at least 10 years beyond the rotation age at which stands may be harvested. The size of each plot should be determined primarily by the number of trees, which should be from 100 to 300 per plot. Plots in older timber will therefore be larger than those in young small timber. Plots will seldom exceed an acre in size unless overmature or decadent timber is included in the range of ages covered.

The shape of plots should be rectangular when possible, though plots with not less than four sides and with no angles of less than 60 degrees are acceptable. Each plot should, if possible, be surrounded by an isolation strip of timber of the same age and character. In dry regions it is very difficult to lay out small plots accurately because the roots spread a considerable distance beyond the crowns. In such stands gaps in the crown canopy are a part of normal stocking.

The sides of the plot may be laid out with compass and tape, plane table, or other surveying methods of equal accuracy. The sides of plots on slopes must be measured horizontally, disregarding the evident fact that an acre therefore has a greater surface. Increased surface area up to a 40 per cent slope increases the yield, but this variation is disregarded in yield tables. Normality must be judged by inspection, with special effort in humid regions to avoid gaps in the crown cover. Too rigid a standard for rejection of plots in the field may make it impossible in some regions to secure the requisite number except at prohibitive expense. In this case, the standard of normality should be lowered sufficiently to secure the necessary data. Since plots will still vary considerably in volume even when they all show 100 per cent of stocking, the introduction of a slightly greater variation by accepting plots with small holes in the crown cover does not lower the utility or value of the standard to any great extent. Too rigid selection and too small size of the plot may result in a standard of normality 25 per cent higher than that obtainable for average normal utilization of the site.

The selection of abnormal plots in the field can be avoided by keeping a running plotted record on graph paper on which the number of trees on the plot is entered on the ordinate scale over the average diameter of the plot on the scale of abscissas. Abnormal plots with too few trees will fall below the band of dots. Those showing overstocking will fall above the average values.

225. Measurements Taken on Plots. The d.b.h. of every tree is taken by 1-in. classes and tallied by species into two crown classes, upper crowns (dominant and codominant), and the intermediate and suppressed trees. The lower limit is 0.6 in., or the 1-in. class, and the usual class limits from 0.6 below to the 0.5 above are observed.

The heights on d.b.h. of enough trees on the plot should be taken to form a reliable curve of height on diameter for each species present on the plot. Fifteen to twenty trees usually should suffice. For plots in the same site class, heights for adjoining plots may be combined for a single curve. The distance to the trees must be measured with a tape, and a standard hypsometer used. The ages of six or more average dominants are determined with increment borer, at the stump or at d.b.h. Separate studies are made to determine the age of free-growing seedlings at stump height or b.h. (Sec. 188). If the trees selected for age determination are average dominant trees and the plot is even-aged within the recognized limits (Sec. 193), their average age is accepted, to which is added the correction for seedling age, to obtain the age of the plot. Occasionally the largest trees are advance growth and will be a few years older than the average dominants. An effort should be made to arrive at an average age from a sample which will represent the entire dominant stand.

A description of the plot should include locality and specific location, date, and names of crew; a note on estimated crown density given in decimals; forest type; slope in per cent; aspect in one of the eight divisions of the compass; altitudes, absolute and relative (elevation above bottomland or stream); soil (type, class, moisture, and drainage); rock; soil cover, including humus and litter; herbaceous ground cover; underbrush and tree reproduction; history, including origin of stand, fires, cutting, grazing if any; and present condition.

Plots are frequently located by a compass tie to some known point and the bearings of the boundaries given, to permit of reexamination, but they are seldom staked permanently with the expectation of remeasurement. However, it is good practice to stake out the plots and blaze their boundaries. This permits remeasurement 5 to 10 years later if opportunity arises.

374 FOREST MENSURATION

YIELDS—EVEN-AGED STANDS									Date				Crew			Plot No.
Type		Locality			Situation				Aspect				Slope			Dimensions
Soil, type		Humus, depth			Ground Cover %				Stand, origin				Ages, test trees Diam. Age Diam. Age			Area
" class		" kind			" species				Fires, etc.							Shape
" moisture drainage		Litter, depth			Brush %			Hts.	Reprod. %			Hts.				Multiple
Rock		" kind			" species				" Species				Seedling Age			Age
D.B.H.	Sp. Dom. Int. Sup.	Sp. D. I. S.	Sp. D. I. S.	Tot. dom.	Tot. I.&S.	Tot. All.	Hts. taken	Ht. curve	Basal areas	Cubic volume	Board feet rule rule
1											
2											
3											
4											
5											

226. Office Work—Plot Computations. It is a great advantage to record on the original sheet, along with the tally of the trees in each d.b.h. class, crown class, and species, the totals for the "dominant" tree class that will be used to determine the site index; the height curve; local volume tables used; and the basal areas, cubic volume, and volume in board feet by the International $\frac{1}{4}$-in. and at least one other log rule; all under separate d.b.h. classes. Computations from the tally thus recorded should then be totaled for the plot and entered at the foot of each column.

The next step is to convert the plot values to terms of 1 acre by use of the requisite factor (for $\frac{1}{2}$-acre plots, 2, etc.). These plot summaries can be recorded on the reverse side of the blank, which may also show the plotted heights and resultant height curve and may have a space for any further data such as plot location and the lines to permanent monuments. Plots are numbered serially and dated.

An example of such record form is shown. All calculations are thus recorded on the original form.

A summary of the data expressed in terms of 1 acre, usually computed for each plot, includes:

Total stand:

1. Basal area of trees 1 in. and over.
2. Number of trees. Subordinate totals may be made for each species and crown class, for basal area, and number of trees.
3. Average d.b.h. for the entire stand, obtained by dividing the total basal area by number of trees, getting the average basal area, and converting to d.b.h. by means of a table of basal areas.
4. Average height of all trees, from average d.b.h. and height curve.
5. Cubic volume of entire stem i.b. for all trees.

YIELDS, PLOT SUMMARIES YEAR ____

Plot No.	Age		Av. Diam.	Av. Height	Site Index
Entire Stand	Plot	Multiple	Acre	Remarks:	
No. of Trees					
Basal Area					
Diam. Av. Tree					
Cubic Volume					
Dominants	Plot	Multiple	Acre		
No. of Trees					
Basal Area					
Diam. Av. Tree					
Ht. Av. Tree					
Cubic Volume					
Above ____ inches	Plot	Multiple	Acre		Height Curve
No. of Trees					
Basal Area					
Diam. Av. Tree					
Cubic Volume					
Bd. ft., rule ____					
Bd. ft., rule ____					
Above ____ inches	Plot	Multiple	Acre		
No. of Trees					
Basal Area					
Diam. Av. Tree					
Bd. ft., rule ____					
Bd. ft., rule ____					

Heights / Diameters

Merchantable stand:

6. Cubic volume of merchantable wood to given top diameters, with bark, or peeled.

7. Board-foot volume by International log rule, either to a fixed (5-in.) top or to average utilized top diameters for each d.b.h. class.

8. Board-foot volume by other log rules in common use, to average utilized top diameter for each d.b.h. class. The two latter figures include only trees above the minimum merchantable d.b.h. limits.

There remains the computation of the site index for each plot, as shown in Secs. 227 and 228.

227. Construction of a Site Index Graph or Table. The volume on the plot, in cubic-foot and board-foot units, has been computed with the aid of its height curve and its resulting local volume table, and adjusted to volume per acre. Before the classification of plots on the basis of site, which is an office computation, is undertaken, the site class curves must be prepared from the dominant heights and ages from all plots.

Height, for the purpose of classifying site, is obtained solely from the dominant (and codominant) trees, ignoring intermediate and suppressed trees. This height is not the same, therefore, as the average height of all the trees on the plot.

1. For each plot or stand, compute the total basal area at b.h. of the dominant class.

2. From this, compute the average basal area by dividing by the number of dominants.

3. Find the diameter corresponding to this average basal area in Table 13, Sec. 27.

4. Look up the height of a tree of this diameter on the curve of height on diameter drawn for the plot. This is the height that will later be used in determining the site index for this plot.

5. Classify these heights by 10-year age classes, and plot an average curve of dominant height over age. Table 53 shows a set of such averages. A weighted curve is balanced among these points, which is the *guide curve* of progressive increase in average height of dominant trees on age. It is also an advantage to place a dot for the height of each plot on the graph of height on age, which will show the distribution and upper and lower limits of the height range at each age.

The plots must be well distributed over the range of sites on which a species occurs in the region being studied. *Otherwise this average curve will be distorted. If the distribution of plots by site is not representative within each age class, further bias will take place.* In no case will the average curve show accurately the development of the average height of the dominant trees at each age. Since the successive average heights on age are found from the total number of dominant trees at each age, and since this number constantly diminishes, each successive average excludes some of the slower growing trees found in the preceding average (Sec. 200).

6. As discussed before, a standard age is chosen at which to make the fixed or arbitrary divisions of height upon which site classes will be

based and by means of which the plots will be classified by sites. If this is taken at 50 years and for 10-ft. intervals, the ordinate of height at the 50-year abscissa is divided into these 10-ft. gradations regardless of where the guiding or weighted average curve falls. The 10-ft. height intervals, 40, 50, 60, etc., are at the center of each band; *viz.*, at 50 years the 70-ft. band will include plots whose average heights lie between 66 and 75 ft.

7. The problem now consists of drawing a set of curves through these fixed 50-year heights, each curve representing one 10-ft. site index class and showing the development in average height of dominant trees for that site class throughout the age range. The basis for the shape of these curves is the single weighted curve of step 5. So far the plots have not been and cannot be separated into sites by field examination but must be so classified on the basis of the guide curve and the set of site index curves that will be shaped to it. Each of these curves is assumed to have the same shape as the guide curve, but separated from it by distances that are proportional to the intervals between the site index heights at 50 years and the average height of the guide curve at this age. This means that the distances apart on each ordinate will be equal; but, since all start from zero and reach 10-ft. differences at 50 years, the amount of the difference will start from zero, reach 10 at 50 years, and be greater than 10 above 50 years. The percentage distance of each curve from the guide curve will be the same for all ages.

It is easiest to draw in the site index curves by using these percentages, although more elaborate methods have been suggested. The percentage is found by dividing the height at 50 years by the height of the guide curve at 50 years. For example, referring to Fig. 70, which is based on the data of Table 53, the average dominant height at 50 years as shown by the guide curve is 94 ft. The 70-ft. site index will have a ratio of 70/94, or 74.5 per cent of the guide curve for each 10-year age class. Each reading on the guide curve for ages at 10-year intervals, 40, 30, 20, 10, is then reduced by 74.5 per cent, and a new curve is drawn through points plotted from the adjusted readings. This new curve is the curve for site index class 70 ft. The same procedure is followed for all other desired site classes, which should be sufficient in number to cover the complete range of heights of the individual plots. The completed chart will serve two purposes: first, to classify the plots by site; and, second, to divide any forest land into site classes when the table is being applied in the field on a given property.

The assumption that the true curves of average dominant height on age for different sites are strictly proportional to the shape of the guiding curve is a temporary expedient in the absence of historical data from permanent sample plots, repeatedly measured. In time it may be shown that the shapes of the curves for high and low site qualities differ from the average curve.

The foregoing assumption of proportional shape can be checked in the absence of historical data and an adjustment can be made, provided that the supposition is valid that the basic plot data cover the same range of site classes equally effectively in the entire range of age classes. The procedure is as follows:

1. Draw the guide curve as before.
2. Sort the plots by 10-year age classes, and compute the standard deviation of average dominant heights for each age class based on differences of actual heights from the curved heights for the age classes.
3. Divide the standard deviation for each age class by the average height from the guide curve. This is called the coefficient of variation (Sec. 16d).
4. Plot the coefficients of variation over the average ages of the plots in the age classes.

   a. If the plotted points can be represented by a straight horizontal line, the assumption of proportional shape is supported, *i.e.*, the percentage variation around the guide curve is constant.

   b. If the plotted points fall in a curve, which is apt to be descending with age and concave upward, then proceed with the following steps.

5. Select a 10-ft. site index value, and compute an adjustment factor as follows:

$$\text{Adjustment factor} = \frac{\text{site index height} - \text{guide curve height at standard age}}{\text{guide curve height} \times \text{coefficient of variation at standard age}}$$

6. Use this factor to determine the distance in per cent from the guide curve at all other ages for the site index curve in question as follows:

Difference in per cent from guide curve at x age
$$= \text{adjustment factor} \times \text{coefficient of variation at } x \text{ age}$$

7. Convert percentages to feet by multiplying by guide curve height at x age and plot the differences on the chart. Connect the points, which gives the site index curve.
8. Repeat steps 5 to 7 for all desired site index classes.

In the loblolly pine example, a straight horizontal line was obtained under step 4, and the foregoing steps were not necessary.

TABLE 53. CLASSIFICATION OF LOBLOLLY PINE PLOT DATA BY AGE CLASS

Age class, years	Average age, years	Number of plots	Average height of dominants, ft.	Average number trees per acre	Basal area per acre, sq. ft.	Volume per acre, cu. ft.
20–24	23	16	51	530	121	2,332
25–29	27	15	62	431	138	3,390
30–34	32	11	70	381	152	4,147
35–39	37	19	78	255	146	4,328
40–44	43	42	86	170	137	4,539
45–49	46	43	88	166	135	4,589
50–54	52	24	94	149	132	4,884
55–59	56	34	95	139	125	4,657
60–64	61	31	98	149	138	5,204
65–69	66	16	101	137	133	5,199
70–74	70	19	102	123	138	5,383
75–79	77	5	113	97	145	6,404

228. Determination of Site Index for Each Plot. The lines of the graph (Fig. 70) correspond to the exact 10-ft. site index classes, 50, 60 ft., etc. Lines are then drawn halfway between so that the plots or stands can be classified easily by 10-ft. classes. Each plot is given a site index to the nearest foot, by noting the position of its plotted average height of dominants in reference to the nearest site index curve. For example, a plot in a 25-year-old stand has a height of 70 ft. This plotted point lies slightly above the 110-ft. curve. At 25 years, the distance between two adjacent curves is 8 ft. The 70-ft. height falls 1 ft. above the curve, or at $\frac{1}{8}$, or 0.125, of the 8-ft. interval. In site index units this is equal to 0.125 times 10, or 1.25, rounded off to 1. Hence the site index for the plot is 110 plus 1, or 111 ft. With a little practice, the site index can be interpolated ocularly. The assumption here is that the plot in question will remain at site index 111, attaining this average height at 50 years and becoming still taller at more advanced ages. A second plot, 42 years old, has a present average height of 63 ft. By the above process its site index rating is found to be 69 ft.

229. Construction of a Table of Total Basal Area Based on Age. To supplement the initial field selection and reject plots that are abnormal to a degree which might seriously affect the purposes of the table, a method is needed which can be applied in the office. Total basal area per plot furnishes this basis.

A table of total basal areas based on age for each site is prepared from curves by a method similar to that used for height on age (Table 53). First, the basal areas of all plots regardless of site are averaged by 10-year age classes and a weighted curve is plotted of basal area on age. Next, the plots for each separate 10-ft. site class are averaged on the basis of age. For each site class a curve must be drawn of basal area based on age. It is usually assumed that for total basal area, each of these curves will have a shape similar to the guiding or average curve for the entire set of plots. But the separate curves will not necessarily be spaced at equal intervals. In fact, the distance between curves usually decreases progressively from poor to better sites, and there is far less total difference in basal area between poor and good sites than there is in height and volume. The plot averages, too, are apt to be irregular except at heavily weighted points.

An attempt may be made to draw separately the curve for each site class, on the original graph, by following the trend of the guiding curve but weighting each point on the curve by the number of plots in the plotted averages. The averages for each curve may be plotted with different symbols or colors. When all curves are plotted, they must be harmonized by determining the spacing between them, whether it be equal or progressive, and readjusting the curves to this regular arrangement.

A simpler yet more rigorous method than the foregoing one is available, which has the advantage of establishing the curve shape and balancing the system of curves in one operation. The steps are as follows:

1. Compute the percentage relationship that the basal area of each plot bears to the basal area on the guide curve corresponding to the age of the plot. For example, a plot in a 32-year-old stand is found to have 120 sq. ft. of basal area per acre. The guide curve at this point reads 128 sq. ft. The value is $120/128 \times 100$, or 93.8 per cent.

2. Collect together all plots within the same 10-ft. site index class regardless of age, and find the average of the percentages. Compute also the exact average site index.

3. Plot the average percentages over the average site indexes, and balance a curve among the points. This graph is commonly called the "intercept" chart (see Fig. 71).

4. From this intercept curve, read the curved percentages at even 10-ft. site index intervals.

5. Apply each percentage to the basal areas read from the guide

curve for ages and 10-year intervals to obtain the basal areas for each site index curve. Plot these values. Each series will fall in a smooth curve. These curves show the increase in total basal area with age for each site class.

The percentages plotted on the intercept chart do not necessarily

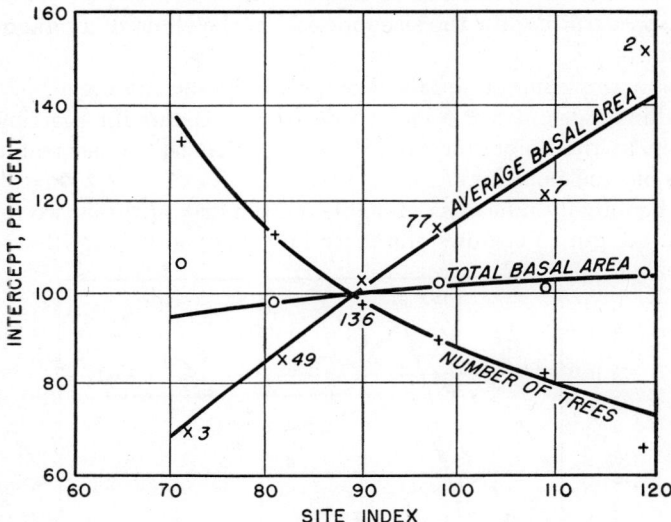

FIG. 71. Intercept curves of total basal area, number of trees, and average basal area for loblolly pine.

fall in a straight line but sometimes are best fitted by a curve which is concave downward yet tends to rise with site index.

This method is valid only when the coefficients of variation of basal areas around the guide curve form a horizontal straight line as in the case of heights. To test for this condition, use the same method as for dominant height. If the fitted line is rising or falling, or if it is curved, however, this is direct evidence that the basal area curves by site index do not bear a constant percentage relationship to the guide curve. In this case, a second method of establishing the curves by site index is called for. It consists of the following steps:

1. Plot standard deviations of basal area over age, and balance a curve among the points.
2. Divide the deviation of each plot from the guide curve by the standard deviation corresponding to the plot age. This ratio is called a "standard unit."
3. Assemble the plots by site index classes, and compute an average standard unit for the class.

382 FOREST MENSURATION

4. Plot the average standard unit over the exact site index average for the class, and curve out, a process analogous to the construction of the intercept chart for dominant heights.

5. Each site curve is then plotted on the original chart containing the guide curve as follows:

   a. Take a reading from the guide curve.
   b. Take a reading for the site index from the chart of average standard units.
   c. Take a reading of standard deviation at the same age.
   d. Add the product of b and c to a to get a value for plotting.
   e. Repeat the above for each decade and each site class, and curve out the plotted points.

For the loblolly pine data of Table 53, the first of these two methods sufficed. Figure 71 shows the intercept curve and Fig. 72 the guide

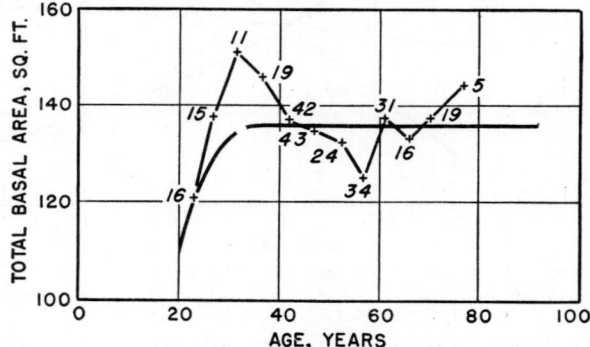

FIG. 72. Guide curve for total basal area for loblolly pine.

curve. The plotted points for the guide curve show clearly the effect of overstocked plots, particularly in the young age classes, and considerable judgment was necessary to establish the position of the curve. The sharp downward trend at the extreme left is necessary because the basal area at age 0 must be 0 sq. ft. The intercept curve is usually horizontal or slightly concave downward. Headings for even 10-ft. site index intervals show the following:

Site index	Intercept percentage
70	95
80	97.5
90	100
100	101.5
110	103
120	104

The basal area curve for site index 70 is prepared by multiplying the guide curve basal areas by 95 per cent; that for site index 80 by 97.5 per cent, etc. Figure 73 shows the final chart, from which the total basal area of normal stands can be read directly for any age and site index.

The same method is repeated for number of trees and for average basal area and later for volume per acre in cubic feet.

Since the total basal area, number of trees, and average basal area curves were drawn independently, they may not cross-check with each other at each 10-year point. For example, the number of trees divided

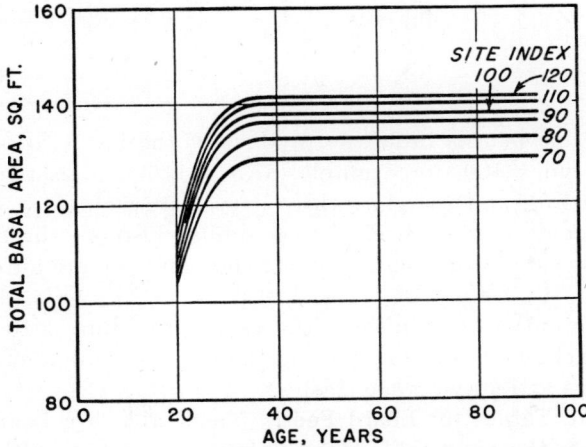

Fig. 73. Harmonized total basal area curves by site index for loblolly pine.

into total basal area may not give the average basal area for each 10-year age class as read from the curve drawn independently. For this reason they must be cross-checked as follows: For each site-age reading the relationship should hold true that number of trees times average basal area equals total basal area. Since this is an essential requirement, the three guide curves are drawn simultaneously and checked throughout their lengths by the above relationship. If the cross checks are not perfect, the curves must be adjusted until perfect agreement is obtained. One difficulty encountered here will be that the average basal area computed from the individual plots will not be equal to the average basal area compiled by dividing the basal area of all plots in the age class by the number of trees on all plots. The second calculation for basal area should be the basis for the guide curve of average basal area. After the guide curves are made to check, and the intercept curves are drawn, a similar check is made of these intercept curves by using the following relationship: average basal area in per cent times number of trees in per cent equals basal area in per cent. The following data from Fig. 71 illustrate the procedure:

Site index	Intercept percentages		
	Total basal area	Number of trees	Average basal area
70	95	138.5	68.5
80	97.5	114.7	85
90	100	98	102
100	101.5	87.5	116
110	103	79.5	129.5
120	104	73.2	142

In each line in this table the product of the last two percentages (with site index 100, for example, 87.5 × 116) equals the first percentage (101.5), and therefore three curves cross-check at this point. With the guide curves checked in a similar fashion, the full sets of curves with values for each site index class and age are automatically in balance.

After the curves are finished, the usual procedure is to prepare a table for each class of data covering the complete range of 10-ft. site index classes and 10-year age classes.

230. Yield Tables for Board Feet. Yield tables for board feet are derived from the yield tables for total cubic feet by multiplying the yields in cubic feet by board foot–cubic foot ratios, curved over the average diameter of the stand. To obtain this curve, the board-foot volume for each plot is divided by its total cubic-foot volume. The ratios are averaged and plotted over the average diameters of the plots. This curve is very well defined. If not, the plot records and calculations should be carefully checked. Figure 74 is the board foot–cubic foot curve for loblolly pine.

To compute the board-foot yield for a given age-site class, obtain the average stand diameter for this class from the diameter table and read the board foot–cubic foot ratio for this diameter from the curve. Next multiply the yield in cubic feet for this class by the ratio and record the computed board feet in the board-foot yield table.

231. Merchantable Yields for Partial Stands. Tables for merchantable yields in cubic feet and cords are usually also prepared for a given species. This is also true for board feet. For example, a table show-

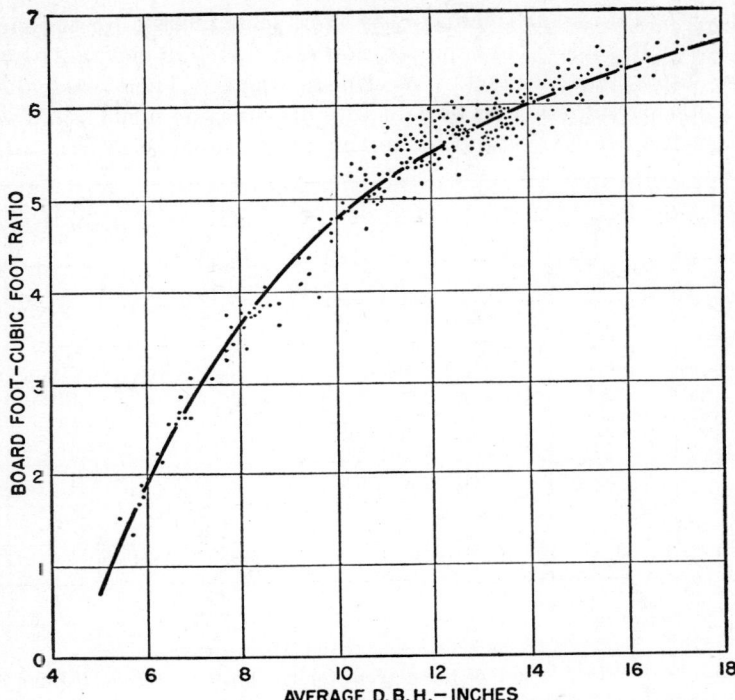

Fig. 74. Board foot–cubic foot ratios plotted on average stand diameter for loblolly pine.

ing the yields in cubic feet, basal area, number of trees, and average diameter for the stand 12 in. and over in diameter may be compiled. The basic tables for the entire stand are used as the starting point for constructing the merchantable tables. As in the case of the board-foot ratio, strong correlation holds for the percentage relationships between the corresponding stand data for the entire and merchantable stand when these are plotted over stand d.b.h.

On the original plot records the number of trees, basal area, and cubic-foot yields of the merchantable stand are divided by the corresponding data for the entire stand and expressed as percentages. These percentages are classified and plotted over average diameter of the stand. Figure 75 is an illustration. If the cubic-foot yield percentages had been plotted on the same chart they would lie slightly above the curve for the basal area percentages. The curve trend is usually well defined, and the plotted points (not shown here) form a narrow band. To compute the tables for the merchantable stand, the curved percentages are read, for the average diameters, for the specific ages and site classes, and the appropriate values in the table for the entire stand are multiplied by these percentages. The average diameter of the partial stand is obtained by dividing the new basal area by the new number of trees and converting this average basal area to d.b.h.

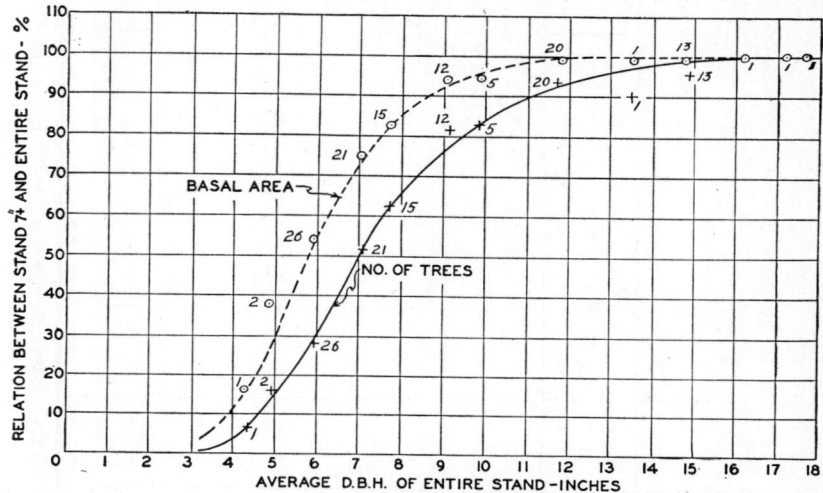

FIG. 75. ------ Derivation of basal areas for the portion of stands above 7 in., as a percentage of the basal area of entire stand. ——— Derivation of number of trees above 7 in., in percentage of number of trees in entire stand. Plotted on average d.b.h. of stand. [*From Donald Bruce, A Method of Preparing Timber Yield Tables, Jour. Agric. Res.,* **32**, 543–547 (1926).]

Yield tables for cordwood, rough or peeled, are expressed in the standard cord unit of 128 cu. ft. of stacked wood. Their construction follows the same principles as above.

232. Stand and Stock Tables for Fully Stocked Even-aged Stands. Stand tables are defined as those showing the number of trees per acre by diameter class for given stands. Stock tables show the volume in each diameter class. In yield work, stand and stock tables for fully stocked stands are exceedingly useful because they give a more complete picture of the distribution of the number of trees and volume in the stand. With these tables it is possible to find the distribution and proportion of volume in any diameter class, group of diameter classes, or above or below any diameter limit. The construction of a stand table involves constructing frequency distributions or curves (Sec. 13). For even-aged stands the frequency curve of number of trees by diameter classes is usually skewed especially for the younger stands. A very young stand starts out with a J distribution; *i.e.*, the greatest number is found in the smallest class, with rapidly decreasing numbers towards the right. Gradually the curve shifts to the right with advancing age, and a definite peak (Sec. 15g) appears. The peak of the curve is still in the lower diameter classes but not in the smallest

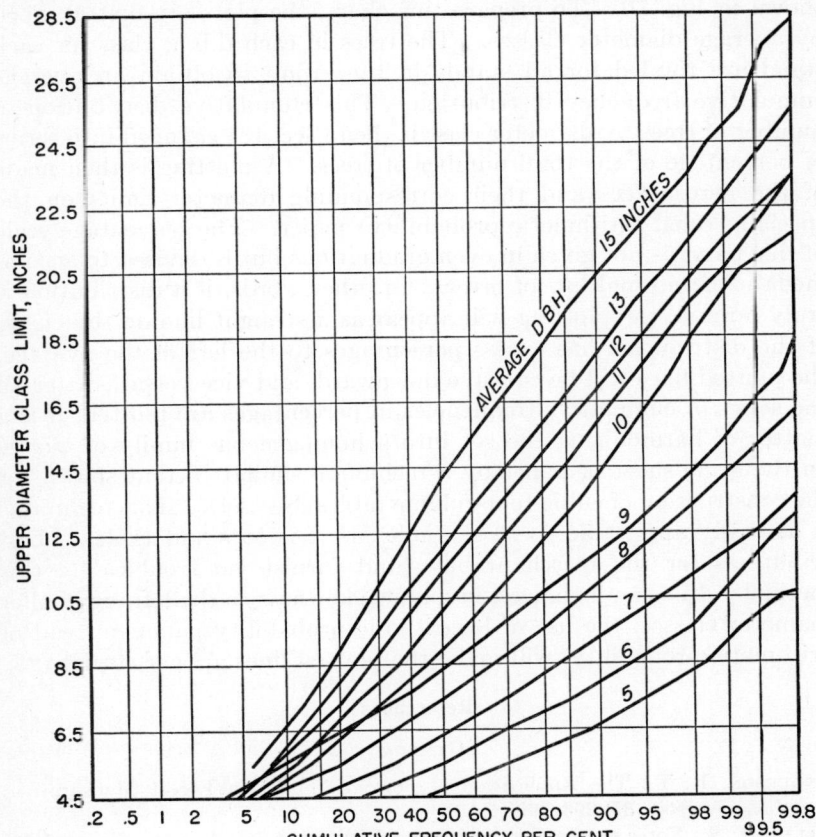

FIG. 76. Cumulative stand distributions for loblolly pine plotted on arithmetic probability paper.

class. As the stand gets older, the peak shifts closer and closer to the center. For some species it even moves beyond the center of the distribution, resulting in skewness of the opposite character. The frequency curves for the stock tables are even more extreme and have the added complication that, for the merchantable stand, the lower end of the distribution is cut off sharply by the limiting diameter.

The construction of stand and stock tables is therefore not an easy matter. At least two graphical methods, and several mathematical formulas, have been used for this purpose. An adequate discussion of these advanced methods is beyond the scope of this book. The method of using arithmetic probability paper is the simplest and is shown in Fig. 76. To prepare this chart, the plots are first grouped by average diameter classes. The trees in each d.b.h. class for each stand are totaled for all stands in the group to obtain an average cumulative frequency distribution. This cumulative distribution of number of trees by diameter class is then (Sec. 13) changed into terms of percentage of the total number of trees. A plotting is then made of the percentages and their corresponding diameter limits on the specially ruled (arithmetic probability) paper. The percentage scale of this paper is not given in even graduations but is devised to accommodate the normal law of error. In other words, if a distribution is truly normal, the plotting will appear as a straight line on the chart. If the distribution has excess percentages to the left of the average, the plotted lines will be concave downward, and vice versa. After all the series of cumulative frequencies in percentages are plotted, it is a matter of harmonizing the set into a homogeneous family of curves on the same sheet of paper by a technique similar to that shown for the construction of cubic-foot volume tables (Sec. 84). This technique is as easily applicable to stock tables as it is to stand tables. Distributions for the merchantable stand furnish no troubles by this method. In some instances, involving skewness to the left, or smaller diameter classes, the use of logarithmic probability paper, instead of arithmetic probability, will make the plotted lines appear straight.

References

General

Anderson, R. T.: The Application of Fourier's Series in Forest Mensuration, *Jour. Forestry*, **35,** 293–299 (1937).

Baker, F. S.: Notes on the Composition of Even-aged Stands, *Jour. Forestry*, **21,** 712–717 (1923).

Behre, C. E.: Preliminary Normal Yield Tables for Second-growth Western Yellow Pine in Northern Idaho and Adjacent Areas, *Jour. Agric. Res.*, **37,** 381–397 (1928).

British Columbia Forest Service: Volume, Yield, and Stand Tables for Some of the Principal Timber Species of British Columbia, Research Division, 1936, 53 pp.

Bruce, Donald: Anamorphosis and Its Use in Forest Graphics, *Jour. Forestry*, **21,** 773–783 (1923).

———: A Method of Preparing Timber-yield Tables, *Jour. Agric. Res.*, **32,** 543–557 (1926).

―――: Preliminary Yield Tables for Second Growth Redwood, *Univ. Calif. Agric. Expt. Sta. Bul.* 361, 1923, 43 pp.
――― and L. H. REINEKE: The Use of Alinement Charts in Constructing Forest Stand Tables, *Jour. Agric. Res.*, **38**, 269–308 (1929).
Douglas Fir Second-growth Management Committee: Management of Second-growth Forests of the Douglas Fir Region, U.S. Forest Service, Pacific Northwest Forest and Range Experiment Station, processed, 1947, 151 pp.
DUNNING, D., and L. H. REINEKE: Preliminary Yield Tables for Second-growth Stands in the California Pine Region, *U.S. Dept. Agric. Tech. Bul.* 354, 1933, 23 pp.
FORBES, R. D., and DONALD BRUCE: Rate of Growth of Second-growth Southern Pines in Full Stands, *U.S. Dept. Agric. Cir.* 124, 1930, 77 pp.
FRICKE: Einheitliche Schätzungstafel für Kiefer, *Ztschr. f. Forst u. Jagdw.*, **46**, 325–342 (1914). Review, *Forestry Quart.*, **12**, 629–631 (1914).
FROTHINGHAM, E. H.: White Pine under Management, *U.S. Dept. Agric. Bul.* 13, 1914, 70 pp.
GERHARDT, E.: Ertragstafeln für reine und gleichartige Hochwaldbestände von Eiche, Buche, Tanne, Fichte, Kiefer, grüner Douglasie und Lärche, 2d ed., Verlag Julius Springer, Berlin, 1930, 73 pp.
GEVORKIANTZ, S. R., and R. ZON: Second-growth White Pine in Wisconsin, *Univ. Wis. Agric. Expt. Sta. Res. Bul.* 98, 1930, 40 pp.
Great Britain Forestry Commission: Growth and Yield of Conifers in Great Britain, *Bul.* 10, 1928, 187 pp.
HAIG, I. T.: Comparative Timber Yields, *Jour. Forestry*, **30**, 575–578 (1932).
―――: Second Growth Yield, Stand, and Volume Tables for the Western White Pine Type, *U.S. Dept. Agric. Tech. Bul.* 323, 1932, 68 pp.
HOUGH, A. F.: Some Diameter Distributions in Forest Stands of Northwestern Pennsylvania, *Jour. Forestry*, **30**, 933–943 (1932).
ILVESSALO, Y.: Methods for Preparing Yield Tables, *Silva Fenn.*, 5, 1927, 32 pp.
―――: Possibilities of Finding a Uniform Basis for the Study of Growth and Yield in Different Countries, *Proc. Internatl. Cong. Forestry Expt. Sta*, 1930, 24 pp.
KITTREDGE, J., JR., and S. R. GEVORKIANTZ: Forest Possibilities of Aspen Lands in Lake States, *Univ. Minn. Agric. Expt. Sta. Tech. Bul.* 60, 1929, 84 pp.
LONG, H. D.: A Proposed Method of Estimating the Growth of Pulpwood Stands, Pulp and Paper Research Institute of Canada, 1947, 8 pp.
MCARDLE, R. E., and W. H. MEYER: The Yield of Douglas Fir in the Pacific Northwest, *U.S. Dept. Agric. Tech. Bul.* 201, 1930, 64 pp.
MCINTYRE, A. C.: Growth and Yield in Oak Forests of Pennsylvania, *Pa. Agric. Expt. Sta. Bul.* 283, 1933, 28 pp.
―――: Virginia Pine in Pennsylvania, *Pa. State Col. School Agr. Expt. Sta. Bul.* 300, 31 pp.
MACKINNEY, A. L., and L. E. CHAIKEN: Volume, Yield and Growth of Loblolly Pine in the Mid-Atlantic Coastal Region, *Appalachian Forest Expt. Sta. Tech. Note* 30, 1939, 30 pp.
MEYER, W. H.: Diameter Distribution Series in Even-aged Forest Stands, *Yale Forest School Bul.* 28, 1930, 105 pp.
―――: Yield of Even-aged Stands of Loblolly Pine in Northern Louisiana, *Yale Forest School Bul.* 51, 1942, 39 pp.

———: Yield of Even-aged Stands of Ponderosa Pine, *U.S. Dept. Agric. Tech. Bul.* 630, 1938, 60 pp.

———: Yield of Even-aged Stands of Sitka Spruce and Western Hemlock, *U.S. Dept. Agric. Tech. Bul.* 544, 1937, 86 pp.

———: Yields of Second-growth Spruce and Fir in the Northeast, *U.S. Dept. Agric. Tech. Bul.* 142, 1929, 52 pp.

MUNGER, T. T.: Watching a Douglas Fir Forest for Thirty-five Years, *Jour. Forestry*, **44**, 705–708 (1946).

OSBORNE, J. G., and F. X. SCHUMACHER: The Construction of Normal-yield and Stand Tables for Even-aged Timber Stands, *Jour. Agric. Res.*, **51**, 547–564 (1935).

REINEKE, L. H.: A Modification of Bruce's Method of Preparing Timber Yield Tables, *Jour. Agric. Res.*, **35**, 843–856 (1927).

———: Yield Tables—How Many Plots? *Jour. Forestry*, **25**, 448–451 (1927).

SCHNUR, G. L.: Diameter Distribution for Oldfield Loblolly Pine Stands in Maryland, *Jour. Agric. Res.*, **49**, 731–743 (1934).

———: Yield, Stand, and Volume Tables for Even-aged Upland Oak Forests, *U.S. Dept. Agric. Tech. Bul.* 560, 1937, 87 pp.

SCHUMACHER, F. X.: Concerning Normal Stocking of Even-aged Stands, *Jour. Forestry*, **26**, 608–617 (1928).

———: A New Growth Curve and Its Application to Timber Yield Studies, *Jour. Forestry*, **37**, 819–820 (1939).

———: Yield, Stand, and Volume Tables for Douglas Fir in California, *Univ. Calif. Col. Agric. Bul.* 491, 1930, 41 pp.

———: Yield, Stand, and Volume Tables for Red Fir in California, *Univ. Calif. Col. Agric. Bul.* 456, 1928, 29 pp.

———: Yield, Stand, and Volume Tables for White Fir in the California Pine Region, *Univ. Calif. Col. Agric. Bul.* 407, 1926, 26 pp.

SHOW, S. B.: Yield Capacities of the Pure Yellow Pine Type on the East Slope of the Sierra Nevada Mountains in California, *Jour. Agric. Res.*, **31**, 1121–1135 (1926).

SPAETH, J. N.: Growth Study and Normal Yield Tables for Second Growth Hardwood Stands in Central New England, *Harvard Forest Bul.* 2, 1920, 21 pp.

TAYLOR, R. F.: Yield of Second Growth Western Hemlock–Sitka Spruce Stands in Southeastern Alaska, *U.S. Dept. Agric. Tech. Bul.* 412, 1934, 29 pp.

TROREY, L. G.: A Mathematical Method for the Construction of Diameter Height Curves Based on Site, *Forestry Chron.*, **8**, 121–132 (1932).

U.S. Forest Service: Volume, Yield and Stand Tables for Second-growth Southern Pines, *U.S. Dept. Agric. Misc. Pub.* 50, 1929, 202 pp.

WACKERMAN, A. E., R. ZON, and F. G. WILSON: Yield of Jack Pine in the Lake States, *Wis. Agric. Expt. Sta. Res. Bul.* 90, 1929, 23 pp.

WILLIAMSON, A. E.: Cottonwood in the Mississippi Valley, *U.S. Dept. Agric. Bul.* 24, 1913, 62 pp.

WOODWARD, K. W.: A Comparison of Yields in the White Mountains and Southern Appalachians, *Forestry Quart.*, **11**, 503–508 (1913).

SITE DETERMINATION

BATES, C. G.: Concerning Site, *Jour. Forestry*, **16**, 383–388 (1918).

CAJANDER, A. K.: The Theory of Forest Types, *Acta Forest. Fenn.*, 28, 1926, 108 pp.

CLAMENTS, F. E.: Plant Formations and Forest Types, *Soc. Amer. Foresters Proc.*, **4**, 50–63 (1909).

COILE, T. S.: Relation of Soil Characteristics to Site Index of Loblolly and Shortleaf Pines in the Lower Piedmont Region of North Carolina, *Duke Univ. School Forestry Bul.* 13, 1948, 78 pp.

DUNNING, D.: Site Classification for the Mixed Conifer Selection Forests of the Sierra Nevada, *Calif. Forest and Range Expt. Sta. Res. Note* 28, 1942, 21 pp.

FROTHINGHAM, E. H.: Classifying Forest Sites by Height Growth, *Jour. Forestry*, **19**, 374–381 (1921).

———: Height Growth as a Key to Site, *Jour. Forestry*, **16**, 754–760 (1918).

———: Site Determination and Yield Forecasts in the Southern Appalachians, *Jour. Forestry*, **19**, 1–14 (1921).

HAIG, I. T.: Colloidal Content and Related Soil Factors as Indicators of Site Quality, *Yale Forest School Bul.* 24, 1929, 33 pp.

HANZLIK, E. J.: The Determination of Site Qualities for Even-aged Stands by Means of a Site Factor, *Soc. Amer. Foresters Proc.*, **9**, 229–234 (1914).

HEIMBURGER, C. C.: Forest-site Classification and Soil Investigation on Lake Edward Forest Experimental Area, *Canadian Forest Serv. Silvicultural Res. Note* 66, 1941, 60 pp.

HOLMAN, H. L.: Ground Vegetation: An Index of Site Quality, *Forestry Chron.*, **5** (1), 7–9 (1938).

KITTREDGE, J., JR.: Interrelations of Habitat, Growth Rate, and Associated Vegetation in an Aspen Community of Minnesota, *Ecol. Monog.*, **8**, 151–246 (1933).

KORSTIAN, C. F.: The Indicator Significance of Natural Vegetation in the Determination of Forest Sites, *Plant World*, **20**, 267–287 (1917).

———: Native Vegetation as a Criterion of Site, *Plant World*, **22**, 253–261 (1919).

PARKER, H. A.: Dominant Height and Average Diameter as a Measure of Site, *Canadian Forest Serv. Silvicultural Res. Note* 72, 1942, 19 pp.

RAY, R. G.: Site-types and Rate of Growth, *Canadian Forest Serv. Silvicultural Res. Note* 65, 1941, 56 pp.

RIGG, G. B.: Using the Vegetation Cover as an Aid in Studying Logged-off Lands as Forest Sites, *Jour. Forestry*, **27**, 539–545 (1929).

ROTH, F.: Another Word on Site, *Jour. Forestry*, **16**, 749–753 (1918).

———: Concerning Site, *Forestry Quart.*, **14**, 3–13 (1916).

SISAM, J. W. B.: Site as a Factor in Silviculture—Its Determination with Special Reference to the Use of Plant Indicators, *Canada Dept. Mines and Resources, Lands, Parks, and Forests Branch, Silvicultural Res. Note* 54, 1938, 88 pp.

Society of American Foresters, Committes on Site Standardization: Classification of Forest Sites, *Jour. Forestry*, **21**, 139–147 (1923).

SPILSBURY, R. H., and D. S. SMITH: Forest Site Types of the Pacific Northwest, A Preliminary Report, *Brit. Columbia Forest Serv. Tech. Pub.* T. 30, 1947, 46 pp.

SPRING, S. N.: Site and Site Classes, *Jour. Forestry*, **15**, 102–103 (1917).

TAYLOR, R. F.: Site Prediction in Virgin Forests of Southeastern Alaska, *Jour. Forestry*, **31**, 14–18 (1933).

WATSON, R.: Site Determination, Classification and Application, *Jour. Forestry*, **15**, 552–563 (1917).

CHAPTER 28

PREDICTION OF THE PERIODIC ANNUAL GROWTH AND YIELD OF "AVERAGE" EVEN-AGED STANDS BY USE OF NORMAL YIELD TABLES

233. The Application of Normal Yield Tables to Average Stands. The primary purpose of yield tables for even-aged stands is to predict the future growth of actual stands over extensive forest areas. Unmanaged stands over large areas will never produce the normal yields shown by the yield table. Hence, without reduction to actual densities, normal yield tables will always overestimate yields by percentages which vary with the density of the actual stands and forest. This, however, in no way invalidates the usefulness of normal yields. The most practical and accurate way to predict the growth of even-aged stands is by means of a normal yield table. The prediction is made by converting the normal growth values from the yield tables to actual growth.

234. Empirical, or Subnormal, Yield Tables. From time to time efforts have been made to construct yield tables for the average run of understocked stands by the same method used for normal yield tables; *viz.*, to measure a representative sample consisting of sufficient plots in stands of different ages but not limiting them to fully stocked stands; and with the definite purpose of constructing tables to show existing actual average yields over large areas rather than fully stocked or normal yields. Since actual stocking at any age may vary from zero to 100 per cent of normal, such tables express merely the accidental averages obtained from the plots measured and therefore represent only the conditions existing at the moment on the areas included in the sample. They must be adjusted to individual stands and forests just as are normal yield tables. They fail to give as complete satisfaction as the normal tables, since the closer a yield table comes to average full production per acre, the more accurately can reductions be made from this standard to obtain actual yields. Any standard less than normal would result in a yardstick that would be difficult to associate with the normal development of a stand. Because of the great range of yields for the same species, age, and sites found under natural subnormal conditions of stocking, not only would a much larger sample be required for an empirical yield table, but such a table would be less valuable for indicating site and possible yields from thinnings and would also fail to indicate the normal potential growing stock in a normal forest, the attainment of which is the goal of management.

235. Understocked Stands. Since the radius of expansion of a normally shaped crown is strictly limited for most species, more so for conifers than for hardwoods, two conditions are required if stands are to be fully stocked throughout their life. First the area should be uniformly and completely stocked with seedlings or sprouts with a spacing that will permit the crowns to form a closed canopy at latest in the earlier years of maturity. This close stocking is necessary if the best bole form and quality are to be attained by suppression of limb growth.

The second condition is even more important. No accident or injury must occur which kills groups of trees on an area large enough to exceed the normal crown radius for a given final or mature age. This would leave a hole in the crown canopy and therefore permanently reduce the yield of the age class from the year of damage to maturity, in spite of expansion of surrounding crowns or growth of seedlings springing up in the opening. Because of the relatively long period of from 50 to 120 years during which stands pass through the successive stages of seedling growth, increasing current growth, and later decreasing current growth (Sec. 194) to the year of maximum average (mean) annual growth per acre, several accidents or injuries may occur to a stand. As a result, the stands in a given age class that escape such depletion and remain fully stocked are usually scarce and their area small. All the remaining depleted stands will be understocked to a greater or less degree and will fall short of the fully stocked condition of the normal yield table.

The standard of stocking set for a normal yield table is that the stand shall be fully stocked at the time a plot is taken, regardless of earlier or later degrees of stocking. The density of understocked stands may then be expressed in terms of this standard as an average maximum.

236. Numerical Measure of Stocking—Density Ratio. The degree of stocking as compared with the fully stocked stands of the yield table may be measured by the ratio of the basal area per acre of the actual stand to the basal area of the equivalent normal stand. It can also be based on the actual and normal number of trees or on volume. However, basal area has been generally accepted as the basis for computing a stocking index, because it gives the most consistent results. Basal area per tree has a direct linear relationship to crown spread, consistent within each species but differing between species possessing different degrees of tolerance. When stocking is complete, and the stand has reached merchantable size, the ratio of total basal area to crown cover for any given species tends to become stabilized. Total basal area thus bears a close relationship to the total land area which produces it. It has been definitely shown for loblolly pine and ponderosa pine in fully stocked stands that, when this ratio is once attained in a stand, the average basal area per acre neither increases nor decreases despite fluctuation of the basal area on individual plots, and the curves level off to flat lines. For other species there may be a slight increase with advancing age. For these reasons, relative basal area constitutes a sound basis for determining the percentage of full stocking. Hence for the stands in a given site and age class, an average of the total basal areas per acre for the group is computed and divided by the basal area per acre for a normal stand of the same site and age to obtain the relative degree of stocking. For example, if for a western white pine stand on 70-ft. land (site index 70 ft. at 50 years), the basal area actually measures 150 sq. ft. per acre, and the basal area in a fully stocked stand is 244 sq. ft., the density ratio of this stand is $150/244 = 0.615$, or in other words the stand is 61.5 per cent stocked.

When degree of normality is measured in terms of number of trees or volume, the several percentages of stocking may differ. For example, in a study of second-growth Douglas fir,[1] the percentages of stocking computed for 81 large tracts varied as follows:

Basis	Average percentage of normal stocking
Total number of trees	78.2
Total basal area	82.0
Cubic-foot volume	82.1
Board-foot volume, International rule	84.3
Board-foot volume, Scribner rule	85.7

From the above, it appears that the assumption of basal area density equaling that of volume is exact for cubic-foot volume but leads to an underestimate of from 2 to 4 per cent for board-foot volume.

237. A Measure of Stocking Independent of Age and Site—Stand Density Index. In the above applications, stocking is always a function of age and site. A method which eliminates these two factors in the determination of stocking has considerable advantage. Gevor-

[1] MEYER, W. H., A Study of the Relation between Actual and Normal Yields of Immature Douglas Fir Forests, *Jour. Agric. Res.*, **41**, 635–665 (1930).

kiantz[1] and Reineke[2] found that, if the number of trees per acre is plotted over stand diameter, this number decreases systematically as the diameter increases, independent of age and site. For example, full stocking for a certain species may be represented by 300 trees per acre averaging 9 in. in d.b.h. This condition may be reached at 27 years on a 120-ft. site, at 32 years on a 100-ft. site, and at 37 years on 80-ft. land. As a fully stocked stand gets older, the crowns and the diameters continue to increase, but trees drop out, usually by death of suppressed trees. The total number of trees therefore decreases, and since the smaller trees disappear the average diameter increases. Young fully stocked stands have large numbers of trees per acre and a small average diameter. Old fully stocked stands have relatively few trees per acre and a large average diameter. By plotting the trees per acre in fully stocked stands of various ages and sites over average diameter on logarithmic paper, a straight line is formed which can be accepted as an expression of normal stocking or as a guide curve for understocked stands. When the average diameters and trees per acre of actual understocked stands are plotted on this chart, the relative percentage of stocking is shown.

A normal guide curve is necessary for each species, since the ratio between width of crown and diameter of the bole varies by species. Hardwoods, for example, have a consistently greater crown spread for given diameters than do conifers. Reineke's formula for the general relationship is

$$\text{Number of trees} = KD^{-1.605}$$

or, in logarithmic form,

$$\log N = -1.605 \log D + k$$

where N = number of trees per acre
D = stand diameter
k = a constant which varies with species = $\log K$

The *stand density index* is the number of trees per acre at the arbitrarily selected average diameter of 10 in. Thus, when k is 4.605, the formula and therefore the guide curve shows 1,000 trees at 10 in. The entire guide curve is then given the designation of "stand density index 1,000." If this were actually the guide curve for a given species,

[1] GEVORKIANTZ, S. R., and R. ZON, Second Growth White Pine in Wisconsin, Its Growth, Yield and Commercial Possibilities, *Wis. Agric. Expt. Sta. Res. Bul.* 98, 1930, 40 pp.

[2] REINEKE, L. H., Perfecting a Stand-density Index for Even-aged Forests, *Jour. Agric. Res.*, **46**, 627–638 (1933).

lines parallel to it could be drawn intersecting the 10-in. abscissa at 900, 800, or 700 trees per acre, etc., and the stand density index for any actual stand of the same species could be easily determined. Furthermore, a stocking ratio can be derived by dividing the actual stand index by the stand density index of a fully stocked stand.

Reineke found, for the species investigated, that the stand density index for the 10-in. average tree in fully stocked unmanaged stands varied from about 400 for the southern pines to 1,000 for red fir and redwood.

Neither the stand density index nor the stocking ratios and percentages remain constant as the stand gets older. Diameter growth in understocked stands is more rapid than in dense stands (Sec. 202). Understocked stands therefore make comparatively rapid progress toward a fully stocked density because fewer trees die and the diameter growth is more rapid. Since it is possible to predict, with a practical degree of accuracy, the future number of trees and average diameter of an understocked stand, the future degree of stocking may be read from the guide curve for the species, whatever its present or future age or site class. This cannot be done from the number of trees or average diameter alone, but it can be predicted with a practical degree of accuracy, using both.

238. Field Procedure for Predicting Growth of Even-aged Stands by Use of Normal Yield Tables. The field procedure for obtaining the basic data to predict the growth of even-aged stands is essentially a forest survey or timber estimate. For this purpose, however, the forest survey must obtain the total age and site class of each stand in addition to the census of the trees within each stand by size classes, which would suffice in the case of an ordinary timber estimate in which growth is not considered. On the map, areas in each site and age class within each forest type must therefore be separated in addition to the usual segregation of blank areas, burns, and barrens (Secs. 150 and 155).

The boundary of each site-age class is mapped in the field, and separate tally sheets for the estimate and stand table are kept for each area. This method results in more subdivisions than are necessary for timber estimating alone but will improve the accuracy of the estimate of volume and is indispensable in predicting future growth, since the accuracy of the growth predictions cannot be greater than that attained by the volume cruise (Sec. 5).

The total age of each stand is found during the cruise by borings plus separately determined seedling growth.

The site class for each stand is then found by obtaining the average height of a sample of dominant trees and locating this height over total age of the plot on the site index graph (Sec. 219).

For management purposes for large areas, where the three factors of age, site, and density vary widely, each of these factors should be combined into broad classes. Age is first determined to the nearest year, and the plots are then grouped and averaged in 5-year, 10-year, or 20-year groups. For example, age may be classified only in 20-year groups, for species with rotations of over 100 years, and in 10-year groups for shorter rotations. Site divisions should be confined to four or five broad site quality classes. Density of stocking can be designated as poor, medium, and good. Greater refinements in classification will add greatly to the field and office computations without a compensating increase in accuracy of average yield estimation.

239. Prediction by Use of Present Stocking Percentage. Previous to the decade 1930-1940 no satisfactory method had been generally recognized for measuring the rate of approach of understocked stands toward full stocking and their increased rate of growth. The pioneer work of Carter[1] in 1914 had been largely ignored. The known fact that trees in these subnormal stands grew faster as individuals than they did in normal stands was accepted as giving a factor of safety in using yield tables for growth prediction, serving to offset future possible losses from destructive agencies. It was therefore the practice to apply the present percentage of normality, as determined by basal area or other standard, to the normal yields to obtain future yield of understocked stands for any future periods. A stand now 70 per cent stocked would be assumed to yield but 70 per cent of the normal yield at 10, 20, 30, or more years in the future.

Carter showed that, for white pine at 60 years, the comparisons in the accompanying table held. The last values are those of a normal yield table.

Number of trees	Per cent of normal number	Total basal area, cu. ft.	Per cent of normal basal area	Board-foot volume	Per cent of normal volume
136	36	175	74	39,200	83
176	46	208	88	45,830	97
380	100	236	100	47,400	100

[1] CARTER, E. E., The Use of Yield Tables in Predicting Growth, *Soc. Amer. Foresters Proc.*, **9**, 177–188 (1914).

It is evident that in this case a stocking of 36 per cent of the normal number of trees for 60-year age has yielded 130 per cent more than would a prediction based on the ratio shown with normal tree number. The same discrepancy, to a lesser degree, would be shown had the relative basal areas of the stand at earlier ages been taken as the basis of future yields.

Belatedly, it was realized that understocked stands of second growth show a greater growth ratio than is predicted by straight proportion. Improved methods allowing for this increased growth are now available for several of our important timber species. Studies of additional species will extend the range of data available for predictions. One such method is based on repeated remeasurements of even-aged stands of various degrees of understocking to determine the trend toward normality (Sec. 240). This is the surest method but involves much time and effort. A second method is based on empirical formulas, such as the one proposed by Gerhardt (Sec. 241). Gevorkiantz has adapted this method to Lake states conditions. A third method is derived from analysis of growth from cumulative stand tallies (Sec. 243). With data on the approach to normality available, it is a mistake to ignore the trend and to assume that present degree of stocking will remain unchanged in the future.

240. Prediction Based on Normality Changes Measured from Permanent Sample Plots. At least two samples of data based on permanent sample plot measurements remeasured every 5 years, one for Douglas fir[1] and one for loblolly pine,[2] have been used to determine the trend of understocked stands toward normality. Normality percentages were computed for each 5-year period, and the changes in stocking during the period were determined. The approach of the actual basal area of understocked stands to the basal area of a fully stocked stand for these two species is shown in Fig. 77. When the periodic changes are added cumulatively from an initial stocking of 60 per cent, the continuous curve of Fig. 78 is produced. Both species show that basal area stocking per acre of understocked stands increases as the stands get older; the lower the degree of stocking, within limits, the faster the change. Consequently, as a stand advances in age, the

---

[1] MEYER, W. H., Approach of Abnormally Stocked Forest Stands of Douglas Fir to Normal Conditions, *Jour. Forestry,* **31,** 400–406 (1933).

BRIEGLEB, P. A., Progress in Estimating Trend of Normality Percentage in Second-growth Douglas Fir, *Jour. Forestry,* **40,** 785–793 (1942).

[2] MEYER, W. H., Yield of Even-aged Stands of Loblolly Pine in Northern Louisiana, *Yale Forest School Bul.* 51, 1942, 39 pp.

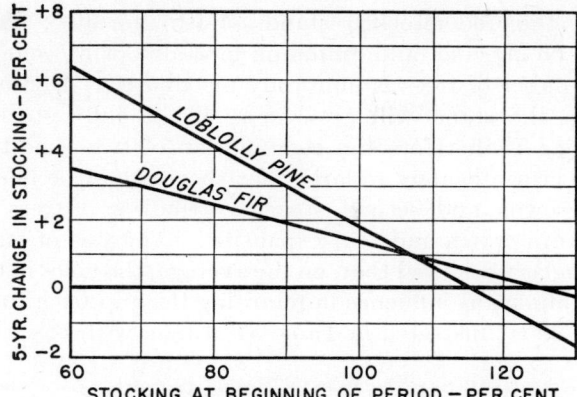

FIG. 77. Five-year change in basal area stocking related to the percentage of normal stocking at the beginning of the period.

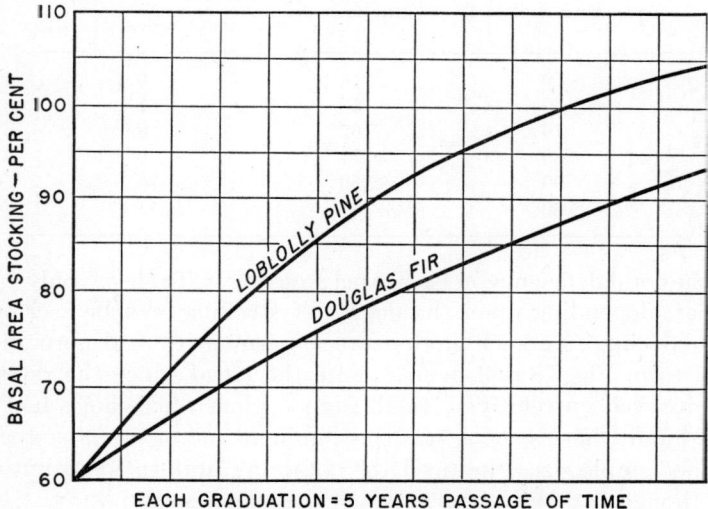

FIG. 78. Increase in basal area stocking percentage with advancing age.

average stocking percentage increases more and more slowly but keeps on developing sometimes even beyond the 100 per cent, or normal, condition. This can be taken as evidence that the yield tables from which the comparisons have been drawn are slightly less than normal standard.

It is evident that the future yield, as a percentage of the normal yield for a given age, will depend on the rate of growth in basal area of all trees of the understocked stand. Both mortality and average growth will be affected, and minimum mortality will occur where the optimum number of trees is uniformly distributed. Dense groups of trees within the stand will grow more like a fully stocked stand. Understocking is therefore the combination of two conditions: first, holes in the crown canopy so large that they cannot be filled later by expanding crowns; and, second, sparse distribution of trees within the range of future crown and root expansion. Analysis of stand conditions in Douglas fir showed that, on the average, the holes in the canopy exerted an important influence in retarding the approach to normality of stocking, as is illustrated in Table 54 dealing with basal area.

TABLE 54. EFFECT OF SPACING VERSUS HOLES IN STAND ON PERCENTAGE OF STOCKING. DOUGLAS FIR, PACIFIC NORTHWEST

Net area of stand = total area minus holes, %	Basal area stocking, %	Difference attributable to spacing exclusive of holes, %
50	44	6
60	53	7
70	62	8
80	71	9
90	80	10
100	89	11

Of the total deficiency in basal area stocking in Table 54, only 6 to 12 per cent, depending upon the degree of stocking, can be ascribed to sparse spacing. In such circumstances, growth toward normality as judged from Fig. 78 will overestimate the trend, since the rates had been observed on relatively small sample plots where holes had been avoided and where spacing was the dominant factor in understocking. This fact emphasizes the need for obtaining uniform distribution of trees, though it may be sparse, when the stand is reproduced.

241. Prediction by Use of a Density Percentage Based on Gerhardt's Empirical Formula. A method of adjusting the periodic growth estimate as obtained from normal yield tables to take care of growth toward normality was developed by E. Gerhardt in 1930.[1] The formulas for the adjustment are as follows:

[1] GERHARDT, E., Ertragstafeln für reine und gleichartige Hochwaldbestände von Eiche, Buche, Tanne, Fichte, Kiefer, grüner Douglasie und Lärche, 2d ed., Verlag Julius Springer, Berlin, 1930, 73 pp. [Review, S. R. Gevorkiantz, *Jour. Forestry*, **32**, 487–488 (1934).]

For tolerant trees: $G = dZ(2 - d)$
For intolerant trees: $G = dZ(1.7 - 0.7d)$
where G = actual periodic growth in cubic volume for 10 years
Z = normal periodic growth for 10 years
d = percentage that the actual basal area is of the fully stocked basal area

The term dZ in the equation reduces the normal periodic growth as taken from a yield table to conditions of present stocking; the following terms in parentheses, either $(2 - d)$ or $(1.7 - 0.7d)$, make an adjustment for growth to normality in the ensuing period. In application, let us assume that in a stand of tolerant northern hardwoods normal stocking gave a periodic annual increment Z of 300 bd. ft. per acre for the next 10 years. The degree of stocking of a subnormal stand is found, by relative basal areas, to be 0.5. Then, for this stand,

$$G = 0.5(300)(2 - 0.5) = 150(1.5) = 225 \text{ bd. ft.}$$

A stand of Norway pine with a stocking of 0.5 and a normal periodic annual growth of 500 bd. ft. would by the formula for intolerant species have a growth of

$$G = 0.5(500)[1.7 - (0.7 \times 0.5)] = 250(1.35) = 337 \text{ bd. ft.}$$

The poorer the density and hence the lower the value of d, the greater will be this correction. Table 55 lists a series of periodic

TABLE 55. GERHARDT GROWTH FACTORS

Present degree of stocking	For tolerant trees		For intolerant trees	
	Percentage comparison between actual and normal periodic growth	Ratio between (2) and (1)	Percentage comparison between actual and normal periodic growth	Ratio between (4) and (1)
(1)	(2)	(3)	(4)	(5)
90	99	1.10	96.3	1.07
80	96	1.20	91.2	1.14
70	91	1.30	87.4	1.21
60	84	1.40	76.8	1.28
50	75	1.50	67.5	1.35
40	64	1.60	56.8	1.42
30	51	1.70	44.7	1.49
20	36	1.80	31.2	1.56
10	19	1.90	16.3	1.63

growth values, to show the effect of these corrections. The first column, or density percentage, would be identical with dZ and would correspond with the density percentage of Sec. 239, where the increase in growth is neglected. The second column gives the results of the formula for tolerant species. Thus, if the stand is 90 per cent stocked, the growth for the next 10-year period instead of 90 per cent is expected to be 99 per cent of normal; for stand 80 per cent stocked, it is 96 per cent of normal, etc. The third column shows the relation between the corresponding values of the first two columns and indicates that there is a consistent 10 per cent increase for each 10 per cent decrease in density. The formula for tolerant species gives a consistent 7 per cent increase for each 10 per cent decrease in density, as is shown in the fourth and fifth columns.

A general expression of these formulas used by Gevorkiantz and Duerr[1] is as follows:

$$g = dG(1 + K - Kd)$$

where g = 10-year growth of the understocked stand
G = 10-year growth of the normal stand of same age and site
d = density, or percentage of stocking
K = a constant for the species

The values of K are fixed as follows:

Intolerant species	0.6–0.7
Intermediate species	0.8–0.9
Tolerant species	1.0–1.1

These authors found that the factors K for cubic feet applied without change to growth in board feet for normal yield tables. For yield tables not based on fully stocked stands, the value of K is correspondingly lower. Although applied to gross growth, it was also found applicable to net growth, *i.e.*, growth minus mortality and defect. For northern hardwoods Gevorkiantz[2] used a value of 1 for the factor K irrespective of age and density of stocking, and applicable even to stands of mixed ages. It can therefore be used to predict the growth of both even-aged and many-aged stands, provided that the basic normal growth of a corresponding even-aged stand can be applied to the stand in question.

[1] DUERR, W. A., Comments on the General Application of Gerhardt's Formula for Approach towards Normality, *Jour. Forestry*, **36**, 600–604 (1938).

[2] GEVORKIANTZ, S. R., The Approach of Northern Hardwood Stands to Normality, *Jour. Forestry*, **35**, 487–489 (1937).

242. Prediction Based on Growth Variables—Gevorkiantz.[1] Gevorkiantz analyzed the age distribution in a selection stand of Lake states northern hardwoods by determining the ages of all trees on sample plots, and found that most of the stands were composed of a main crop of trees consisting of a single age class, regenerating over a short period of years. Usually a few veterans of an older age class were present, also a group of younger trees. He therefore concluded that growth prediction could be based largely on the "even-aged" main crop (Fig. 79*A*).

Site classes were determined by plotting curves of average age of main stand as the dependent variable over average diameter of main stand as the independent variable. A guiding curve of age based on diameter served as the basis for a set of five curves in which the ages for given diameters were respectively less or more than those of the guiding curve (Fig. 79*B*). On the best sites (5) the time required to attain successive average diameters was less than for each lower set down to the poorest site (1). Such average ages indicate the periods of time which the stand has taken to develop from smaller sized material and the number of years it will require to develop into larger sizes. The immediate past and the future period of years required for an increase of a given number of inches in average diameter is the fact sought from the table, and this period determines the site. The site index chart is similar to that for height, but the shape of the curve is concave upward as for volume on d.b.h. Charts *B* and *C* are prepared for each cover type from plots taken in well-stocked stands of that type.

Density of stand, the third variable considered for these stands, was based on total basal area per acre at successive ages and was presented for five density classes on a chart (Fig. 79*C*). Separate basal area charts were required for each site class shown in Fig. 79*B*.

The fourth variable of growth is termed the "merchantability index." This is the number of board feet per square foot of basal area in the stand and is analogous to the board foot–cubic foot ratio used in preparing yield tables for even-aged stands (Sec. 230). Again an index chart is constructed (Fig. 79*D*) with the merchantability ratios plotted over age, with separate curves to represent merchantability classes. This ratio is largely affected by the density of the stand and the variation of diameters of average tree with changes in density, and is the reflection of the effect of the number of trees which make up the basal area. The fewer (and larger) the trees, the greater will be the mer-

---

[1] GEVORKIANTZ, S. R., and W. A. DUERR, Growth Predictions and Site Determination in Uneven-aged Timber Stands, *Jour. Agric. Res.*, **56**, 81–98 (1938).

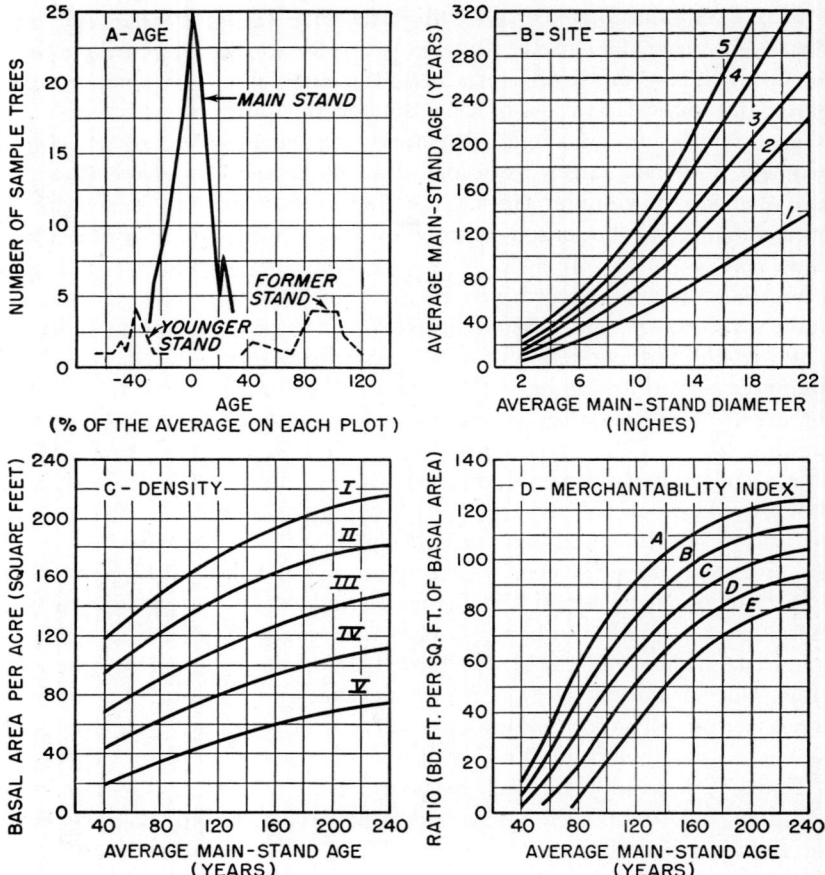

Fig. 79. The four main variables which affect growth, illustrated by the northern-hardwood cover type in the Lake states. (*From S. R. Gevorkiantz and W. A. Duerr, Methods of Predicting Growth of Forest Stands, Lake States Forest Expt. Sta. Economic Notes 9, 1938.*)

chantability index, and vice versa. These charts, in turn, were prepared separately for each site class shown in Fig. 79B.

Growth of stands is predicted from these four charts, one for each variable for a short (10-year) period as follows:

1. Determine the average main-stand age on the basis of experience supplemented by borings of selected trees covering the range of diameters in the main stand.

2. Determine the site from the site index chart (Fig. 79B) by locating age above average diameter of the main stand.

3. Select the density chart (Fig. 79C) for the site indicated, and determine the density class by locating basal area per acre over age of main stand. Five density curves are shown, guided by the growth predictions of the Gerhardt formula.

4. Compute the merchantability ratio by dividing the total volume per acre by the basal area per acre, and find the index curve by locating the ratio over main-stand age in Fig. 79D.

5. Future basal area density and merchantability ratio can be taken from the projections of the curves, and the future volume can be found by multiplying the two.

A specific example of the detailed procedure follows:

1. Age of stand, 120 years.
2. Present volume, 4,000 bd ft.
3. Average main-stand diameter, 14 in.
4. Indicated site, class 2 (Fig. 79B).
5. Basal area of stand, 80 sq. ft.
6. Density, class IV (Fig. 79C).
7. Merchantability ratio, 4,000/80 = 50 bd. ft. per square foot of basal area.
8. Merchantability class for 50 ratio, D (Fig. 79D).
9. Prediction of basal area at 130 years from chart (Fig. 79C), 87 sq. ft.
10. Prediction of merchantability ratio, 63 at 130 years, from chart (Fig. 79D).
11. Future volume 87 × 63 = 5,481 bd. ft.
12. Growth, 10 years = 5,481 − 4,000 = 1,481 bd ft.
13. Periodic annual growth = 148 bd. ft.

This general method is thus dependent on empirical standards for volume on age, density on basal area, site on average diameter at given ages, and merchantability index for dealing with the effect of variable numbers of trees. Obviously the effect of these variables on understocked stands makes it impossible to predict growth for more than about 10 years, by contrast to growth predictions for even-aged normal stands, which set a standard for an entire rotation.

243. Prediction Based on Comparative Cumulative Tallies of Stocking by Numbers of Trees per Acre. A method adopted by H. H. Chapman, unpublished, in 1945 for immature longleaf pine in Alabama was based on 32-year-old fully stocked stands at Urania, La., growing on the same site quality, 80-ft. land. Measurements of each tree made at three previous dates permitted an analysis of growth in diameter, height, and volume of groups of trees. The trees in each plot were grouped by rates of diameter growth. The 50 trees showing the most rapid growth were averaged. To this group the second 50 trees were added, and an average was thus obtained for the fastest growing 100 trees. By adding additional groups of 100 trees and computing new averages, for 200, 300 trees, etc., growth rates were obtained for these respective numbers of trees representing a descending scale of dominance and of growth rate for increasing numbers of the larger and faster growing trees per acre in the fully stocked stands. The final figure gave the growth rates for the complete stand. From the successive measurements at 5-year intervals, points were obtained for curves of growth in volume per acre for the largest 50, 100, 200 trees, etc., based on age of stand up to 32 years, which curves were then projected to 42 years (basis of Fig. 80).

These curves do not show the growth which the indicated number of trees per acre would make if they constituted the entire stand. If only 100 trees were present, the stand would be understocked. The proportion of understocking could be referred to basal area, but in this case the basis of comparison had been worked out as the growth that the same number of trees would show in fully stocked stands. All that was necessary now was to determine how much faster this number of trees would grow when instead of being the fastest growing group in fully stocked stands they constituted the entire understocked stand. This comparison was obtained from 2 plots at Urania and 25 plots in Alabama. The points obtained from these stands were plotted over the age of the stand, against the standard curves for trees of the same number in the fully stocked stands, and corrected curves were drawn. These curves, for yields of unpeeled cords of pulpwood, are shown in Fig. 80. The original guiding standard, or guide curves, are not shown. The comparison agreed closely with Gerhardt's assumptions. The greatest excess of growth over the normal occurred with the smallest number of trees in the actual stand and with the younger ages. Approach to normality was rapid, and full stocking was reached at different ages depending upon the number of trees in the initial stocking. It was attained always with fewer trees than the average numbers shown in normal yield tables for the same age.

The yields of pulpwood in cords for the total area at the end of the next 10-year period were determined by applying the values shown in Fig. 80 to areas obtained from aerial photographs. When the age classes of longleaf pine which originated in past seed years were known,

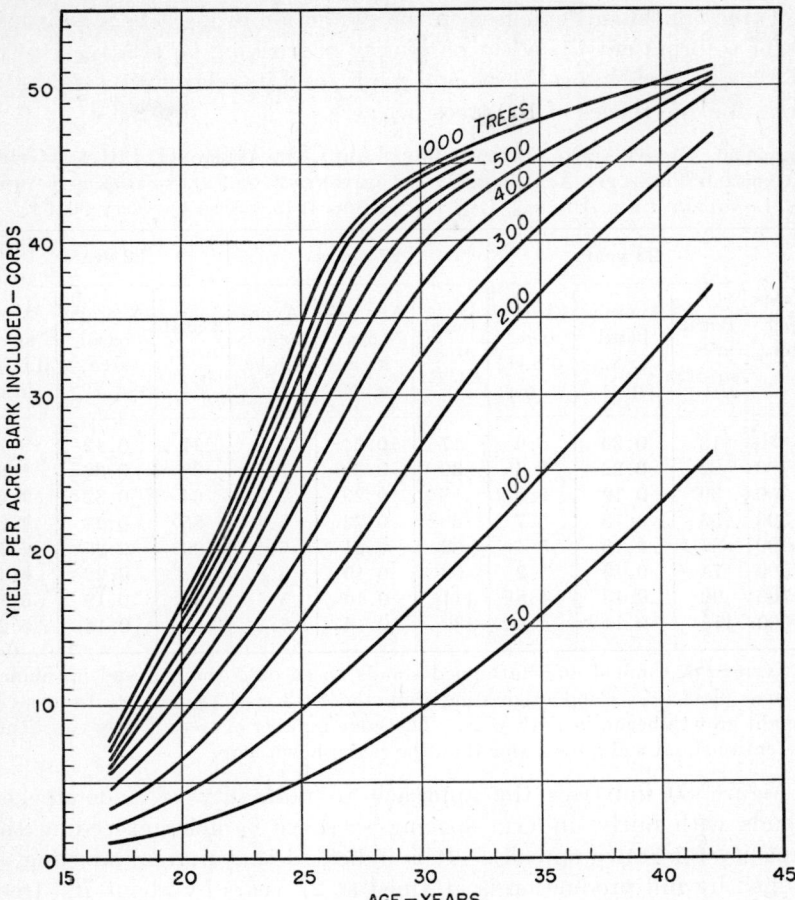

FIG. 80. Longleaf pine, southern Alabama. Yields in rough standard cords for understocked stands. For 50 trees per acre and for each cumulative 100 trees to 1000 trees of established reproduction. Site class 80 ft. at 50 years.

they could be separated on the map by differences in height, and the areas mapped. Density classes differing by 100 trees per acre could also be distinguished and separated on the photographs. The future yields, for stands 10 years older, were then applied to each area. Mortality in understocked longleaf pine saplings is practically negligible.

The differences in relation to site, fire relationships, and intolerance among longleaf, slash, and loblolly pine are such that mixtures exceeding the 75 per cent limit for pure stands did not occur.

Table 56, though not used in the prediction of growth, is shown in order to bring out the effect of density of stocking on relative growth of average trees in basal area and d.b.h., and its relation to total basal area, for differences of 100 trees per acre.

TABLE 56. BASAL AREAS, AVERAGE BASAL AREA, AND AVERAGE D.B.H. FOR GIVEN NUMBER OF TREES PER ACRE, GROWN AS COMPONENTS OF FULLY STOCKED STANDS. LONGLEAF PINE, URANIA, LA., DECEMBER, 1945. SITE QUALITY 80 FT.

Trees per acre	22 years			27 years			32 years		
	Basal area, sq. ft.	Average basal area, sq. ft.	Average d.b.h., in.	Basal area, sq. ft.	Average basal area, sq. ft.	Average d.b.h., in.	Basal area, sq. ft.	Average basal area, sq. ft.	Average d.b.h., in.
50	13	0.26	6.9	17	0.34	7.8	21	0.42	8.7
100	22	0.22	6.4	30	0.30	7.4	38	0.38	8.3
200	39	0.19	5.9	58	0.29	7.3	64	0.32	7.6
300	53	0.18	5.7	69	0.23	6.4	85	0.28	7.2
400	65	0.16	5.4	82	0.21	6.1	100	0.25	6.7
500	74	0.15	5.2	93	0.19	5.8	112	0.22	6.4
700	90	0.13	4.8	111	0.16	5.4	129	0.19	5.8
1,000	115	0.11	4.5	139	0.14	5.1	151	0.15	5.2

Average of thinned and unthinned stands based on 5 thinned and unthinned ¼-acre plots in a stand originating from the 1913 seed crop. Age from seed. Height growth began in sixth year. The same number of trees if they constitute the entire stand will grow faster than the rates shown here.

Figure 80 indicates the approach to normality of understocked stands with fairly uniform spacing (absence of holes exceeding the capacity for crown spread). It is indicated that normal stocking as judged by full production is attained at 27 years by about 700 trees per acre; at 32 years by about 500 trees; at 37 years by about 300 trees; and at 42 years by about 200 trees per acre. The number of trees required by these species for complete utilization of the site continues to diminish as the stands get older.

These objectives could be better attained by the establishment and repeated measurements of plots with varying percentages of subnormal stocking, thus creating an empirical yield table classified by number of trees per acre of established stocking, and showing the effect on future yields, approach to normality, and quality as well as volume of the yields. The next advance in growth predictions will come from the establishment of numerous permanent plots in stands of all degrees of stocking and different silvicultural methods of management.

In general, yield tables not only enable the forest manager to predict future growth in volume but are also the best ultimate means of indicating the character of thinnings and silvicultural treatment; the present worth of a forest property as affected by future yields; the justified extent and value of all such improvements as roads, logging, and manufacturing equipment; and the intensity of measures that should be taken to protect the forest from damage by fire, insects, and diseases. Only by balancing present expenditures against present value as derived from expected future yields can a forest property be brought to the highest state of profit for the owner.

References

ASHE, W. W.: Determination of Stocking in Uneven-aged Stands, *Soc. Amer. Foresters Proc.*, **9**, 204–206 (1914).

AUGHANBAUGH, J. E.: Yield of the Oak-chestnut–Hard Pine Forest Type in Pennsylvania, *Jour. Forestry*, **32**, 80–89 (1934).

BALDWIN, H. I.: A Method for Computing the Proper Density for Maximum Increment, *Jour. Forestry*, **31**, 346–347 (1933).

BARNES, G. H.: The Importance of Average Stand Diameter in Forecasting Timber Yields, British Columbia Forest Service, 1931, 24 pp.

BLASCHECK, A. D.: Method of Calculating Yield in Working Plans in India, a Correction, *Forestry Quart.*, **8**, 332–334 (1910).

BRIEGLEB, P. A.: Progress in Estimating Trend of Normality Percentage in Second-growth Douglas Fir, *Jour. Forestry*, **40**, 785–793 (1942).

———— and J. W. GIRARD: New Methods and Results of Growth Measurement in Douglas Fir, *Jour. Forestry*, **41**, 196–201 (1943).

CARTER, E. E.: The Use of Yield Tables in Predicting Growth, *Soc. Amer. Foresters Proc.*, **9**, 177–188 (1914).

CHAIKEN, L. E.: The Approach of Loblolly and Virginia Pine Stands toward Normal Stocking, *Jour. Forestry*, **37**, 866–871 (1939).

CHAPMAN, H. H.: Application of Yield Tables in the Measurement of Growth on Large Areas, *Jour. Forestry*, **20**, 581–588 (1922).

————: The Measurement of Increment on All-aged Stands, *Soc. Amer. Foresters Proc.*, **9**, 189–203 (1914).

CHISMAN, H. H., and F. X. SCHUMACHER: On the Tree-area Ratio and Certain of Its Applications, *Jour. Forestry*, **38**, 311–317 (1940).

COWLIN, R. W.: Sampling Douglas Fir Stands by the Stocked Quadrat Method, *Jour. Forestry*, **30**, 436–439 (1932).

DUERR, W. A.: Comments on the General Application of Gerhardt's Formula for Approach towards Normality, *Jour. Forestry*, **36**, 600–604 (1938).

DWIGHT, T. W.: The Use of Growth Charts in Place of Yield Tables, *Jour. Forestry*, **24**, 358–377 (1926).

EYRE, F. H., and P. ZEHNGRAFF: Red Pine Management in Minnesota, *U.S. Dept. Agric. Circ.* 778, 1948, 70 pp.

GEVORKIANTZ, S. R.: The Approach of Northern Hardwood Stands to Normality, *Jour. Forestry*, **35**, 487–489 (1937).

————: The Approach of Understocked Stands to Normality, *Jour. Forestry*, **32**, 487–488 (1934).

———: Comments on the "Approach of Loblolly and Virginia Pine Stands toward Normal Stocking," *Jour. Forestry*, **38**, 512–513 (1940).
——— and W. A. DUERR: Methods of Predicting Growth of Forest Stands, *Lake States Forest Expt. Sta. Economic Notes* 9, 1938, 59 pp.
——— and N. W. HOSLEY: Form and Development of White Pine Stands in Relation to Growing Space, *Harvard Forest Bul.* 13, 1929, 83 pp.
——— and H. F. SCHOLZ: Determining Site Quality in Understocked Oak Forests, *Jour. Forestry*, **42**, 808–811 (1944).
HAIG, I. T.: The Application of Normal Yield Tables, *Jour. Forestry*, **22**, 902–906 (1924).
———: The Stocked Quadrat Method of Sampling Reproduction Stands, *Jour. Forestry*, **29**, 747–749 (1931).
HORNIBROOK, E. M.: A Preliminary Yield Table for Selectively Cut Lodgepole Pine Stands, *Jour. Forestry*, **38**, 641–643 (1940).
———: Yields of Cutover Stands of Engelmann Spruce, *Jour. Forestry*, **40**, 778–781 (1942).
MACKINNEY, A. L., and L. F. CHAIKEN: A Method of Determining Density of Loblolly Pine Stands, *Appalachian Forest Expt. Sta. Tech. Note* 15, 1935, 3 pp.
———, F. X. SCHUMACHER, and L. F. CHAIKEN: Construction of Yield Tables for Non-normal Loblolly Pine Stands, *Jour. Agric. Res.*, **54**, 531–545 (1937).
MEYER, H. A.: A Simplified Increment Determination on the Basis of Stand Tables, *Jour. Forestry*, **33**, 799–806 (1935).
———: Increment Determination on the Basis of Stand Tables, *Jour. Forestry*, **34**, 948–950 (1936).
MEYER, W. H.: Approach of Abnormally Stocked Forest Stands of Douglas Fir to Normal Condition, *Jour. Forestry*, **31**, 400–406 (1933).
———: A Study of the Relation between Actual and Normal Yields of Immature Douglas Fir Forests, *Jour. Agric. Res.*, **41**, 635–665 (1930).
———: Understocked Stands, *Jour. Forestry*, **26**, 786–789 (1928).
——— and P. A. BRIEGLEB: Forest Growth in the Douglas Fir Region, *Pacific Northwest Forest Expt. Sta. Forest Res. Notes* 20, 1936, 8 pp.
MOORE, B.: Yield in Unevenaged Stands, *Soc. Amer. Foresters Proc.*, **9**, 216–228 (1914).
MULLOY, G. A.: Empirical Stand Density Yield, *Canadian Forest Serv. Silvicultural Res. Note* 82, 1947, 13 pp.
———: A Practical Measure of Stock Density in White and Red Pine Stands, *Forestry Chron.*, **19**, 108–118 (1943).
REINEKE, L. H.: Perfecting a Stand-density Index for Even-aged Forests, *Jour. Agric. Res.*, **46**, 627–638 (1933).
SCHUMACHER, F. X.: Misuse of Yield Tables, *Jour. Forestry*, **33**, 438–439 (1935).
SIMMONS, E. M., and G. L. SCHNUR: Effect of Stand Density on Mortality and Growth of Loblolly Pine, *Jour. Agric. Res.*, **54**, 47–48 (1937).
SOVULESKI, R., and T. H. HARRIS: A Method for the Classification of Sugar Pine Land According to Expected Yield, *Jour. Forestry*, **46**, 432–437 (1948).
WELLWOOD, R. W.: Trends toward Normality of Stocking for Second Growth Loblolly Pine Stands, *Jour. Forestry*, **41**, 202–209 (1943).
WESTVELD, M.: Yield Tables for Cutover Spruce-fir Stands in the Northwest, *Northwestern Forest Expt. Sta. Occas. Paper* 12, 1941, 16 pp.
WILSON, F. G.: Numerical Expression of Stocking in Terms of Height, *Jour. Forestry*, **44**, 758–761 (1946).

CHAPTER 29

PERIODIC AND CURRENT GROWTH IN DIAMETER BASED ON STAND TABLES

244. Substitution of Diameters for Age Classes in All-aged Stands. By any method of growth prediction, the volume of the *present stand* is obtained from a stand table giving the number of trees per acre, by species, for each diameter class, with average heights for each class obtained from a curve of height on diameter. Yield tables based on age merely show these data for fully stocked stands at each age, from which the predicted volumes of the existing partially stocked stands for similar ages and sites can be derived. The general method used is that of *comparison* of the assumed future condition of the existing stand with the volume which a normal stand of the same age and site class will have in this future year. It is often impossible to obtain a significant or useful average age for stands of tolerant species because of the great differences in the rates of growth of the trees composing the stand, brought about by the influence of shade and tolerance. In Sec. 179 it was shown that the maturity of trees, especially of tolerant species, is more closely related to size than to age. It follows that, for a short future period, usually 10 years, the diameter and volume growth of individual trees and of diameter classes is more consistently related to their present *diameters* than to the average *ages* of the trees within each diameter class, which may vary over a wide range.

Even with intolerant species such as longleaf and ponderosa pines, early attempts to separate the virgin forest into three or four rough age classes, such as veterans, mature, young merchantable, and immature, were found to be impracticable. The age classes could be approximately distinguished by relating them to size and appearance (blackjack versus yellow pine), but no two maps of the areas occupied by each age class coincided. Without area, yields per acre of different age classes cannot be found, and the method was abandoned.[1]

Under these circumstances, the stand table of individual trees, grouped and averaged by diameter classes, can be substituted for the

[1] CHAPMAN, H. H., A Method of Studying Growth and Yield of Longleaf Pine Applied to Tyler County, Texas, *Soc. Amer. Foresters Proc.*, **4**, 207–220 (1909).

ages either of the stand as a whole or of the separate trees in the stand, as the basis for growth predictions. The prediction of growth is made separately for the average tree in each diameter class. Past growth is measured by borings at b.h., for definite past periods, such as 5, 10, 15 and 20 years. From the data thus obtained, average growth is predicted for a future period of 10 years for each d.b.h. class.

This general method of substituting average trees of each diameter class for age of stand is not confined to all-aged stands but can be applied to any stand, whether even-aged or many-aged. For instance, when existing even-aged fully stocked stands of second growth such as old field loblolly pine cannot be found beyond the age of 70 years, the yield tables can give no comparisons for stands 80 years old. In this case, the 70-year-old stands on a given site are shown in the form of stand tables. Past growth for 5, 10, 15, and 20 years is obtained by increment borings. Growth for each diameter class is determined for these respective past periods, and separate curves are plotted for each class, showing the past and present diameters on the vertical scale, and the respective years on the horizontal scale. In even-aged stands, the growth rate of the larger trees is always faster than that of smaller diameters (which is why they are now larger). The growth rate of the smallest diameter classes approaches zero, and the curve tends to level off. When growth ceases, trees of these classes die. From such a set of curves of diameter growth, it is in fact possible to predict closely the mortality that will occur within the next 10 years. This method is treated in detail in Chap. 30. The greatest advantage of the stand table method is that it can be directly applied to stands of any degree of subnormal stocking and hence requires no discount from a normal standard of yield.

Like any other prediction (such as future prices on the stock market), the longer the period involved, the more uncertain the results and the greater the possible error. When stand tables and diameter classes are substituted for volume per acre and yield tables based on age, prediction of the future behavior of the *trees which compose* the stand, instead of the stand itself is in turn made the basis of the study. The stand is separated into its component parts, the trees. Future growth in diameter and height and future changes in form give future tree volumes. The trees that die or are killed or cut must be dealt with as individuals and not merely ignored in the prediction of future "net" volume. The character and behavior of average trees of different diameter classes, and even of trees of different vigor and crown spread within each diameter class, are the tools with which to work.

In Chap. 22 it was shown that the rate of growth in diameter and volume of individual trees is directly related to the space available for expansion of roots and crowns. The more crowded the stand, the less the growth per tree. This loss is more than offset by the greater number of trees per acre which when multiplied by the average growth per tree of the respective diameter classes gives the total growth per acre. By intensive management of stands, consisting of well-timed thinnings, the average rates of diameter growth can be controlled so as to produce the largest yield of timber of required qualities, density, closeness of rings, clearness, and strength, which is affected by thickness of summerwood.

The differences in average rates of diameter growth caused directly by varying densities of stocking or number of trees per acre make it necessary to conduct separate studies of average growth rates for each diameter class, not merely for different species but for each separate area on which growth is to be predicted. The purpose is to obtain the past growth, by species, type, and site class, which fits the average density of stocking to be found on the tract in question, and from this solid basis to venture the prediction for that specific area for the future. Large errors, from any cause, may result in serious losses in management, from either over- or underinvestment in mills and equipment dependent on future yields. For any one property, however, average results, with not too much detail or hairsplitting, will answer these questions.

245. Objectives of Stand Table Method. The problems presented in deriving growth predictions from past diameter (and height) growth of tree classes, as applied to a stand table of present diameters are

1. To determine which, or what proportion, of the present number of trees in each diameter class and species will survive to the end of the future period, hence the future mortality (Secs. 247 and 257).

2. To make reasonably certain that the trees bored for the growth sample are representative of the entire stand and grow at the average rate of the stand, neither faster nor slower (Sec. 248).

3. To include in the tally of diameters for the stand table a range of sizes low enough to account for all small trees that may grow into the merchantable sizes within the period of growth prediction (Sec. 249).

4. To interpret past diameter growth by a method which will forecast actual future growth in diameter, for each d.b.h. class. This is not accomplished by the mere assumption that future growth will equal past growth (Sec. 250).

5. To apply the results obtained by sampling to the entire area of

the forest and thus to predict the yield for the next cutting period, as the basis of management (Sec. 251).

246. The Stand Table as Related to Forest Types, Site Classes, and Age Classes. As shown, species, or groups of species with similar growth habits, are always considered separately in growth predictions. But it is also important to decide whether, for different site qualities, there may be consistent differences in average growth rates for the same species that should be recognized.

On large holdings, for an over-all prediction of total growth for purposes of regulating the amount of the future cut, it may be sufficient to adopt the total area of a given forest type as the unit for determining *average* growth of each separate species. This can be done safely if the sample taken for growth is properly representative (Sec. 248). On small areas, differences in site may not be serious, and an average growth for each separate species on the entire area suffices. But where two or more forest types exist, such as bottomland hardwoods and upland mixed pines and hardwoods, the growth rates of the component species must not be averaged together for these separate types. Red oak or red gum on hardwood bottomlands may grow at much faster rates and reach larger sizes than the same species on the pine-hardwood uplands. Even such pines as survive among bottomland hardwoods may grow faster than the average of the same species on the upland type. On a large holding in Arkansas, the species for the upland pine-hardwood type which were separately studied were loblolly pine, shortleaf pine, white oak, and the red oak group. On bottomlands, the groups were red gum, overcup oak, red oaks, hickory, and miscellaneous species.

Forest types which clearly produce marked differences in average growth rates per tree must always be separated both by mapping and in the tally taken for the stand tables. But separation of the tally for any other reason is necessary only when diameter growth and stand tables are applied to even-aged stands. Otherwise one composite stand table, in which the trees are tallied separately for all important species, will suffice for the forest type in question.

On a very large area which is divided for purposes of regulating the amount of the annual cut into smaller subdivisions perhaps containing 50,000 acres each, a separate stand table is required for each division. Otherwise the average growth of the subarea cannot be coordinated with the rate at which the timber from this area should be cut. A thinly stocked area may be overcut, and vice versa.

The accuracy of growth predictions cannot exceed that of the stand table. As this table except on very small areas is obtained from a sample, not a total tally, it incurs the normal error of estimate. To this initial unavoidable, but controlled, error of estimate is added the error inherent in predictions of future mortality and future growth rates. For a 10-year period, the total error of growth predictions can hardly be less than twice the error in the estimate of the present stand; hence the first consideration in growth prediction is to control the initial error of estimate in the present stand table.

247. Mortality as Affecting Future Stand Tables. Offsetting the future growth in diameter, height, and volume of the trees making up the present stand we have the inevitable and necessary reduction in number of trees surviving to the end of the future period. When a tree dies or is cut, the entire volume of the tree is lost to the stand—not merely its future growth but all its past growth as well. Hence the loss of one tree cancels the future growth on many trees. When these dead or dying trees can be salvaged, their volume is included in the yield for the future period. If logging is annual and continuous, they will have averaged 5 years' growth in a 10-year period, and this growth can be added to their present volume in the prediction of total future realization of yield.

If trees are suddenly killed by such agencies as wind, fire, or bark beetles, growth to the year of death remains normal. Not so if these trees are slowly dying of suppression from shading. In this case, the growth for the last 10-year period of fraction thereof before they die will be negligible.

Mortality thus occurs from two sources, *viz.*, competition by dominant trees, leading to shading and ultimate death unless release occurs, and outside factors such as insects, diseases, wind, abnormal drought, and fire. Here a sharp distinction must be made between the widespread and often total destruction by fire, hurricanes, or imported enemies and the normal endemic losses due to an annual toll from bark beetles, an occasional loss by wind, the slow disintegration of the strength of a tree from rot in bole or roots, or the gradual enlargement of fire scars in southern pine, until the tree finally falls in a wind. In the prediction of the mortality of mature and overmature timber, the line is drawn between these normal endemic losses and catastrophic destruction such as occurred in the great forest fires of the past, which may be repeated, the hurricane damage of 1938 in New England, or the extinction of the chestnut by Asiatic blight. Prediction of mortality in mature timber is based, instead, on the average past losses that normally serve to reduce the numbers of trees in all stands of whatever form.

Stand tables when projected to a future year will therefore show fewer live trees than the same stand holds at present. How can this loss be predicted? Only by data showing past mortality for these stands, and an estimate of future losses, based on this record.

Expressed in number of trees that will die before the end of the future period, either for each diameter class or for the stand as a whole, the losses are always greatest during the first year of the seedlings. From then on, fewer and fewer, but larger and larger trees, for any one age class, die yearly under the influence of shading by the expanding crowns of the survivors or in many-aged stands by the overhead shade. For either even-aged or many-aged stands, and for the forest as a whole, there must be many times the number of live trees in the younger and smaller classes than in those that are mature. When a stand table is constructed for an entire forest including all its age classes, the form of the curve of diminishing numbers with increasing diameters is termed J-shaped. For single all-aged stands, the same form of curve is found. For single even-aged stands, the curve is not of this form but resembles a normal curve of error (Sec. 168) with a skewness toward the left. But when even-aged stands of all existing age classes in the forest are combined in one stand table, again the curve will be J-shaped, with large numbers of trees in the smaller diameters, flattening out as diameter increases, until for large old overmature trees, only a few lone survivors remain.

Death by shading and competition continues as long as the crown canopy remains fairly complete. But with large trees and crowns, death of dominant trees caused by other agencies leaves gaps incapable of closing which are then captured by reproduction or by surviving suppressed trees, and the stand tends toward a many-aged form. This also happens in dry regions with species such as ponderosa pine, where soil moisture is the controlling factor of survival. In this case an extremely intolerant species is found growing in stands of many ages.

There are three possible methods for predicting future mortality of the trees shown in an existing stand table.

1. By inspection of live trees, separate the tally into those that will probably die within the next 10 years and those that will live. This method applies chiefly to young and dense stands. The rate and probability of death are conditioned by the relative tolerance of the species. Small hemlocks die chiefly when they grow under the shade of their

own species, seldom when under hardwoods. Small intolerant pines are quickly killed by expanding crowns of hardwoods. In this separation of the tally for the "future" stand table, future growth on the suppressed trees is not measured, for it would reduce the average growth of the surviving trees of a given diameter class below their actual attainment. Since this method is applicable chiefly to stands containing large numbers of small trees, increased accuracy may be attained by confining the tally of trees that will die within 10 years to a sample smaller than that used in obtaining the general stand table, and using better trained men for this purpose. Where growth borings are entrusted to a separate crew, this tally can be made by them and will include only a small percentage of the area covered for the stand table.

In this case the mortality percentages for each diameter class will be applied to the larger number of trees in the stand table as follows:

FOR 12-INCH CLASS

Stand table	Growth sample	Live trees that will die in 10 years	Per cent dying	Trees to be deducted from stand table	Net future stand table
418	16	1	6.25	26	392

2. The most reliable method consists in recording actual mortality over long periods of years on permanent sample plots and noting the diameter of each tree that dies, the year or period within which death occurs, and the cause of death. From these cumulative records, percentage figures applicable to trees of each diameter class for given future periods can be applied to stands having similar characteristics and distribution of sizes. Such records for old-growth timber measure chiefly the cumulative effects of external agencies, such as wind, insects, and diseases, and of continuing loss from suppression of the smaller trees. On 240 acres of longleaf pine in Louisiana, the net losses in 15 years were 30 per cent in volume of the original stand, of which 8 per cent occurred from a 2-year drought and the remainder from bark beetles and from wind, with a few trees felled by fire. Records for ponderosa pine reveal that a constantly increasing percentage of loss occurs from external causes as trees get larger. From such records, when classified by d.b.h., percentage deductions are made from future gross volumes to fit other stands.

3. In the absence of historic records, the more widely used and standard method of arriving at future probable mortality is to tally all trees that have recently died within each diameter class. From these data the percentage of mortality by number of trees is found. In discussing this third method we may first consider what happens in even-aged stands, for which a "normal" number of trees for each site and age class is usually tabulated. Table 57 gives a sample of the data for loblolly pine grown on old fields in Louisiana.

TABLE 57. PAST LOSSES IN TOTAL STAND, FOR 10 YEARS, EXPRESSED AS PERCENTAGE OF STAND AT BEGINNING AND END OF THE PAST PERIOD. FULLY STOCKED EVEN-AGED STANDS, OLD FIELD LOBLOLLY PINE, LOUISIANA. IN STANDS 3.6 IN. AND OVER*

Age, years	Number of trees	Loss for 10 years, number	Loss in percentage of initial number	Loss in percentage of final or "present" number
Site class 70				
20	810			
30	528	282	34.8	54.4
40	289	239	45.3	82.7
50	211	70	24.2	32.0
60	198	21	9.6	10.6
70	186	12	6.1	6.5
80	176	10	5.4	5.7
Site class 100				
20	580			
30	333	227	40.5	68.2
40	183	150	45.0	30.0
50	138	45	24.6	32.6
60	125	13	9.4	10.4
70	117	8	6.4	6.8
80	111	6	5.1	5.4

* From MEYER, W. H., Yield of Even-aged Stands of Loblolly Pine in Northern Louisiana, *Yale Forest School Bul.* 51, 1942, Table 4.

The increase in mortality at 40 over 30 years is due to the omission from the table of trees 3 in. and under, in which mortality percentages are much greater than in the 4-in. class and over.

An example[1] of classification of mortality, expressed as percentage

[1] DEEN, J. L., A Survival Table for Even-aged Stands of Northern White Pine, *Jour. Forestry*, **31**, 42–44 (1933).

survival for trees of each initial diameter class, for even-aged plots of white pine, was worked out by Deen and is shown in Table 58, *viz.*, for a 40-year-old stand in the next 10 years, 30 per cent of the trees now 3 in. in diameter will survive; 55 per cent of the 4-in. trees, etc.

TABLE 58. WHITE PINE SURVIVAL, NEW ENGLAND. PERCENTAGE OF TREES IN EACH DIAMETER CLASS WHICH WILL SURVIVE THE ENSUING 10-YEAR PERIOD AT DEFINITE AGES

Age, years	Diameter class, in.													
	1	2	3	4	5	6	7	8	9	10	11	12	13	14
20	6	84	100											
25	..	54	88	100										
30	..	30	68	87	100									
35	..	11	49	72	87	96	100							
40	..	..	30	55	73	85	93	98	100					
45	..	..	16	40	61	75	86	93	98	100				
50	..	..	...	30	51	64	77	88	93	97	99	100		
55	..	..	...	23	43	59	71	80	87	93	96	99	100	
60	..	..	...	16	37	53	66	74	82	88	93	96	99	100

Although Tables 57 and 58 are derived from even-aged stands, the same general tendencies for reduction of numbers by suppression apply to stands of mixed ages. How, then, shall the percentages of past mortality in each diameter class be used in order to predict future mortality of trees of the same present diameters? When past mortality based on age is applied to past numbers in the stand as a whole (giving the "future" mortality of this past stand), the percentage is obviously less than if this past loss of numbers is applied to the present reduced number of live trees. But even this present basis for percentage loss is excessive, for the percentage loss of the future continues to decrease.

The answer lies in the fact that size, not age, is also the basis of mortality records. Mortality in trees of a given *diameter class*, which has occurred over a past period, is assumed to apply to trees of the same diameter class for a future period. By the same basis of *comparison, not of prediction*, the loss in numbers and percentage in a past even-aged stand 30 years old would be applied by comparison to a present stand of the same density, now 30 years old. In stand tables based on diameter classes, it is thus assumed that the rate of loss in trees of given diameters for an immediate or measurable past period will apply, within a 10-year future period, to the trees which now have this same diameter.

Since percentage of numbers is the basis to be used, and the losses in percentage will decrease as the trees grow larger, the one precaution is to derive the mortality percentage for the future period from the total number of living and dead trees, and not from living trees as the percentage basis. Whatever the past period may be within which dead trees are tallied, whether it is 2 or 5 years, the annual and not the periodic loss must be found. This current annual loss is the figure used to predict average annual losses, in each diameter class, for the future period.

Whatever the length of the past period in years that is chosen, it must be covered by a complete tally, in the area of the sample, of the trees which have died within these years, with a rigid exclusion of those whose deaths occurred earlier. For different species and regions the length of this period will be determined by progressive changes that take place in the appearance and condition of the dead trees with each passing year since death occurred. With pines, the dead needles fall after the first year. Next, the smaller twigs drop, followed by successively larger limbs until only the skeleton trunk remains. Bark begins to slip during the second year, and its progressive disappearance is noted in terms of years. For hardwoods, similar standards are derived from records on tagged trees. Girdled trees are a source of information. In the Southern Forest Survey, trees that have died within the last 4 years are tallied. The guides used are:

1. All trees 5 to 7 in. in diameter and standing will be tallied as dead 4 years or less provided that they do not fall over when leaned against. Down trees 5 in. in diameter will be tallied regardless of condition; down trees 6 and 7 in. will be tallied only if the wood is not soft, pulpy, and easily broken apart.

2. All trees 8 in. and over, whether standing or down, will be tallied if they have branches or branch stubs of the following sizes or smaller:

a. 3 in. or under for woods of high to intermediate durability, such as oaks, black walnut, locust, and hickory.

b. 3½ in. or under for woods of low durability, such as pine, gum, cottonwood, and willow.

With such yardsticks, the cruiser makes a separate tally of timber already dead, covering a definite past period, separately for each diameter class.

248. Requirements of Sampling for Diameter Growth. The measurement of past current growth for the determination of increment on a specific area requires the boring or chipping of trees. Although growth data can be taken from stumps and corrected for taper to convert to d.b.h. outside bark (Secs. 203 to 206), such measurements can seldom if ever fulfill the requirements of sampling for obtaining average future growth on the area to be evaluated; hence stump measurements are not used for this purpose. The sample taken for growth will always be much smaller than that covered by the stand table. The trees chosen for growth measurement must then represent either a true random or a systematic sample of the growth rates of the trees in the stand table. It is possible to obtain these average growth figures from the smaller sample, since the number of trees to which the growth data are to be applied is correctly given by separate diameter classes in the stand table. Within the limit of error in the timber cruise, all that is further needed is enough borings for each diameter class to give a statistically sound estimate of past growth on the separate classes, which number can be approximated, during the accumulation of the borings, from the standard error of the mean (Secs. 17 and 171).

Since each diameter class constitutes in effect a separate sample, it might be assumed that an equal number of trees for each class would satisfy the requirements of sampling. It is true that the sample should cover the entire range of diameters measured. But two other considerations are of greater importance in view of the wide range and high standard deviation of diameter growth even in separate 1-in. classes.

1. The purpose of sampling is to obtain an average of a large enough number of individuals to confine the standard error of the mean within prescribed limits. This could be accomplished separately for each diameter class by boring equal numbers of trees in the several classes, provided that the standard deviation for each class were about the same. But identical errors are more serious when applied to large numbers of trees than when applied to small numbers. Since the error diminishes as the number of units in the sample increases, the practice described in 2, rather than that of sampling equal numbers for each class, is standard.

2. It does not follow that the selection of single trees at fixed intervals will give an accurate sample of growth rates per diameter class. Diameter growth per tree increases as density of stocking diminishes. One tree bored for 10 acres of strip may represent 500 trees or less on stands with poor stocking, and 2,000 trees or more on the denser stands. Yet each tree would be given equal weight. This results in an overestimate of average growth rate, a biased error which cannot be

permitted. When, instead, *all* the trees are bored on small plots equally spaced, the sample should reflect accurately the average growth rate for each diameter class on the entire area, within the controlled limits of error of sampling on plots.

By systematic unbiased sampling, the number of trees measured in each diameter class will be in rough proportion to the total number of such trees in the stand table. Diameter classes containing the larger numbers of trees will have a greater number, though not a greater proportion, measured for growth.

249. Modifying the Stand Table for Use in Growth Studies— Ingrowth. A stand table used solely for a timber estimate includes only merchantable trees. For saw logs, trees too small to make lumber would be omitted. With closer utilization, the limit for saw timber has been progressively lowered but remains about 7 in. and over for small and medium-sized trees in eastern regions, and 12 or 16 in. or even higher for the West Coast. The development of pulpwood markets and other uses has set a minimum d.b.h. for timber estimates at 4 ins.

The stand table required for growth prediction should include trees below these merchantable limits. The number of diameter classes smaller than the minimum merchantable diameter limit that should be included in the stand table must cover the range of diameters capable of growing into the merchantable classes during the period to be covered by the estimate of future growth. For fast-growing species or for periods longer than 10 years, the range of lower diameter is increased and the minimum diameter for the stand table set lower than for those of slower growth for the shorter period.

For a 10-year period, three additional 1-in. classes, depending on the maximum rate of growth, are usually adequate for a 7-in. limit for board feet. When a 4-in. limit is originally used, the tally below this limit should be taken on a smaller sample. Growth rates on these smaller diameters are measured by the same method as for the remaining stand.

The term *ingrowth* is defined as the entire volume of trees, now below the minimum merchantable diameter limit, that will grow into and above this limit during a definite future period of years, plus the growth that these trees make after reaching the merchantable size during the period predicted. Hence ingrowth consists of the total merchantable volume, at the end of the future growth period, of all trees entering the merchantable diameter classes within the period.

Total growth on an area, when measured by cubic volume regardless of merchantability, would include all diameters down to 1 in. (0.6 to 1.5 in.) and would require a tally of these small trees. As shown, stand tables used in timber estimating do not include these unmerchantable sizes. In this case the entire growth, from seed, of the ingrowth trees will remain unaccounted for until the period within which the trees become merchantable. At the end of the future period, all the previous growth during the entire life of these smaller diameter classes that attain merchantable sizes is entered as growth for that period alone.

If we could visualize a forest in which the number of trees in each diameter class is such that the ultimate yield of all classes on attaining maturity will be about equal, it would follow that the volume of the ingrowth at the end of each future period would also be equal. The inclusion, as current growth of the next future period, of the entire growth which the ingrowth tree classes have put on since they were seedlings is roughly offset by not including in the total the growth that the still smaller and younger diameter classes are making. If this condition of balance between larger and smaller trees does exist in the forest, there is no essential difference between the future volume growth of trees now merchantable plus the total volume of ingrowth, on the one hand, and the total actual volume growth on all tree classes down to seedlings, on the other.

This equalization of growth capacity for the different tree groups, young and old, small and large, practically never exists in the actual forest. Instead, the extent of the young group of trees which produce ingrowth may be greatly in excess or the reverse. When growth is predicted in this or any other manner, for a short future period on a forest area, it is the current *periodic* growth of the stands and forest as a whole, not the average, or *mean*, annual growth of single stands, that is determined (Secs. 194 and 195).

When, as is usually the case on lands from which only the larger timber has been cut, the remaining stands and trees are those on which current growth is increasing rapidly, the volume of ingrowth for the next 10 years, on well-stocked areas, may be far in excess of the rate that can be maintained permanently. On stands of second growth just approaching merchantable sizes, the proportion of total periodic growth represented by ingrowth may for a time exceed the growth on the remaining merchantable stand. This condition can endure only for a relatively short time and must not be made the basis of a sustained annual cut. When a proper distribution of age and size classes is attained, the spurt in current growth is over, and the mean annual yield that the property can produce continuously is approached.

Since a minimum d.b.h. is always used, ingrowth is always a factor and may be either a large or a small addition to growth of the present merchantable stand. It must therefore be included in periodic growth predictions based on stand tables.

250. Application of Past Diameter Growth to Stand Tables. The boring and measurement at b.h. of ring growth for a past period of years merely gives a record of past performance of the tree, similar to that kept on the books of a business. How shall this past record be interpreted and the "business trend" of growth per acre be determined for the stretch ahead?

The simplest and at the same time the least dependable assumption is that the past rates of diameter growth will not change in the future. If a tree now 10 in. d.b.h. had a d.b.h. of 9 in. 10 years ago, we assume that it will grow to be an 11-in. tree in the next 10 years. Past growth is thus merely projected into the future. A table of growth in inches for each present d.b.h. class for the past 10 years is merely labeled "growth" and is assumed to apply equally well to the past and the future, for this 10-in. tree. Until recently, this method, known as *straight-line projection*, has been widely used. It ignores both past and future trends of growth and as a method is obsolete. By this and the next method to be described, a single measurement of growth in inches for the past 10 years is the only one considered necessary.

The second method, which was substituted for straight-line projection, rests on a different assumption, which is that trees of the same d.b.h. tend to grow at similar rates in a many-aged forest, not only in the present decade, but in any past or future period. When this premise is accepted, if the past 10-year growth o.b. at b.h. of a tree now 10 in. amounts to 1 in. in diameter, this constitutes the future growth of a tree 9 in. in diameter at the beginning of the decade just past. It is then assumed that any tree *now* 9 in. d.b.h. should grow 1 in. in the next 10 years. The past 9-in. tree is compared with the present tree of the same size. In this way, by determining *past d.b.h.* for the trees bored, a table of "future" growth for trees of each d.b.h. class is prepared and applied to the present stand table. This is termed the *method of straight-line comparison* and is the standard method used widely by the Forest Service in the Forest Survey. It differs from the first method only by the shift from present to past diameters for the basis of the growth rate used in predictions.

Neither of these two methods actually studies past *trends* of growth;

hence both may fail of a close prediction as to the continuation of these trends into the future. They take no account of the almost inevitable lessening of the future rate of diameter growth due to increasing cross-sectional area of the tree, the constant increase of competition of other trees, or the attainment of physical maturity in the life cycle, any or all of which causes would slow down the future rate of diameter growth (Sec. 202). These changes may be quite swift, as when crowns are rapidly shortened at about 20 years of age in old field pine stands. Their effect is not automatically recorded by measuring merely the resultant total net volumes produced, as is done by the method of comparison, by using yield tables for even-aged stands. When the tree and the diameter class are the units, the trends of future growth for each d.b.h. class, and the mortality suffered, are the individual elements which make up the total net growth, and each must be considered as carefully as the separate sums in a ledger.

Past trends, it is assumed, should indicate the future behavior, unless a radical change occurs such as release by cutting or other cause. Hence past trends must be measured. For fast-growing species and regions this is done by separating past growth into cumulative 5-year periods, and measuring two or preferably three of these periods. For slow-growing conditions, two to three 10-year periods may serve better. When the average past growth is obtained separately for each d.b.h. class and is plotted (Sec. 256, Fig. 81), the three or four points will locate a curve of past growth, which will indicate whether that class is slowing down, maintaining its rate of growth, or increasing it. Any well-based curve of past performance gives a fairly accurate basis for projection over a future 10-year period. By this extension the future d.b.h. is found for the average tree of each d.b.h. class. This is termed the *method of curved projection* (3) as contrasted with straight-line projection (1) and straight-line comparison (2).

By any one of these three methods of applying past diameter growth to the future, it may be assumed that the average tree dealt with lies at the exact mid-point of the diameter class. For 1-in. classes, a 10-in. tree represents all trees 9.6 to 10.5 in., and technically the mid-point would be 10.05 in. For 2-in. classes, the 10-in. tree is the "average" of trees 9.1 to 11.0 in., with the same small discrepancy. By any of the three methods then, the rate and amount of future growth when found is added to the present d.b.h. For 1 in., of growth, the average tree becomes 11 in. For 0.9 in. it reaches only 10.9, etc. This rule applies whether the future growth is found by straight line or by curved projection.

426   FOREST MENSURATION

But it is not necessary to stop here in seeking to improve the utility and accuracy of these diameter growth measurements. If it is a convenience to classify or bunch trees in 1- or 2-in.-diameter classes now, it is equally convenient to do the same thing for the future stand table, 10 years hence. This may permit of a further projection to 20 years.

The movement of trees from one diameter class to those above based on the assumption of even distribution in $\frac{1}{10}$-in. classes and the same growth rate for all trees in each single diameter class is best shown by diagram. For 10 trees now in the 10-in. class, with 3 smaller and 3 larger trees, all growing 1.3 in. in 10 years, the present and future distribution by classes will be as shown on page 427.

When we assume that every tree in each tenth of a 1-in. d.b.h. class grows at the same rate, it makes no difference in the growth prediction whether we use the straight average or split the class into tenth inches. Since the latter choice enables us to construct the future stand table in terms of d.b.h. classes, it has been accepted as the standard method of applying future diameter growth to the present stand table. The method of application is as follows:

The upward movement of numbers of trees into higher classes is seen to be proportional to the ratio which growth in inches for the period bears to the diameter class interval.

Let e = number of trees which will enter the next higher diameter class

t = total number of trees in the present class

g = diameter growth in inches

i = diameter class in inches

Then $e = t(g/i)$, in which g/i is the fraction of the number of trees t in the class i which will go up one class during the period for which g is measured. When g exceeds i, the ratio is greater than 1. If it is 1.3, all trees, or 100 per cent, go up one class, and 0.3 of these trees advance two classes, leaving 0.7 in the lower of the two classes into which the original class has grown. As stated, the total growth predicted by this method is practically identical with that obtained by using a single average. See example in Tables 66 and 67. Where the stand is producing piece products in which minimum diameter is a limiting requirement, this method, and not that of growth of the *average* tree for the class, will indicate the number of trees that will represent ingrowth into the minimum merchantable class and yield the required dimensions, whereas the average diameter method assumes that either all or none of a certain smaller diameter class will be included as reaching merchantable size.

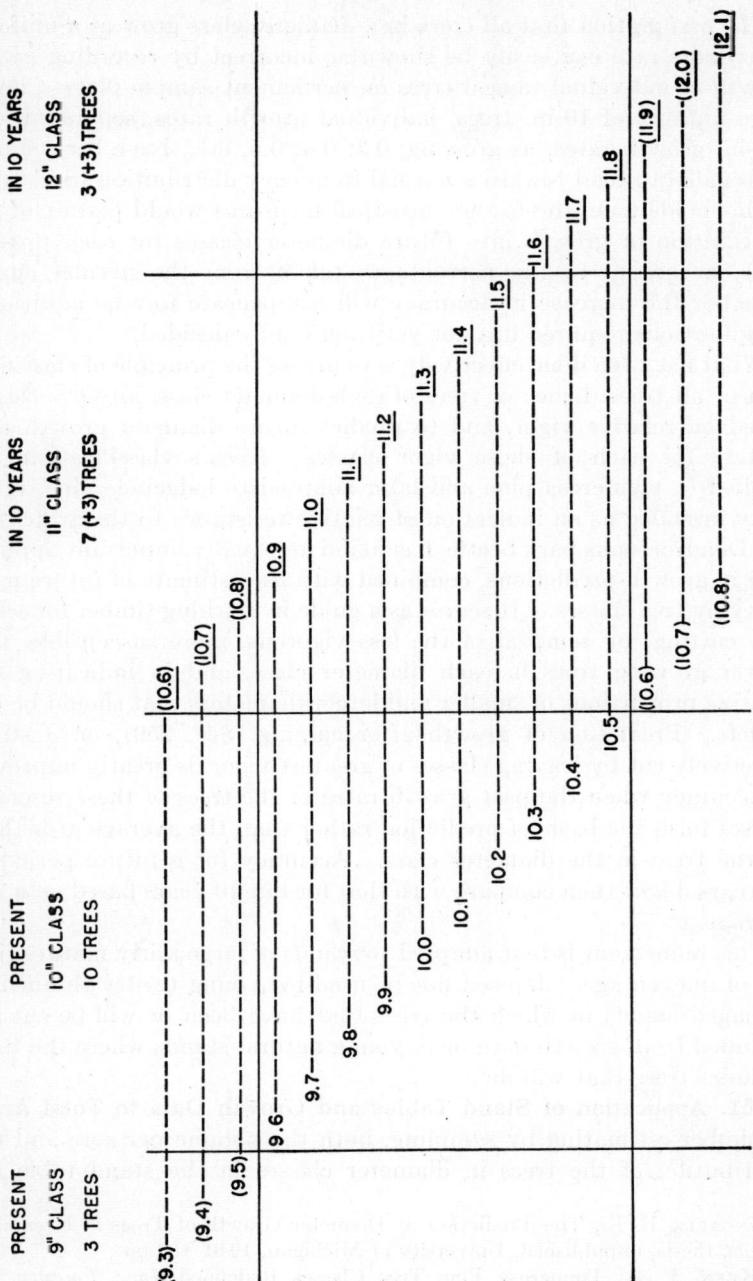

The assumption that all trees in a diameter class grow at a uniform or average rate can easily be shown as incorrect by recording actual growth of individual tagged trees on permanent sample plots. For a large number of 10-in. trees, individual growth rates, separated into $\frac{1}{10}$-in. growth rates, as growing, 0.3, 0.4, 0.5, etc., have been shown by Ingalls[1] to tend toward a normal frequency distribution (Sec. 168). If this holds true, the use of statistical measures would permit of the distribution of growth into future diameter classes for each present class, according to the percentages taken from the normal curve. Whether the increase in accuracy will compensate for the additional computations required has not yet been finally decided.

What has been done effectively is to utilize the principle of classification of all trees, hence of trees of each diameter class, into *tree classes* based on relative vigor, and to predict future diameter growth separately for each of these vigor classes. Keen's classification,[2] as applied to ponderosa pine and later adapted to lodgepole pine, originally intended as an indication of relative resistance to the attacks of the Dendroctonus bark beetle, has found an equally important application in growth predictions, combined with the estimate of future mortality by tree classes. It serves as a guide in marking timber for selective cutting, by removal of the less vigorous, more susceptible, and slower growing trees in each diameter class, and in indicating the relative proportions of smaller and larger diameters that should be cut or left. Prediction of growth after cutting (Sec. 259), on a stand selectively cut by leaving classes of greatest vigor, is greatly improved in accuracy when the past growth rates of the trees of these reserved classes form the basis of prediction rather than the average growth of all the trees in the diameter class. Accuracy for a future period of 20 years might then compare with that for but 10 years based on a flat average.

This refinement is best adapted to stands of large fairly mature timber of uneven age. It need not be used for young thrifty thinned or managed stands in which the trees that have been or will be cut are excluded from growth data or in young natural stands where the tally excludes trees that will die.

251. Application of Stand Tables and Growth Data to Total Area. In timber estimating by sampling, both the volume per acre and the distribution of the trees in diameter classes in the stand table are

---

[1] INGALLS, E. E., The Prediction of Diameter Growth of Trees in Even-aged Stands, thesis, unpublished, University of Michigan, 1940, 188 pp.

[2] KEEN, F. P., Ponderosa Pine Tree Classes Redefined, *Jour. Forestry*, **41**, 249–253 (1943).

assumed to apply to the larger areas sampled, within the limits of the standard error of estimate as controlled by the number of plots taken for the sample. The volume of growth predicted is likewise assumed to apply to the average acre on the area sampled. This area may be the entire holding. More frequently it is a subdivision, consisting primarily of a large block within which types have been separated.

Should it happen that growth is sought for an area on which the sample taken for present volume has for any reason been corrected (Sec. 158), it is assumed that growth in volume will be adjusted by this same correction. If volume per acre on the sample estimate has been raised 10 per cent, volume growth is assumed as 10 per cent larger than on the sample. Although such ocular corrections in volume may prove possible in timber estimating for units as small as 40 acres, growth is usually found for large units, in which case the volume per acre from the sample is always taken as indicating that on the total area.

Local volume tables applied to both present and future stand tables, as reduced by mortality and estimated commodity drain, are used to obtain the total estimated volume increment for the future 10-year period. The volume growth for the period is then reduced to volume growth for the present, or current, year.

A method commonly used for this process is to determine the ratio which future volume bears to present volume and from this ratio to find the percentage of compound interest that would give the future volume from the present volume. The flaw in this procedure is to assume that the growth rate in terms of absolute volume for the current year can be obtained by multiplying present volume by compound interest per cent. For an entire forest region and even for a single stand, this does not accord with the facts. It is true that for a 10-year period a calculation of increased volume involving compound interest will give the same result as one involving simple interest, provided that the two rates are correlated, but for a single year the growth in absolute volume should be based on a simple interest calculation. For example, in a stand which increases from 10,000 to 15,000 bd. ft. in 10 years, the rate of increase is 50 per cent, which corresponds to a simple interest rate of 5 per cent and a compound rate of about 4.1 per cent. The current annual growth rate of this period is 500 bd. ft., which is obtained by multiplying 10,000 by 0.05. The use of the compound rate of 4.1 will give only 410 bd. ft., or an underestimate of 18 per cent. Compound interest does not apply to a going concern in which growth, drain, and mortality are on a current annual basis. Nor does it apply as a measure of either current or mean annual growth in a forest or a single stand. In view of the error cited above, it destroys at one stroke the standards of accuracy set up for all other steps in the forest survey.

Volume (and value) growth may also be obtained from current annual diameter growth by determining the percentage relationship between diameter growth and volume growth. This would include adjustment for height growth and form change. An approximation formula for the volume growth percentage in use in the Lake states is as follows:

$$P = f\frac{d}{D}$$

where P = annual growth percentage in volume
d = estimated 10-year diameter growth
D = present d.b.h.
f = a factor which adjusts for height growth, form change, and the relation between diameter and volume

If f/D is called k, the formula is reduced to

$$P = kd$$

and if a table of k values is available, the percentage can be readily computed for any diameter growth rate in inches. Table 59 is such a computation developed for Lake states conditions.

TABLE 59. k FACTORS FOR CONVERTING DIAMETER INCREMENT INTO VOLUME GROWTH PER CENT*

D.b.h., in.	Bd. ft. volume, International ¼ in.	Cordwood or merchantable cu. ft. volume, all species
6	....	6.73
8	....	4.05
10	....	2.80
12	....	2.15
14	2.34	1.74
16	1.90	1.46
18	1.60	1.24
20	1.36	1.07
22	1.17	0.95
24	1.03	0.84
26	0.92	0.77
28	0.84	0.72
30	0.77	0.67

* Taken from a larger table by S. R. Gevorkiantz and L. P. Olson, A Simplified Method for Predicting Growth of Forest Stands, *Lake States Forest Expt. Sta.* RE-LS *Forest Survey*, Sept. 15, 1947.

By use of Table 59, the annual volume growth percentage based on board-foot volume for a 20-in. tree which is expected to grow 1.50 in. will be

$$P = 1.36 \times 1.50, \text{ or } 2.04 \text{ per cent}$$

since the factor k for a 20-in. tree is 1.36.

252. Reproduction Surveys—Stocked Quadrats. With the excessive mortality of seedlings during the first season, caused by their failure to establish themselves and secure contact with ground moisture, or through shade or other competition, a count of the seedling class is useless. The presence of numerous seedlings following a seed year may be noted and their later establishment watched. Survival of seedlings, in turn, depends on further competition with grass, brush, and other trees of the same age of their own or of different species, and finally on freedom from overhead shade and root competition of larger trees. For these reasons, a tally, or count, of reproduction in itself, on the basis of the number counted, yields no information of value in predicting its future importance as a forest crop. Instead the objective is to predict the proportion of the total *area* which will ultimately be stocked with surviving trees, capable of yielding wood products.

The *stocked-quadrat* method has been widely used for this purpose. It consists in taking small plots regularly spaced along equally distant strips and noting the presence or absence of *one* thrifty seedling on the plot capable of developing to maturity through absence of shade or other deterring influences. More than one such seedling does not alter the plot classification. Area, not numbers, is the basis of the tally. The plots can be circular or square, ranging in size from 1 to 4 mil-acres. The 1-mil-acre size is 6.6 ft. square; the 4-mil-acre size is 12.2 ft. square. A 10-ft. square plot is also used. A common practice is to take a block of four quadrats at each station. If only one quadrat is stocked, a tally of 1 is given the station; if two are stocked, a tally of 2, etc. Upon the completion of the reproduction survey, the tally numbers are added and divided by the total quadrats observed, thus giving the stocking percentage. Owing to the effect of size of plots on sampling in irregular distributions, the larger the plot, the greater will be the percentage shown as stocked. Choice of size is related to species, growth habits, and products to be grown.

Where tallies are made of stems down to and including the 1-in. class (0.6 to 1.5 in.) the objections to a numerical rather than an area and survival basis for growth predictions still hold good. Except on permanent sample plots for special studies, little use is found for the extension of a stand table below 4 in. at b.h., and conclusions as to future stocking drawn from such numerical counts are invalid when applied to area and ultimate suvival.

The stocked-quadrat method is chiefly applicable to species growing in even-aged stands. Reproduction of tolerant trees in all-aged stands, below the minimum diameter limit of the stand table, is seldom tallied or counted, any more than it is under the stocked-quadrat method. Its presence, absence, or relative abundance may be noted and briefly described. Survival of a tree depends largely on its relative tolerance and the future history of the stand. Until released, its comparatively slow growth is a reason for avoiding any expense for recording or enumerating the individual seedlings present.

A method has lately been devised by which the future yields for different percentages of stocked quadrats, for Douglas fir, can be predicted.[1]

References

BRIEGLEB, P. A.: Calculating the Growth of Ponderosa Pine Forests, Pacific Northwest Forest Experiment Station, 1945, 60 pp.

BRIGHT, G. A.: The Relation of Crown Space to the Volume of Present and Future Stands of Western Yellow Pine, *Forestry Quart.*, **12**, 330–340 (1914).

CHAPMAN, H. H.: The Causes and Rate of Decadence in Stands of Virgin Longleaf Pine, *Lumber Trade Jour.*, **84** (6), 11, 16–17 (1923).

———: The Problem of Forecasting Timber Growth in Irregular Stands, *Jour. Forestry*, **40**, 805–806 (1942).

DUERR, W. A., and S. R. GEVORKIANTZ: Growth Prediction and Site Determination in Uneven-aged Timber Stands, *Jour. Agric. Res.*, **56**, 81–98 (1938).

DUNNING, D.: Predicting the Second Cut in National Forest Management Plans, *Jour. Forestry*, **24**, 785–790 (1926).

GEVORKIANTZ, S. R., and L. POISEN: An Improved Increment-core Method for Predicting Growth of Forest Stands, *Lake States Forest Expt. Sta. Paper* 12, 1948, 19 pp.

GRAVES, H. S.: Practical Forestry in the Adirondacks, *U.S. Dept. Agric. Div. Forestry Bul.* 26, 1899, 84 pp.

HAIG, I. T.: Accuracy of Quadrat Sampling in Studying Forest Reproduction on Cutover Areas, *Ecology*, **10**, 374–381 (1929).

HERVEY, D. E.: A Method of Measuring the Current Mortality of a Timber Stand, *Jour. Forestry*, **34**, 1003 (1936).

HERRICK, A. M.: Multiple Correlation in Predicting the Growth of Many-aged Oak-hickory Stands, *Jour. Forestry*, **42**, 812–817 (1944).

HORNIBROOK, E. H.: Preliminary Yield Tables for Selectively Cut Ponderosa Pine in Black Hills, *Jour. Forestry*, **37**, 807–812 (1939).

---

[1] Management of Second-growth Forests in the Douglas-fir Region, Douglas Fir Second-growth Management Committee Progress Report, Pacific Northwest Forest and Range Experiment Station, 1947, 151 pp.

KAKASAI, M. A.: Increment Determination on the Basis of Stand Tables, *Jour. Forestry*, **34,** 628–631 (1936).

KORSTIAN, C. F.: Growth on Cutover and Virgin Western Yellow Pine Lands in Central Idaho, *Jour. Agric. Res.*, **33,** 501–541 (1924).

KRAUCH, H.: Mortality in Cutover Stands of Western Yellow Pine, *Jour. Forestry*, **28,** 1085–1097 (1930).

LEXEN, B.: Growth Following Partial Cutting in Ponderosa Pine, *Jour. Forestry*, **37,** 943–946 (1939).

———: Space Requirements of Ponderosa Pine by Tree Diameter, *Southwestern Forest and Range Expt. Sta. Res. Note* 63, 1939.

McCARTHY, E. F.: Accelerated Growth of Balsam in the Adirondacks, *Jour. Forestry*, **16,** 304–307 (1918).

——— and W. M. ROBERTSON: Volume Increment on Cutover Pulpwood Lands, *Jour. Forestry*, **19,** 611–617 (1921).

MEYER, E. A.: Methods of Forest Growth Determination, *Pa. State Col. Agric. Expt. Sta. Bul.* 435, 1942, 93 pp.

STAEBLER, G. R.: Predicting Stocking Improvement in Reproduction Stands of Douglas Fir, *Pacific Northwest Forest Expt. Sta. Res. Note* 41, Apr. 1, 1948, processed, 5 pp.

WAHLENBERG, W. G.: Methods of Forecasting Timber Growth in Irregular Stands, *U.S. Dept. Agric. Tech. Bul.* 796, 1914, 55 pp.

CHAPTER 30

PREDICTING THE PERIODIC ANNUAL GROWTH OF STANDS BY THE STAND TABLE METHOD

253. Measurement of Diameter Growth at B.H. The three steps in predicting the growth of trees are to determine:
1. Growth in d.b.h. outside bark.
2. Growth in total, or merchantable, height.
3. Growth in volume correlated with 1 and 2 by use of volume tables based on d.b.h. and height.

After the exact point $4\frac{1}{2}$ ft. above the ground is determined, the d.b.h. of each tree is measured twice, at right angles, and averaged to the nearest $\frac{1}{10}$ in.

Growth in d.b.h. means increase in diameter outside, *not* inside bark, at $4\frac{1}{2}$ ft. from the ground and not at the stump (Sec. 110). Growth outside bark includes growth in double thickness of bark, plus growth in double radius, *i.e.*, in diameter of wood. Growth of wood alone, taken from increment borings, fails to indicate the growth in d.b.h. by the amount of bark "growth," an error amounting roughly to 10 per cent or more.

Radial measurement of wood growth taken from increment borings at b.h., and width of bark, measured separately (Sec. 65), must be doubled to express diameter growth, since a radius is one-half the diameter of a circle. But the radial measurements must be recorded, for both wood and bark, directly as taken. Doubling each measurement not only increases the labor but is a source of error. After the data have been averaged and plotted, the scale of radial growth is then doubled to read in terms of diameter. Increase in bark thickness will be treated in Sec. 254.

The measurements of radial growth of wood are listed under each separate present d.b.h. class. The average past growth is obtained by dividing the total growth of all trees in the class by the number of trees. This gives a table which is the basis for a curve of past growth of wood (but not of wood and bark, hence not of d.b.h.) based on present d.b.h.

In obtaining the radial growth by the increment borer, the boring should be taken along an average radius, if possible, which would mean placing it at one of the two end points of an average diameter. This assumes that the pith is in the exact center of the tree. The cross section of practically every tree, however, is more or less eccentric. The growth along different radii may be even more eccentric because the pith is usually found off center. The object is to obtain a measurement which is equivalent to the growth on the average radius. This may be best accomplished by observing the following rules:

1. Do not bore leaning trees unless required by rule 3 below.
2. Locate the average diameter, and bore at that point, holding the borer at right angles to the vertical axis of the tree, or
3. Make all borings on a given plot on the side of the tree facing the center, or the center line, of the plot. This method of sampling overcomes any bias that slope or lean may have on the rate of growth.

When the increment borer is used, the core is counted and measured as it lies in the extractor. When cores are difficult to count because of slow growth or indistinct rings, they may be labeled with an indelible pencil, recording the d.b.h., species, and number, and later examined under a magnifying glass or microscope, or tested with dyes to bring out the details. For most species the measurement is completed in the field when the boring is made. When it is feasible to substitute the ax for the borer (the notch will come off as a slab or edging in the mill) a horizontal cut is made at b.h. deep enough to include the growth for the period required. Rings are measured directly on this notch.

The total period for which growth is to be determined is first carefully counted inward, marking each subperiod of 5 or 10 years with a soft pencil, preferably indelible. For the purpose of determining the rate at which growth slows down (Sec. 184), at least three 5-year, or two 10-year intervals should be measured, depending on species, region, and rate of growth. The radial growth is measured from the inmost ring to the cambium with a ruler graduated and read to the nearest 0.05 in. The measurement should be cumulative in order to reduce fractional errors caused by shifting the ruler, as shown below.

Species	D.b.h., in.	Current radial growth, in.	
		10 years	20 years
Loblolly pine......................	10.1	1.25	2.00

This tree grew 1.25 × 2, or 2.5 in., in d.i.b. in the first 10 years and only 0.75 × 2, or 1.5 in., in the last decade. Hence its rate of diameter growth is decreasing.

It has been common practice in the past to measure the number of rings in the last ½ in. of radius, thus determining the number of years required to grow 1 in. in d.i.b., and, ignoring bark growth, to assume that the average tree of each 1-in. class will in the future grow 1 in. in diameter in this period of years. This practice is wrong, since it makes growth the independent variable and years the dependent variable, while actually the reverse is true. Only when there is perfect correlation between the two variables (Sec. 175) can this interchange be made, which in this case would mean that the only influence affecting diameter growth is the age of the tree!

The average growth in diameter must therefore be based on past periods of years, not on past inches of diameter. The error incurred in using number of years to grow ½ in. in diameter as the basis of the averages of growth, instead of number of inches grown in a fixed period, is to reduce appreciably the indicated average rate of growth per decade. Since years are averaged instead of growth, the slower a tree grows the more years it has in the last inch of diameter, and the greater will be the weight in determining the average growth rate. One slow-growing tree may have a weight equal to three trees growing rapidly, as in the example below.

Tree number	Rings in last ½ in.	Rate of growth, in. per year
1	5	0.200
2	15	0.067
Total.........	20	0.267
Average.......	10	0.133

The data in column 2 would indicate that it took 10 years for the average tree to grow 1 in. in diameter (Sec. 185). The true average rate of growth of the two trees is found in column 3 and indicates a period of about 7 years and an error of 33 per cent.

254. Correction for Growth in Thickness of Bark. Bark becomes thicker as trees increase in diameter. The difference between the average thickness of bark at b.h. at the beginning and end of a period of years gives the second of the two elements of growth in d.b.h. This increase in thickness is for most trees a net, not a total, growth of bark. Although some smooth-barked trees appear to retain all their original

PREDICTING GROWTH BY THE STAND TABLE METHOD 437

bark, most pines and hardwoods may lose portions of the outer bark by scaling off, as is seen in sycamore. Bark growth forms no discernible annual rings. The average thickness of bark is more closely related to the diameters of the trees than to their ages. Diameter growth of wood for a past period is shown for different diameter classes, not age classes. The average increase in double bark thickness can be found for trees growing 1 in. in d.b.h. without reference to the period of years required. This net "growth" of bark is proportional to any fraction or multiple of 1 in. of tree growth in d.b.h. for a past period. If growth of tree is ½ in. in 10 years and growth of bark for 1 in. of tree growth is 0.1 in., then growth of bark for ½ in. of tree growth equals 0.05 in. in 10 years.

Tree volume tables may or may not include bark. The volume of bark may be shown separately from that of wood. When bark and wood are both consumed as fuel the table will show volume outside bark. Separate volume of bark is sought when it is used for tannin, insulation, fillers, or other products. Volume of wood without bark may include the entire peeled cubic contents or only the portions merchantable. But in every case, regardless of the units of measurement used for the table, the size class into which the tree falls in the volume table, and hence its volume, is determined by its d.b.h. *outside bark*. Growth of these trees must therefore be found for outside bark dimensions, or no correlation is possible between growth in diameter and growth in volume for the tree. An inch of growth should always mean an inch of *both* wood and bark growth.

Although the growth sought is that of the volume of whatever products are to be grown, this objective is most easily obtained by correlating the volumes based on d.b.h. with growth outside bark. Otherwise *volume tables* as well as growth would have to be based on measurements taken *inside* the bark at b.h., with much greater labor and no appreciable increase in accuracy.

Single width of bark can be measured rapidly by the use of the Swedish bark gauge (Sec. 66) or by notching; two measurements, on opposite sides of the tree should be taken; for very large trees with thick bark four should be made (Sec. 65). When large numbers of trees are measured, a single bark reading per tree is sufficient.

Diameter growth at b.h. equals present d.b.h. minus past d.b.h. with both measurements taken outside bark. The latter must be computed. The steps are

1. Obtain average single bark thickness, and average past radial growth of wood for trees of each present d.b.h. class.

2. Double the average figures for bark thickness and wood growth to put them in terms of diameter.

3. On cross-section paper graduate both the horizontal and vertical axes to the same scale, in inches.

4. Draw a straight line at 45-degree angle to represent d.i.b.

5. From 2, find the d.i.b. for each d.b.h. class, and plot the d.b.h. above the proper d.i.b. point on line 4.

6. These points may fall in a line which is straight or slightly curved. Draw a smooth line through them, harmonizing any irregularities. This is the curve of d.b.h. The vertical space between the curve and the 45-degree line is the average double width of bark for trees of given d.b.h. and d.i.b.

To find growth in double bark thickness corresponding to a past period of years:

7. Subtract the double radial growth of wood from the d.i.b. of the average tree for each d.b.h. class to get the past d.i.b.

8. On the chart (6) find this smaller d.i.b. and read the corresponding d.b.h. above it.

9. Subtract past d.b.h. from present d.b.h., to obtain the past growth at b.h. for trees of each present diameter class, and tabulate. These points can be used for predicting growth by either the obsolete method of straight-line projection when but one past period is measured, or by curved projection when two or more periods have been taken. It will not serve for the method of comparison (see Sec. **250** for these methods), in which past diameters should be the basis of classifying and predicting diameter growth.

In this case

10. From the chart (6) read the past d.b.h. for each present d.b.h. as indicated by growth, for the period required, or take this figure from the table (9).

11. On a chart similar to (6) represent past d.b.h. by a 45-degree line, and for each actual past d.b.h. plot above this point the present d.b.h.

12. From points on the past d.b.h. curve corresponding exactly with the mid-point of each past d.b.h. class read the present d.b.h. for this past class. Trees of present d.b.h. corresponding to this past d.b.h. will, by comparison, be assumed to have this rate of future growth.

Double bark width may be expressed as a percentage of the diameter of wood inside bark. When found for each d.i.b. class it has been shown that the percentage of bark width is approximately the same for trees of each d.i.b. class, hence, by reversal, for each d.b.h. class.

The percentage by which d.i.b. is increased to equal d.b.h. is termed the *bark factor*. It differs with species but gives consistent factors for each separate species. Bark factors for species in the Lake states, as determined by the Lake States Forest Experiment Station, are given in Table 60. If adopted, the curve of bark thickness can be plotted above the 45-degree curve of d.i.b. in a straight line by fixing a position of two points, each of which is the required percentage above the d.i.b. value at that point on the curve. If bark thickness falls off in percentage of d.i.b. for larger classes, the d.i.b. curve would be modified from a straight line.

TABLE 60. BARK FACTORS FOR TREE SPECIES IN THE LAKE STATES*
(Obtained from original list)

Ash, green	1.117	Oaks	1.122
Aspen	1.096	Hickories	1.087
Basswood	1.104	Pine, Jack	1.107
Beech	1.046	White	1.117
Birch, paper	1.075	Norway	1.112
Hemlock	1.124	Spruce	1.065
Maple, sugar	1.096	Tamarack	1.056

* Methods of Predicting Growth of Forest Stands in the Forest Survey of the Lake States, *Lake States Forest Expt. Sta. Economic Notes* 9, Table 28, April, 1938.

255. Periodic Growth in Height, Based on D.B.H. Classes. The purpose of measuring the diameter and height growth of individual trees is to determine past growth and from this to predict future growth in volume for trees and stands and finally for entire forest areas. When prediction is based on stand tables showing number of trees per acre by d.b.h. classes, growth in height determines the future average heights which are used to obtain the local volumes per tree from a standard table, for application to the future stand (Sec. 86).

The prediction of future height growth for trees in the present diameter classes requires a knowledge of past height growth. The alternative methods for obtaining past height growth are:

1. Use of the total height growth based on ages of trees, applicable chiefly to even-aged stands. This method serves as a check on other methods and is useful in dealing with older timber and with trees of larger sizes released by cutting. It should not ordinarily be substituted for the methods described below.

2. Actual determination of the current height growth for a past (10-year) period by cutting back the tips of a sample of felled trees

until the age of the cross section shows the required number of rings, in this example, 10 rings. For intolerant trees in even-aged stands, the trend of height growth is more closely related to age than to diameter, at least for the dominant or final crop trees. For method 2 the trees are always classified by diameter.

3. Preparation of d.b.h.-height curves for even-aged stands at different ages, and their use to derive a similar curve of height on diameter for the stand at the end of the period of prediction. Both methods 1 and 2 are true growth curves of height, but based respectively on age (1) and on diameter (2). Neither method directly answers the question, what average height will trees of each present diameter class have at the end of the period? A curve is needed to show the average height for each d.b.h. class, in other words, a "height curve" in which growth, or age, is not shown. Such a curve, for example, is used in estimating the present volume of a stand. A similar d.b.h.-height curve is needed for the future stand.

A curve of height on diameter is not a growth curve but shows the average height of trees of different diameters at a given time, present or future, and therefore cannot be used as a means of determining true height growth. Just as in the case of growth percentage (Sec. 260), actual past and probable predicted periodic growth in height must first be determined by other means, after which the curve of future height on future d.b.h. can be plotted. When future growth is based on diameter classes, the ease and comparative simplicity of cutting the Gordian knot by assuming that trees in the future stand will have the same average heights for each diameter class as they have in the present stand has caused this short cut to be used almost exclusively. By this method, if a 7-in. tree grows to 8.2 in. during the period and the difference between the heights read from the same d.b.h.-height curve is 8 ft., the growth in height is assumed to be 8 ft.

But heights do not normally follow the same growth trends as diameters (Sec. 182). The former maintains a more or less consistent relation to age of tree or stand, *i.e.*, to the typical curve of height growth (Sec. 200). Diameter growth, by contrast, varies far more with the tree class, such as dominant, intermediate, or suppressed, and responds to release to a far greater extent than does height growth (Sec. 183). It follows that in young even-aged stands, 10-in. dominant trees in the present year will be only slightly taller than the rest of the stand but will have diameters considerably greater than the average. Ten years from now, the trees that are *then* 10 in. may be intermediate or suppressed trees. Such trees, with 10 years more of height growth, will be taller than 10-in. trees in the present stand. Thus the new curve of height on diameter for even-aged stands falls at a higher level instead of coinciding with the present curve, as is easily shown by plotting successive curves of height on diameter for remeasured sample plots in even-aged stands (see Table 61).

A reverse relationship will be shown for trees that have completed most of their height growth but are released as seed trees or by stand improvement. In this case diameter growth increases relatively faster than height growth; hence heights are shifted further to the right, and the curve of height on diameter for the future 10-year period will fall below, not above, the present curve of height on diameter to the degree by which the ratio of height growth to diameter growth is lessened. This does not mean that the actual heights are shortened. For any single diameter class, the larger future diameter will also show a greater height, but not so great as for the original curve. Where these actual growth relationships are ignored, as is done when identical height-on-diameter curves are used for present and future local volume tables, large errors may be incurred, and growth may be underestimated for young stands and overestimated for older timber released by cutting.

TABLE 61. HEIGHT ON DIAMETER AT DIFFERENT AGES
Even-aged stand, loblolly pine, forest grown, Louisiana, site class 86 ft. at 50 years. Thinned at each indicated age

D.b.h., in.	Age, years			
	18	23	28	33
	Height, ft.			
4	42	40	39	
5	43	45	45	
6	45	49	52	
7	46	52	56	
8	47	56	59	60
9	48	58	61	63
10	48	60	63	67
11	48	61	65	70
12	..	62	67	73
13	..	..	70	75

In this table, as is evident, actual height growth is not shown, but merely the existing relationship of total height on diameter for the same stand at different ages. For instance, the smaller heights of 4-in. trees at 28 years apply to trees that were less than 4 in. in previous years and have been suppressed, growing more slowly than trees in the 4-in. class in previous years. The 4-in. trees in the young stand were dominant trees, in the older stand intermediate or even suppressed trees. The magnitude of the errors inherent in applying curves of height on diameter obtained from an existing stand to the stand at any future period is obvious.

The relationship of these d.b.h.-height curves to the true height curves plotted over age for the same stand reveal the essential difference between basing height growth on age and determining height based on diameter.

TABLE 62. HEIGHT GROWTH BASED ON AGE, SEPARATED BY ORIGINAL DIAMETER CLASSES AT 18 YEARS OF AGE

Original d.b.h., in.	Age, years			
	18	23	28	33
	Height, ft.			
6	45	52	59	64
7	46	55	62	70
8	49	57	66	74
10	51	60	69	76

In this table, the diameters for subsequent years are not revealed. There will be an increasing spread of d.b.h. for each of the initial classes, passing up into successive d.b.h. classes at different rates of growth even for the same initial d.b.h. class. Since growth is to be based on d.b.h. and static height for each age, Table 61, not Table 62, is required.

4. There remains the choice of predicting future heights as maintaining the fixed ratio to diameters indicated in the present d.b.h.-height curve. Although shown to be inaccurate for small areas, it may be permissible for uneven-aged stands of tolerant species on large areas of forest. The assumption is merely a device for making allowance for growth in height that has not actually been determined. It is applied by using the single curve of present heights on diameter for the prediction of future height on future diameters. The change in heights as read from the height curve for the present and future d.b.h.

is assumed to be the growth in height. The same volumes per tree for each d.b.h. class will thus be used for the future stand as applied to the present stand, since they are derived from the same height curve. For single stands the error as usual increases as the area diminishes. Since tolerant trees in all-aged forests do not follow the laws of growth applicable to trees in even-aged stands, the method has greater justification for such forests. It is universally adopted for regional forest surveys covering very large areas, simply because of convenience and reduced cost and the assumption that the results are reasonably accurate.

256. Prediction of Growth in Volume, Based on Growth in D.B.H., Height, and Use of Volume Tables—Influence of Changing Form Factors. In the construction of conventional volume tables, trees in each d.b.h. and height class are averaged regardless of the variation in form of the individual trees (Sec. 149). The form of all trees tends to change with age, the trees becoming more full-boled (Sec. 81). Dominant trees maintain a lower form than codominant and suppressed trees, although showing the same trend toward fuller form. For any one even-aged stand, the use of the same volume table uncorrected for changing future form, will tend progressively to underestimate the actual growth in volume for a future period. The only way of correcting for this error is by determining *stand form classes* for both the younger and the older or future stand and then applying the appropriate form class volume tables. Even this is an approximation. Where stands of different average sizes are combined in a single stand table for a large area, and growth is based on d.b.h., the average volume, taken from a standard volume table for each d.b.h. class, should reflect the effect of average changes in form as well as in diameter and height.

Once the future stand table, corrected for mortality, and the future curve of height on diameter have been prepared, the future volume of the stand is computed by the same simple method employed for computing present volumes in timber estimating. The number of trees per acre in the future stand table for each d.b.h. class is multiplied by the future volume of a tree of this class. The total volume is obtained by multiplying the volume per acre by the total area covered, to obtain the future stand per acre. Periodic growth in volume for the period is the difference between present and future volumes; average growth per year is this total divided by the years in the period. Future growth percentage in simple interest equals the periodic annual growth divided by the present volume. Expressed as compound interest, it is the rate by which the present volume increases to the future volume as shown by compound interest tables.

An illustration can now be given of the use of some of these methods of predicting periodic annual growth based on present diameter classes, allowance for bark, and projected future growth in d.b.h., height, and volume. In the basic data for a yield table for unthinned old field loblolly pine in Louisiana, the oldest stand measured was 80 years. Since these 80-year-old stands were to be cut and an estimate of volume at 90 years was desired, a prediction was made by two alternative methods, both based on projecting past growth for three 5-year periods to obtain future growth in d.b.h. outside bark for the next 10 years. Future height based on diameter was predicted from past plot records of diameter-height relations, plotted for diameter classes, and projected 10 years. It could have been found from felled trees, by similar projection (method 2 above). The same volume table was used, expressed in board feet by International $\frac{1}{4}$-in. rule for each age.[1]

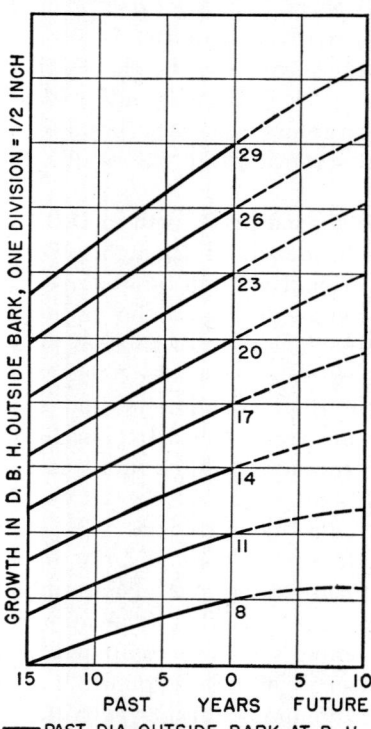

FIG. 81. Predicted growth in d.b.h. outside bark by 3-inch diameter classes for 9 acres of 80-year-old, natural fully stocked old field stands of loblolly pine, Louisiana.

The prediction of diameter growth at d.b.h., including bark, is shown in Fig. 81 for 3-in. classes. As usual, the projection for even as short a period as 10 years involves judgment and may be in error.

Table 64 shows the prediction of height growth by 3-in.-diameter classes based on heights now and 15 years ago, for each d.b.h. class.

The heights 10 years in the future were obtained by projection from curves of height on diameter for 15, 10, and 5 years in the past, for which only the data from the 15-year past curve are given in the table.

[1] *U.S. Dept. Agric. Misc. Pub.* 50, Table 5 reduced from $\frac{1}{8}$-in. rule by a factor of 0.905.

TABLE 63. PAST, PRESENT, AND FUTURE DIAMETER O.B. AT B.H. BY DIAMETER CLASSES FOR 9 ACRES OF 80-YEAR, OLD FIELD LOBLOLLY PINE IN NATURAL FULLY STOCKED STANDS, LOUISIANA*

15 years ago, in.	Present, in.	10 years in future, in.
7.5	8	8.1
10.4	11	11.2
13.3	14	14.3
16.2	17	17.4
19.1	20	20.5
22.1	23	23.6
25.0	26	26.6
27.8	29	29.7

* Derived from curved averages shown in Fig. 81. Each third 1-in. class shown.

TABLE 64. PAST, PRESENT, AND FUTURE HEIGHT FOR 9 ACRES OF 80-YEAR, OLD FIELD LOBLOLLY PINE IN NATURAL FULLY STOCKED STANDS, LOUISIANA, BASED ON D.B.H. AT 80 YEARS

D.b.h., in.	Height, 15 years ago, ft.	Height, present, ft.	Height 10 years in future, ft.
8	80	85	87
11	92	98	101
14	100.5	105	108
17	106	110	113
20	111	115	117
23	114.5	118	120
26	118	121	122
29	121	122.5	123.5

Computation of the present volume at 80 years on 9 acres is given in Table 65.

In Tables 66 and 67 the volume before deduction for mortality has been predicted for the loblolly pine stand at 90 years of age. Two methods of prediction are used (Sec. 250). In the first, that of finding future average diameters for each present d.b.h. class (Table 65) the following steps are taken:

1. Obtain these future d.b.h.'s by adding to the class diameter the average growth for the decade expressed to the nearest $\frac{1}{10}$ in.
2. Obtain the heights, corresponding to the future diameters.
3. Interpolate to the nearest foot in height and to the nearest $\frac{1}{10}$ in. of future d.b.h. in the volume table to obtain the volume per tree. This is done in two steps, as is shown below for the present

TABLE 65. PRESENT VOLUME OF OLD FIELD LOBLOLLY PINE 80 YEARS OLD ON 9 ACRES, BOARD FEET BY INTERNATIONAL $\frac{1}{8}$-IN. RULE (FINALLY REDUCED TO $\frac{1}{4}$-IN. RULE BY A FACTOR OF 0.905)

D.b.h., in.	Trees, number	Height, ft.	Volume per tree, $\frac{1}{8}$-in. rule, bd. ft.	Total volume, bd. ft.
7	6	78	34	204
8	9	85	57	513
9	18	90	83	1,494
10	31	94	118	3,658
11	45	98	161	7,245
12	60	101	210	12,600
13	50	103	262	13,100
14	58	105	319	18,502
15	60	107	378	22,680
16	75	108	436	32,700
17	69	110	504	34,776
18	76	112	577	43,852
19	54	114	658	35,532
20	52	115	731	58,012
21	39	116	805	31,395
22	35	117	877	30,695
23	27	118	967	26,109
24	22	119	1,053	23,166
25	12	120	1,134	13,608
26	3	121	1,226	3,678
27	2	121.5	1,316	2,732
28	2	122	1,406	2,812
29	2	122.5	1,496	2,992
30	2	123	1,586	3,172
Total volume by $\frac{1}{8}$-in. rule International....................				405,197
Total volume by $\frac{1}{4}$-in. rule...............................				366,703
Average volume per acre..				40,745

17-in. tree, which in 10 years will be 17.4 in. in diameter and 113 ft. in height. The interpolation lies between 110 and 120 ft. for the 17- and 18-in. classes.

a. Interpolation for future height of 113 ft. of a 17-in. tree:

	Bd. ft.
Volume at 110 ft...	504
Volume at 120 ft...	559
Difference for each foot....................................	5.5
Difference for 3 ft...	16.5
Volume adjusted to 113 ft.................................	520

TABLE 66. FUTURE VOLUME OF OLD FIELD LOBLOLLY PINE, 80 YEARS OLD, ON 9 ACRES, BY METHOD OF INTERPOLATION FOR AVERAGE FUTURE DIAMETERS, NO CORRECTION FOR MORTALITY

Board feet by International ⅛-in. rule

Present d.b.h., in.	D.b.h. 10 years in future, in.	Present stand table, trees	Height 10 years in future, ft.	Volumes per tree adjusted for		Future volume, bd. ft.
				Height	Diameter	
7	7.0	6	80	35	35	210
8	8.1	9	87	58	61	549
9	9.1	18	93	88	92	1,656
10	10.1	31	97	124	129	3,999
11	11.2	45	101	169	179	8,055
12	12.2	60	104	220	231	13,860
13	13.3	50	106	274	291	14,550
14	14.3	58	108	332	349	20,242
15	15.3	60	109.5	390	408	24,480
16	16.4	75	111	451	479	35,925
17	17.4	69	113	521	546	37,674
18	18.4	76	114.5	593	625	47,500
19	19.5	54	116	673	710	38,340
20	20.5	52	117	747	788	40,976
21	21.5	39	118	828	869	33,891
22	22.55	35	119	910	957	33,495
23	23.55	27	120	1,005	1,043	28,161
24	24.55	22	121	1,082	1,143	24,860
25	25.6	12	121.5	1,169	1,220	14,640
26	26.6	3	122	1,254	1,307	3,921
27	27.6	2	122.5	1,342	1,398	2,796
28	28.7	2	123	1,435	1,502	3,004
29	29.7	2	123.5	1,530	1,579	3,194
30	30.7	2	124	1,630	1,700	3,400
Total volume ⅛-in. rule International.................						439,378
Total volume ¼-in. rule (factor 0.905).................						397,674
Average volume per acre on 9 acres....................						44,186

Similarly for the 18-in. tree, giving 593 bd. ft.

    b. Interpolation for future diameter of 17.4 in. 113 ft. high.

	Bd. ft.
Volume of a 17-in. tree.................................	520
Volume of an 18-in. tree................................	593
Difference for each ¹⁄₁₀ in.............................	7.3
Difference for 0.4 in....................................	29
Volume adjusted for height and diameter...............	546

Standard procedure in rounding off figures falling at the dividing point (0.5) between classes is to raise the odd values to the class above and place the even values in the class below as

520.5 is given above as 520, but 521.5 would be 522

In the second choice, method 2, or the stand table projection procedure, the trees in the future are reclassified in the diameter classes into which they would then fall. The assumption is made that equal numbers of trees fall within each $\frac{1}{10}$ in., within a diameter class, all having the same rate of diameter growth. This permits the same pro-

TABLE 67. FUTURE VOLUME OF OLD FIELD LOBLOLLY PINE, BY METHOD OF TREE MOVEMENT. SAME STAND AS IN TABLE 66

D.b.h., in.	D.b.h., 10 years in future, in.	Present stand table, trees	Future stand table			Local volume in 10 years, bd. ft.	Future volume, bd. ft.
			Stationary	Moving up from below	Resultant		
7	7.0	6	6	0	6	35	210
8	8.1	9	8	0	8	58	464
9	9.1	18	16	1	17	88	1,496
10	10.1	31	28	2	30	124	3,720
11	11.2	45	36	5	41	169	6,929
12	12.2	60	48	9	57	220	12,540
13	13.3	50	35	12	47	274	12,878
14	14.3	58	41	15	56	332	18,592
15	15.3	60	45	17	62	390	24,180
16	16.4	75	45	15	60	451	27,060
17	17.4	69	41	30	71	521	36,991
18	18.4	76	46	28	74	593	43,882
19	19.5	54	27	30	57	673	38,361
20	20.5	52	26	27	53	747	39,591
21	21.5	39	19	26	45	828	37,260
22	22.55	35	16	20	36	910	32,760
23	23.55	27	12	19	31	995	30,845
24	24.55	22	10	15	25	1,082	27,050
25	25.6	12	5	12	17	1,169	19,873
26	26.6	2	1	7	8	1,254	10,032
27	27.6	2	1	1	2	1,342	2,684
28	28.7	2	1	1	2	1,435	2,870
29	29.7	2	1	1	2	1,530	3,060
30	30.7	2	1	1	2	1,630	3,260
31				1	1	1,730	1,730

Total volume $\frac{1}{8}$-in. rule International.................... 437,718
Total volume $\frac{1}{4}$-in. rule (factor 0.905).................. 396,135
Average volume per acre.. 44,015
Difference based on Table 66, −0.39 per cent

portion of the numbers of trees in the class to advance into the next class as the growth in diameter bears to the number of tenth inches in the d.b.h. class, in this case, 10 for 1-in. classes (Sec. 250). Thus, for the class width of 1 in. and for the 17-in. trees, growing 0.4 in., four-tenths of the present stand of 69 trees, or 28 trees, will move up to 18 in., and the remainder, 41 trees, will stay in the 17-in. class. When this method is used, two advantages accrue. First, the future stand is expressed in a stand table similar to that of the present stand; and, second, only the first of the two interpolations, that of the nearest foot of height, is required; hence the local volume table is the one shown under "adjustment for height" in Table 67.

257. Correction for Mortality. The correction for mortality in this sample must now be made. It has an excellent basis. Practically none of the trees below 7 in. grew fast enough to enter this class, and in fact 19 out of 33 died; no ingrowth had occurred. A comparison of the stand tables of 1939 and 1947 showed a loss of 69 trees in 7 years on 9 acres. In the tally for 1947, 65 dead trees were tallied by diameter. The four remaining lost trees were added to the 7-in. and 8-in. classes. By computing the volume lost for each diameter class, the total mortality loss was 9,834 bd. ft. for 8 years. This is past mortality but is computed as if each tree had lasted to the final year. But mortality in the future is sought. For this old stand, mortality will tend to increase rather than decrease, but since the stand is still healthy, most of the future losses will be suppressed trees, except for the unpredictable losses from wind, bark borers, fire, etc.

Mortality is then predicted for the past 10 years on 9 acres as follows:

	Bd. ft.
Average past mortality, 8 years	9,834
Average past mortality, annual	1,229
Average past mortality, per acre	136.5
Margin for increase, 13.5 ft. per acre	150
Total deduction per acre, 10 years	1,500
Gross yield per acre in 10 years	44,186
Net yield per acre, 44,186 − 1,500	42,686

A further deduction will eventually be necessary to take care of losses from red rot, which has begun to destroy the merchantable contents of the timber. This deduction may finally be obtained by scaling the logs when the stand is cut. It is greater for stands that have never been thinned and have consequently grown slowly in diameter, so that the knots from dead branches have not healed quickly, than for stands with normal diameter growth.

A check on the prediction of growth for 10 years by this method was obtained by computing the site class of each of the nine stands, using the stand data for the dominant trees to obtain average basal area, average d.b.h., and the average height of the dominant stand from the height curves. This gave 112 ft. at 80 years corresponding to a site class of 97.5 ft. at 50 years. Interpolating in the yield table for old field loblolly pine[1] for 80 years, a value of 37,925 bd. ft. by International ¼-in. rule is shown as against 40,745 bd. ft. on these plots. Extending the curves of this table to 90 years and interpolating, the indicated yield is 39,825 bd. ft., as against 42,686 bd. ft. in the above example. For the 80-year-old stand the yield is 107.4 per cent of the table. For the predicted 90-year-old stand, it is 107.2 per cent. By contrast, a set of early yield tables for southern pine issued by a public agency, in which original data on the older age classes were lacking, would give a yield at 80 years for this site class as 47,625 bd. ft. or difference of +16.9 per cent, while the prediction for 90 years would be 53.395 bd. ft. or difference of +25.1 per cent.

258. Method of Comparison of Present with Past Growth. As shown (Secs. 250 and 251) the most reliable method of predicting growth in diameter is based on projecting average past growth on trees of each diameter class into the future for the same trees and class. The method of comparison (3, Sec. 250) does not do this. Instead, it seeks to predict the growth of trees of a given d.b.h. class by determining the growth that trees, formerly of this same d.b.h. class, have made in the past period, and through assuming that by this comparison the future growth of the present trees of this class can be predicted.

This method is subject to large errors when applied to even-aged stands, or even to entire forests composed of such stands and of intolerant trees, where diameter growth tends to slow down and the larger trees in individual stands have the faster growth rates. Nor is it reliable for single stands of small extent anywhere. But many growth studies deal with large areas, where the entire forest (type) is the unit, and all age classes are present and overlap. When in addition the forest may be composed of mixtures of tolerant species and the stands tend to be many-aged, it may be found that for the whole area the growth rate is associated more closely with diameter than with age. With tolerant species, such as hemlock and some hardwoods, the growth tends to increase as the trees become larger and secure more light. Under these conditions the method of basing diameter growth, by comparison, on the past growth shown by trees which then had the same diameters may more closely reflect actual conditions than

[1] *Yale Forest School Bul.* 51.

PREDICTING GROWTH BY THE STAND TABLE METHOD 451

other methods, although even in this case projected curves of past growth may show *acceleration* and their use may be equally effective.

A simple method of referring the past growth of trees of present diameter, as measured on radii, to the past d.b.h. outside bark is shown for hemlock in Connecticut, for which it was found that growth in bark thickness was 11.1 per cent of growth of wood for all diameters. The steps are:

1. Plot the average growth at b.h. outside bark over present d.b.h., and draw a growth curve as in Fig. 82. If it were now wrongly

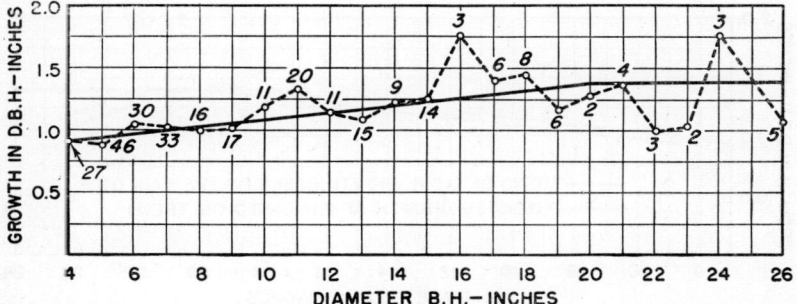

FIG. 82. Curve of growth on d.b.h., for 10-year period, hemlock, Connecticut.

assumed that this past growth could be projected as a straight line into the future, the data on past growth would be tabulated as indicating future growth for each present d.b.h. class. It is obvious from the form of the curve that growth increases with diameter until maturity of size is reached at 20 in. and over, and hence it would be wrong to assume that past growth of a given diameter class would be equal to its future growth.

2. The prediction must therefore be made by referring the growth back to the *past* sizes of the trees on which it was laid. Since the period was 10 years and the growth shown on the ordinates gives this deduction from d.b.h. for each diameter class, the curve shown in Fig. 82 can be directly used as the basis for a new curve based on these past diameters. The points on this new curve are obtained as follows: For the present 12-in. tree the past growth was 1.15 in. The tree 10 years ago was then 10.85 in. in d.b.h. On the ordinate of 12 − 1.15, or 10.85, in. a point is plotted at 1.15 in. Points corresponding to each present diameter class are thus plotted and are connected by a new curve. The dots in Fig. 83 indicate the points plotted from the original curve (Fig. 82) for all the diameter classes as an example of the process. The rate of diameter growth for each present

diameter is read from the modified curve. On such a rising curve as this, it is seen that the *prediction* of growth for 10 years for each diameter class will be larger if taken from the derived curve; *viz.*, for 10-in. trees, the original curve predicts a growth of 1.11 in. and the derived curve, 1.14 in. In the same manner it is seen that, where no upward or downward trend occurs between diameters, the two curves are identical. To demonstrate the effect of a falling curve the values between 21 and 26 in. in Fig. 82 have been arbitrarily lowered in Fig. 83. The derived curve as indicated then predicts less diameter

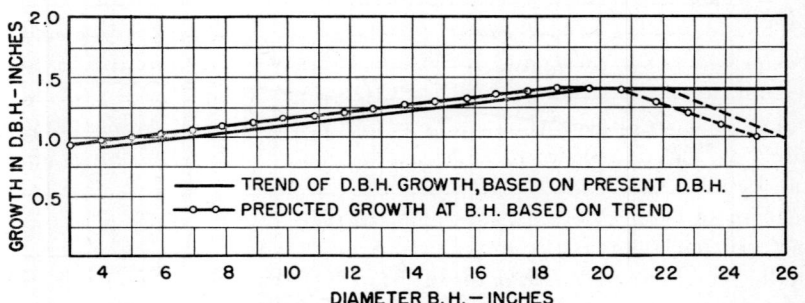

FIG. 83. Method of predicting growth at b.h. by comparison with growth for last 10 years on past d.b.h. classes.

growth than the original curve for these diameter classes; *viz.*, the growth of the 24-in. tree from the original curve is 1.20 in., and from the derived curve, 1.10 in.

Since this modification of the method of predicting diameter growth is identical with the original method in case no change in rate occurs, and when such change does occur it predicts the indicated changes, the modified curve and method should always be employed when growth is predicted by diameter classes by the method of comparison.

Since the method of calculating present and future stand volume is identical to the case of loblolly pine in Tables 65 to 67, once the diameter growth rate is determined, this example of hemlock need not be followed further.

259. Measuring Directly the Effect of Release Cutting. The methods of predicting diameter growth so far described have been based on the premise that the stands have not been partly cut or disturbed in any way, or that the acceleration due to release by partial cutting can be averaged in with the growth on the trees which have not been affected. Whenever conditions affecting future growth on the tract to be studied are now or will soon be altered radically by cutting a part of the stand, which will change the growth rate on at least some of the remaining trees, it follows that no measurements or averages of growth obtained from the forest before these changes occurred can be used to predict what will happen in the immediate future. Ultimately after perhaps 10 to 30 years or more (depending on the character of the present cut) the conditions of crown and root competition and the resultant slowing down of diameter growth will again cause the average growth of the trees to approach the rate before the stand was cut over. Meanwhile the important phase of growth lies in the next 10, 20, or 30 years.

The chief cause of a change in growth rate consists in recovery and acceleration of growth due to release by removal of dominant trees by blowdowns or cutting. Where cutting is selective with the intent to leave the smaller thrifty trees, removing the mature timber, release may be very extensive and affect a high percentage of the residual stand. The extent to which trees are released is determined by the location of each tree with reference to its competitors. "Clear cutting" in early days was not so close as it became later. Scattered small trees below saw-log size left in the South, being relieved of all competition, displayed unusually fast rates of growth, in sharp contrast to their often stunted condition in the original stand. Where "seed trees," which are in fact old but retarded trees, are left, growth in diameter jumps ahead while height growth does not increase proportionately.

Confronted by such changes, predictions based on past growth of uncut stands would seriously underestimate the growth on the residual stand after logging. Yet future growth must not be merely guessed at. The basis of prediction must be actual measurements. Under these circumstances, two courses are possible. First, in the absence of old cuttings of similar character, an effort should be made to find trees in the forest which have been released by accident or which have grown free from excessive competition and under circumstances resembling those left on the cutover areas and to determine their growth rate. The second and more reliable choice, if available, is to make the measurements on past cuttings, if any can be found that are 5, 10, or more years old.

In either of these cases it is safe to expect little if any subsequent slowing down in diameter growth for from 5 to 10 years in the future, provided that the greater portion of the remaining stand is released. For this reason, the method of comparison is preferred over that of curved projection. The growth rate would be found on trees measured on past cuttings, some of which have been growing at accelerated rates. It can then be applied to trees in stands recently cut whose past growth, *before* the stand was cut, was much slower and is no longer applicable. It is intended to predict, as usual, the *future* growth of these trees in turn. Since only a portion of the trees in a selectively cut stand may be released and show accelerated diameter growth, the growth rate applied to obtain future growth of a given stand or forest must be computed as a composite or proportion of the rates shown on released trees and those retained in the portion of the stand not benefited by release. This can be done by separating these two classes of trees in a sample stand table, taken for this purpose, on recently cutover lands. The composite rate for each d.b.h. class is obtained from a weighted percentage of diameter growth for trees released and not released in each diameter class. The method of weighting the average growth of released trees and those not released is shown in Table 68 taken from the first published systematic study of tree growth in this country, "The Adirondack Spruce," by Gifford Pinchot in 1898. Although taken on stumps and not corrected for growth of bark, the arrangement of the data is similar to that now employed in such studies.

TABLE 68. DIAMETER GROWTH AT STUMP, BEFORE AND AFTER SELECTION CUTTING, ON 115 RED SPRUCE, SANTA CLARA, N.Y.

Stump diameter, etc., in.	All trees, number	5 years before cutting, in.	5 years after cutting, in.	Trees released, number	Periodic annual growth, in.	Trees not released, number	Periodic annual growth, in.
(1)	(2)	(3)	(4)	(5)	(6)	(7)	(8)
6	11	0.085	0.150	6	0.205	5	0.084
7	24	0.085	0.130	9	0.190	15	0.094
8	18	0.115	0.150	7	0.180	11	0.131
9	21	0.125	0.185	12	0.240	9	0.112
10	28	0.145	0.170	9	0.215	19	0.149
11	11	0.130	0.140	2	0.210	9	0.124
12	2	0.175	0.255	1	0.280	1	0.230
Totals............	115	.....	.....	46	.....	69	.....
Weighted averages...	...	0.118	0.158	..	0.211	..	0.123

The stand had been cut 5 years previous to measurement. The table shows one of the plots taken. Columns 3 and 4 give the average growth before and after cutting, for the entire residual stand. In column 6 is shown the growth of that portion of the stand which was released by cutting, and in column 8 that which showed no release.

Growth on the stand would be based on the average diameter growth shown in column 4 reduced to b.h., o.b., average increase in height for each diameter class, and the local volumes tables for present and future diameters and heights.

References

BARNES, C. H.: The Development of Uneven-aged Stands of Engelmann Spruce and Probable Development of Residual Stands after Logging, *Forestry Chron.*, **13**, 417–457 (1937).

BARRETT, L. I., Increased Growth of Longleaf Pine Left after Logging, *Southern Lumberman*, **135** (1751), 39–40 (1929).

BUELL, J. H.: The Prediction of Growth in Uneven-aged Stands on the Basis of Diameter Distribution, *Duke Univ. School Forestry Bul.* 11, 1945, 70 pp.

DAVIS, V. B.: Growth in a Selectively Logged Stand in Louisiana Bottomland Hardwoods, *Jour. Forestry*, **33**, 610–615 (1935).

DEEN, J. L.: A Survival Table for Even-aged Stands of Northern White Pine, *Jour. Forestry*, **31**, 42–44 (1933).

GALLAHER, W. H.: Second Growth Yellow Pine, *Forestry Quart.*, **11**, 531–536 (1913).

GARVER, R. D., and R. MILLER: Selective Logging in the Shortleaf and Loblolly Pine Forests of the Gulf States Region, *U.S. Dept. Agric. Tech. Bul.* 375, 1933.

KRAUCH, H.: The Determination of Increment in Cut-over Stands of Western Yellow Pines in Arizona, *Jour. Agric. Res.*, **32**, 501–541 (1928).

———: The Determination of Increment in Cut-over Stands of Western Yellow Pine in the Southwest, *Jour. Forestry*, **28**, 978–986 (1930).

LUTZ, H. J.: Increased Growth of Released Hemlock, *Jour. Forestry*, **26**, 1047–1049 (1928).

MACKINNEY, A. L.: Increase in Growth of Loblolly Pines Left after Partial Cutting, *Jour. Agric. Res.*, **47**, 807–821 (1938).

MARSHALL, R.: The Growth of Hemlock before and after Release from Suppression, *Harvard Forest Bul.* 11, 1927, 43 pp.

MERRILL, P. H., and R. C. HAWLEY: Hemlock, Its Place in the Silviculture of the Southern New England Forest, *Yale Forest School Bul.* 12, 1924, 68 pp.

MEYER, H. A.: A Simplified Increment Determination on the Basis of Stand Tables, *Jour. Forestry*, **33**, 799–806 (1935).

MEYER, W. H.: Growth in Selectively Cut Ponderosa Pine Forests of the Pacific Northwest, *U.S. Dept. Agric. Tech. Bul.* 407, 1934, 64 pp.

———: A Method of Constructing Growth Tables for Selectively Cut Stands of Western Yellow Pine, *Jour. Forestry*, **28**, 1076–1084 (1930).

PEARSON, G. A., and A. D. FOLWEILER: Acceleration of Growth on Western Yellow Pine after Cutting, *Jour. Forestry*, **25**, 981–988 (1927).

STETSON I. G.: Some Suggestions on Predicting Growth for Short Periods, *Forestry Quart.*, **8**, 326–331 (1910).

THOMSON, R. B.: Methods of Determining Increment over Large Areas, *Jour. Forestry*, **30**, 47–57 (1932).

WEIDMAN, R. H.: A Study of Windfall Loss of Western Yellow Pine in Selection Cuttings 15 to 30 Years Old, *Jour. Forestry*, **18**, 616–620 (1920).

CHAPTER 31

GROWTH PERCENTAGE

260. Definition and Character of Growth Percentage. The growth of a tree has been compared with that of a money capital invested at interest, in which no withdrawal of a part is possible until the entire sum has remained for a certain period, when all must be taken at once. In a forest composed of stands or trees of all ages, this is not true, and it is possible to take from the area each year an amount which may be equal to the "interest" earned or may be greater or less than that sum. If greater, the capital shrinks in amount; if less, it increases. Withdrawals are also made by natural processes, or "losses," independent of the owner's will but subject to a certain amount of salvage if the capital is capable of being converted into cash, as in the case of mature trees killed by fire.

The normal condition of a forest capital is that permitting the greatest percentage of annual returns, not for any short period but for the entire rotation. This requires a considerable stock of wood of various ages. Forest areas composed of cutover or bare land have suffered an almost complete withdrawal of wood capital and may even have lost their natural productiveness as an investment by inability to reproduce valuable species of trees. The outstanding difference between a depleted forest capital and similar depletion in most other lines of business is that it is impossible to accelerate the rate of restoration or to shorten the period required to build up this capital to the normal state again, beyond the limits of time imposed by the natural processes of tree production and growth as speeded up by thinnings and prompt reforestation.

In the case of sums of money that are allowed to accumulate in this manner, interest is computed at a compound rate. It is always possible, by dividing the total interest earned on an investment by the number of the years in the period, to express these earnings in terms of simple interest, but the rate thus shown is of course higher than that representing an accumulation of compound interest in which the interest (tree growth) is each year put to work to earn new interest.

In expressing growth of trees or stands it is necessary to keep these relations in mind. At the same time, growth percentage refers primarily to the volume of wood laid on to the wood capital, leaving financial calculations of investment in land, cost of production, and value to the subject of forest valuation. These additional considerations of finance bring in as an indispensable factor the increase in value of each unit of volume with increasing size and quality of the timber. They deal also with market prices and with the changing value of the dollar, factors which may greatly increase the percentage being laid on by a stand over that shown by increase in cubic volume alone. Growth percentage varies with the unit of product used in the measurement, since the increase in the board foot–cubic foot ratio with larger diameters is reflected in this percentage relation. The combined and total growth percentage of a stand is therefore the sum of the percentages represented by, (a) volume in terms of specific products, (b) quality (and relative increase in quantity) of product, and (c) the price and value of the product in terms of dollars, or

$$1 + (a + b + c)$$

For a given period, growth percentage in volume is found by comparing the volume of the tree or of the stand in the initial year with its final volume at the end of the period. After these two figures are determined, growth percentage may be computed either as simple interest by dividing the average annual net growth for the period by the initial volume, or as compound interest by finding for the entire period the ratio of increase without reference to growth trend during this period, looking up the equivalent of this ratio in a table for compound interest rates. The choice of either of these methods is optional. For a forest in which the yield is harvested annually, the current percentage for the year, *i.e.*, simple interest, is actually laid on each year on the forest as a whole. For separate stands, the increase is in terms of the curve of current annual growth. For a normal forest in which cutting can approximate actual growth, current growth and its percentage will also approximate the annual cut, unless the growing stock should be built up to a higher state of productiveness, which is usually the case. In a forest or stand which is immature and in which current growth exceeds the average annual growth attained for the rotation, there is some advantage in computing growth percentage on a compound interest basis.

In no case should it be assumed that a curve of compound interest percentage will represent the curve of actual annual growth. This would be true only of immature stands before the period of entry into merchantable size classes, and this period is not usually measured for current growth.

261. Base of Growth Percentage. A growth percentage indicates the growth for a single year, but it is rarely possible to compute a growth rate from values established at the beginning and end of the same year, since mensurational techniques do not allow a sufficiently high degree of accuracy for such a short period and annual growth fluctuates with rainfall and average temperature. It is necessary and customary to use a longer period, commonly 10 years, and to compute the average annual rate by dividing the periodic growth by the number of years in the period.

If diameter growth alone is considered, the rate of increase may be fairly constant for a decade. This rate may be used to compute the current growth percentage on the tree diameter at the beginning of the present year, without too great an error. When volume growth for a period is determined, this tends to increase annually on trees growing in height and with uniform diameter growth, since growth in volume varies as $\pi D^2/4 \times$ height $\times$ form factor. In stands, growth is complicated by the losses occurring, which require deduction of the entire volume of the trees lost as an offset against the growth on the surviving trees alone (Sec. 247).

The first outstanding fact is that growth percentage is a relation between two quantities, *viz.*, growth in volume and the initial volume on which this growth is laid. To determine the percentage, both these quantitites must first be measured. Percentage is a relation or ratio, not a quantity, and it is dangerous to substitute percentage as a means of predicting growth in volume. It changes with each change in either the volume laid on or the wood capital. As growth occurs, the capital becomes constantly larger. The increase in the size of the capital would in turn require a larger growth, if the rate or percentage is to be maintained.

Actually, in the development of either trees or stands, the rate of annual volume growth never increases rapidly enough to maintain the rate of current growth percentage. In the very early life of a tree, growth percentage is at its highest. It constantly falls until at maturity it drops below the current rate of interest even on a financial basis. The growth percentage of stands shows the same tendency even more pronounced, by reason of the loss of trees. It stands to reason, therefore, that the current growth percentage of a tree or stand should be the growth for a single year, the current year, applied to the volume of the stand at the beginning of said year.

262. Methods of Determining Growth Percentage. The utility of growth percentage is primarily to determine, for timber which is fairly mature, whether or not the tree is increasing at a satisfactory rate. For young timber, the growth percentage is known to be high and is without significance, because the wood capital is too inadequate for a satisfactory basis.

In mature trees, height growth has decreased or has practically ceased, and changes in form are inconsiderable. Diameter growth is therefore the most significant phenomenon.

For this purpose the formula derived by Prof. Schneider of Eberswalde, Germany, is the standard method.

Let n = number of rings in the last inch of radius
D = d.b.h. outside bark
P = current growth percentage

Then

$$P = \frac{400}{nD}$$

The formula is based on the theory that the average rate of diameter growth for the last inch will hold good for both the past year and the year to come, hence is the most equitable basis for this ratio, or percentage. Height and form factor are regarded as constant for the two years. Diameter at the beginning of the past year is $D - (2/n)$, and at the end of the current year is $D + (2/n)$; viz., a 10-in. tree contains 10 rings in the last inch of radius; hence $n = 10$. Its growth in the past year in diameter is therefore 2 in. divided by 10 years, or 0.2 in., and its diameter 1 year ago is 9.8 in.

Changes in volume are in this formula correlated with basal area alone, which is

For past year,

$$\frac{\pi}{4}\left(D - \frac{2}{n}\right)^2$$

$$\text{Growth} = \frac{\pi}{4}D^2 - \frac{\pi}{4}\left(D - \frac{2}{n}\right)^2$$

$$= \frac{\pi}{4}\left(\frac{4D}{n} - \frac{4}{n^2}\right)$$

Growth percentage P is found by the proportion

$$\frac{P}{100} = \frac{\pi/4[(4D/n) - (4/n^2)]}{\pi D^2/4}$$

$$P = \frac{400}{nD} - \frac{400}{n^2 D^2}$$

For the current year, the above derivation is started with $D + (2/n)$, giving

$$P = \frac{400}{nD} + \frac{400}{n^2 D^2}$$

The average of the two percentages is

$$P = \frac{400}{nD}$$

The results of this formula as applied to d.b.h. outside bark are conservative, first, because the diameter outside bark is greater than that inside bark; second, because the growth of bark is ignored and this alone may make 10 per cent of difference in growth; and, finally, because the tree may still be growing in height. It also ignores increase in form factor, quality, and price. These facts do not diminish its usefulness as a quick indicator of relative increase in volume as between different trees and different stands, to permit of intelligent selection of those which should receive priority in cutting, when other factors are equal.

263. Current versus Mean Annual Growth Percentage. Mean annual growth percentage is a relation of simple interest, in which the period is the life of an even-aged stand, the capital is the total volume produced, and the interest is the average, or mean, annual growth. The percentage is this mean annual growth divided by the total volume.

True current growth percentage, based on the division of the growth of the year by the total volume at its beginning, and mean annual growth percentage become equal in the year that the current growth falls to the level of the maximum average, or mean (Sec. 195). Neither involves compound interest. Previous to this year, the percentage of current annual increase exceeds this mean, and the stand should obviously be held for growth. After this year the reverse will be true if maximum average volume growth is desired for the stand.

264. Periodic Growth Percentages. When the principles of simple interest are adhered to, as in the above cases, the growth percentage is always based on the volume and growth of the present year. But volume growth is measured in periods of 10 years. For trees, the period is the past 10 years, though this growth may, by methods set forth in Sec. 250, be applied to the future decade. For stands, the past growth is of little moment, and the growth of the future decade may be determined either from yield tables or by predicting the volume growth of trees. From either basis, the net volume growth of stands may be found.

Once the growth for the period is known it can be expressed as growth percentage. But if the volume at the beginning of the period is taken as the basis and the total periodic growth divided by the number of years in the period is taken as the annual increment, the resultant percentage is on the basis of simple interest on this minimum initial volume. It will exceed the true average percentage laid on each year on the volume of that year, or the average of the actual current percentage.

The percentage of compound interest is easily found by obtaining the ratio between the volume at the end of a period and that at the beginning and using a table of compound interest. For example, if a stand now has 10,000 bd. ft. per acre but had only 5,000 bd. ft. 10 years ago, hence 500 bd. ft. per acre per year average growth, the ratio is 10,000/5,000 = 2. Compound interest tables show this to be equivalent to an annual rate of 7.2 per cent laid on the original volume of 5,000 bd. ft. In terms of simple interest, the rate is 10 per cent. If it is now assumed that growth will continue at the rate of 500 bd. ft. per acre per year for 10 more years, the ratio will be 1.5, equivalent to a compound rate of slightly more than 4 per cent and a simple rate of 5 per cent. This example serves to show the disparity between the two forms of interest and particularly the rapid change in rate under equal conditions of growth per acre per year.

To avoid the use of compound interest tables, Pressler based a simple rate of interest on the average volume for the period, which had the effect of reducing the rate by increasing the base of comparison. The derivation of the Pressler formula is as follows:

Let V = volume at end of period
v = volume at beginning of period

$$\text{Average volume} = \frac{V + v}{2}$$

$V - v$ = growth for period of n years

$$\frac{V - v}{n} = \text{annual growth for period}$$

$$\text{Growth per cent} = \frac{(V - v)/n}{(V + v)/2} \, 100$$
$$= \frac{V - v}{V + v} \frac{200}{n}$$

Table 69 shows several interesting characteristics of the interest curve. An understanding of these is fundamental to the correct use of growth percentages. Here the condition of a constant diameter

TABLE 69. COMPARISON OF SIMPLE INTEREST, COMPOUND INTEREST, AND PRESSLER GROWTH RATE IN BASAL AREA, UNDER THE CONDITION OF CONSTANT DIAMETER INCREMENT OF 1 IN. PER DECADE

Initial d.b.h., in.	Annual growth rate in basal area		
	Simple interest, %	Compound interest, %	Pressler rate, %
2	12.52	8.48	7.70
4	5.58	4.53	4.36
6	3.62	3.14	3.07
8	2.66	2.39	2.35
10	2.11	1.93	1.91
12	1.74	1.62	1.61
14	1.48	1.39	1.38
16	1.29	1.22	1.21
18	1.14	1.08	1.08
20	1.02	0.97	0.97
22	0.93	0.89	0.89
24	0.85	0.82	0.82
26	0.78	0.76	0.75
28	0.73	0.71	0.70
30	0.68	0.66	0.65

increase of 1 in. per decade has been assumed without regard to the size of the tree, yet all three growth percentages start from relatively high levels in the smallest diameter classes and decrease rapidly at first and then more and more slowly with continued increase in size. Thus it would be a bad mistake to use a simple growth percentage of 12.52 per cent for a 2-in. tree for more than one decade since in the first future decade it would be much smaller and in the second decade it would be only 5.58 per cent. The relationship between simple and compound interest is also shown. Both give the same value at the end of the decade, but at intermediate years, the compound interest calculation would give slightly smaller values than the simple interest calculation. The relationship between compound interest and the Pressler formula shows strikingly good agreement except for the smallest diameter classes, where the Pressler rate tends to be low. This is of additional interest, since the Pressler computation is in effect a simple interest computation, but based upon the mid-period value instead of the initial value.

The shape of the growth percentage curve is fundamental not only to the diameter and basal area relationships but also to cases where growth is related to diameter, age of tree or stand, and volume per acre. The growth percentages for cubic-foot volume tend to run slightly higher than those for basal area under equal conditions of diameter growth, since height increment becomes involved. Those for board-foot volume are substantially greater, especially in the smaller merchantable size classes because of the rapidly increasing ratio between board feet and cubic feet in these trees, plus the height increment. In the extreme case where a submerchantable tree suddenly enters the merchantable class, its growth percentage as calculated for the first future decade would be infinity.

265. Utility of Growth Percentage in Growth Predictions. In the use of either the Pressler rate or the straight compound interest formula, the percentage if calculated for a past period, from yield tables or for trees, applies only to this past period and not to a future period, when it will inevitably be greatly reduced.

Growth percentage is never constant but is continually diminishing. Thus its use as a means of predicting the growth of a tree or stand for any future period greater than 1 year is certain to give exaggerated results unless it has actually been placed on a true future basis with regard to the volume to which the percentage is applied. Current annual growth percentage will not give a true prediction for the next 10 years for a tree or stand, nor will the growth percentage of a past period.

On the other hand, if the growth percentage has been determined on a true future basis it can be applied to stands or trees now younger in age or smaller in diameter by the method of comparison (Sec. 250), and thus it may be a convenient means of computing growth. Being a ratio, it would in theory apply to stands of any total volume, provided that they had the requisite age or diameter or contained about the same average tree and distribution of diameter classes as the sample measured. Applied in this way it has had quite an extensive use; *viz.*, for the stand above cited (Sec. 264), the simple growth percentage on the basis of present volume was found to be 5. A similar stand of the same age now contains only 6,000 bd. ft.; 5 per cent per year simple interest on this amount is 3,000 bd. ft. for the next 10-year period, which is the same as would be obtained by the use of a yield table reduced to 0.6 density.

If, however, the initial calculation of past growth percentage were applied to the same stand for the future, it would greatly exceed the actual growth; *viz.*, if the stand, (Sec. 264) at 15,000 bd. ft. continues to grow at the rate of 5,000 bd. ft. for 10 years, this gives but $3\frac{1}{2}$ per cent per year instead of 5 per cent. If 5 per cent is again applied to 15,000 bd. ft., the growth is predicted at 7,500 bd. ft., an error of 50 per cent.

Growth percentage obtained from trees cannot with safety be applied to stands because of the losses of numbers in the stand as a whole. The growth percentage of average trees, even when properly figured, will therefore be too high for the growth of the stand. When these stand percentages can be determined directly for the stand, either from yield tables or from studies of growth based on trees diameters corrected by tree losses, they may be capable of service.

One illustration of the use of current growth percentage is that of a company in Texas, which adds 10 per cent annually to its initial estimates of young second-growth timber, on the basis that for this specific class of timber the actual increase in merchantable volume roughly approximates this figure as the result of ingrowth plus the high percentage of increase laid on the relatively low merchantable volume of the trees composing such stands. Actual percentages for specific stands may vary widely from this figure, but it saves the cost of annual reestimates and can suffice for a short period as an approximation to prevent undervaluing the stands.

Generalizations of this character are frequently substituted for actual determination and prediction of growth, thus avoiding expensive field and office calculations, but with correspondingly decreased accuracy. Future growth cannot be assumed to equal past growth; and, when to this assumption is added the fundamental error of projecting the same growth percentage for the increased volume on which this percentage will be based, the prediction has no relation to fact.

Both European and American authorities consider growth percentage an unsafe tool with which to predict the growth of stands for a future period. The error involved is always plus, and large in amount, unless it is applied by comparison to stands which are practically identical in diameter distribution and in density, site, and species to those from whose past growth the percentage was derived. In order to determine these facts, the present stand table and volume must be found for the stand in question, and it will practically always be different from that possessed "10 years ago," from which the percentage to be used was derived. Use of growth percentages for growth prediction is like putting the cart before the horse. It can be found accurately only after both present and future volume are determined; hence it should not be substituted as a short cut for finding one of the two basic factors (*i.e.*, future volume) from which it is derived. As a means of expressing the static present relation between growth and volume, for a stand or tree, it can be used effectively to indicate the condition of maturity reached and the advisability of cutting or reserving the tree or stand, but for no other purpose. Growth must be predicted quantitatively in the first instance, using percentage, as usual, to express the relationship between two known, or appraised, quantities. (Future volume is always an appraisal.) The use of the Gerhardt percentage relation for understocked stands is based not on prediction but on comparison of the relative growth percentages of stands of given ages on identical sites.

References

BARRETT, L. I.: Growth in Mixed Hardwood, *Jour. Forestry*, **24,** 574–578 (1926).
BELYEA, H. C.: Current Annual Increment of Red Spruce and Balsam Fir in the Western Adirondacks, *Jour. Forestry*, **20,** 603–605 (1922).
BENTLEY, J., JR., and A. B. RECKNAGEL: Accelerated Growth of Spruce after Cutting in the Adirondacks, *Jour. Forestry*, **15,** 896–898 (1917).
CHANDLER, B. A.: Notes on a Method of Studying Current Growth Percent, *Forestry Quart.*, **14,** 453–460 (1916).
GEVORKIANTZ, S. R.: A New Growth Per Cent Formula, *Jour. Forestry*, **25,** 44–49 (1927).
HANZLIK, E. J.: More about Growth Per Cent, *Jour. Forestry*, **25,** 443–447 (1927).
———: Predicting Future Volumes by the Growth Per Cent, Western Yellow Pine Region, Oregon, *Univ. Wash. Forest Club Quart.*, **6** (2–4), 5–11 (1928).
RUDOLPH, P. O.: A Comparison of Several of the Growth Per Cent Methods of Predicting Growth, *Jour. Forestry*, **28,** 28–33 (1930).

CHAPTER 32

SAMPLE PLOTS

266. Growth Predictions Based on Historical Records—Recurring Inventories. The effect of all causative factors on the growth of trees and the resultant yields of stands of either even or dispersed ages may be determined for specific stands by remeasurement at intervals of time. In this way the effect of deceleration, mortality, thinnings, wet and dry cycles, tolerance, and soil are expressed in their final form. Predictions based on historical records may take the form of comparison of the observed results with expectations for younger stands of similar character. Or they may be attempts to project the future growth of the stands so measured from their past performance.

Historical records of past growth may be obtained in at least two ways. The first is by small permanent sample plots in which the d.b.h. of every tree is recorded to $\frac{1}{10}$ in. every 5 or 10 years and either a complete tally or a sample tally of heights is made to obtain a height curve. The second method is an expansion of the first and is termed the "recurring inventory method." Stand tables, either 100 per cent or obtained by sampling, are taken at intervals, usually of 5 to 10 years, and the past movement of trees from lower to higher diameter classes used to compute the growth for the period. The end results of either method are identical; *viz.*, that the difference in volume at the beginning and end of the period, plus the amount cut, or salvaged from loss by mortality, equals the realized growth of the area for the past period.

Permanent sample plot records, after 10 to 20 years, permit scientific comparison of different methods of predicting period growth, thus enabling the forester to choose and to modify methods for application to present stands for the prediction of future growth. The inclusion in permanent plots of stands not fully stocked and differing in degrees of stocking is as important a procedure for growth prediction as is the measurement of fully stocked stands. These two methods have the advantage that the growth is measured directly outside the bark at b.h., hence avoiding the requirement for computing bark thickness and growth which is needed in increment borings. If all trees on a permanent plot are numbered with metal tags, or preferably by painting, taking care to avoid removal of bark at b.h., any problem of dispersion of growth rates, slowing down, effect of thinnings, relation of crown to growth, relative height growth, and relative growth of competing species, is capable of solution.

Where the recurring inventory method is used, a 100 per cent tally is in effect a permanent sample plot. But if the diameter distribution is obtained from a sample composed of strips or plots, a valid comparison can be obtained only if in the succeeding surveys the identical areas of the original sample are again traversed. When the centers of circular or square plots are permanently marked for reference for the second survey, there is little difference between this work and permanent sample plots. For American conditions, therefore, the benefit of historical records can be best obtained (1) by a system of representative sample plots of different densities, small enough not to constitute too great a burden for initial record or for remeasurement or (2) by the repetition of an over-all survey on large areas, in which the area covered for the stand table can be reduced to the required percentage of accuracy, comparable with the initial survey, and using the same units of measurement, the same merchantable products, stump height, and top diameters, and the same volume table. Corrections in total volume can then be made if necessary to bring the second result into harmony with changed conditions involving closer utilization and different products.

The recurring inventory method is practically identical with the method of prediction by movement of diameter classes (Secs. 250 and 256), except that it has as its basis the actual past happenings in the forest area concerned. It serves only as the usual *past* basis for predicting future increment. What it amounts to is making two successive estimates of the stand on identical areas by the use of stand tables. The basis for prediction of growth is no better nor any worse than that from sample plots except for the wider distribution of the basic sample from which the stand tables are arrived at. Future yields may be greater, less, or consistent with those of the immediate past period according to the condition of the stand, its ages and sizes, and the amount of cuttings. For these reasons, the method of recurring inventory, in itself, is merely a historical record of changes in the diameter distribution of forests or parts of forests, and of consequent changes in volume, and derived past net growth. Its application in predicting future growth in unmanaged many-aged stands, based on size classes, may be less accurate than methods based on actual attempts to derive future diameters by projection or comparison with past growth. As experience accumulates in the recording of yields per acre of these forests under systems of thinning and selective cutting, the future possible yields per acre become better known. On the basis of experience, the future yields can be approximated more closely whatever the method used, since only past growth furnishes a basis for specific attempts at prediction of future diameter distribution and yields. In the study of growth by the method of recurring inventory, plots may be as large as 40 acres, on which the same accuracy of diameter measurement must be observed as for other permanent plots, but in this case the tally can be by diameter classes and the trees need not be numbered.

Problems of diameter growth are not solved merely by stating that trees "grow" at a certain rate. This usually, under straight-line projection, assumes that the past and the future growth are identical, and tables then show past growth only. By distinguishing between actual past growth on the one hand and predicted future growth on the other, and by using the first solely as a means of appraising the amount of the second, an intelligent choice of the proper method for a growth study can be made. Physical laws govern growth, and the results are not controlled purely by mathematics. In previous sections the most important of the physical factors influencing growth have been discussed. Growth predictions have the same importance in forestry as do predictions in the economic field and require an equally thorough grasp of each separate case. They cannot be left to amateurs.

267. Purpose of Sample Plots. A sample plot is an area small enough to permit complete measurement, to an established standard of scientific accuracy, of the vegetation occupying the plot, with notes on environmental factors such as soil and exposure. In forestry, sample plots are especially concerned with trees; in grazing studies, with forage plants. Where sample plots are measured only once, the objective is obtained through comparison of large numbers of plots of different ages and sites as in the construction of yield tables (Chap. 27). If it is certain that they will have no further use, their location needs only temporary markings. Permanent sample plots are those established for the study of physical and ecological changes in the tree vegetation by growth and mortality: the relationship of crown character to volume growth; the effect of different methods and degrees of thinning; cleaning; methods of final cutting and reproduction; different silvicultural systems; effects of prescribed fire on success of reproduction and on fire hazard; progressive changes in the forest floor, litter, and humus; changes in soil conditions; in fact, any and all changes brought about by the dynamic forces of nature or man, whose character must be understood and controlled in the direction of greater benefit to man himself, through improvement of the forest for the ends which serve him best.

The two important features by which the measurement of sample plots differs from timber cruising or forest surveys are, first, the standard of accuracy employed and, second, the detailed character of the record. Diameter measurements are standardized at the nearest 0.1 in. (with the b.h. point carefully measured) instead of rounded off into 1-in. or 2-in. classes. Heights are recorded to the nearest foot. Records are made on permanent adequate forms, which for permanent plots may provide for repeat measurements on the same record blank. Permanent sample plots have great advantages over temporary plots in the study of changes in the forest. The only time at which an accurate appraisal and record of conditions on a plot can be made is the present moment. Inferences may be drawn as to past conditions and future changes, but both are obscure and uncertain. The speed with which natural processes alter and conceal past events is astonishing, and only by deduction from fragmentary evidence can the past history be pieced together. This is especially true of mortality in young stands. The small dead trees, either fire-killed or suppressed, are mostly sapwood and fall down within 2 or 3 years. In 6 to 10 years, no trace of them remains except where fragments are lodged in a dry situation as over a log or rock. Unless the changes in composition, growth, reproduction, ground cover, litter, and soil can be based on repeated accurate observations systematically recorded and compared, important and controlling facts may escape notice and wrong conclusions be drawn.

268. Choice and Location. Sample plots are not the solution of all problems facing the forest manager, and their establishment and maintenance are expensive, a fact which can be offset by relatively small size and reduced cost per plot. The choice and location of permanent plots should be the last, not the first, step in a preliminary study of the problems to be investigated. Otherwise, from failure to evaluate an important controlling factor, many years of effort may prove worthless. The preliminary requirement is therefore as intensive a study as possible from general observation, based on a broad and thorough background of experience, in order to grasp the essential factors whose influence should be determined.

With the broad aspects of the problem formulated, two principles are invoked in selection of plots, *viz.*, isolation of specific factors, and contrast. A specific question would be, for instance, the age and size of stands requiring thinning, the degree and class of thinning to be used, and for what objectives or products. For this purpose only fully stocked stands are suitable. The lessons learned here can be applied to stands of lesser density. But for growth, plots are also needed in understocked stands. Contrast is the key to measurement of results of varying treatment, and this is secured by the establishment of *check plots*. Such plots must be as nearly identical as possible with the treated plots, be located in their immediate vicinity, and be allowed to develop without the factor imposed on the treated plot. So important are check plots that the recommended procedure is to establish one check plot for each treated plot, especially in thinnings.

Since uniformity of the conditions to be studied or treated gives the most certain and accurate results, it follows that areas of a size suitable for the establishment of a series of adequate numbers of duplicated treated and check plots are not readily available. Hence the search for and location of the best areas for the experiment are more important than their premature establishment. Any permanent plot will yield results of value, but these may be wholly negative in character.

Once established, the treatment itself must be founded on sound appreciation of the workings of natural forces and their guidance by man. In one instance a 10-year experiment in methods of reproduction of longleaf pine produced only negative results, and the successful natural method was observed only as the result of accidental occurrence of fire on an adjoining area. Plots to be treated must be freed from the influence of untreated areas as far as possible. In thinning, this requires an *isolation strip* surrounding the plot, on which strip the same treatment is carried out. The width of the strip should be sufficient to ensure that adverse influences, shade, and root competition from the untreated stands, cannot affect the measured portion, the permanent plot. In thinning, 25 ft. is usually adequate. For control of brown spot disease in longleaf pine even $\frac{1}{4}$ mile may not be enough to prevent reinfection.

The danger of premature fixation of plans for plot experiments lies in the freezing both of methods and of area involved into a controlled procedure which excludes subsequent modifications by either the lack of suitable additional area or the impossibility of retracing the steps already taken on the plots. The experiment then proceeds in a groove, ending in futility, while other agencies or individuals with a new outlook and start, or merely by close observation over a period of years, arrive at sound conclusions. Adequately planned and amenable to changes, there is no real substitute for permanent plot records for determining quantitatively and scientifically the results of the operation of natural and of controlled environmental factors on the production of forest crops.

Although the cost of far-flung programs of plot research may run into six figures, the method of accurate observations at successive intervals of time on the same area is open to every forester. If consistently and carefully pursued, and reinforced by photographs taken from identical marked points and by permanent plot records, no matter how few or small the plots, the forester or landowner is on the way to becoming a practical silviculturist with a sure grasp of crop problems and methods of management. The ability to observe contrast in existing plots, and changes at successive periods, is what distinguishes the successful forest manager from the mere office administrator. On such men the future productiveness of our forests will depend. An attitude fatal to the procurement of reliable results from permanent plots is to start with a preconceived postulate and then to endeavor to prove that this thesis is correct. This attitude introduces bias both in the planning of the experiment and in interpreting the results. It leads, and has led, to deliberate efforts to suppress "unfavorable" results, or to explain them away, instead of exposing them to scientific analysis in the effort, fundamental to all honest investigators, to ascertain the true trends of natural forces in order to work with and not against them. Nature makes no attempt to conceal or camouflage her operations and will defeat all efforts by investigators to divert or disguise fundamental facts.

269. Size and Number. Size of the plot usually varies directly with the size of the tree and the age of the stand under investigation. Number of plots often has the inverse relation, owing principally to the ease of measuring small plots as compared with measuring large plots. Statistical methods place great emphasis on the repetition of plots, since measurements made on a single plot or comparisons drawn from a single contrasting pair cannot be considered indicative of a true development. Only when an event occurs time and again on different plots does the result become significant. The number of plots required can be determined statistically, but only if some advance knowledge of variation in behavior is available, either through experience in similar fields of research or through past experience on the series of plots in question. While small plots have the advantage of ease in measurement and, hence, of replication, as well as of uniformity within the plot, they exaggerate small differences in environmental conditions, which should better be averaged out in favor of major environmental factors by the use of larger plots.

The mil-acre, 6.6 ft. square, is the accepted unit of area in establishing sample plots. The plot may vary in size from 1 to 1,000 mil-acres, depending upon the objective. Reproduction plots and plots established for the study of surface vegetation may contain but a single mil-acre, sometimes only one-fourth of a mil-acre. On the other hand, sample plots established for the study of the rate of growth and volume yield over intervals of time may contain a full acre (1,000 mil-acres) or in special cases 5 or even 10 acres.

270. Establishment of Plots. Temporary sample plots can be established without regard to the status of land ownership. Permanent sample plots must be under a stable and permanent ownership, sympathetic with the purposes of the enterprise; otherwise the owner during a period of years will almost certainly cause damage or loss of results by overt acts such as grazing, cutting of timber, or even carelessness with fire. Sample plots are best established on public or institutional forests. Best results are obtained when the same agency which is conducting the research owns the land.

Rectangular permanent sample plots should be clearly marked at the corners with indestructible posts, preferably of iron pipe. The location of all permanent plots must be placed on a map of the forest area, with their boundaries, as well as corners, marked on the ground. Otherwise, sooner or later, woods operations conducted by persons not familiar with the research program will be apt to result in the cutting of timber on these plots, ruining many years of work. Plots are usually rectangular, preferably square, to facilitate relocation in later years and the assembling of field data. The bearing of the boundary lines should be shown on a large-scale map of the project, also the bearing and distance of a tie line to some permanent survey corner or other monument.

It has frequently happened that plots originally considered temporary, but which were marked and located, have since proved of great value for remeasurement. The boundaries of all rectangular plots, temporary or permanent, can best be marked by blazing each tree outside of the boundaries, the blaze facing the boundary and marked by a horizontal black crayon line. These marks will endure for many years, and the scar is permanently indentifiable. Another method is to spot all trees along the boundary, inside the plot, with white paint.

Recently a method of locating permanently large numbers of circular plots $\frac{1}{5}$ acre in size has been advocated by Stott[1] as a means of acquiring rapidly the necessary data on the growth of stands of different degrees of stocking, age, and species. In this case the center of the plot is permanently marked by painting a tree or setting a stake. This is the only permanent corner needed. The radius of a $\frac{1}{5}$-acre plot is 52.6 ft., and all border trees are located by a tape to determine whether more than one-half of the bole lies within the plot. The increasing recognition of the need for data on growth of understocked stands and the relative speed and cheapness of the method give promise of its wide adoption. Plots are located on definite courses and at fixed intervals, which permit relocation and remeasurement.

271. Identification of Trees. Each tree on a permanent plot above a specified d.b.h. (usually 1.5 in.) is identified by a serial number. These numbers may be painted on the bark, but if bark is smoothed off for this purpose there is danger of diminishing the d.b.h.; hence the numbers must be placed well above or below this height. A good quality of paint is necessary, and the number must be renewed at 5-year intervals. Where the plot is safe from vandalism by removal of numbers, metal tags, preferably strips, numbered serially for each plot, may be attached at b.h. by a copper nail, No. 11 gauge, $2\frac{1}{2}$ in. long. The exact position of b.h. from the ground is permanently marked by the nail or by a spot of paint so that remeasurement will be made at the original point. On rapidly growing trees the nails may become embedded and should be pulled and reset at 5-year intervals with a claw hammer. On smooth-barked species, such as spruce, balsam fir, or beech, a rubber stamp may be used for numbering trees. Rubber stamps are made for this purpose with figures about 2 in. high by $1\frac{1}{2}$ in. in width. Such stamps are used exclusively in some European countries for numbering trees on permanent plots.

On plots established for the study of reproduction at intervals over a period of time, each seedlings is best identified by means of a metal tag attached to the seedling by a fine copper wire wound loosely around the stem. Where no disturbance is anticipated, a marker of stiff wire bearing a numbered tag may be thrust into the soil beside the seedling. For short-time studies, small gardener's stakes may be used. It is not always necessary to number *all* the trees on a plot. In thinning

[1] STOTT, C. B., Permanent Growth and Mortality Plots in Half the Time, *Jour. Forestry*, **45**, 669–673 (1947).

experiments in fully stocked stands, most of the trees falling originally in the unmerchantable suppressed classes will die, both in unthinned stands and in those properly thinned (but not too heavily). These trees are not tagged, but a complete tally by d.b.h. classes should be made in each measurement, to determine their behavior, fate, and any influence which they may have on the stand. This is especially needed where an understory, such as hardwoods under pine, is permitted to develop, which may interfere with moisture supply and growth of the pine overstory. This effect can be measured by periodically removing the hardwood on some of the control and check plots by either cutting or fire.

272. Measurement of Trees. The d.b.h.'s of all trees, from 1.6 in. up, are recorded by species. The numbered trees are recorded to the nearest 0.1 in., and the unnumbered trees in their respective inch classes. Either the diameter tape or the calipers may be used. With the former, consecutive measurements of individual trees are more consistent but are apt to include a small plus error. With calipers, the individual errors are greater but are compensating for the plot as a whole. Repeat measurements should always be made with the same type of instrument as originally used. Trees may be separated in the tally on the basis of crown class as dominant, codominant, intermediate, and suppressed. If required in the investigation, the age of the stand may be determined by increment borings supplemented by seedlings studies (Chap. 23).

Measurement of heights involves the choice between data on every tree, which will make a complete scientific record, or measuring only a sufficient number to permit the construction of a curve of height on diameter. In every case, however, a separate curve is required for each plot. If the curves of two or more plots agree, one composite curve can be drawn, having the advantage of increased number of height measurements. This cannot be assumed in advance. From 10 to 20 heights per plot are a minimum requirement.

Heights of small trees are best measured by a pole graduated in feet. One form of jointed pole is effective up to 32 ft.[1] With a pole the operation is rapid and comparatively inexpensive. Where the hypsometer is required, for greater heights, a crew of two men is required independent of the caliper record and numbering, and the time required to measure all trees is three to five times that for the

[1] LIMING, F. G., A Sectional Pole for Measuring Trees, *Jour. Forestry*, **44**, 512–514 (1946).

diameter record and tagging, which is a strong inducement to substitute the height curve.

Length of green crown should be measured on the bole to the point of attachment of the lowest whorl of live limbs. It can be done readily when height is measured, by one additional observation. Sampling rather than complete records may suffice, in which case the measurement of 25 heights per plot, distributed among all crown classes and diameters, may be adequate. Unless the height of each tree is measured successively from the same point, the error incurred for leaning trees will distort the growth records, but for the plot as a whole, these errors should compensate, making this difficult requirement unnecessary.

273. Listing Sheets versus Source Tables. The data derived from current and periodic measurements on sample plots are tabulated on listing sheets and form the source tables relating to the particular investigation. Source tables may be developed for all trees above the specified diameter, also for the intermediate growth from the specified tree diameter down to vegetation only 3 ft. high, and for the ground cover including reproduction up to 3 ft. in height. In the source table for the trees above a specified diameter they may be listed under their respective numbers by species or variety, d.b.h. in inches, height in

feet, height to base of crown, crown form, crown class, bole form, physical conditions, and miscellaneous features.

The data relative to the intermediate or shrubby vegetation and the ground cover including reproduction are usually tabulated as to species and their abundance and frequency.

A sample form for listing sheet is given on page 475. This sheet provides for four successive records of individual numbered trees and forms the basis for subsequent compilations and for studies of behavior of tree classes.

Information relative to the nature and condition of the humus, litter, and soil is usually noted. In certain studies, samples of each might be obtained from the plots for scientific analysis in the laboratory. Consecutive analyses of soil at definite periods would show the changes taking place under the conditions of the individual plots.

274. The Charting of Sample Plots. In addition to the data as recorded in source tables, charts of various kinds are useful for the graphic presentation of the data assembled. Cross-section paper is used in developing the chart in the field.

The scale of the chart varies between wide limits, depending upon the particular data charted. Thus, in charting the distribution of the numbered trees over the sample plot, a chart on the scale of 10 in. to 66 ft. is usually acceptable. On the other hand, where there is a dense ground cover to chart on a single mil-acre, a scale of 10 in. to 6.6 ft. may be none too large.

Charts are of two general types, horizontal and vertical. Horizontal charts refer chiefly (1) to the location of each tree on paper, as to its position in the plot, or (2) to the position of the crown of each tree.

The charting of sample plots is greatly facilitated by dividing them into small units by running heavy string across the plot in two directions, dividing it into squares. In charting the numbered trees on the plot, the size of the quadrats should vary with the age and size of the timber but should be some multiple of 6.6 ft. When the squares are sufficiently small, the location of the trees on the cross-section paper can be made ocularly or by employing two poles marked in feet and placed at right angles to each other on the quadrat.

Vertical charts are graphic presentations of forest vegetation in vertical planes in which the crowns and boles in a single plane or narrow belt are charted to scale, usually on a scale of 1 to 60 or 1 to 120, depending on the height of the canopy. Such a chart shows the crown position and crown form, also the form, length, position, and taper of the bole. It also shows the successive stories in the crown cover.

By charting a sample plot at successive intervals of time the progressive changes in the various factors are graphically shown.

The crowns are best mapped if a man is ined in with a plumb bob underneath the edge of the crown and then the distance is measured from the bole of the tree to the point he occupies.

References

BALDWIN, H. I.: The Strip Survey Adapted to Sample Plots, *Jour. Forestry*, **36**, 41–43 (1938).

CAMPBELL, A. S.: An Error Which Arises in the Use of Sample Plots, *Scot. Forestry Jour.*, **48**, 65–66 (1934).

CLAPP, R. T.: Painted Numbers on Trees in Permanent Sample Plots, *Jour. Forestry*, **34**, 139–140 (1936).

International Union of Forest Research Organizations: Outlines for Permanent Sample Plot Investigations, *Forestry Comm. Gt. Brit.*, 1936, 33 pp.

KORSTIAN, C. F.: Growth on Cutover and Virgin Western Yellow Pine Lands in Central Idaho, *Jour. Agric. Res.*, **28**, 1139–1148 (1924).

KRAUCH, HERMANN: A New Method of Measuring Tree Heights on Sample Plots, *Jour. Forestry*, **20**, 220–222 (1922).

MEYER, W. H.: Rates of Growth of Immature Douglas Fir as Shown by Periodic Remeasurement of Permanent Sample Plots, *Jour. Agric. Res.*, **36**, 193–215 (1928).

MOREY, H. F., and P. W. STICKEL: Numbering Trees on Permanent Sample Plots with Rubber Stamps and Paint, *Jour. Forestry*, **33**, 422–425 (1935).

ROBERTSON, W. M.: Silvicultural Research Committee, Canadian Society of Forest Engineers, Outline for the Study of Cutover Lands, *Forestry Chron.*, **13**, 490–502 (1937).

—— and G. A. MULLOY: Sample Plot Methods, Canadian Dominion Forest Service, 1946, 69 pp.

U.S. Forest Service, Division of Silvics, Branch of Research: Sample Plots in Silvicultural Research, *U.S. Dept. Agric. Cir.* 333, 1935, 90 pp.

CHAPTER 33

THE FOREST SURVEY OR INVENTORY

275. Objectives of the Forest Survey. Forest properties or estates are managed for one or more major purposes; *viz.*, commercial wood production, protection and utilization of fish and game, watershed protection, grazing, parks, or aesthetic enjoyment. Public properties, such as national forests, will as a rule be managed for a greater variety of uses than private forests, since the intangible benefits derived from noncommercial uses constitute a vital factor in our public economy. The holding of properties under private ownership, however, demands that the commercial uses be stressed. Under any circumstances, the forester must seek out all possibilities of use and must plan for their highest coordinated development within the scope of the owner's desire and ability.

If the forest property is not yet under management, the forester's business is frequently of an advisory nature. The owner's general objectives in the future management of the property should be ascertained even prior to investment in the venture if possible. Even so, there may be possibilities for development along supplementary lines or even along lines completely different from those the owner intended, and he should be so advised.

If the property is already under management, the business of the forester is to maintain and build up the forest capital so that the area may become increasingly productive. In addition such further uses must be sought as have not yet been developed.

In any case, whether the property is publicly or privately owned, or unmanaged, one of the first needs is an inventory of the supply of raw material, an estimate of the rate of growth and production, and an investigation of the markets for the finished products. Constant effort in the future will be required to keep these three quantities in balance, in accordance with the needs of the owner and good business practice. The inventory, or forest survey, using the term in its broadest sense, thus becomes the basis for all planning and future operation.

The present discussion will be confined chiefly to the inventory of the wood capital itself, including its capacity to maintain and replenish itself by growth. Forest mensuration thus creates the foundation for forest management. There is no distinct cleavage between the two, since forest mensuration constitutes an essential part of forest management. In the following discussion, however, an arbitrary division will be made at the point where the inventory is concluded. From that point on, only general statements as to the use of the inventory for management purposes will be made.

Owing to the varied desires of the owner, the new possibilities detected by the forester, the range in utility of a property, the degree of intensity of management, and the scope of the silvicultural pattern of the forest itself, no single schedule of forest survey can be recommended. Each property must be handled as a unit by itself. The most suitable mensurational techniques must be selected from those available and must be applied by the most efficient field methods. It is possible, however, to set forth a list of general principles which will serve as a guide and will clarify and coordinate the processes through which the forester must go. Following this, a discussion will be made of the case of a hypothetical large property, in which the selection and application of specific techniques will be demonstrated.

Although the inventory of the present condition of the growing stock and the survey of its possibilities for growth may be considered as two separate activities, in most instances they can be coordinated into a common pattern of field work. Such procedure is in fact highly recommended, since the accuracy of the growth predictions depends not only upon the character of the growth data itself but also upon that of the basic inventory data.

276. General Outline of Procedure to Be Followed in a Forest Survey

A. Preliminary Analysis

1. Analyze the present uses of the property. Is it or can it be handled for wood production, grazing, recreation, protection of watersheds, game protection, aesthetics, or any combination of these?
2. Ascertain future objectives and desires of the owner and compare with present uses.
3. Determine the nature of the local and accessible markets for different classes of forest products or forest uses. Estimate the approximate future needs for specific products and the demand for recreational or other uses.

4. Decide upon the classes and character of data that must be collected to provide the desired information.

B. Initial Field Work

5. Obtain map of the property, if available. Convert it to a scale which will lend itself to mapping of details and determination of subareas, such as forest types (Sec. 155). If no maps are available, locate boundaries, conduct the survey, and make the map (Sec. 141).
6. Make a quick preliminary examination of the tract, including general notations on:
    a. Areas apparently available for different uses.
    b. Approximate areas and location of forest cover types, swamps, open and waste lands, immature and merchantable timber.
    c. Tree species present, their relative abundance and importance by cover types.
    d. Estimated ranges of height and diameter, volumes per acre, and ranges of volume per acre within each forest type.

If the area is large and the expense involved in a forest survey is considerable, this preliminary examination may be accompanied by a report indicating certain measures that should be immediately adopted, such as fire protection, crude silviculture, consolidation of areas by further acquisition, and essential improvements. The detailed forest survey is then undertaken as soon as possible, to give the basis for more systematic future management.[1]

C. Intensive Field Work

7. Based upon the preliminary field work, select stock-taking method best adapted to the situation, the owner's needs, and the apparent possibilities of the tract. For description of different methods of sampling forest areas and methods of laying out a plan of sampling, see Chap. 17. For methods of cruising, see Chaps. 18 to 20.
8. Decide on a specific plan for the intensity of sampling to obtain a desired degree of accuracy (Sec. 171).
9. Determine method to use in predicting future yields.

[1] CHAPMAN, H. H., Forest Management, Chaps. IX, X, Williams Press, Inc., Albany, N.Y., 1931.

a. Even-aged stands (Chaps. 27 and 28).
   b. Uneven-aged stands (Chaps. 29 and 30).
10. Coordinate the field techniques required for the growth study with those of the forest inventory.
11. Obtain necessary field equipment for the collection of the data. For description and use of instruments for measuring distances, tree diameters, heights, and radial growth, see Chaps. 14, 15, and Sec. 190.
12. Determine the method of calculating wood volume based on the use to which the timber will be put.
    a. Volume tables available.
       i. Total or merchantable height (Sec. 80).
       ii. Products (Sec. 2 and Chap. 7).
       iii. Proper utilization (Sec. 77).
       iv. Check accuracy of volume tables (Sec. 106).
    b. Volume tables not available.
       i. Prepare local tables (Chaps. 9 and 12).
       ii. Prepare piece products tables and estimates (Chap. 20).
       iii. Plan on estimating by log rule (Chap. 18).
13. Prepare field tally sheets for recording the data with regard to proper analysis later (Sec. 10).
14. Collect field data that will furnish the basis for:
    a. Present quantity and location of timber.
    b. Growth data or yields by forest types and forest condition classes. Also locate and map roads, trails, rivers, and creeks, and areas of possible special value, such as for recreation, park, or protection.
15. Calculate volume of timber by species, forest types, and area subdivisions as required (Secs. 155 to 157, inclusive). Deduct for defect (Sec. 154). Determine stand per acre by desired type and area subdivisions.
16. Prepare stand and stock tables per acre for the desired subdivisions (Sec. 232).
17. Apply the growth data to the inventory or the stand tables in accord with the principles dictated by the method chosen for making the growth and yield estimate.

D. Application of the Results

At this point, planning for future management begins. The development of such a plan lies outside the scope of forest mensuration and is within that of forest management, which in itself constitutes a major field of activity.[1] A few steps will be shown, however, to indicate briefly the use to which the mensurational data will be put.

18. Determine the silvicultural policy to be followed for each forest type, including methods of harvest cutting, cleaning, thinnng, stand improvement, planting, etc.
19. With aid of the stand tables and growth data predict the allowable annual or period cut and show the effect of optional cutting practices on the future development of the growing stock.
20. Prepare a preliminary plan of management for the property based on:
    a. Objectives of the owner.
    b. Recommended added objectives.
    c. Economic conditions.
    d. Silvicultural conditions and necessary silvicultural practices.
    e. Business objectives.
21. Obtain the owner's approval of the plan of management.
22. Proceed with putting the plan into effect.
23. Collect more data; make detailed observations with the intent of improving and correcting the initial plan.

277. Aerial Photographs as an Aid in Forest Survey. For many years aerial photographs have been used as an aid in the mapping of forest areas. Within recent years the techniques of photography from the air and the methods of interpreting the photographs have become so well developed that they now have an assured position in forest inventory work itself, *i.e.*, in the measurement of volumes of timber as well as the areas of forest land. There is good reason to believe that substantial further improvement will be made and hence that the photographs will become still more useful and valuable. In no event, however, should it be assumed that an intensive forest survey can be based in its entirety on aerial photographs alone. Ground survey or checks of one kind or another will always be essential. Although the novice can gain much information directly from photographs without good training in photogrammetric techniques, the best job of interpretation will be made by those who have had an adequate and intensive training. The subject has become so involved that there can be no attempt in this text to cover it adequately. It is worth separate consideration and study. Only a few general principles will be set forth to indicate the usefulness of the method.

[1] CHAPMAN, H. H., Forest Management, Williams Press, Inc., Albany, N.Y., 1931.

Aerial photographs are divided into two classes, obliques and verticals. The vertical photograph is taken with the axis of the camera perpendicular to the ground and the oblique with the axis at an angle departing from the vertical. Low obliques do not take in the horizon; high obliques do. Although oblique photographs are used fairly extensively for reconnaissance purposes, the vertical photographs have become more and more popular, particularly when they are so taken as to include stereoscopic coverage. Stereoscopic coverage means that successive photographs on a line of flight have overlapping areas so that when adjoining pairs are put under the stereoscope the ground and all vegetation on the ground will be seen three-dimensionally. Under such viewing it becomes possible to measure differences in elevation and heights of objects, such as trees. The recommended overlap is about 60 per cent along the line of flight and 20 to 30 per cent along the edges of pictures between the lines of flight. Such overlapping permits the integration of a complete set of photographs into an accurately controlled mosaic by the radial-line method. With stereoscopic coverage it becomes possible to produce a complete contour map by the use of special instruments that have been developed for this purpose. The U.S. Geological Survey and the military forces make extensive use of aerial photographs for topographic mapping.

Experimentation in aerial photography has brought out the fact that differences in forest vegetation or forest type may be distinguished if suitable combinations of film and filter are used. The bulk of aerial photographs now available in the United States have been made with panchromatic film and a minus blue filter. Differentiation between forest types and even between a few species can be made, but no general rule can be drawn. Much depends upon the season of photography, the time of exposure, the developing of the negative, the printing of the positives, and the quality of the paper. This is also the case with any film and filter combination, but less so with some than with others. The recent use of infrared film with a filter which admits some of the visible light shows much promise in revealing differences in foliage types. Infrared film with the red filters produces a photograph with too much contrast and with areas of blackness impenetrable under the stereoscope. In surveys in the South, photography with infrared film and filter other than red has shown differences in tone as between various pines, such as loblolly, longleaf, and slash pine, and between conifers as a whole and hardwoods. Even with the hardwood class itself a few species can be distinguished from the rest, particularly the gum. In a sense photographs from infrared film may seem unrealistic since they do not give a portrayal as the human eye would see it but introduce shadings which are not humanly perceptible. If, however, these shadings bring out distinctions that assist in subdividing the forest stands into classes of forest types or species, the result is desirable. In forestry work, except where pure conifer types are involved, winter photography is not recommended, since the leafless crowns of the hardwoods are poorly defined and not subject to measurement.

As a rule the larger the scale of the photographs, the better are the results. However, a limit to scale is dictated by cost. Four or five inches to the mile in the case of vertical stereoscopic coverage is considered to be a suitable compromise between scale and cost. The scale of the photograph is determined by the focal length of the camera, the height of the airplane above ground level, and the variations in topographic elevation. Omitting the last factor, $RF = f/H$ where RF is the representative fraction or scale, f is the focal length of the camera, and H is the height of the plane above the ground. Thus if the camera has an $8\frac{1}{4}$-in. (0.6875-ft.) lens and is flying at 10,890 ft. above level ground the representative fraction is

$$0.6875/8{,}250 = 1/15{,}840.$$

This means that 1 in. on the photograph is equal to 15,840 in., or 1,320 ft., on the ground. Should the level of the ground vary, for example, if one portion of the photograph covers a plateau 400 ft. above the ground level, the scale for this portion will be

$$\frac{0.6875}{10{,}890 - 400} = \frac{0.6875}{10{,}490} = \frac{1}{15{,}258}$$

and the images on the ground will appear somewhat larger than equal images located on the plain below. The same scale of 1:15,840 can be obtained by a camera of longer focal length, flying at a higher altitude. Thus if a 12-in. (1-ft.) lens were used, $RF = 1/15{,}840 = 1/H$ and the necessary altitude of the plane would be 15,840 ft. instead of 10,890 ft. There is a decided advantage in using a camera of short focal length since under stereoscopic viewing the displacement of vertical objects on the photographs is more pronounced and measurements of differences between bottom and top are more easily and accurately read. However, in areas of rugged topography, the short focal length is not desirable, since the distortion within a single photograph becomes too pronounced.

When the scale of the photographs is once determined it becomes possible to take measure of area, distances, and even heights under stereoscopic viewing and to relate these measurements in terms of absolute units. On the single photograph heights of objects which throw a clear shadow can be measured by their shadow length related to the angle of the sun, the latitude, and the scale of the photograph. However, this method has distinct limitations on irregular terrain and has given way in large part to the measurement of differential parallax obtained from stereoscopic pairs. A number of instruments have been devised to make the parallax measurement quick and satisfactory.

The density of stands can be measured either by tree counts or by more general photographic rendition.

The following data are thus made available as basic mensurational information convertible into terms of tree, stand, and forest volume:

1. Accurate areas within general forest types.
2. Heights of trees, which can be related to volume and site classification.
3. Crown widths of trees, which can be related to volume, particularly in the case of conifers.
4. Number of trees, which can be related to density and to volume.
5. Density of stand as revealed by the character of the canopy.
6. General structure of the stand in regard to its being even- or uneven-aged.

Information taken from aerial photographs must always be checked by ground measurements. A tree count on the photograph will include only those trees which compose the main canopy and will not take in the understory trees. Neither will such a count include small trees or reproduction which have very narrow crown widths (less than 3 ft.), since such trees may not register. Hence ground check is necessary to supply a correction factor for the number of trees. The measurement of crown width on a photograph will include only that width which can be seen from above and will not include overlap. Correlative studies are again indicated. The measurement of tree height will in the usual case be somewhat scant, since the visible point of the photographs is somewhat below the true tip and since vegetation or ground cover may form an apparent but untrue ground level. In each case therefore the measurements must be checked on the ground with the purpose of establishing correction factors which will have a general application.

The ground checks are best made on areas that are identical with those which have been scanned and measured on the photographs.

Varying combinations of the above named types of data will be devised to give the best interpretation of the volume existing on the ground. No one combination has yet been found to give satisfactory estimates for all forest conditions. In some instances, the volumes of individual trees or even of fully stocked even-aged stands may be closely correlated with total height. In other cases heights and crown width, height and density, or height, crown width, and density will be needed.

The final test of success of the aerial photographic method is the extent to which it will furnish information that is as accurate and as complete, if not more so, than that furnished by a ground survey, without the aid of photographs, and at a reasonable cost. The combination of the two types of survey into a pattern controlled by a sampling technique designed to produce the best results has been found satisfactory and promises to be more so with further improvement and development of photographic, photogrammetric, and ground correlating techniques.

278. Survey Procedure on a Large Area

A. Determining the Percentage of Area for the Sample

As stated previously, each forest area is an entity in itself, to which the most appropriate methods of survey are adapted. The percentage of the area to be sampled depends intimately upon the size of the unit for which accurate estimates are desired and the variability of conditions within this unit. If the unit is 40 acres and contains scattered highly valuable trees, nothing less than a 100 per cent cruise will suffice, regardless of whether the unit constitutes the entire forest area or is only a minute fraction of the forest holdings. If, on the other hand, the 40-acre unit is relatively homogeneous, a 10 or 20 per cent cruise may be sufficient. If the forest tract is large and accurate estimates are needed not by small subdivisions but merely by major blocks, the sampling percentage may be only a fraction of 1 per cent. The decision as to the intensity of sampling is based entirely on the required limit of accuracy (Secs. 17 and 171). In many instances, a standard error of $1\frac{1}{2}$ per cent, or a maximum error of twice this quantity, 3 per cent, may be a feasible goal. In other instances, where the variability of the forest is great and the area unit for which an estimate is desired is comparatively small, this limit of accuracy is too close and must be expanded, owing to the inordinate amount of field work and cost involved in securing the higher standard of accuracy. In all cases, where the limit of accuracy is based on the smallest area subdivision, that required for the forest as a whole is substantially lower, depending upon the size of the total area.

The estimates obtained in step 6d of Sec. 276 furnish the starting point for determining the intensity of sampling. These are the preliminary determinations of average volumes and standard deviations by forest types and conditions. It is planned that the forest will be stratified into such classes during the course of the survey. Strata may be based on:

1. Forest type in broad classifications according to prevailing regional types such as conifers, upland hardwoods, bottomland hardwoods.
2. Age classes, when even-aged stands prevail.
3. Stand-size classes as a substitute for age classes, including seedlings and saplings, poles, and merchantable timber, with diameter limits set for each.
4. Stand density classes, based on:
    a. Volume per acre in board or cubic feet.
    b. Crown density in not over four classes, obtainable from ground estimates or aerial photographs.
5. Condition classes, such as virgin, cutover, burned.
6. Site classes (optional).
7. Ownership: public or private, which may be further subdivided.

Except for item 7, these stratifications decrease the standard deviation of volume per acre and permit the lowest percentage of area to be covered in timber surveys. Aerial photographs perform a valuable service in stratification (Sec. 277). In the usual case, the number of strata should be held down, since too fine a stratification leads to inefficiency and finally even to loss of accuracy.

A hypothetical area of 400,000 acres in the loblolly pine–hardwood type is now chosen to illustrate some of the above points. A survey is made so that a broad long-term management plan can be framed. Consequently the inventory need not be accurate by 40-acre subdivisions but merely by major management subdivisions, here chosen as grossing 50,000 acres, or 8 divisions to the entire forest. Preliminary examinations have suggested stratification into relatively few major forest types; *viz.*, pine-hardwoods, upland hardwoods, and bottomland hardwoods. Age and condition of stand add the further stratification into old fields, virgin timber, selectively cut areas, and clear cut areas. These strata will constitute the basis of the growth predictions, as well as the forest inventory.

One of the 50,000-acre blocks has approximately the following acreages, estimated volumes, and standard deviation of volumes within forest type and condition class:

Type and condition	Area, acres	Volume per acre, bd. ft.	Standard deviation of volume, bd. ft.
Virgin pine–hardwood............	20,000	12,000	8,000
Bottomland hardwood............	15,000	7,000	4,500
Selectively cut pine–hardwood....	15,000	6,000	3,500
Total or average................	50,000	8,700	5,600

A standard error of the estimate for a major type area is tentatively set at 2 per cent (maximum 4 per cent). In the virgin pine–hardwood type this limit of accuracy will require 1,111 plots, as obtained in the following manner (Sec. 17):

$$s_M = \frac{s}{\sqrt{N}}$$

where s_M should be 2 per cent of the average volume of 12,000 bd. ft. or 240 bd. ft. and s is 8,000 bd. ft., giving

$$240 = \frac{8,000}{\sqrt{N}} \quad \text{or} \quad N = \left(\frac{8,000}{240}\right)^2 = 1,111$$

If the plots are $\frac{1}{4}$ acre in size, this will mean $1,111/4 = 277.75$ acres covered by the cruise, or 1.39 per cent of the total area in the type.

Similar calculations for bottomland hardwoods and selectively cut pine give 987 and 850 plots, respectively, or 246.75 and 212.50 acres, or in turn 1.64 and 1.42 per cent of the type areas.

The combined acreage in the total sample is 738 acres, or 2,948 plots, which when applied to the total area in the block, the average volume, and standard deviation without regard to type, gives a standard error of 1.2 per cent, obtained by

$$s_M = \frac{5,600}{\sqrt{2,948}} = 103 \text{ bd. ft.} = 1.2 \text{ per cent of 8,700 bd. ft.}$$

The average percentage of cruise is 1.48 per cent. A line-plot survey (Sec. 142) in which the lines are spaced at 20-chain intervals and $\frac{1}{4}$-acre circular plots are taken at 10-chain intervals along the line gives a 1.25 per cent cruise and is therefore adopted as being a reasonable cruising procedure. The other blocks are analyzed similarly, and it is assumed that no change is indicated in the percentage of area to be sampled.

B. Subsampling for Growth Data

The method of growth prediction chosen for this loblolly pine forest is the stand table method (Sec. 250) except for old field pine stands which will be handled by the yield table method (Sec. 233). The growth data for application in the stand table method will be collected on a sample smaller than that covered by the inventory of standing timber. Determination of the number of trees to be bored in the stratified sample is contingent upon the standard deviation of growth rate in d.b.h. outside bark and the limit of accuracy desired. This accuracy cannot ordinarily be set at a close figure, because of the nature of a growth prediction, which applies past growth to future periods and makes allowances for mortality. In this case an accuracy of about 10 per cent for the 10-year period of prediction is expected, although no advance check is available. The number of trees sampled in each diameter class will depend upon the variation of radial growth within the diameter class and not upon the number of trees as indicated in a stand table. Hence a small percentage of trees will be bored in the small diameter classes, where variation in growth rate is small and numbers of trees are large, and a higher percentage will be bored in the larger diameter classes, where variation in radial growth is apt to be large and numbers of trees small.

A second factor in sampling for growth is the need for avoiding bias introduced by density of stand. If one sample tree is selected at fixed intervals along a survey line, the effect of density on diameter growth is disregarded, resulting in selecting a proportionately greater number of trees in less dense stands and obtaining an average growth rate which is too high. The recommended practice is to sample all trees on plots of standard size at fixed distances along the line. But for different tree-size groups, the size of the plot can be reduced by taking smaller plots within the larger ones (Sec. 248). Standards in use are:

Region	Tree diameter range and size of plot in acres
Pacific Northwest	Over 40 in., $\frac{1}{10}$; 11 to 39 in., $\frac{1}{20}$; 3 to 10 in., $\frac{1}{200}$
Southeast	Over 17 in., $\frac{1}{4}$; 9 to 16 in., $\frac{1}{40}$; 3 to 8 in., $\frac{1}{60}$
Northern Rockies	Over 25 in., 1; 3 to 25 in., $\frac{1}{10}$
Central hardwoods	Over 15 in., $\frac{1}{10}$; 7 to 15 in., $\frac{1}{50}$; 4 to 6 in., $\frac{1}{100}$

It is advisable to mark growth plots permanently for remeasurement.

For this reason circular plots are becoming widely accepted. The size divisions are laid out as concentric circles.

At the beginning of the hypothetical contemplated survey, one plot out of every four survey plots will be considered a growth plot. The growth data will be recorded and plotted as assembled. When the growth trends by species and types become stabilized and the standard error of estimate well known, fewer plots will be taken, but they will never be dropped out entirely. Continued summarizing and plotting will show whether any new conditions are being uncovered.

C. Measuring Periodic Growth in D.B.H.

The trees will always be bored at b.h. with the increment borer. Wherever possible, the age of the tree will be determined. One boring per tree will be made, facing the center of the circular plot to overcome bias created by lean of tree or slope. One measurement of bark width will be taken, by notching or by use of the Swedish bark gauge. Accuracy is to $\frac{1}{20}$ in. of radius, or, on very slow-growing trees, $\frac{1}{100}$ in. The growth will be recorded by two 5-year periods to detect changing growth rate. Increment cores will be measured on the spot, except where ring identification is difficult, in which case the cores will be numbered and measured in the office by use of stains and magnification. A maximum of 10,000 trees is anticipated.

D. Mortality

Mortality will be recorded by species and diameter class. Past mortality will be taken as the basis of current annual mortality and used to reduce present stand tables for future dates. The dead trees will be tallied on the total area. Guides will be furnished to the cruisers to enable them to identify trees which have been dead for not more than 2 years.

E. Defect

Reduction of the future growth estimate for advancing decay is based on accumulated knowledge of the percentage allowances for defect of trees of increasing age and diameter. Flat reduction factors may be applied, or separate deductions by diameter class and age may be made, depending upon the evidence collected during the survey. Defect deductions constitute as important a factor in growth predictions as do mortality adjustments.

F. Equipment

With the above technical decisions on the character of the cruise and growth survey made, the most suitable equipment is assembled, the crew organization established, and the most appropriate forms prepared.

The field party will consist of three men, one acting as a compassman, the second as estimator, and the third as growth and height measurer.

The equipment for each party will consist of
1 standard compass and staff
1 tape for measuring plot diameters
1 Abney
1 ax for blazing and marking
1 calipers
1 diameter tape
1 Swedish bark gauge
2 increment borers
Maps, forms, pencils, tally board, etc.
Since the distance between plots will be measured by pacing, a chain is not needed.

G. Forms

The character of the form always depends upon the classes of data required. For volume alone, the categories are:

1. Saw timber, above a given d.b.h. to utilized top diameters. The minimum d.b.h. is here set at the 12-in. class. Since a local volume table will be derived from a regional table by means of a study of the run of timber in the area to be surveyed and of the current utilization practices, it will not be necessary to set up a form including log-height classes, but only by d.b.h. and major groups of species, such as loblolly pine, shortleaf pine, white oaks, red oaks, red gum, bottomland hardwoods, and other hardwoods.

2. Pole timber, covering a lower range of diameter classes, 5 to 11 in., inclusive. Volumes for these sizes will be computed in cubic feet.

3. Reproduction and saplings, below 5 in., will be estimated by the stocked quadrat system (Sec. 252), one set of four quadrats being located at each plot center.

4. Mortality tally.

5. Defect observations.

The tally for each plot will be kept separate, since they must later be sorted and combined by type-condition classes for the purposes of computing the inventory, the stand tables, and the growth predictions.

Growth measurements will be carried on a separate form, on which the sample trees will be described in detail.

H. Field Work

The crews will operate from maps which are furnished to them. The survey lines will be run through the middle of the forties and not along the edges since many type changes occur along forty lines. Line location will be checked and corrected if necessary at recognizable map features. Car transportation will be available to drop off the crews at the nearest starting point of a day's run and to pick them up at a designated point at the end of the day.

Each crew will organize its own routine of taking measurements, dividing the tasks among themselves so that each member will be kept busy all the time. This is necessary to obtain maximum crew efficiency. In order to get complete coordination between crews, it will be necessary to have an intensive training period at the beginning of the job. Standards of performance and methods of notation must be made consistent, if reliable accuracy is desired.

Computations on the tally sheets will be kept up to date by the crew. Weekly shipments to the main office will be made for purpose of checking and compilation. If the computations prove a burden on the crew and cause a loss of time spent in the field, it may be advantageous to hire computing clerks in the main office and to send in the tally forms daily.

I. Computations

1. The plot volumes are computed from the tallies as the survey proceeds. Study of the average volumes and standard deviations within the recognized strata and for the area as a whole may indicate a change in the intensity of sampling, but this is hardly anticipated.

2. Plots are assembled by strata within each 50,000-acre block. For each stratum, a stand table is computed showing the number of trees per acre for each species group by diameter classes. Average volumes per acre are also found. The distribution of plots on a percentage basis by strata is found.

3. It is then assumed that the total acreage in a stratum is equal to the total area of the tract, multiplied by the percentage of the total

number of plots found in the stratum. Total volume per stratum is obtained by multiplying acreage by average volume. Total volume for the entire block and for the entire forest is found by adding the stratum totals within a block. The volume of the entire forest is that of all the blocks.

4. The future 10-year growth for each forest and condition class is obtained by the stand curved-projection method. The 2-year mortality multiplied by 5 is deducted to get net growth for the period.

5. Ingrowth to merchantable saw-log size should be computed and kept separate from growth of the present saw-timber stand.

6. Periodic annual growth is one-tenth of the periodic growth.

7. The growth estimates are applied to acreages of types, blocks, and total forest area, similarly to the method used for volume estimates. The estimate at this stage makes no allowance for cuttings which will be made in the future period.

8. Different cutting practices will be studied to determine their effect on total growth for the next 10-year period and consequently to analyze the maintenance, improvement, or deterioration of the growing stock.

   a. One-half of the periodic growth on the volume removed by cutting is lost, but one-half is realized.

   b. The future stand table, as of 10 years hence, is computed.

   c. Assumed periodic cuts and resultant net periodic growth are studied to determine necessary changes in cutting practice.

9. The prediction should be extended over a second 10-year period, still assuming similar alternative methods of cutting.

10. Finally, on the basis of the above calculations and analysis, the method of cutting and the volume of the annual and periodic cut will be recommended for each block and for the forest as a whole. Timber may have to be purchased from the outside to supplement the supply in order to keep the mill operating efficiently.

279. Survey Procedure on a Small Area. The case developed in Sec. 278 was that of a relatively large area. It represents a much more complex situation than that met on a small area, such as a wood lot or a small group of forties. As a result the general procedure of Sec. 276 is stripped of many of its details. One of the chief distinctions is that the intensity of sampling must be increased, even up to 100 per cent coverage if variation of forest condition is great or if distribution of volume and value is erratic. In deciding upon intensity of sampling by computation of the standard error of the mean, one may use a revised version of the formula, owing to the restriction in the population. It is no longer infinite or comparatively so, as in the case of the 50,000-acre blocks previously treated; it is definitely limited, or finite. A 40-acre tract, for example, contains only 160 ¼-acre plots and not an infinite number, as the statistical theory behind the formula previously used demands. The new formula is

$$s_m = \sqrt{\frac{s^2}{n}\left(\frac{N-n}{N}\right)}$$

where s_m is the standard error of the mean, s the standard deviation, n the number of sample plots taken, and N the total possible number of plots in the area. For example, since a 40-acre tract has 160 ¼-acre plots, N will be 160. To find the number of plots to be taken, the formula is changed to

$$n = \frac{s^2}{s_m^2 + (s^2/N)}$$

To illustrate the calculation, assume that an estimate of average volume per acre in a 40-acre tract is 4,000 bd. ft.; the standard deviation is 2,000 bd. ft., and a standard error of the mean of 10 per cent, or 400 bd. ft., is desired. Then,

$$n = \frac{2{,}000^2}{400^2 + (2{,}000^2/160)} = \frac{4{,}000{,}000}{160{,}000 + 25{,}000} = \frac{4{,}000{,}000}{185{,}000} = 21.6 \text{ plots}$$

280. Case Examples of Volume and Growth Surveys in Various Forest Regions

A. For an Area of 200,000 Acres, Northeastern Spruce, Balsam, and Hardwoods

1. Control

Base lines, traverse lines, property lines, or township lines are to be used as control lines. Whenever there are large lakes, traverse shore lines by transit stadia both for control of area and for starting and closing points of cruise lines. Base lines are to be run at 1½-mile intervals; if there are existing base lines, use these lines. Starting and closing points on township, property, or base lines to be located by chaining, and to be scribed on nearest tree plainly. Stations in traverse to be marked as on base lines. Prominent roads may be used as base lines; if so, to be traversed by transit stadia and stations marked as on base lines and traverses of lakes. When running traverses, base lines, etc., sketch maps are to be made indicating types and topographic features as under vi and vii below.

Squared cedar posts are to be set at the intersection of property or base lines with roads or other property, base or town lines, these to replace old posts if present or to be set now if old ones cannot be found.

Maps to be plotted by latitudes and departures and detail filled in with protractor. Errors of closure to be distributed pro rata through the control map, by both latitudes and departures.

2. Cruising

a. Cruise line interval:

On tracts over 5,000 acres use ¼ mile.

On tracts less than 5,000 acres use ⅛ mile.

Except that in special cases, where special conditions warrant, special instructions will prevail.

b. Cruise line direction. In one of the cardinal directions, or approximately so, but at right angles to topography in general.

c. Cruise line running. Hand compass for direction and pacing for distance. Circular ¼-acre plots to be established every 5 chains. Starting point and closing point on base line or traverse to be placed on nearest tree of reasonable size, blazed and scribed. Trees nearest the center of each ¼-acre plot to be blazed and scribed with cruise line number and plot number.

d. Tallying.

i. Species: Spruce, fir, hemlock, poplar, white pine, Norway pine, white ash, maple, beech, white birch, yellow birch, cedar.

ii. Size: All trees 5 in. (4.6 in.) d.b.h. and over and in addition cedar ties on basis of minimum 10-in. top 8 ft. long.

iii. Merchantability: No trees to be tallied unless merchantable (spruce, fir, hemlock, poplar, and the pines for pulpwood, hardwoods for bolt wood).

iv. Cruisers will caliper sufficient trees per plot to make sure ocular estimates are correct. Plot radius measurements to be taken four times at least. Doubtful trees near the edge of the plot to be checked as necessary.

v. Tally sheet data: Tally sheet and map sheet shall show beginning point, point where line is tied, cruise line bearing, and total length.

vi. Types: The following types shall be recognized and marked on the map sheet:

(*a*) Productive forest (each sample plot to be put on map sheet as a circle, if productive).

(1) Softwood type: 75 per cent or more spruce, fir, hemlock, or white pine.

(2) Mixed wood type: Less than 75 per cent softwood and less than 75 per cent hardwood.

(3) Hardwood type: More than 75 per cent hardwood species by number.

(4) Cedar swamp: Similar to mixed softwood type, but cedar to comprise more than 50 per cent of the softwood species, in general on low swampy land.

(5) Cutover: To be typed as hardwood cutover (HCO), mixed wood cutover (MCO), softwood cutover (SCO), cedar swamp cutover (CCO).

(6) Land on which hardwood has recently died, changing type in the direction of softwood type, should generally be typed as mixed wood.

(b) Wastelands (designated on map sheet at each sample plot by a cross (X) if wasteland).

(1) Burn: To be designated and typed with approximate date if possible.

(2) Flowage: Along edges of lakes, or beaver flowage on brooks.

(3) Water: Such as lakes or ponds.

(4) Heath.

(5) Alders.

vii. Topographic features:

(a) At each sample plot indicate by arrow direction of slope. Indicate heights of land between larger watersheds by a single heavy dot-dash line.

(b) Woods roads.

(c) Brooks. Show by double line if drivable, and in all cases show width in feet approximately.

(d) Camps or old campsites.

(e) Other features which would be useful in laying out logging operations, such as cliffs, localized large boulders, etc.

(f) Write on each tally sheet general condition of reproduction, considering age and species and condition of thrift.

3. Growth.

For the purpose of determining growth rate in the future, permanent sample plots are to be established at least at intervals of 20 chains on the cruise lines, spacing ½ mile apart. Cruise lines upon which permanent plots are to be established will be followed by a second crew and painted from beginning to end and the location of each sample plot more clearly marked with paint.

The plot center will be marked and scribed as indicated above under "Cruising," but the outline of the plot will be painted to indicate all trees inside the plot. On each permanent plot careful measurement will be taken of d.b.h. of each tree with calipers, measuring the trees to the nearest $\frac{1}{10}$ in., but tallying them to the nearest inch. All trees are to be calipered at right angles and the average d.b.h. taken.

At each plot, or oftener if necessary, in order to secure 100 trees per township of each species, the cruiser will take the total height of one each of the following species: spruce, fir, poplar, hemlock, white pine, and one of the hardwoods.

At each plot a smaller sample plot ($\frac{1}{10}$ acre) with radius 11.78 ft. (17.85 links), tally d.b.h. of each tree from 1 in. (0.6 to 1.5 in., etc.) to 4 in. (3.6 to 4.5 in.).

B. *For an Area of 20 to 500 Acres in the Central Hardwood Region, Containing Mature Timber*

1. Tally all hardwoods of commercial size or cover the area by a sampling scheme.
2. Get diameter tally by species. For commercial sizes, estimate log length of each tree tallied.
3. Use Girard form class volume tables.
4. Tally trees in cordwood or pole sizes on a smaller sample down to 4 in. d.b.h.
5. Note or tally reproduction and saplings on the basis of plots or by general observation.
6. Take growth borings in merchantable timber, but make use of standard growth rates if obtainable. Small sizes may be bored with less damage to the trees. Do not bore veneer logs. Adjust the number to be bored to get a fair sample of each diameter class.
7. Use method of comparison, rather than projection of diameter growth.
8. Distinguish types and site quality only if marked differences in growth rates are evident.
9. Apply actual current height growth whenever necessary, instead of the height-diameter curve. Obtain this from felled trees or from outside sources.
10. Estimate growth by the usual method of determining future minus present volume. Ingrowth of poles to merchantable size must be included.

C. *For an Area of 100,000 Acres of Even-aged Second-growth Longleaf Pine on the Gulf Coastal Plain*

1. From aerial photographs controlled by a radial-line plotting, prepare a type map and get areas by:

a. Forest types: longleaf pine, loblolly pine, slash pine, bottomland hardwoods, oak brush, farm lands.

b. Age classes based on height classes for the various types.

c. Density of stocking by gradations of 100 trees per acre.

2. Predict future yields for each pine species for 10 to 20 years by use of yield tables adjusted for understocked stands.

a. Compare yield data from permanent sample plots with available normal yield tables.

b. By reduction for present relative density of stocking by number of trees per acre, to which correction factor is added for more rapid proportionate growth of trees in incompletely stocked stands. Site classes are distinguished.

c. Get curves of yield on age, and extend them 20 years beyond the age of the oldest existing stands.

D. *For a Wood Lot, Second-growth Hardwoods, Managed for Saw Logs, Piling, and Cordwood*

1. Correlate diameter and total age of individual dominant trees from stumps, stump tapers, and ring counts, for different growing sites to determine the products which can be grown on this site and the approximate length of the rotation required.

2. Test growth per cent by Schneider's formula (Sec. 262) for trees now in or approaching merchantable class for saw logs or piling.

3. Determine total basal area per acre, separately by diameter classes. If between 100 and 150 sq. ft., determine the basal area which can be removed in trees which are mature, deformed, or crowded, without reducing the total basal area below 60 to 70 sq. ft.

4. Apply Gevorkiantz method (Sec. 242) for predicting growth for the next 10 years to the stand after cutting, or apply method of comparison of diameter growth (Sec. 250) to predict diameters at the end of a 10-year period.

5. Compare growth for this period on the remaining stand with proposed cut, and try to balance the two on a 10- or 20-year cutting cycle basis. Reduction of cut will increase periodic growth, and vice versa.

6. If stand is young and even-aged, predict future volume on the basis of conservative thinnings for cordwood, which will not be heavy enough to reduce the final yield. Apply yield tables for second-growth hardwoods to predict the future cubic volume. As stands get older, they will be measured as indicated for saw-log class.

E. For an Area of 300,000 Acres, Ponderosa Pine

1. Classify the stand table by diameters (2-in. classes) and vigor classes (Keen).
2. From existing data on mortality, response to release by well-formed codominant trees, need for thinning in crowded blackjack stands, and duration of period of increased growth after selective cutting, decide on the shortest length of the cutting cycle permitted by economic and transportation factors. Basis is soil moisture per tree as the limiting factor of growth.
3. From measurements of growth based on diameter taken on cut-over areas, or on trees with adequate growing space, determine the diameter growth of each vigor class for the past 10, 20, and 30 years, and its past d.o.b. Apply these growth rates to present diameter of trees on areas to be cut selectively to obtain by stand table projection the future stand table (Sec. 250).
4. Estimate future height, preferably from current height growth projected, otherwise (on large areas) from height-diameter curves.
5. Compute present and future volumes from local volume table or tables. Compare periodic growth with present proposed cut and length of proposed cutting cycle.

F. For a Wood Lot, Southern Pine

1. Decide on silvicultural system which will ensure reproduction after cutting.
   a. For longleaf pine, clear cutting with adequate seed trees.
   b. For loblolly or slash pine, clear cutting of even-aged stands leaving seed trees; group selection cutting in large groups of 1 to 2 acres minimum; or shelterwood cutting removing about one-half the stand, with complete removal in about 10 years. In all cases, reduction of hardwood growth just previous to seed fall.
2. For stands previous to reproduction cutting. Thinnings and stand improvement at proper intervals. Avoidance of too heavy thinnings, which would reduce total yields per acre. Adequate thinnings in codominant crown class to prevent stagnaton of growth.
3. For even-aged stands.
   a. Yield tables, reduced first by percentage of actual to normal stocking (crown density or number of trees) and then increased by factor representing faster growth of individual trees in understocked

stands (using Gerhardt's formula or other data) (see longleaf pine, Sec. 243).

b. Projection of past diameter growth for each d.b.h. class by curves based on three 5-year or two 10-year past periods, applied to present diameters. Mortality eliminated by thinnings.

4. Do not use continuous all-aged selection system, which results in replacement of pines by hardwoods of low value. Stands of mixed ages managed previous to reproduction by improvement cuttings and thinnings; for reproduction, by the group selection, seed trees, or shelter wood systems.

References

CLARK, K. McR.: Method of a Forest Survey and Estimate in Nova Scotia Forestry Quart., **11**, 201–208 (1913).

ELDREDGE, I. F.: The Forest Survey in the South, Jour. Forestry, **33**, 406–411 (1935).

HAMMATT, R. F.: Winter Reconnaissance in Californian Mountains, Forestry Quart., **9**, 557–562 (1911).

KERR, A. F.: Notes on Strip Mapping for Intensive Reconnaissance, Forestry Quart., **12**, 341–346 (1914).

MASON, G. Z.: Winter Reconnaissance in the Rocky Mountains, Forestry Quart., **11**, 516–518 (1913).

MUDGETT, B. D., and S. R. GEVORKIANTZ: Reliability of Forest Surveys, Jour. Amer. Statis. Assoc., **29**, 257–281 (1934).

RECKNAGEL, A. B.: Forest Survey of a Parcel of State Land, N.Y. Conserv. Comm. Bul. 11, 1915, 19 pp.

SCHUMACHER, F. X., and H. BULL: Determining the Error of Estimate for a Forest Survey, with Special Reference to the Bottomland Hardwood Region, Jour. Agric. Res., **45**, 741–756 (1932).

SEWALL, J.: A Canadian Forest Survey, Forestry Quart., **9**, 400–405 (1911).

U.S. Forest Service: Instructions for Making Timber Surveys in the National Forests; Including Standard Classification of Forest Types, 1925, rev., 45 pp.

WATSON, R.: Forest Surveys on the Michigan State Forests, Jour. Forestry, **16**, 567–575 (1918).

WIJKSTROM, S.: Sample Plot Method of the Minnesota Land Economic Survey for Determining Growth and Yield, Jour. Forestry, **28**, 734–738 (1930).

WILSON, E.: Survey Methods and Costs for a Large Area, Forestry Quart., **8**, 287–293 (1910).

AERIAL PHOTOGRAMMETRY

ABRAMS, T.: Essentials of Aerial Surveying and Photo Interpretation, McGraw-Hill Book Company, Inc., 1944, 289 pp.

American Society of Photogrammetry: Manual of Photogrammetry, Pitman Publishing Corp., New York, 1944, 841 pp.

ANDREWS, G. S.: Air Survey and Forestry: Developments in Germany, Forestry Chron., **10**, 91–107 (1934).

———: Tree Heights from Air Photographs by Simple Parallax Measurements, *Forestry Chron.*, **12**, 152–197 (1936).
EARDLEY, A. J.: Aerial Photographs: Their Use and Interpretation, Harper & Brothers, New York, 1932, 203 pp.
FOSTER, E.: The Use of Aerial Photographs in Mapping Ground Conditions and Cruising Timber in the Mississippi River Bottom Lands, *Southern Forest Expt. Sta., Occas. Paper* 37, 1934, 7 pp.
GARVER, R. D.: Aerial Photographs in Forest Surveys, *Jour. Forestry*, **46**, 104–106 (1948).
GRUMBINE, A. A.: Aerial Phototimber Cruising, *Southern Lumberman*, **173** (2177), 220–222 (1946).
KELSH, H. T.: The Slotted Templet Method for Controlling Maps Made from Aerial Photographs, *U.S. Dept. Agric. Misc. Pub.* 404, 1940, 29 pp.
KRAMER. P. R., and E. E. STURGEON: Transect Method of Estimating Forest Area from Aerial Photograph Index Sheets, *Jour. Forestry*, **40**, 693–696 (1942).
KRUEGER, T.: Use of Aerial Photos for Timber Surveys in the Rocky Mountain Region, *Jour. Forestry*, **39**, 922–925 (1941).
LOSEE, S. T. B.: Air Photographs and Forest Sites, *Forestry Chron.*, **18**, 129–144, 169–181 (1942).
MOESSNER, K. E.: A Crown Density Scale for Photo Interpreters, *Jour. Forestry*, **45**, 434–436 (1947).
———: Forest Stand–Size Class Keys for Photo Interpreters, *Jour. Forestry*, **46**, 107–109 (1948).
———: The Value of Aerial Photographs to the Forest Manager, *Proc. Soc. Amer. Foresters Meeting*, 377–383 (1947).
Northeastern Forest Experiment Station: Forest Survey Field Manual, 1947, 92 pp.
ROGERS, E. J.: Use of the Parallax Wedge in Measuring Tree Heights on Vertical Aerial Photographs, *Northeastern Forest Expt. Sta. Forest Survey Note* 1, 1946, 17 pp.
SEELY, H. E.: Determination of Tree Heights from Shadows in Air Photographs, *Canadian Forest Serv., Aerial Forest Survey Res. Note* 1, 1942, 10 pp.
SISAM, J. W. B.: The Use of Aerial Survey in Forestry and Agriculture, *Imperial Agric. Bureaux Joint Pub.* 9, 1947, 59 pp. + 67 figs.
SPURR, S. H.: Developments in Aerial Photography as Related to Forest Education, *Proc. Soc. Amer. Foresters Meeting*, 38–42, 1947.
———: Aerial Photographs in Forestry, The Ronald Press Company, New York, 1948, 333 pp.
——— and C. T. BROWN: Tree Height Measurements from Aerial Photographs, *Jour. Forestry*, **44**, 716–722 (1946).
———: Specifications for Aerial Photographs Used in Forest Management, *Photogrammetric Eng.*, **12**, 131–141 (1946).
STANDISH, M.: The Use of Aerial Photography in Forestry, *Jour. Forestry*, **43**, 252–257 (1945).
U.S. War Department: Advanced Map and Aerial Photograph Reading, Basic Field Manual FM 21–26, 1941, 190 pp.
———: Air Corps Aerial Photography, Training Manual 2170-6, 1930, 126 pp.

APPENDIX

TABLE I. SCRIBNER DECIMAL C LOG RULE

Diameter, in.	Length, ft.											Diameter, in.
	6	7	8	9	10	11	12	13	14	15	16	
	Contents, bd. ft.											
6	0.5	0.5	0.5	0.5	1	1	1	1	1	1	2	6
7	0.5	1	1	1	1	2	2	2	2	2	3	7
8	1	1	1	1	2	2	2	2	2	2	3	8
9	1	2	2	2	3	3	3	3	3	3	4	9
10	2	2	3	3	3	3	3	4	4	5	6	10
11	2	2	3	3	4	4	4	5	5	6	7	11
12	3	3	4	4	5	5	6	6	7	7	8	12
13	4	4	5	5	6	7	7	8	8	9	10	13
14	4	5	6	6	7	8	9	9	10	11	11	14
15	5	6	7	8	9	10	11	12	12	13	14	15
16	6	7	8	9	10	11	12	13	14	15	16	16
17	7	8	9	10	12	13	14	15	16	17	18	17
18	8	9	11	12	13	15	16	17	19	20	21	18
19	9	10	12	13	15	16	18	19	21	22	24	19
20	11	12	14	16	17	19	21	23	24	26	28	20
21	12	13	15	17	19	21	23	25	27	28	30	21
22	13	15	17	19	21	23	25	27	29	31	33	22
23	14	16	19	21	23	26	28	31	33	35	38	23
24	15	18	21	23	25	28	30	33	35	38	40	24
25	17	20	23	26	29	31	34	37	40	43	46	25
26	19	22	25	28	31	34	37	41	44	47	50	26
27	21	24	27	31	34	38	41	44	48	51	55	27
28	22	25	29	33	36	40	44	47	51	54	58	28
29	23	27	31	35	38	42	46	49	53	57	61	29
30	25	29	33	37	41	45	49	53	57	62	66	30
31	27	31	36	40	44	49	53	58	62	67	71	31
32	28	32	37	41	46	51	55	60	64	69	74	32
33	29	34	39	44	49	54	59	64	69	73	78	33
34	30	35	40	45	50	55	60	65	70	75	80	34
35	33	38	44	49	55	60	66	71	77	82	88	35
36	35	40	46	52	58	63	69	75	81	86	92	36
37	39	45	51	58	64	71	77	84	90	96	103	37
38	40	47	54	60	67	73	80	87	93	100	107	38
39	42	49	56	63	70	77	84	91	98	105	112	39
40	45	53	60	68	75	83	90	98	105	113	120	40
41	48	56	64	72	79	87	95	103	111	119	127	41
42	50	59	67	76	84	92	101	109	117	126	134	42
43	52	61	70	79	87	96	105	113	122	131	140	43
44	56	65	74	83	93	102	111	120	129	139	148	44
45	57	66	76	85	95	104	114	123	133	143	152	45
46	59	69	79	89	99	109	119	129	139	149	159	46
47	62	72	83	93	104	114	124	134	145	155	166	47
48	65	76	86	97	108	119	130	140	151	162	173	48
49	67	79	90	101	112	124	135	146	157	168	180	49
50	70	82	94	105	117	129	140	152	164	175	187	50

TABLE II. THE INTERNATIONAL LOG RULE FOR SAWS CUTTING A ¼-INCH KERF
Standard scale for seasoned lumber with 1/16-in. shrinkage per 1-in. board, and saws cutting a ¼-in. kerf, or for green lumber, for saws cutting a 5/16-in. kerf

Diameter, in.	Length of log, ft.												Diameter, in.
	8	9	10	11	12	13	14	15	16	17	18	19	20	
	Contents, bd. ft.													
4	....	....	5	5	5	5	5	5	5	5	5	10	10	4
5	5	5	5	5	10	10	10	10	10	15	15	15	15	5
6	10	10	10	10	15	15	15	20	20	20	25	25	25	6
7	10	15	15	15	20	20	25	25	30	30	35	35	40	7
8	15	20	20	25	25	30	35	35	40	40	45	50	50	8
9	20	25	30	30	35	40	45	45	50	55	60	65	70	9
10	30	35	35	40	45	50	55	60	65	70	75	80	85	10
11	35	40	45	50	55	65	70	75	80	85	95	100	105	11
12	45	50	55	65	70	75	85	90	95	105	110	120	125	12
13	55	60	70	75	85	90	100	105	115	125	135	140	150	13
14	65	70	80	90	100	105	115	125	135	145	155	165	175	14
15	75	85	95	105	115	125	135	145	160	170	180	195	205	15
16	85	95	110	120	130	145	155	170	180	195	205	220	235	16
17	95	110	125	135	150	165	180	190	205	220	235	250	265	17
18	110	125	140	155	170	185	200	215	230	250	265	280	300	18
19	125	140	155	175	190	205	225	245	260	280	300	315	335	19
20	135	155	175	195	210	230	250	270	290	310	330	350	370	20
21	155	175	195	215	235	255	280	300	320	345	365	390	410	21
22	170	190	215	235	260	285	305	330	355	380	405	430	455	22
23	185	210	235	260	285	310	335	360	390	415	445	470	495	23
24	205	230	255	285	310	340	370	395	425	455	485	515	545	24
25	220	250	280	310	340	370	400	430	460	495	525	560	590	25
26	240	275	305	335	370	400	435	470	500	535	570	605	640	26
27	260	295	330	365	400	435	470	505	540	580	615	655	690	27
28	280	320	355	395	430	470	510	545	585	625	665	705	745	28
29	305	345	385	425	465	505	545	590	630	670	715	755	800	29
30	325	370	410	455	495	540	585	630	675	720	765	810	860	30
31	350	395	440	485	530	580	625	675	720	770	820	870	915	31
32	375	420	470	520	570	620	670	720	770	825	875	925	980	32
33	400	450	500	555	605	660	715	765	820	875	930	985	1,045	33
34	425	480	535	590	645	700	760	815	875	930	990	1,050	1,110	34
35	450	510	565	625	685	745	805	865	925	990	1,050	1,115	1,175	35
36	475	540	600	665	725	790	855	920	980	1,045	1,115	1,180	1,245	36
37	505	570	635	700	770	835	905	970	1,040	1,110	1,175	1,245	1,315	37
38	535	605	670	740	810	885	955	1,025	1,095	1,170	1,245	1,315	1,390	38
39	565	635	710	785	855	930	1,005	1,080	1,155	1,235	1,310	1,390	1,465	39
40	595	670	750	825	900	980	1,060	1,140	1,220	1,300	1,380	1,460	1,540	40
41	625	705	785	870	950	1,030	1,115	1,200	1,280	1,365	1,450	1,535	1,620	41
42	655	740	825	910	995	1,085	1,170	1,260	1,345	1,435	1,525	1,615	1,705	42
43	690	780	870	955	1,045	1,140	1,230	1,320	1,410	1,505	1,600	1,695	1,785	43
44	725	815	910	1,005	1,095	1,195	1,290	1,385	1,480	1,580	1,675	1,775	1,870	44
45	755	855	955	1,050	1,150	1,250	1,350	1,450	1,550	1,650	1,755	1,855	1,960	45
46	795	895	995	1,100	1,200	1,305	1,410	1,515	1,620	1,730	1,835	1,940	2,050	46
47	830	935	1,040	1,150	1,255	1,365	1,475	1,585	1,695	1,805	1,915	2,030	2,140	47
48	865	975	1,090	1,200	1,310	1,425	1,540	1,655	1,770	1,885	2,000	2,115	2,235	48
49	905	1,020	1,135	1,250	1,370	1,485	1,605	1,725	1,845	1,965	2,085	2,205	2,330	49
50	940	1,060	1,185	1,305	1,425	1,550	1,675	1,795	1,920	2,045	2,175	2,300	2,425	50
51	980	1,105	1,235	1,360	1,485	1,615	1,745	1,870	2,000	2,130	2,265	2,395	2,525	51
52	1,020	1,150	1,285	1,415	1,545	1,680	1,815	1,945	2,080	2,215	2,355	2,490	2,625	52
53	1,060	1,195	1,335	1,470	1,605	1,745	1,885	2,025	2,165	2,305	2,445	2,590	2,730	53
54	1,100	1,245	1,385	1,530	1,670	1,815	1,960	2,100	2,245	2,395	2,540	2,690	2,835	54
55	1,145	1,290	1,440	1,585	1,735	1,885	2,035	2,185	2,330	2,485	2,640	2,790	2,945	55
56	1,190	1,340	1,495	1,645	1,800	1,955	2,110	2,265	2,420	2,575	2,735	2,895	3,050	56
57	1,230	1,390	1,550	1,705	1,865	2,025	2,185	2,345	2,510	2,670	2,835	3,000	3,165	57
58	1,275	1,440	1,605	1,770	1,930	2,100	2,265	2,430	2,600	2,770	2,935	3,105	3,275	58
59	1,320	1,490	1,660	1,830	2,000	2,170	2,345	2,515	2,690	2,865	3,040	3,215	3,390	59
60	1,370	1,545	1,720	1,895	2,070	2,250	2,425	2,605	2,785	2,965	3,145	3,325	3,510	60

Formula: $[(D^2 \times 0.22) - 0.71D] \times 0.904762$ for 4-ft. sections.
Taper allowance: ½ in. per 4 lin. ft.

TABLE III.—CUBIC CONTENTS OF CYLINDERS AND MULTIPLE TABLE OF BASAL AREAS

Length, ft., or number of trees	Diameter, in.																		
	2	3	4	5	6	7	8	9	10	11	12	13	14	15	16	17	18	19
	Contents of cylinders, cu. ft., or basal areas, sq. ft.*																		
1	0.02	0.05	0.09	0.14	0.20	0.27	0.35	0.44	0.55	0.66	0.79	0.92	1.07	1.23	1.40	1.58	1.77	1.97
2	0.04	0.10	0.17	0.27	0.39	0.53	0.70	0.88	1.09	1.32	1.57	1.84	2.14	2.45	2.79	3.15	3.53	3.94
3	0.07	0.15	0.26	0.41	0.59	0.80	1.05	1.33	1.64	1.98	2.36	2.77	3.21	3.68	4.19	4.73	5.30	5.91
4	0.09	0.20	0.35	0.55	0.79	1.07	1.40	1.77	2.18	2.64	3.14	3.69	4.28	4.91	5.59	6.31	7.07	7.88
5	0.11	0.25	0.44	0.68	0.98	1.34	1.75	2.21	2.73	3.30	3.93	4.61	5.35	6.14	6.98	7.88	8.84	9.84
6	0.13	0.29	0.52	0.82	1.18	1.60	2.09	2.65	3.27	3.96	4.71	5.53	6.41	7.36	8.38	9.46	10.60	11.81
7	0.15	0.34	0.61	0.95	1.37	1.87	2.44	3.09	3.82	4.62	5.50	6.45	7.48	8.59	9.77	11.03	12.37	13.78
8	0.17	0.39	0.70	1.09	1.57	2.14	2.79	3.53	4.36	5.28	6.28	7.37	8.55	9.82	11.17	12.61	14.14	15.75
9	0.20	0.44	0.79	1.23	1.77	2.41	3.14	3.98	4.91	5.94	7.07	8.30	9.62	11.04	12.57	14.19	15.90	17.72
10	0.22	0.49	0.87	1.36	1.96	2.67	3.49	4.42	5.45	6.60	7.85	9.22	10.69	12.27	13.96	15.76	17.67	19.69
11	0.24	0.54	0.96	1.50	2.16	2.94	3.84	4.86	6.00	7.26	8.64	10.14	11.76	13.50	15.36	17.34	19.44	21.66
12	0.26	0.59	1.05	1.64	2.36	3.21	4.19	5.30	6.55	7.92	9.42	11.06	12.83	14.73	16.76	18.92	21.21	23.63
13	0.28	0.64	1.13	1.77	2.55	3.47	4.54	5.74	7.09	8.58	10.21	11.98	13.90	15.95	18.15	20.49	22.97	25.60
14	0.31	0.69	1.22	1.91	2.75	3.74	4.89	6.19	7.64	9.24	11.00	12.90	14.97	17.18	19.55	22.07	24.74	27.57
15	0.33	0.74	1.31	2.05	2.95	4.01	5.24	6.63	8.18	9.90	11.78	13.83	16.04	18.41	20.94	23.64	26.51	29.53
16	0.35	0.79	1.40	2.18	3.14	4.28	5.59	7.07	8.73	10.56	12.57	14.75	17.10	19.63	22.34	25.22	28.27	31.50
17	0.37	0.83	1.48	2.32	3.34	4.54	5.93	7.51	9.27	11.22	13.35	15.67	18.17	20.86	23.74	26.80	30.04	33.47
18	0.39	0.88	1.57	2.45	3.53	4.81	6.28	7.95	9.82	11.88	14.14	16.59	19.24	22.09	25.13	28.37	31.81	35.44
19	0.41	0.93	1.66	2.59	3.73	5.08	6.63	8.39	10.36	12.54	14.92	17.51	20.31	23.32	26.53	29.95	33.58	37.41
20	0.44	0.98	1.75	2.73	3.93	5.35	6.98	8.84	10.91	13.20	15.71	18.44	21.38	24.54	27.93	31.53	35.34	39.38

* To use this table as a multiple table for finding total basal area for several trees, the left-hand column may be used as the multiple, *e.g.*, the total basal area of 16 eight-in. trees is 5.59 sq. ft. The total basal area of 43 eight-in. trees is $2 \times 20 + 3$ trees or $2 \times 6.98 + 1.05 = 15.01$ sq. ft.

TABLE III.—CUBIC CONTENTS OF CYLINDERS AND MULTIPLE TABLE OF BASAL AREAS.—*(Continued)*

Contents of cylinders, cu. ft., or basal areas, sq. ft.*

Length, ft., or number of trees	20	21	22	23	24	25	26	27	28	29	30	31	32	33	34	35	36
1	2.18	2.41	2.64	2.89	3.14	3.41	3.69	3.98	4.28	4.59	4.91	5.24	5.59	5.94	6.30	6.68	7.07
2	4.36	4.81	5.28	5.77	6.28	6.82	7.37	7.95	8.55	9.17	9.82	10.48	11.17	11.88	12.61	13.36	14.14
3	6.54	7.22	7.92	8.66	9.42	10.23	11.06	11.93	12.83	13.76	14.73	15.72	16.76	17.82	18.92	20.04	21.21
4	8.73	9.62	10.56	11.54	12.57	13.64	14.75	15.90	17.10	18.35	19.63	20.97	22.34	23.76	25.22	26.73	28.27
5	10.91	12.03	13.20	14.43	15.71	17.04	18.44	19.88	21.38	22.93	24.54	26.21	27.93	29.70	31.53	33.41	35.34
6	13.09	14.43	15.84	17.31	18.85	20.45	22.12	23.86	25.66	27.52	29.45	31.45	33.51	35.64	37.83	40.09	42.41
7	15.27	16.84	18.48	20.20	21.99	23.86	25.81	27.83	29.93	32.11	34.36	36.69	39.10	41.58	44.14	46.77	49.48
8	17.45	19.24	21.12	23.08	25.13	27.27	29.50	31.81	34.21	36.70	39.27	41.93	44.68	47.52	50.44	53.45	56.55
9	19.63	21.65	23.76	25.97	28.27	30.68	33.18	35.78	38.48	41.28	44.18	47.17	50.27	53.46	56.75	60.13	63.62
10	21.82	24.05	26.40	28.85	31.42	34.09	36.87	39.76	42.76	45.87	49.09	52.41	55.85	59.40	63.05	66.81	70.69
11	24.00	26.46	29.04	31.74	34.56	37.50	40.56	43.74	47.04	50.46	54.00	57.66	61.44	65.34	69.36	73.49	77.75
12	26.18	28.86	31.68	34.62	37.70	40.91	44.24	47.71	51.31	55.04	58.90	62.90	67.02	71.27	75.66	80.18	84.82
13	28.36	31.27	34.32	37.51	40.84	44.31	47.93	51.69	55.59	59.63	63.81	68.14	72.61	77.21	81.97	86.86	91.89
14	30.54	33.67	36.96	40.39	43.98	47.72	51.62	55.67	59.86	64.22	68.72	73.38	78.19	83.15	88.27	93.54	98.96
15	32.72	36.08	39.60	43.28	47.12	51.13	55.31	59.64	64.14	68.80	73.63	73.62	83.78	89.09	94.58	100.22	106.03
16	34.91	38.48	42.24	46.16	50.27	54.54	58.99	63.62	68.42	73.39	78.54	83.86	89.36	95.03	100.88	106.90	113.10
17	37.09	40.89	44.88	49.05	53.41	57.95	62.68	67.59	72.69	77.98	83.45	89.10	94.95	100.97	107.18	113.58	120.17
18	39.27	43.30	47.52	51.93	56.55	61.36	66.37	71.57	76.97	82.56	88.36	94.35	100.53	106.91	113.49	120.26	127.23
19	41.45	45.70	50.16	54.82	59.69	64.77	70.05	75.55	81.24	87.15	93.27	99.59	106.12	112.85	119.80	126.95	134.30
20	43.63	48.11	52.80	57.71	62.83	68.18	73.74	79.52	85.52	91.74	98.17	104.83	111.70	118.79	126.10	133.63	141.37

*To use this table as a multiple table for finding total basal area for several trees, the left-hand column may be used as the multiple, *e.g.*, the total basal area of 16 eight-in. trees is 5.59 sq. ft. The total basal area of 43 eight-in. trees is $2 \times 20 + 3$ trees or $2 \times 6.98 + 1.05 = 15.01$ sq. ft.

INDEX

A

Abney hand level, 10, 196
  errors in use of, 209
Abnormal mortality, 306
Abscissas, 20
Acceleration, growth, 451
  after cutting, 453
Accidental errors, 10
Accuracy, of measurement, 9
  scientific, in log measurement, 50
Adaptation of volume tables, 167
Addition charts, 139
Aerial photographs, as an aid in forest survey, 482
  data from, 485
  forest types in, 483
  and ground measurements, 485
  oblique, 483
  scale, 484
  vertical, 483
Age, basal area based on, 379
  basis of height growth, 335
  even-aged stands, 320
  form of trees based on, 356
  growth variable, 403
  normal, for suppressed trees, 322
  plots for yield tally, 373
  relations to size of trees, 304
  seedling, 319
  suppressed trees, 322
  timber estimating, 263
  trees and stands, 318
Age classes, stand table as related to, 414
Aggregate difference, alinement chart, 154
Airplane maps in timber estimating, 261
Alienation, coefficient of, 294
Alinement chart volume tables, for board feet, 165
  for cubic feet, 144

Alinement charts, 139
All-aged stands, 325
  diameters substituted for age classes, 411
Annual increment, current, 326
  mean, 330
Annual rings, 347
Appearance of data, 12
Application, of results, forest survey, 481
  of stand tables and growth data to total area, 428
Area, choice of percentage to be covered, 234
Area determination, 213, 224, 242
Area measurement, 232
Area method of fitting normal curve of error, 283
Areas, circles, 46
  small, forest survey procedure on, 493
  strip survey for, 238
Arithmetic mean, 22
  computation, 280
  weighted, 25
Assumed mean, 23
Average (*see* Mean)
Axis graduation, 20

B

Bark, 105
Bark factors for tree species in Lake states, 439
Bark gauge, Swedish, 94
Bark measurement, 52
Bark measurements and scaling diameter, 52
Bark thickness, correction for growth in, 436
  and form class, 118
  measurement, 93
Bark volume, 43

Bark width, forest survey, 490
Basal area, basis for mean diameter, 25
  and crown spread, 393
  as indicator of stocking, 393
  per acre as standard, 364
  total, based on age, 379
    intercept curves, 381
Basal areas, measurement, 48
  table, 46
Base of growth percentage, 458
Base line, 215
Bearing objects, 214
Biased error, 10
Biltmore stick, 187
  as dendrometer, 271
  graduation, 189
  use of, 191
Blodgett log rule, cubic, 53
Board feet, alinement chart volume tables for, 165
  measurement in the log, 56
  standard volume tables, construction, 160
  timber estimating, by log rules, 247
    by use of volume tables, 251
  yield tables for, 384
Board foot–cubic foot ratio, 74, 75
  in yield tables, 384
Board-foot equivalents of cordwood, 43
Board-foot log rule, 57, 113
Board-foot standard, 57
  standard volume tables, checking the accuracy of, 167
Board-foot unit, 113
Board-foot volume, felled trees, 103
Board-foot volume tables, basis, 160
Boards, length, minimum, 72
Bole, upper diameter, growth in, 355
Bolts, measurement of, 45
Boundary survey, 213
Branch wood, 92, 105
Breast-high form factor, 132
Brereton log rule, 74
Butt log form class, determination, 164
Butt log form quotients, 161
Butt rot, 62
Butt swell, measurement, 93

C

Calamities, 306
Calculation, proficiency in, 15
Calipers, 185
  use of, 189
Cardboard hypsometer, 200
Case examples of volume and growth surveys, 494
Catastrophic destruction, 415
Center rot, 62
Central hardwood region, forest survey for small area, 497
Central tendency, measurement, 22
Chaining, slope, 238
Chapman, H. H., 78
Chapman hypsometer, 204
Chapman method of growth prediction for long-leaf pine, 405
Chart corrections, alinement, 147
Chart drawing, 21
Charting sample plots, 476
Charts, addition, 39
  alinement, 139
  multiplication, 142
  vertical, sample plots, 476
Check plots, 470
Check scaling, 69
Checking accuracy, of curves, 127
  of estimates, 256
Checks, surface, 64
Christen hypsometer, 201
  errors in use of, 212
Clark, Judson F., 77
Class range, 25
Class width, 25
Classification of forest soils into site classes, 365
Climatic cycles, effect on periodic growth, 315
Codominant trees, 308
Coefficient, alienation, 294
  correlation, 295
  of variation, 31
Commercial methods of log measurement, 50
Compensating errors, 10
Compound interest, growth expressed as, 429

INDEX

Compound interest, and Pressler's growth rate, 462
Computations, elementary, 8
   forest survey, 492
Cone, 48
Constantine log rule, 74
Construction, log rules, 72
   normal yield tables, 361
   site index graph, 376
Control, normal mortality, 304
Conversion of volume tables, 167
Converting factors, cubic feet to cords, 132
Coordinates, cartesian, 20
   rectangular, 20
Cord, long, 36
   running, 35
   short, 35
   standard, 35
Cord foot, 35
Cord measure, 34
   timber estimating for, 227
   units, 35
   utility, 34
Cords, converting factors for cubic feet, 132
   variation in solid contents, 38
Cordwood, board-foot equivalents, 43
   cull table, 41
   deduction for defect, 41
Cordwood volume tables, 131
Cordwood yield tables, 386
Corners, marking, U.S. land survey, 218
   reestablishment of, 219
Correction for mortality, loblolly pine example, 449
Correlation, 294
Correlation coefficient, 295
Cost of timber estimating, factors, 229
Cover types, effective percentages in strips, 266
Crook or sweep, deductions, 65
Crossties, specifications, 86
   tie log rules, 87
Crown canopy, even-aged, 320
Crown classes, relation to diameter growth, 343
Crown ratio, effect on diameter growth, 343

Crown spread on basal area, 393
Cruisers, timber, 227
Cubic contents, felled trees, 96
   log rules, 53
   scaling for, Quebec, 69
   stacked wood, 36
Cubic feet, alinement chart volume tables, 144
   to cords, converting factors, 132
   standard volume tables, construction, 120
Cubic foot–board foot ratio, 74
Cubic form factors, basis, 131
Cubic log rule, 53
   Humphrey caliper, 51
Cubic log rules, based on cylinders, 52
   market, 53
Cubic measure, timber estimating for, 227
Cubic meter, 53
Cubic volume, 1
   felled trees, graphic determination, 98
   logs, 45, 48
   squared timbers, 54
Cubic volume tables for partial utilization, 158
Cull deduction in estimating, 250
Cull logs, 60, 66
Cull table, cordwood, 41
Cumulative distribution, 19
Current increment, 326
Current and mean annual growth percentage, 460
Current and periodic annual increment for even-aged stand, 326
Curve, height on diameter, 205
   S-shaped, 20
Curved projection, in diameter growth, 425
Curves, checking the accuracy of, 127
   harmonized, 121
   reading and rechecking, 111
Curvilinear relationships, 296
Cutting, release, measuring effect of, 452
Cylinder, 48
   as basis for cubic volume of logs or bolts, 43

D

Data, appearance of, 12
D.b.h. (*see* Diameter breast high)
Deceleration of diameter growth, 310
Declination, magnetic, 222
Deduction for defect in cordwood, 41
Deductions, from gross scale, 60
  from Scribner Decimal C rule, 61
Defect, cordwood, deduction, 41
  and cull, deduction in estimating, 250
  deduction, in scaling, 60
    in timber estimating, 259
    forest survey, 490
Defects, hidden, 61
  interior, deduction, 61
  piece products, 84
  side or exterior, 63
Dendroctonus beetle, 428
Dendrometer for top diameters of piling or poles, 269
Density of stand, as a growth variable, 403
  in timber estimating, 264
Density percentage, Gerhardt's formula, 400
Density ratio, 393
Derivation of local from standard volume tables, 128
Determination of areas, 224
Deviation, average percentage, 156
  mean, 25
  standard, 30
Diagram log rules, 73
Diameter, average, of stand, 365
  normal and actual, 181
  outside and inside bark, 93
  scaling of, 52
Diameter breast high (d.b.h.), 89
  correlation with diameter growth on stump, 349
  errors in measuring, 237
  measurement of, 89
  stump tapers, 349
Diameter classes, basis of growth rates, 412
  basis for periodic height growth, 439
Diameter growth, acceleration, 451
  average tree, 344
Diameter growth, b.h. curve, 351
  based on stand tables, 411
  characteristics of, 309
  comparison of present with past, method of, 450
  current, on stump, derivation of b.h. growth from, 353
  curved projection in, 425
  after cutting, 312
  deceleration of, 310
  effect on, of crown ratio, 343
    of logging and mortality, 312
    of site quality, 343
  influences affecting, 343
  measurement at b.h., 434
  past, application to stand tables, 424
  recovery from suppression, 312
  relation to crown classes, 343
  requirements of sampling, 420
  and silviculture, 342
  straight-line comparison, 424
  straight-line projection, 424
  stump, 345
    compilation and averaging, 349
    correlation with d.b.h. outside bark, 349
  tree movement in, 426
  trees, 341
  upper bole, 354
Diameter instrumentation, errors of, 189
Diameter limit in logging, 314
Diameter limits, merchantable, 114
Diameter tape, 186
  use of, 191
Diameters, scaling logs, 57
  standing trees, measurement, 185
  substitution for age classes in all-aged stands, 411
  top of piling in tables, dendrometer for, 269
  upper, growth in, purpose of determining, 354
Diameters breast high, Douglas fir, 13, 14, 131
Difference, standard error of, 292
Discrete series, 277
Dispersion, measures of, 28
  relation between measures of, 31

Distance, measurement by pacing, 240
Distribution, bimodal, 28
  cumulative, 19
  curtailed, 18
  J-shaped, 18
  normal, 18, 276
  skewed, 18
  tree volume, by logs, 164
  unimodal, 28
Dominance, expression of, 309, 370
Dominant and codominant trees, heights of, basis of site, 366
Dominant trees, 308
Douglas fir, stocking percentages, 400
Doyle log rule, 78, 113
Duerr, use of Gerhardt formula, 402

E

Elementary computations, 8
Elements of forest mensuration, 1
Empirical yield tables, 361–392
Equipment, forest survey, 491
Error of the mean, standard, 31, 284
Errors, 9
  accidental, 10
  biased, 10
  compensating, 10
  in diameter instrumentation, 189
  in height instrumentation, 209
  in measurement on strips, 236
  random, 9
  systematic, 10
  in timber estimating, 257
  in use of Abney hand level, 209
  in use of Biltmore stick, 191
  in use of Christen hypsometer, 212
  in use of diameter tape, 191
  in use of improper volume tables, 237
  in use of Merritt hypsometer, 209
Establishment of plots, 472
Estimate, by plots, 243
  total, determination, 242
Estimating, by sampling, 275
  of timber for cubic measure, 227
    (See also Timber estimating)
Even-aged stand, definition, 321
Even-aged stands, age of, 320
  and all-aged stands, 325

Even-aged stands, growth prediction by yield tables, 392
  height growth in, 307
  normal yield tables for, 361
  predicting growth from yield tables, procedure, 396
  stand tables for, 386
  stock tables for, 386

F

False rings, 318
Faustmann hypsometer, 196
Felled trees, board-foot volume, 103
  cubic content, 96
  graphic determination of cubic volume, 98
  measurement procedure, 94
  volume, 89
Field notes, 214
Field work, forest survey, 492
  initial, 480
  intensive, 480
Film, infrared, 483
Fire scars, 64
Forest capital and growth percentages, 456
Forest cover in timber estimating, 263
Forest management, dependence on mensuration, 479
Forest mensuration, elements of, 1
  scope, 1
Forest property, uses of, 478
Forest Service, cubic log rule, 53
  hypsometer, 196
  standard log rule, 78
Forest soils classified by sites, 335
Forest survey, aerial photographs as aid in, 482
  application of results, 481
  bank width, 490
  computations, 492
  defect, 490
  equipment, 491
  field work, 492
  forms, 491
  growth data, subsampling for, 489
  initial field work, 480
  intensive field work, 480

512 FOREST MENSURATION

Forest survey, inventory, 478
  longleaf pine, 100,000 acres, 497
  measuring periodic growth in d.b.h., 490
  mortality, 490
  ponderosa pine, 300,000 acres, 499
  preliminary analysis, 479
  procedure, general outline, 479
    on a large area, 486
    on a small area, 493
  small area, central hardwood region, 497
  southern pine wood lot, 499
  spruce, balsam, and hardwoods, 494
  standard error of estimate, 488
  stratification, 487
  volume and growth, case examples, 494
Forest types, in aerial photographs, 483
  stand table as related to, 414
  as units for timber estimating, 259
Form, growth in, purpose of determining, 354
  for short periods, 358
  of trees, 341
  of logs, 48
  quotient, 90
    butt log, 161
  tree, as variable in tree volumes, 90
  trees based on age, graph, 356
Form class, 90, 118
  butt log, determination, 164
  measurement, 90
  taper tables, 180
  volume tables, 119
Form factor, breast-high, 132
  cubic, basis, 131
  influence on volume growth, 443
Forms, forest survey, 491
Formula, Huber, 49
  log rules, construction, 74
    Doyle, 78
    International, 77
  Newton, 50
  Smalian, 49
Formulas, log rules based on, 76
Forty, 232

Frequency, cumulative, stand distribution, 387
  curve, 18, 277
  distributions, 17
    cumulative, 19
    normal, 276
  polygon, 18
  skewed, 278
Frustums, 48
Fundamental statistical techniques, 275
Future stand tables, mortality as affecting, 415

G

Geometric mean, 26
Gerhardt growth factors, 401
  formula for growth prediction in understocked stands, 400
Gevorkiantz, prediction based on growth variables, 403
  use of Gerhardt formula, 402
Girard method, volume tables, 161
Graph paper, 21
Graphic determination of cubic volume of felled trees, 98
Graphic plotting of averages, 107
Graphs, 20
Great Britain, d.b.h. in., 89
Gross scale, deductions from, 60
Ground checks with aerial photographs, 485
Growth, bark thickness, 436
  from circular plots, forest survey in New England, 496
  diameter, characteristics of, 309
    (See also Diameter growth)
  expressed as compound interest, 429
  height, 336
    characteristics, 307
    (See also Height growth)
  measurement, 5
  vs. mortality, 315
  periodic, annual, prediction by normal yield tables, 392
    stand table method, 434
  effect of climatic cycles on, 315
  of stands, five stages, 328
  vs. tolerance, 308
  of trees, character, 298

Growth, of trees, in diameter, 341
  in dimension and form, 341
  influences affecting, 298
  purpose of studying, 298
  and stands, 6, 324
  in total height based on age, 335
  value, from diameter growth, 430
  volume, of average tree, 358
    based on d.b.h., height and volume tables, 443
    correlation with volume estimates, 429
    from diameter growth, 430
    (*See also* Increment)
Growth data, application to total area, 428
  in forest survey, subsampling for, 489
Growth factors, Gerhardt, 401
Growth percentage, base, 458
  current versus mean annual, 460
  definition and character, 453
  methods of determining, 459
  periodic, 460
  Pressler's formula, 461
  Schneider's formula, 459
  unsafe for predictions, 464
  utility in growth predictions, 463
  in volume, 430
  quality, and value, 457
Growth prediction, accuracy, 6
  based on growth variables, 403
  based on historical records, 466
  based on normality changes, 398
  in even-aged stands, by use of normal yield tables, 396
  from present stocking percentage, 397
  in understocked stands, Gerhardt's formula, 400
  utility of growth percentage in, 463
Growth rates, by diameter classes, 312
  of tree species, 303
Growth studies, purpose of, 332
Growth variables, prediction based on, 403

H

Hardwoods, Connecticut, normal diameter, 181

Hardwoods, log length, 104
  second-growth, wood lot, forest survey of, 498
Harmonized curves, 121
Heart checks, 63
Height, on age, curves of, 337
  as basis of site, 365
  on diameter, curve of, preparation, 205
Height curves, red oak, 129
Height-diameter ratios as basis of height growth, 442
Height growth, based on age, 335
  based on height-diameter ratio, 442
  characteristics, 307
  even-aged stands, 307
  periodic, based on d.b.h. classes, 439
  single tree, 199
Height instrumentation, errors in, 209
Height measurement, on felled trees, 90
  principle of sampling, 207
  on sample plots, 474
Heights, errors in sampling, 237
  leaning trees, 195
  merchantable in timber estimating, 254
  rules for measuring, 92
  on plots for yield table, 373
  standing trees, measurement, 194
  timber estimating, 253
  total and merchantable, 91, 117
  tree, and standard volume tables, 115
Histogram, 18
Historical records, growth predictions based on, 466
Hoppus log rule, 54
Horseshoe method, timber estimating, 231
Huber formula, 49
  for branch wood, 98
Humphrey caliper log rule, 51
Hypsometer, Abney, 196
  cardboard, 200
  Chapman, 204
  Christen, 201
  Faustmann, 196
  Forest Service, 196

514   FOREST MENSURATION

Hypsometer, Merritt, 201
  staff, 199
Hypsometers, 194

I

Identification of monuments, 213
Increment, current and periodic annual, 326
  many- or all-sized stands, 331
  mean annual, 326, 330
Increment borer, 321
Index, site, 367
Influences, affecting diameter growth, 343
  affecting the growth of trees, 298
Ingalls, E. E., 428
Ingrowth, effect on periodic growth, 423
  in stand table, 422
Instrumentation, diameter, errors of, 189
  height, errors of, 209
Intercept, 287
Interior defects, deductions, 61
Intermediate trees, 308
International log rule, 74–76
  for ¼-inch kerf, 78, 504
  for ⅛-inch kerf, 78
Inventories, recurring, 466
Inventory, 1
  or forest survey, 478
  permanent sample plot method, 467
  purpose, 299
Isolation strip, 470

K

Keen, F. P., 428

L

Land measurement, units, linear, 232
Land value and tree growth, 301
Leaning trees, avoidance, in growth, 435
  errors in height of, 209
  heights of, 195
Least squares fit, 290
  method of, 133
Lengths, scaling, 58

Life pattern of tree species, 303
Lightning scars, 64
Linear units of land measurement, 232
Lines, retracement of, 220
Listing sheets, sample plots, 475
Loblolly pine, changes in degree of stocking, 399
  current height growth, 445
  diameter growth, 445
  future volume, 447
    by tree movement, 448
  height on diameter at different ages, 441
  height growth based on age in diameter classes, 442
  site index chart, 366
  total basal area curves, 383
  volume as basis of growth, 446
  yield table, in board feet, 362
Local application of volume tables, correction for, 169
Local volume table, construction, 106
  derivation from standard tables, 128
  use in growth predictions, 429
Log as unit in timber estimating, 247
Log diameters, ocular estimates, training in, 249
  in scaling, 57
  tally of, 247
Log estimates, rough methods of, 247
Log lengths, 91, 114
  conifers, 91
  hardwoods, 91, 104
  scaling, 58
  timber estimating, 253
Log measure, log rules as standard for, 56
Log measurement, commercial methods, 50
  necessity for, 56
  scientific accuracy in, 50
Log rule, 247
  Blodgett, 53
  board-foot, 57
    universal, 81
  Brereton, 74
  Constantine, 74
  cubic, 53
  Doyle, 78, 113

Log rule, Hoppus, 54
  Humphrey caliper, cubic, 51
  International, 47, 76
    ⅛-in. kerf, 78
    ¼-in. kerf, 78, 504
  Maine, 74
  Massachusetts, 81
  New Hampshire, 76
  Quarter Girth, 54
  Scribner, 73
  Scribner Decimal C, 74, 113, 503
  Spaulding, 74
  Tiemann, 75
  Vermont, 76
  volume tables, 103
Log rules, based on cubic contents of logs, 53
  based on formulas, International log rule, 77
  based on mill tables, 81
  construction of, 72
    by formula, 74
  cubic, based on cylinders, 52
  diagram, 73
  Forest Service standards, 78
  metric, 53
  as standard for log measure, 56
  tie, 87
  timber estimates for board feet by use of, 247
Log scaling, 56
  sound, straight, 57
Log tally, 4
Logarithmic paper, 20
Logging, effect on diameter growth, 312
Logs, cubic contents of, log rules for, 53
  cubic volume of, 45
  cull, 66
  form and cubic volume of, 48
  measurement of, 2
  merchantable, 66
Long logs, scaling, 59
Longleaf pine, forest survey, 100,000 acres, 497
  method of predicting growth, 405
  tapers, 99
Lumber, thickness, 57, 72, 73
  width, 72

M

Magnetic declination, 222
Maine log rule, 74
Many- or all-aged stand, increment, 331
Maps, stand per acre, 267
  stocking, 267
  topographic, 267
Market, cubic standard, 53
Markets, 479
Massachusetts log rule, 81
Maturity of trees, 304
Mean, arithmetic, 22
  weighted, 25
  assumed, 23
  geometric, 26
  standard error of, 31, 284
Mean annual increment, 330
Mean versus current annual growth percentage, 460
Mean deviation, 29
Mean diameter, calculation, 23
Meander surveys, 216
Measure, cord, 34
  stacked, 34
Measurement, accuracy, 9
  area, 232
  bark, 52
  board feet in the log, 56
  butt swell, 93
  cord, on sloping ground, 35
  diameter of slanting trees, 185
  diameter growth, at b.h., 434
    at upper sections, 355
  distance, by pacing, 240
  felled trees, procedure, 94
  growth, 5
    in bark thickness, 436
  height, felled trees, 90
    principle of sampling, 207
    standing trees, 194
  land, linear units, 232
  logs, necessity for, 56
    at small end, 51
  piece products, 84
    in the tree, 269
  plots for yield table, 373
  products and logs, 2
  radius, stump, error in, 436

Measurement, stands of timber, 4
  strips, errors in, 236
  sweep in standing trees, 271
  tree, 4
    on sample plots, 474
    standard, 89
    weight, 34
Measures, of central tendency, 22
  of dispersion, 28
    relation between, 31
Measuring, the effect of release cutting, 452
  periodic growth in d.b.h., forest survey, 490
Median, 27
Mensuration, definition, 8
Merchantable and cull logs, 66
Merchantable diameter limits, 114
Merchantable heights, rules for measuring, 92
  versus total heights, 91, 117
Merchantable index as a growth variable, 403
Merchantable yield tables for partial stands, 384
Merritt hypsometer, 201
  errors in use of, 209
Metric, cubic, 53
Metric log rules, 53
Mill tallies, log rules based on, 81
Mill tally, volume tables, 103
Mine timbers, 86
  estimation of, 273
Mode, 28
Monuments, identification of, 213
Mortality, abnormal, 306
  as affecting future stand tables, 415
  correction for loblolly pine, example, 449
  effect on diameter growth, 312
  forest survey, 490
  method of predicting, 416
  normal, its control, 304
  physical standards to judge date of, 420
  recent, tally of, 417
  sample plot method, 417
  Southern Forest Survey standards for date of, 420

Multiple and curvilinear relationships, 296
Multiplication charts, 142

N

Neiloid, 48
New Hampshire log rule, 76
Newton formula, 50
Normal age for suppressed trees, 322
Normal curve of error, areas and ordinates, 279
  fitting of, by area method, 283
    to observational data, 281
    by ordinate method, 282
Normal diameter, 181
Normal distribution, 18
Normal frequency distribution, 276
  of diameter growth, 428
Normal or fully stocked stands, 368
Normal law of error, 18
Normal mortality, control of, 304
Normal standard of stocking for yield tables, 361
Normal yield tables, application to average stands, 392
  for even-aged stands, construction, 361
  field procedure for predicting growth by use of, 396
  need for, for thinned stands, 371
Normality, progress of subnormal stands toward, 369
Normality changes measured from permanent sample plots, 398
Northeast, forest survey for 200,000 acres, 494
Numerical measure of stocking, 393

O

Occurrence, probability of, 284
Ocular estimates, log diameters, 249
Ocular tally, 255
Ogee, 20
Ordinate method of fitting normal curve of error, 282
Ordinates, 20
Overrun, 59

INDEX

Overrun, Doyle log rule, Ontario, 79
Owner's objectives, 478

P

Paces, natural, table, 241
Pacing, measurement of distance by, 240
Paraboloid, 48
Parallax, 485
Partial stands, average diameter, 386
  merchantable yield tables for, 384
Pen for cord measure, 36
Percentage deviation, average, 156
Percentile papers, construction of tables, 182
Periodic annual growth, prediction from normal yield tables, 392
  prediction by stand table method, 434
Periodic annual increment, 326
Periodic and current growth in diameter, stand tables based on, 11
Periodic growth, effect of climatic cycles on, 315
  in form, 358
  in height, based on d.b.h. classes, 439
Periodic growth percentages, 360
Permanent sample plots, 466
  basis of yield table, 367
Photographic method for solid contents, cords, 37
Piece products, classes of, 84
  inspection, 84
  measurement, 3, 84
    in the tree, 269
  specifications, 84
  standards of measurement, 84
Piling, 35
  estimation of, 269
  poles, dendrometer for top diameters, 269
Pitch rings, 63
Pitch seams, 61
Planimeter, polar, 102
Plant indicators of site, 365
Plats, township, 216
Plot computations, yield table, 374
Plots, arbitrarily chosen, 245
  circular, dimensions, 245

Plots, circular, in U.S. Forest Survey, 495
  in estimating, 243
  growth data, forest survey, 489
  measurement for yield table, 373
  method of boring trees on, 435
  requirement for sampling diameter growth, 422
  selection for yield tables, 372
  site index determination for, 379
  standard for normality, 372
Polar planimeter, 102
Pole as height measure, 474
Poles, 85
  and piling, estimation of, 269
Ponderosa pine, forest survey 300,000 acres, 499
Population, and sample, 16
  statistical, 275
Post timber, estimating, 273
Posts and small poles, 85
Preconceived postulates, 471
Predicting periodic annual growth of stands by stand table method, 434
Prediction of growth in volume based on d.b.h., height, and volume tables, 443
Preliminary analysis for forest survey, 479
Pressler's formula for growth percentage, 461
Pressler's growth rate and compound interest, 462
Principal meridian, 215
Probability, law of, 276
  of occurrence, statistical, 284
Probability paper, 21
Procedure, forest survey, 479
  on small area, 493
Products, measurement of, 2
  piece, 84
  round, 84
Proficiency in calculations, 16
Progress of subnormal stands towards normality, 369
Pruning, effect of, 310
  and growth in diameter, 311
Pure stands, 368
Purpose of growth studies, 332

Q

Quadrats, stocked, 431
Quality, growth percentage in, 451
Quarter Girth log rule, 54
Quebec, scaling for cubic contents, 69
Quebec "acre," 232

R

Random error, 9
Random selection, 276
Range, total, 29
Range lines, 215
Records, scaling, 69
Recurring inventories, 466
Recurring inventory method, 467
Red oak, height curves, 129
    standard volume table, 116
Red spruce, diameter growth before and after cutting, 454
    volume table, 157
Reestablishment of lost or obliterated corners, 219
Regression, 287
    coefficient, 287
    curvilinear, 287
    rectilinear, 287
Reineke, L. H., 100
Reineke's formula for stand density, 395
Release cutting, measuring effect of, 452
Reproduction surveys, stocked quadrats, 431
Retracement, of old lines, 220
    of surveys, 213
Ring shake, 61
Rings, false, 318
Rot, butt, 62
    center, 62
    defects from, 61
Rotation, 331
Round products, 84
    measurement of, 2
Rounding off, 11

S

Sample, importance of, 112
    percentage of area for, 486
    and population, 16

Sample, standard deviation from, 30
    stratified, 17, 276
Sample areas in timber estimating, 230
Sample plots, 466
    charting, 476
    choice and location, 469
    establishment, 472
    identification of trees, 473
    for inventory, permanent, 467
    listing sheets, 475
    measurement of heights on, 474
    for mortality records, 417
    permanent, as basis for normality changes, 398
    purpose of, 468
    size and number, 471
    source tables, 475
    standards of accuracy, 469
Sample scaling, 67
    limitations, 68
Sampling, 16
    estimating by, 275
    intensity, 487
    mechanical, 276
    principle, as applied to height measurements, 207
    requirements for diameter growth, 420
Sapwood, rotten, 64
Saw kerf, waste from, 72, 76
Sawed output of logs, variation in, 72
Sawing, method of, 72
    slash, 72
Scale, 57
    net, 60
    sound, 60
Scale book, page, 70
Scale stick, 57
Scaling, 56, 57
    check, 69
    cubic contents, Quebec, 69
    deductions, for crook or sweep, 65
        for defect, 60
        for interior defects, 61
        for side or exterior defects, 63
    diameters, 52, 57
    lengths, 58
        maximum, 58
    logs, 56

Scaling, logs, long, 59
  sample, 67
  sound straight logs, diameter, 57
    lengths, 58
    trimming allowance, 59
Scaling practice, 57
Scaling records, 69
Schneider's formula for growth percentage, 459
Scope of forest mensuration, 1
Scribner Decimal C log rule, 74, 113, 503
  deductions from, 61
Scribner log rule, 73
Seedling age, 319
Selection, random, 276
Semilogarithmic paper, 21
Series, discrete, 277
Shake, 63
Side or exterior defects, deductions, 63
Significant figures, 10
Silviculture and diameter growth, 342
Site, height as basis of, 365
  from plant indicators, 365
Site class differentiation in growth of average trees, 358
Site classes, classification of forest soils into, 365
  as a growth variable, 403
  stand table as related to, 414
Site factors, 363
Site index, 367
  determination for single plots, 379
  graph, 366
    construction, 376
Site quality, effect on diameter growth, 343
  effect on yields, 363
  in timber estimating, 262
Size versus age of maturity in trees, 304
Skewed frequency, 278
Skewness, 18
Slabs, utilization of, 72
  waste from, 76
Slash sawing, 72
Slope correction, 239
Smalian formula, 49
  for felled trees, 96
Soil as a site factor, 364
Solid contents of cords, 38

Sound scale, 60
Sound straight logs, scaling, 57
Source tables, sample plots, 475
Southern pine, wood lot, forest survey, 499
Spaulding log rule, 74
Species in timber estimating, 263
Specifications, piece products, 84
Splits, 63
Spruce, balsam, and hardwoods, forest survey, 494
Squared timbers, cubic volume, 54
Stacked measure, 34
Stacked wood, solid cubic contents, 36
Staff hypsometer, 199
Stagnation of growth, 370
Stand, per acre, maps, 267
  merchantable yield tables, 373
  in timber estimating, 263
Stand density index, 394
Stand distribution, cumulative frequency, 387
Stand table method, objectives, 413
  of predicting periodic annual growth, 434
Stand tables, application of past diameter growth to, 424
  application to total area, 428
  basis for diameter growth, 411
  for fully stocked even-aged stands, 286
  future, mortality as affecting, 415
  modify for use in growth studies, 422
  related to forest types, site classes, and age classes, 414
Standard, board feet, 57
  cubic, Blodgett foot, 53
  market, 53
  meter, 53
  for log measure, log ruler as, 56
  yield table, as, 364
Standard cord, 35
Standard deviation, 30
  computation, 280
  estimated from sample, 30
Standard error, of a difference and of a sum, 292
  of estimate in forest survey, 488
  of the mean, 284
  in scaling by sample, 67

520   FOREST MENSURATION

Standard merchantable heights, rules for measuring, 92
Standard parallel, 216
Standard tree measurements, 80
Standard volume tables, 4
  alinement charts for, 139
  based on butt log form quotients, 161
  for board feet, checking the accuracy of, 167
    construction, 160
  for cubic feet, construction, 120
    by least squares, 133
  and tree heights, 115
Standardizing log rules, diagrams, 73
Standards, of measure, 3
  measurement, piece products, 84
  for sample plots, 469
Standing trees, height measurements, 194
  sweep, measurement of, 271
Stands, age of, 318
  all-aged, substitution of diameters for age classes, 411
  average, application of normal yield tables to, 392
  even-aged, age of, 320
    and all-aged, 325
    growth of, 324
    normal or fully stocked, 368
    overstocked, 368
    pure, 368
    stratified, 365
    subnormal, progress towards normality, 369
    thinned, need for normal tables for yield, 371
    timber, measurement, accuracy of, 4
    understocked, 368, 393
Statistical methods, 16
Statistical techniques, 275
Stem analysis, 302
  by groups of trees, 355
Stock tables for even-aged stands, 386
Stocked quadrats, reproduction surveys, 431
Stocking, density of, effect on diameter growth, 408
  maps, 267
  normal standard, 361

Stocking, by number of trees per acre as method of growth prediction, 405
  numerical measure of, 393
  percentage of, influences affecting, 400
  use in growth prediction, 397
Straight-line comparison, diameter growth, 424
Straight-line projection, in diameter growth, 424
Stratification in forest surveys, 487
Stratified samples, 17, 276
Stratified stands, 365
Stratifying, 17
Strip survey, small areas, 238
Strip width, errors in, 237
Strips, area equivalents, 233
  errors of measurement on, 236
  varying the percentage covered, 266
  width, determination, 236
Stump, diameter growth on, 345
  as basis for current growth at b.h., 353
Stump d.b.h. tapers, 349
Subnormal stands, progress towards normality, 369
Subnormal yield tables, 392
Sum, standard error of, 292
Suppressed trees, 308
  age for, 322
Survey, boundary, 213
Surveys, meander, 216
  retracement and reestablishment, 213
  U.S. public land, 213, 215
Swedish bark gauge, 94
Swedish increment borer, 321
Sweep, 65
  in standing tree, measurement, 271
Systematic error, 10

T

Tagging trees, 473
Tally system, 12
Tape, diameter, 186
Taper, 48
  average upper log, 163
  in logs, 51
Taper tables, construction and application, 180

INDEX 521

Tapers, longleaf pine, 99
  percentile, tables, construction, 183
  stump-d.b.h., 349
Tendency, central, measurement, 22
Thinned stands, longleaf pine, effect on tree growth, 408
  need for normal tables for, 371
Thinning, effects of, on increment, 327
  on yield, 371
Tie log rules, 87
Tie volume tables, 272
Tiemann log rule, 75
Timber cruisers, 227
Timber estimates, checking the accuracy of, 256
  mine timbers, 273
Timber estimating, age of stands, 263
  airplane maps, 261
  for board feet, by log rules, 247
    by use of volume tables, 251
  cost, factors, 229
  for cubic and cord measure, 227
  deduction for defect, 259
  defect and cull, deductions, 250
  density of stands, 264
  forest type, 264
    as units for, 259
  heights, 253
  horseshoe method, 231
  ocular, 249
    tally of dimensions, 255
  one hundred per cent tally, 227
  post timber, 273
  by sample areas, 230
  site qualities in, 262
  small areas, strip survey for, 238
  species, 263
  the stand, 263
  by tally of log diameters, 248
  tree tally, 252
  varying the percentage covered, 266
  volume or density correction, 264
Timber scribe, 218
Timber stands, measurement, 4
Timbers, mine, 86
  squared, cubic volume, 54
Tolerance, relation to growth, 308
Top diameters, 91
Topographic Abney hand level, 198

Topographic maps, 207
Township lines, 215
Township plats, 216
Township subdivision, 215
Training in ocular estimating, 255
Tree calipers, 185
Tree classes, Keen, 428
Tree crops, 302
Tree diameters, 185
Tree form, 118
  local, in timber estimating, 237
Tree growth in volume, 358
Tree heights and standard volume tables, 115
Tree identification, sample plots, 473
Tree measurement, 4
  of piece products, 269
  standard, 89
Tree movement in diameter growth, 426
Tree ring calendar, 347
Tree rings, 298
Tree species, life pattern of, 303
Tree tally in timber estimating, 252
Tree volume, distribution by logs, 164
Tree volume tables for ties by grades, 272
Trees, growth of, influences, 298
  measurement on sample plots, 474
  and stands, age of, 318
    growth of, 324
  suppressed, age for, 322
Triangles, 194
  isosceles, 199
Trimming allowance, 92
Truncated solids, 48
Types, forest, in timber estimating, 259

U

Understocked stands, 362, 393
U.S. public land survey, 215
  corners, marking of, 218
Units, of cord measure, 35
  of measure, 1–2
Universal log rule for board feet, 81
Universal volume table, 105, 115
Universe, finite, 16
  infinite, 16
  statistical, 275
Upper sections, diameter growth, 355

Uses of forest land, 478
Utilization, closeness of, 105
　partial, cubic volume tables for, 158

V

Value, growth percentage in, 457
Value growth from diameter growth, 430
Variable, dependent, 20
　independent, 20
Variation, coefficient of, 31
Vermont log rule, 76
Volume, bark, 43
　board-foot, felled trees, 103
　of felled trees, 89
　growth percentage in, 430, 457
　of logs, cubic, 45
Volume or density correction in timber estimating, 264
Volume-diameter ratio method, 171
Volume growth, of average tree, 358
　from diameter growth, 430
Volume and growth surveys in various regions, 494
Volume table, choice of, in timber estimating, 251
　local, construction, 106
　　volume-diameter ratio method, 172
　red spruce, 122, 157
　second-growth red oak, 116
　standard, volume-diameter ratio method, 173
　universal, 105
Volume table problem, 105
Volume tables, 89
　adaptation and conversion, 167
　alinement charts for, 139
　board feet, alinement chart, 165
　　in timber estimating, 251
　constructed from harmonized curves, 121
　cordwood, 131
　cubic feet, alinement chart, 144
　　for partial utilization, 158
　data which should accompany, 167
　errors in use of, 237
　form class, 119
　local, correction for, 169
　　deviation from standard tables, 128

Volume tables, log rule, 103
　mill tally, 103
　standard, 4
　　based on butt log form quotients, 161
　board feet, checking the accuracy of, 167
　construction, 160
　cubic feet, construction, 120
　least squares method, 133
　and tree heights, 115
　ties by grades, 272
　use in predicting growth, 443
　volume-diameter ratio method, 171

W

Waste, in sawing logs of different sizes, 76
　from saw kerf and slabs, 76
Weight, measurement by, 34
White pine survival, New England, 419
Witness trees, 214
Wood lot, second-growth hardwood, forest survey, 498
　southern pine, forest survey, 499
Worm holes, 64

X

Xylometer, 37

Y

Yield table, alternate methods of construction from plots, 367
　method of comparison, 368
　normal standard of stocking for, 361
　from permanent plots, 367
　plot computations, 374
　plot measurements, 373
　selection of plots for, 372
Yield tables, 6
　board feet, 384
　board-foot–cubic-foot ratio, 384
　cordwood, 386
　definition, 361
　empirical, 361, 392
　merchantable for partial stands, 384
　normal, use in growth prediction, 392
　(*See also* Normal yield tables)
Yields, effect of site quality on, 363